Rare Earth and Critical Mineral Operations and Processing

Richard Skiba

AFTER MIDNIGHT
PUBLISHING

Skiba, Richard (author)

Rare Earth and Critical Mineral Operations and Processing

ISBN 978-1-7641699-8-1 (Paperback) 978-1-7641699-9-8 (eBook) 978-1-7643896-0-0 (Hardcover)

Non-fiction

Contents

Introduction 1

Chapter 1: Introduction to Rare Earth and Critical Minerals 2

    What are Rare Earth and Critical Minerals? 3

    Why they Matter: Applications in Clean Energy, Electronics, and Defence 12

    Key Categories: REES, Lithium, Cobalt, Nickel, Graphite, PGMS, Etc. 14

    Overview of the Global Supply Chain and Criticality Concept 26

    Role of Australia and Other Major Producers 28

    Chapter 1 Review Questions 33

Chapter 2: Geology, Occurrence, and Exploration 35

    How Rare Earth and Critical Mineral Deposits Form 36

    Common Host Rocks and Ore Minerals (Bastnäsite, Monazite, Spodumene, Etc.) 40

    Key Deposit Types and Global Examples 46

    Exploration Methods: Mapping, Sampling, Drilling, and Assay Techniques 69

    Ore Grade and Tonnage Considerations 75

    Chapter 2 Review Questions 94

Chapter 3: Mineral Characterisation and Ore Testing 96

    Physical and Chemical Properties of Ores 97

    Mineral Identification and Liberation Studies 116

    Analytical Tools: XRD, XRF, SEM, MLA, and ICP-MS 122

    Bench-Scale and Pilot Testing for Process Design 142

    Chapter 3 Review Questions 151

Chapter 4: Comminution, Classification, and Particle Preparation 153

    Fundamentals of Crushing and Grinding 154

    Equipment Types: Crushers, Mills, and Classifiers 158

    Energy Efficiency and Particle Size Control 184

    Importance of Liberation Before Separation 186

    Chapter 4 Review Questions 190

Chapter 5: Physical Separation Processes 194

Gravity, Magnetic, and Electrostatic Separation Principles 195

Flotation Fundamentals and Reagent Use 221

Factors Affecting Recovery and Selectivity 231

Examples 233

Chapter 5 Review Questions 239

Chapter 6: Hydrometallurgical Techniques 242

Leaching Principles: Acid, Alkaline, and Pressure Leaching 243

Solvent Extraction and Ion Exchange 255

Precipitation and Purification 268

Case Studies in Rare Earth Separation and Purification 274

Chapter 6 Review Questions 278

Chapter 7: Pyrometallurgy and Thermal Processing 280

Roasting, Calcination, and Reduction 281

Thermal Decomposition and Product Upgrading 291

Equipment and Temperature Control 300

Comparison of Pyro- vs. Hydro-Processing Routes 325

Chapter 7 Review Questions 332

Chapter 8: Plant Design, Process Flows, and Equipment 335

Key Plant Equipment and Instrumentation 336

Layout of a Typical Mineral Processing Plant 338

Process Flow Diagrams for REE, Lithium, and Cobalt 344

Quality Control, Sampling, and Product Specification 347

Chapter 8 Review Questions 356

Chapter 9: Environmental Management and Safety 359

Radiation Hazards (Thorium/Uranium in REE Ores) 360

Tailings and Waste Handling 362

Water Use, Recycling, and Emissions Control 369

Workplace Safety and Regulatory Compliance 371

Introduction to ESG and Community Engagement 372

Chapter 9 Review Questions 379

Chapter 10: Circular Economy, Recycling, and Secondary Sources 383

Recovery of Critical Minerals from E-Waste and Batteries 384

Reprocessing of Tailings and By-Products.. 396

Life-Cycle Assessment and Sustainability Principles 397

Future Trends in Circular Mineral Supply 402

Chapter 10 Review Questions.. 406

Chapter 11: Industry Operations and Global Market Dynamics 409

Major Mining and Processing Operations (Australia, China, USA, Africa) 410

Market Pricing, Supply–Demand Trends, and Value Chain Stages........................... 417

Government Strategies and Critical Minerals Policies 422

Logistics, Exports, and Downstream Manufacturing Links 424

Production Costs 427

Chapter 11 Review Questions.. 436

Chapter 12: Future Skills, Technology, and Career Pathways.................................. 440

Digitalisation, Automation, and AI in Mineral Processing 441

Green Chemistry and Low-Carbon Refining Technologies 449

Training, Education, and Career Development in the Sector 452

Emerging Research and Innovation Priorities 455

Chapter 12 Review Questions.. 459

Review Question Sample Answers.. 462

References 486

Index 516

Introduction

The materials that underpin our modern world are not always visible, but they are everywhere. From the magnets that drive wind turbines and electric vehicles, to the phosphors in smartphones and guidance systems in defence technology, rare earth elements (REEs) and other critical minerals are the silent enablers of the global energy transition and digital revolution. These minerals—small in volume but vast in impact—sit at the intersection of science, technology, and geopolitics, shaping everything from clean-energy innovation to industrial competitiveness and national security.

This book, *Rare Earth and Critical Mineral Operations and Processing*, explores the entire value chain of these essential materials: from their geological origins and extraction to advanced processing, refining, and recycling. It aims to provide students, industry professionals, and policymakers with a comprehensive understanding of how rare earths and critical minerals are discovered, characterised, processed, and brought into the global supply network that powers 21st-century industries.

Across twelve chapters, the book integrates principles of geology, metallurgy, chemistry, environmental science, and economics to explain both the opportunities and challenges of this rapidly evolving sector. Each chapter begins with clear learning outcomes and concludes with review questions designed to consolidate understanding and encourage practical application. Real-world examples, case studies, and comparative analyses highlight the roles of major producing nations, the impact of emerging technologies such as automation and artificial intelligence, and the importance of sustainability and circular-economy approaches in modern resource management.

Ultimately, this book seeks to bridge technical knowledge and strategic context. It equips readers not only with the operational and scientific foundations of rare earth and critical mineral processing, but also with an appreciation of their broader significance in achieving global decarbonisation, technological sovereignty, and supply-chain resilience. Whether you are a student beginning your study of mineral processing, a professional transitioning into the critical-minerals industry, or a policymaker shaping the future of resource security, this text provides the knowledge and framework to understand—and contribute to—the most dynamic and consequential materials story of our time.

Chapter 1

Introduction to Rare Earth and Critical Minerals

The technologies that define the twenty-first century—electric vehicles, wind turbines, smartphones, advanced defence systems—quietly depend on a handful of elements most people rarely hear about. Rare earth elements and other critical minerals sit at the heart of magnets, batteries, catalysts and semiconductors, enabling the global shift toward cleaner energy, faster computation and more resilient infrastructure. Yet the same materials that power innovation also expose economies to new risks: complex chemistries, concentrated reserves, processing bottlenecks and geopolitically sensitive supply chains.

This chapter opens the door to that world. We begin by clarifying what counts as a rare earth element versus a critical mineral, and why the distinction matters—chemically in the lab, economically in the market and strategically for governments. You'll meet the headline materials of the energy transition—lithium, cobalt, nickel, graphite and the platinum group metals—and see how their unique properties translate into real-world applications, from high-strength permanent magnets to high-energy-density batteries and hydrogen technologies.

Because access matters as much as chemistry, we then zoom out to the global supply chain: how deposits are found, how ores are processed, where the chokepoints lie and how materials move from exploration to end-use manufacturing—and, increasingly, back again through recycling. Along the way, we introduce the idea of "criticality": a framework for weighing economic importance against supply risk, and for understanding why national lists differ and evolve over time.

Finally, we situate Australia within this landscape. With world-class geology, established mining capability and a growing processing footprint, Australia is a pivotal supplier to allied markets. We contrast its role with other major producers to highlight complementarities, dependencies and opportunities to build more secure, sustainable value chains.

By the end of the chapter, you'll have a clear map: the key materials and what they do; where they come from and how they are turned into technologies; who produces them and why that matters; and which policies and partnerships can turn resource endowment into long-term strategic advantage.

Learning Outcomes	

This chapter aims to give you the ability to:

1. Define the terms *rare earth elements (REEs)* and *critical minerals*, and explain the distinctions between these groups based on chemical, economic, and strategic criteria.
2. Identify and classify key minerals and elements—such as lithium, cobalt, nickel, graphite, and platinum group metals (PGMs)—that are considered critical for modern industrial and technological applications.
3. Describe the essential roles of rare earth and critical minerals in clean energy technologies (e.g., wind turbines, EV batteries), electronics, telecommunications, and defence systems.
4. Explain the concept of *criticality* in the context of mineral supply chains, including the factors that influence supply risk and economic importance.
5. Illustrate the structure of the global supply chain for rare earth and critical minerals, including stages from exploration and extraction to refining and end-use manufacturing.
6. Analyse Australia's role as a leading producer of several critical minerals and compare it with other key international producers such as China, the United States, and Africa.
7. Discuss the strategic importance of securing critical mineral supply chains for global sustainability, technological innovation, and national security.
8. Recognise how government policies, international cooperation, and technological advances influence global production and market stability for rare earth and critical minerals.

What are Rare Earth and Critical Minerals?

Rare earth and critical minerals are two closely related but distinct categories of strategically important raw materials that are essential to modern technologies, clean energy, and advanced manufacturing. Both play a foundational role in supporting global technological innovation, energy transition, and industrial growth, but they differ in their chemical nature, applications, and supply challenges.

Rare earth elements (REEs) are a group of 17 metallic elements that include the 15 lanthanides, plus scandium (Sc) and yttrium (Y). These elements share similar chemical properties and often occur together in mineral deposits rather than in easily mined, concentrated forms. Rare earths are typically divided into two subgroups based on atomic weight and geochemical behaviour. The light rare earth elements (LREEs) include lanthanum (La), cerium (Ce), neodymium (Nd), praseodymium (Pr), and samarium (Sm), while the heavy rare earth elements

(HREEs) include gadolinium (Gd), dysprosium (Dy), erbium (Er), ytterbium (Yb), lutetium (Lu), and yttrium (Y).

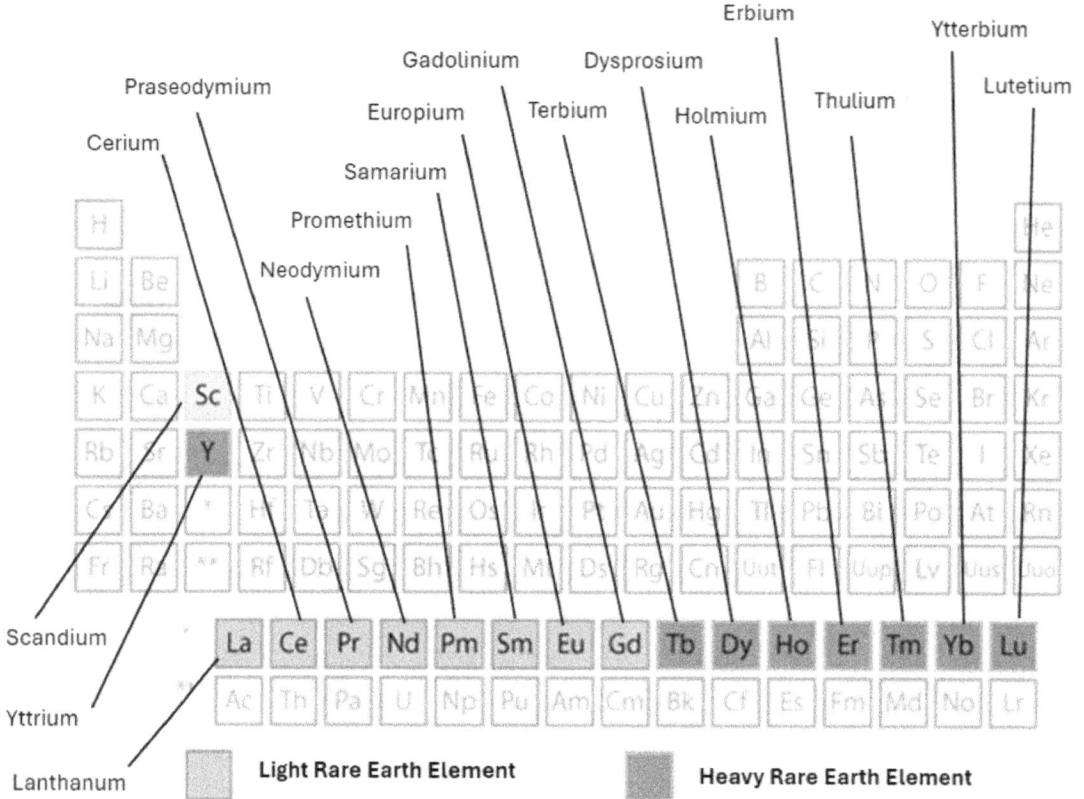

Figure 1: Rare Earth Elements.

Rare earths are notable for their unique magnetic, optical, and catalytic properties. These characteristics make them indispensable in the production of high-performance permanent magnets, electric motors, wind turbines, smartphones, lasers, and defence technologies. Despite the term "rare," these elements are relatively abundant in the Earth's crust; however, they are seldom found in concentrations that are economically viable to mine, making their extraction complex and costly.

Critical minerals, by contrast, are defined as materials that are essential for economic, technological, or national security purposes but are vulnerable to supply disruption. Their criticality arises from limited global production, concentration of supply in a few countries, or challenges in processing and recycling. Common examples include lithium, cobalt, nickel, graphite, and manganese, which are vital for battery production; and niobium, tungsten, vanadium, and the platinum group metals (PGMs), which are key to industrial and defence applications.

Critical minerals are identified through structured criticality assessments that evaluate both economic importance and supply risk. These lists vary between countries—such as Australia, the United States, the European Union, and Japan—reflecting differing industrial priorities and strategic dependencies. They play an increasingly central role in the global energy transition, supporting technologies such as electric vehicles, renewable power systems, and energy storage infrastructure.

Rare earth elements form a subset within the broader category of critical minerals. Together, they underpin the global drive toward decarbonisation, digitalisation, and defence modernisation. Ensuring a sustainable, secure, and transparent supply of these resources through responsible exploration, efficient processing, and recycling has become a strategic priority for many nations. Australia, in particular, plays a leading role in this field, possessing world-class reserves and advanced expertise in both resource extraction and environmental management.

Geological Settings and Deposit Types

The distribution of rare earth elements (REEs) and other critical minerals across the Earth's crust is governed by distinctive geological processes that lead to their concentration in specific deposit types. REEs are not uniformly distributed; rather, they are found in various geological settings shaped by combinations of factors such as rock types, fluid chemistry, and tectonic activities that create economically viable mineral deposits.

One of the principal sources of REEs is hard-rock igneous deposits, particularly within carbonatites and alkaline igneous complexes. These deposits are formations of intrusive rocks characterized by magmas that are rich in carbonates or alkali elements, often leading to concentrations of REEs alongside associated elements like niobium and fluorine. Major global REE deposits, such as Mount Weld in Australia and Bayan Obo in China, are substantially hosted in carbonatite formations [1, 2]. The mining of these deposits involves extensive procedures including crushing, beneficiation, and chemical extraction to retrieve the rare earth elements [3].

Figure 2: Carbonatite breccia from the Precambrian-Cambrian of Norway. James St. John, CC BY 2.0, via Wikimedia Commons.

In addition to hard-rock deposits, weathered or regolith-hosted deposits—commonly referred to as ion-adsorption clays—represent another vital source of REEs. These deposits predominantly occur in subtropical and tropical regions where prolonged weathering processes dissolve REE-rich minerals, allowing REE ions to adsorb onto clay particles. This type of mineralization is notably significant in southern China, which is the world's leading supplier of heavy rare earth elements (HREEs) [4, 5]. The extraction from these weathered deposits typically employs simpler methods such as leaching, which significantly contrasts the complex processing of hard-rock mining [6].

REEs can also be found as accessory minerals within sedimentary or placer deposits. In these scenarios, heavy minerals that contain REEs, such as monazite and xenotime, consolidate through natural processes of erosion, transportation, and sediment accumulation [7]. Such sediments may accumulate within riverbeds, coastal sands, or alluvial plains, often as by-products of other mining endeavours, exemplifying a secondary source of rare earths derived from other mineral extraction activities, such as titanium or zircon mining [8].

Beyond REEs, crucial minerals like lithium, cobalt, nickel, and graphite are also located in diverse geological contexts. Lithium is typically sourced from hard-rock pegmatites or salt lake brines in closed basins, while cobalt and nickel are often found in laterites formed through the weathering of ultramafic rocks. Graphite can emerge in metamorphosed rock formations or within natural vein deposits [7, 9, 10]. Understanding the mineralization and geological settings of these critical minerals is paramount for exploration and subsequent development, as aspects like host rock composition, tectonic history, and the effects of weathering fluids significantly influence the availability and concentration of REEs and critical minerals [2, 11].

Understanding the rarity of economically exploitable rare earth element (REE) deposits requires a multifaceted examination of geological, chemical, and environmental factors. Despite their relatively common occurrence in the Earth's crust, the successful formation of mineable REE deposits is contingent upon several critical factors, including high grade, significant tonnage, suitable mineralogy, and favourable environmental conditions.

To be economically viable, a rare earth deposit must demonstrate a sufficient concentration of rare earth oxides. The geological settings conducive to high-grade REE deposits primarily include carbonatites and alkaline igneous rocks, which have been shown to be associated with specific intracontinental rift-related extensional settings [12]. However, this does not automatically guarantee economic extractability; the mineralogy must also favour extraction processes. Monazite, xenotime, and bastnäsite are among the chief minerals from which REEs are commercially extracted, each varying significantly in REE content and complexity of processing [12].

Moreover, the conditions for effective mining extend beyond pure geological factors to encompass environmental, regulatory, and social dimensions. Mines located in areas with stringent environmental regulations or social opposition often face prohibitive operational challenges. Efficient extraction processes, such as low-cost leaching techniques utilized in ion-adsorption clay deposits—prevalent in regions of southern China and Southeast Asia—demonstrate how geographical and climatic factors can limit accessibility to rare earth resources [5]. While ion-adsorption deposits provide a more economical extraction method, they are geographically restricted, further underscoring the rarity of viable REE deposits [13].

The rarity of economically viable REE deposits can also be attributed to the complexity of processing. The mineralogy of a deposit significantly influences its REE profile, with variations in light and heavy rare earth abundance impacting both the market value and processing difficulties [12, 14]. Studies show that carbonatite and alkaline rocks are typically enriched in light REEs while being relatively depleted in heavy REEs [14]. Additionally, the economic feasibility of REE extraction is compounded by global competition for these resources, with only a few regions, such as parts of Africa, South America, and Southeast Asia, providing conditions conducive to profitable extraction [15].

Major Global Locations of REE and Critical Mineral Resources

The distribution of Rare Earth Elements (REEs) and other critical minerals across the globe is highly concentrated in a few countries. This uneven distribution has significant implications for global supply chains, trade policies, and resource management, especially given the rising demand for these materials.

China is the largest holder of rare earth reserves, estimated at around 44 million metric tons, and accounts for over 86% of the world's total REE production [16]. This dominance is largely due to extensive ion-adsorption clay deposits found in its southern provinces, which are rich in heavy rare earth elements (HREEs) and can be extracted using cost-effective leaching techniques [17, 18]. China's integration of mining, refining, and downstream processing has solidified its central role in the REE supply chain, positioning it as a crucial gatekeeper of these critical resources [19]. Although HREEs sourced from ion-adsorption clays make up a smaller volume of production, they fulfill a substantial portion of global demand, underscoring China's vital importance in this sector [20].

Australia is also a key player, with high-grade hard rock rare earth deposits such as the Mount Weld Mine, operated by Lynas Rare Earths. This site is one of the leading sources of rare earth feedstock globally, supporting both local processing and international exports [21, 22]. Furthermore, Australia's mineral wealth extends beyond REEs to include lithium, nickel, and cobalt, which are essential for the transition to clean energy technologies [23].

In the United States, the Mountain Pass Mine represents one of the few significant REE operations, aiming to revitalize domestic production to enhance self-sufficiency and reduce dependence on imports. Historically, Mountain Pass was a major producer but has seen a revival in recent years to increase its operational capacity [21]. Other countries such as Brazil, Russia, and Vietnam are also recognized for their reserves and are increasing exploration and production efforts. For instance, Brazil's unique carbonatite-hosted deposits and Vietnam's promising resources in the Lai Châu region highlight the potential for these nations to contribute more significantly to the REE supply chain in the near future [23, 24].

Figure 3: An aerial photograph of the Mountain Pass Rare Earth Mine & Processing Facility in San Bernardino County, California. Tmy350, CC BY-SA 4.0, via Wikimedia Commons.

Emerging mining projects in Greenland, Madagascar, and various regions in Africa and Scandinavia illustrate a diversification of REE sources that may decrease the long-standing reliance on China. Projects like Kvanefjeld in Greenland aim to develop rare earth and uranium-bearing resources, reflecting a growing global interest in exploration [25]. These developments are crucial as they signal a move towards a more balanced and secure production of critical minerals, essential for meeting future demand and mitigating supply chain vulnerabilities.

Overall, the global landscape for REE and critical mineral resources is characterized by significant concentration in China, with important contributions from Australia and the United States, along with emerging producers in other regions. This geographic distribution critically shapes economic strategies and trade policies as nations acknowledge the importance of these materials for technological advancement and energy transitions.

The Lanthanides, Scandium and Yttrium

The lanthanides, together with scandium (Sc) and yttrium (Y), make up the group of elements commonly referred to as the rare earth elements (REEs). These 17 metallic elements share many similar chemical and physical properties, which is why they are often found together in nature and are typically discussed as a group rather than individually. Despite their name, most of these elements are relatively abundant in the Earth's crust, but they are rarely found in concentrated, easily mineable forms.

The term "lanthanides" is derived from the Greek verb "lanthanein," meaning "to escape notice." This nomenclature is reflective of the historical circumstances surrounding the discovery of lanthanum (La), the first element of the lanthanide series. Discovered in 1839 by the Swedish chemist Carl Gustav Mosander, lanthanum was initially identified as an impurity within cerium nitrate, thereby epitomizing its "hidden" nature in mineral form [26]. Mosander's finding marked the recognition of the lanthanides as a distinct group of elements, which was further supported as additional elements displaying similar properties were discovered to be associated with lanthanum in various minerals, enhancing the concept of concealment inherent to their chemical and geological contexts [26].

The lanthanides themselves consist of 15 metallic elements that span from lanthanum to lutetium on the periodic table. They are characterized by their closely related electron configurations and comparable chemical behaviours, which complicate their separation and identification in both natural and industrial settings. This shared chemical similarity contributes to their historical "hidden" status, as they frequently occur together within the same mineral deposits, making individual identification challenging [27].

Following their initial grouping as "the hidden ones," the lanthanides have been extensively studied in various domains of chemistry and materials science. A notable aspect of their behaviour is attributed to the phenomenon termed "lanthanide contraction," which explains the progressive decrease in ionic radius across the series, contributing to their distinct properties and interactions in chemical compounds [28]. Thus, the appellation "lanthanides" serves not only as a nod to their origin in discovery but also as a descriptor of their consistent geological behaviour, where their concealed presence in various minerals aligns with their challenging separability and strong interactive characteristics in chemical contexts.

The lanthanides are a series of 15 elements on the periodic table, ranging from lanthanum (La) to lutetium (Lu). They occupy the f-block of the periodic table, beginning with atomic number 57 (lanthanum) and ending with atomic number 71 (lutetium). Each lanthanide atom has a partially filled 4f electron shell, which gives the series its distinctive chemical behaviour.

The lanthanides are typically divided into two subgroups based on atomic weight and behaviour:

- Light Rare Earth Elements (LREEs): Lanthanum (La), Cerium (Ce), Praseodymium (Pr), Neodymium (Nd), and Samarium (Sm).

- Heavy Rare Earth Elements (HREEs): Gadolinium (Gd), Terbium (Tb), Dysprosium (Dy), Holmium (Ho), Erbium (Er), Thulium (Tm), Ytterbium (Yb), and Lutetium (Lu).

LREEs are more common in the Earth's crust and are often found in minerals such as bastnäsite and monazite, while HREEs are rarer and usually occur in minerals like xenotime or in ion-adsorption clays in tropical regions.

Rare Earth and Critical Mineral Operations and Processing

Scandium is a soft, silvery-white metallic element with atomic number 21. Although it is not a lanthanide, scandium is included among the rare earth elements because it shares similar ionic size, valence (+3), and chemical reactivity with the lanthanides. Scandium is relatively scarce and is typically found in trace amounts in minerals such as thortveitite, euxenite, and gadolinite. It is used primarily in high-strength aluminium–scandium alloys, which are valued for aerospace components, sporting goods, and fuel cells.

Yttrium, with atomic number 39, is another element closely associated with the lanthanides due to its similar trivalent oxidation state (+3) and comparable ionic radius. It commonly occurs in the same ores as the heavy rare earths, particularly in xenotime and monazite. Yttrium plays an important role in modern technologies—most notably in phosphors for LED displays, superconductors, and ceramics that can withstand extreme temperatures.

Together, the lanthanides, scandium, and yttrium form a chemically coherent group of rare earth elements that are indispensable to high-tech and clean energy industries. Their shared chemical traits—particularly their ability to form trivalent cations and complex compounds—make them difficult to separate from one another but highly valuable for producing magnets, catalysts, lasers, batteries, and optical materials. Although they are not geologically rare, their economic extraction and refining remain challenging, which contributes to their strategic importance in global markets.

Critical Minerals Other Than REEs

Critical minerals, distinctly important for technologies driving the transition to clean energy, encompass a variety of substances such as lithium, cobalt, nickel, and graphite. Each of these minerals has unique geological settings and formation processes, reflecting the complex chemistry and physics at play. The variety of deposits associated with these minerals not only shows the diversity of their geological origins but also their crucial roles in sustainable technologies like batteries for electric vehicles and renewable energy systems.

Lithium, a key component in lithium-ion batteries, is predominantly derived from two principal types of deposits: hard-rock pegmatites and brine pools. Pegmatites, which are coarse-grained igneous formations, often contain lithium-bearing minerals such as spodumene. Notable lithium pegmatite deposits are found in regions like Australia, particularly in Greenbushes and Pilgangoora, which are recognized as some of the most significant sources globally [29-31]. Conversely, lithium brine deposits—formed through evaporation in closed-basin salt lakes—are prominently located in the "Lithium Triangle" of South America, encompassing parts of Bolivia, Chile, and Argentina, which significantly contribute to global lithium supply [32, 33]. The brine extraction methods are often more environmentally sustainable compared to hard-rock mining, making them increasingly favoured by industry stakeholders [34, 35].

Cobalt and nickel are typically extracted from both laterite and sulphide deposits. Laterite deposits emerge from the weathering of ultramafic rocks in tropical climates, concentrating these metals at the surface. Significant lateritic deposits are found in areas like the Democratic

Republic of Congo and Indonesia [36]. Sulphide deposits, however, originate from magmatic processes and are located at greater depths, with major sources in nations such as Russia and Canada [36, 37]. Additionally, the shift towards recycling used batteries and electronic waste has opened up new avenues for recovering cobalt and nickel, thereby addressing supply risks stemming from geological constraints and geopolitical factors [37, 38].

Graphite, utilized heavily in battery technology, typically forms in metamorphic rocks through heat and pressure, converting carbon-rich sediments into crystalline flake graphite [29]. High-grade deposits are primarily located in China, Mozambique, and Canada, while amorphous graphite can be found in regions like Sri Lanka [39, 40]. The increasing demand for graphite in lithium-ion batteries underscores the need for reliable sources and sustainable supply chains [41].

Like rare earth elements, the global distribution of these critical minerals is often confined to specific countries, creating vulnerabilities in supply chains. This concentration raises significant risk factors related to political instability, environmental concerns, and trade barriers [30, 42]. Ensuring diversified and sustainable sources of these minerals is essential for maintaining technological advancements and energy security [38, 43]. The identification of viable extraction techniques, especially from alternative sources such as geothermal brines and recycled materials, is crucial for enhancing resilience in the supply chain while minimizing environmental impacts [32, 35].

Why they Matter: Applications in Clean Energy, Electronics, and Defence

Rare earth elements (REEs) and critical minerals are increasingly recognized as essential components in the technologies that drive modern society, particularly in sectors such as clean energy, electronics, and defence, due to their unique physical and chemical properties. These materials play a fundamental role across various advanced systems, from renewable energy generation to military hardware.

Rare earth elements (REEs) have a unique set of physical and electronic properties that make them indispensable in many modern technologies. These properties arise mainly from the special behaviour of their 4f electrons, which are shielded by outer electron shells and can interact in highly specific ways with neighbouring atoms. This unusual electronic structure gives rise to a combination of magnetic, luminescent, electrical, and catalytic characteristics that are difficult to replicate with other materials.

The magnetic properties of certain rare earths, such as neodymium (Nd), dysprosium (Dy), and samarium (Sm), make them particularly valuable in manufacturing powerful permanent magnets. These elements possess numerous unpaired electrons whose spins align to generate strong magnetic fields. As a result, neodymium magnets can store up to eighteen times more magnetic energy than ordinary iron magnets. Such magnets are vital for applications like wind turbines, electric motors, aircraft guidance systems, miniature speakers, and computer hard drives. However, in their pure form, these elements lose magnetism before reaching room temperature. To overcome this, they are combined with transition metals such as iron or cobalt to form robust alloys like neodymium–iron–boron ($Nd_2Fe_{14}B$) and samarium–cobalt ($SmCo_5$) magnets, which retain magnetic strength at higher temperatures.

The luminescent properties of rare earths such as europium (Eu), yttrium (Y), erbium (Er), and neodymium (Nd) make them invaluable in lighting and optical technologies. These elements emit light when excited by electromagnetic radiation and are therefore used as phosphors in LEDs and fluorescent lamps. Europium, for instance, produces the red light crucial for early colour televisions, while erbium is used in optical amplifiers that boost signals in fibre-optic communication networks. Additionally, rare earth–doped materials such as yttrium–aluminium–garnet (YAG) crystals form the basis of powerful lasers used in medicine, manufacturing, and military range-finding systems.

The electrical properties of certain rare earths, including cerium (Ce), lanthanum (La), neodymium (Nd), and praseodymium (Pr), are exploited in energy storage technologies. These elements are commonly blended into an alloy known as mischmetal, used in nickel–metal hydride (NiMH) batteries. The alloy improves a battery's energy density, charge retention, and lifespan, making it ideal for use in hybrid vehicles and portable power tools.

Finally, rare earths such as cerium (Ce) and lanthanum (La) demonstrate remarkable catalytic properties due to their ability to easily donate and accept oxygen. Cerium oxide is a key component of automotive catalytic converters, which reduce toxic emissions by converting carbon monoxide into carbon dioxide. Lanthanum plays a major role in fluid catalytic cracking, a refinery process that breaks down crude oil into valuable fuels such as gasoline.

Overall, the unique 4f electron configuration of rare earth elements allows them to interact with light, magnetism, and chemical reactions in extraordinary ways. These properties make REEs indispensable to modern industry, enabling clean energy technologies, advanced electronics, and sophisticated defence systems that define contemporary life.

In the clean energy sector, specific rare earth elements such as neodymium (Nd), praseodymium (Pr), dysprosium (Dy), and terbium (Tb) are crucial for producing high-strength permanent magnets, which are pivotal in wind turbines and electric vehicle (EV) motors. These magnets enhance the efficiency and durability of these technologies compared to traditional materials [44, 45]. Additionally, critical minerals like lithium, cobalt, nickel, and manganese serve as fundamental constituents of lithium-ion batteries, enabling energy storage for electric vehicles and renewable power systems [44, 45]. Furthermore, platinum group metals (PGMs), including platinum and iridium, serve as essential catalysts for hydrogen fuel cells, allowing for efficient energy conversion with net-zero carbon emissions [46]. Without access to these vital resources, achieving global decarbonization objectives would be challenging [47].

In the electronics arena, rare earths facilitate miniaturization and high-performance capabilities critical for developing energy-efficient devices. Elements like lanthanum (La) and cerium (Ce) are utilized in producing optical lenses and polishing powders, while europium (Eu), yttrium (Y), and terbium (Tb) are required for red and green phosphors in LEDs and display technology [44, 45]. Additionally, tantalum and niobium are indispensable for manufacturing capacitors and high-temperature superconductors, whereas tin and indium find applications in soldering and touchscreens. Such materials are integral in fostering the production of compact and powerful devices that propel the information economy forward.

In the defence sector, the strategic importance of rare earths and critical minerals is evident; they underpin technologies essential for national security. Applications of these materials include advanced components in jet engines, radar systems, missile guidance, sonar equipment, and precision-guided munitions, with samarium-cobalt magnets exemplifying their use due to their stability under high temperatures [48]. Moreover, titanium's strength-to-weight ratio and corrosion resistance make it indispensable for aerospace and naval applications, further showcasing the importance of these minerals in maintaining a technological edge [49].

Key Categories: REES, Lithium, Cobalt, Nickel, Graphite, PGMS, Etc.

Rare Earth Elements (REEs)

As described earlier, the rare earth elements (REEs) represent a group of 17 metallic elements that include the 15 lanthanides—from lanthanum (La) to lutetium (Lu)—along with scandium (Sc) and yttrium (Y). These metals share similar chemical behaviours due to their comparable atomic structures, yet each element exhibits subtle differences that give rise to a remarkable range of magnetic, optical, and catalytic properties. These unique characteristics make REEs indispensable in modern technology, renewable energy systems, and advanced manufacturing.

The lanthanides form the core of the REE group, characterised by their partially filled 4f electron orbitals, which influence their magnetic and electronic interactions. The series is commonly divided into two categories: light rare earth elements (LREEs) and heavy rare earth elements (HREEs). This distinction is based on atomic weight, ionic radius, and how these elements occur in nature.

Light REEs (LREEs)—including lanthanum, cerium, praseodymium, and neodymium—are relatively more abundant and are typically found in minerals such as bastnäsite and monazite. These elements are vital in industrial and consumer technologies. For example, lanthanum and cerium are used as catalysts in petroleum refining and in polishing glass and camera lenses, while neodymium and praseodymium are essential for making high-performance permanent magnets used in electric motors and wind turbines. The strong magnetic fields produced by these materials allow for more compact and efficient designs in clean energy and electronics.

In contrast, heavy REEs (HREEs)—such as dysprosium, terbium, and ytterbium—are less common in the Earth's crust and are usually concentrated in more complex mineral deposits like xenotime or ion-adsorption clays. Despite their lower abundance, they are critical for high-temperature and high-performance applications. Dysprosium and terbium, for instance, enhance the heat resistance of neodymium magnets, enabling them to operate efficiently in demanding environments such as electric vehicle drivetrains and aerospace components. Ytterbium and other HREEs are used in laser technologies, fibre optics, and advanced electronics due to their luminescent and conductive properties.

Rare earth elements are embedded in an extraordinary range of technologies that define modern life. They are essential components in wind turbines, electric vehicle motors, smartphones, LED lighting, medical imaging devices, and military guidance systems. Their value, however, lies not in their total abundance but in the difficulty of extracting, separating, and refining them. Because REEs tend to occur together in nature and share similar chemical properties, separating one from another requires complex and energy-intensive chemical processing. Furthermore, environmental and social considerations—such as waste management, water use, and community impacts—add additional challenges to sustainable production.

Lithium (Li)

Lithium (Li) is a soft, silvery-white metal that is the lightest solid element and is known for its highly reactive chemical properties. This element has become a critical component in the global energy transition, particularly due to its applications in energy storage technologies such as lithium-ion batteries. These batteries are essential for electric vehicles (EVs), renewable energy systems, and various portable electronics, including smartphones and laptops [29, 50]. The increasing electrification of transportation and the development of clean energy infrastructures have further underscored lithium's importance, positioning it as a strategically vital resource in the 21st century [51, 52].

Figure 4: Lithium - Elemental lithium is a silvery-gray metal. It occurs in some minerals, like lepidolite mica (= lithium mica) and spodumene (= a feldspar-like lithium aluminosilicate). James St. John, CC BY 2.0, via Wikimedia Commons.

Geologically, lithium is primarily found in two types of deposits: hard-rock pegmatite and brine. In hard-rock deposits, lithium is mainly sourced from the mineral spodumene, typically extracted from coarse-grained igneous rocks known as pegmatites. Australia is the largest producer, significantly dominating the hard-rock lithium mining sector, particularly in regions such as Greenbushes and Pilgangoora [53, 54]. This method involves conventional mining followed by chemical processing to yield lithium concentrates like lithium carbonate and lithium hydroxide [50]. Conversely, brine deposits are usually found in arid areas where lithium-rich groundwater accumulates in evaporated lake basins or salars, such as those in the Lithium Triangle of South America, which includes Chile, Argentina, and Bolivia. These regions are responsible for more than half of the world's known lithium reserves [52, 55]. Extracting lithium from brine requires evaporation and pumping, which, while less energy-intensive than hard-rock mining, raises environmental concerns about water use in fragile ecosystems [56].

Beyond energy storage, lithium is used extensively in various industrial applications. It is integral in producing heat-resistant glass and ceramics, enhancing durability and thermal stability, and is also a key component in lithium-based lubricants and greases that work effectively at high temperatures [29, 50]. Additionally, lithium compounds are utilized in air purification systems and as additives in pharmaceuticals and lightweight metal alloys [34].

However, the surging demand for lithium due to the electric vehicle boom and the renewable energy sector has created challenges for global supply chains. Issues are prevalent regarding the environmental impacts associated with lithium extraction, particularly concerning water consumption in arid regions and the socio-economic implications of mining activities in developing nations [57, 58]. Furthermore, while recycling lithium-ion batteries presents an avenue to mitigate dependence on newly mined resources, the current recycling rate remains low. Development of efficient recycling technologies is essential for achieving a circular lithium economy, vital for sustaining long-term demand [34, 57].

Cobalt (Co)

Cobalt (Co) is a hard, lustrous, silver-grey metal known for its strength, high melting point, and resistance to corrosion. It plays a vital role in both energy storage technologies and advanced engineering materials, making it one of the most strategically important critical minerals in the modern economy. Its unique electrochemical and metallurgical properties allow it to perform functions that few other elements can replicate, particularly in enhancing battery performance and enabling high-temperature industrial applications.

In the battery industry, cobalt is a crucial component of lithium-ion battery cathodes, where it helps to stabilise the battery's chemical structure and improve energy density [59]. Batteries containing cobalt can store more energy, last longer, and operate safely under a wide range of temperatures. These properties make cobalt essential for electric vehicles (EVs), which require batteries that can deliver high power output while maintaining reliability and safety. Cobalt-based cathode materials, such as lithium cobalt oxide ($LiCoO_2$) and nickel-manganese-cobalt (NMC) or nickel-cobalt-aluminium (NCA) chemistries, dominate current EV and portable electronics markets. By preventing overheating and prolonging the lifespan of rechargeable batteries, cobalt plays a critical role in supporting the global transition toward clean energy and sustainable transportation.

Cobalt, a strategically important metal with diverse applications, primarily serves as a by-product of copper and nickel mining. Geologically, it is rarely found in concentrated deposits of its own, instead occurring within sulphide and laterite deposits. Sulphide deposits, formed through magmatic processes typically at greater depths, are commonly found in regions such as Russia, Canada, and Australia. Significant historical cobalt production has occurred in the Idaho Cobalt Belt in the USA and in major mining regions like Sudbury, Canada, and Norilsk, Russia, which are recognized for their nickel-cobalt sulphide ores [60-62]. Cobalt in these contexts is often economically valuable, reflecting its role as a by-product of the extraction of other metals, primarily copper and nickel [60, 62].

In contrast, laterite deposits are more superficial and predominantly found in tropical and subtropical settings, where intense weathering of ultramafic rocks results in the concentration of cobalt and other metals. Countries like Indonesia, the Philippines, and especially the Democratic Republic of Congo (DRC) are noted as significant mining regions for lateritic ores [60, 62, 63]. The DRC, in particular, represents the largest source of cobalt, supplying over 60% of the global demand [61, 64]. Cobalt mining in the DRC is primarily linked to copper and cobalt oxide ores, reflecting a significant portion of the international supply chain [64].

Figure 5: Cobalt is a ferromagnetic, ductile metal, which is very similar to iron, but is much rarer. Hi-Res Images of Chemical Elements, CC BY 3.0, via Wikimedia Commons.

Cobalt is a ferromagnetic, ductile metal, which is very similar to iron, but is much rarer. Hi-Res Images of Chemical Elements, CC BY 3.0, via Wikimedia Commons.

This concentration of cobalt production in the DRC raises serious ethical and environmental issues. Numerous reports have highlighted the negative social implications associated with artisanal and small-scale mining in the country, including child labour and unsafe working conditions [64, 65]. Moreover, the political and economic instability of the region introduces additional risks to the supply chain, prompting governments and manufacturers to explore alternative sources, responsible sourcing, and recycling initiatives to mitigate dependence on this single country [61, 65]. Thus, while cobalt is crucial for various technologies, particularly in the battery sector, the associated risks necessitate a critical re-evaluation of sourcing strategies to ensure sustainability and ethical compliance [60, 64, 65].

Beyond its use in batteries, cobalt is indispensable in a range of industrial and high-technology applications. It is a key ingredient in superalloys, which are used in jet engines, gas turbines, and aerospace components where materials must withstand extreme heat and stress. Cobalt's high melting point (1,495°C) and excellent corrosion and oxidation resistance make it ideal for these demanding environments. It is also used in magnets, cutting tools, and wear-resistant coatings, as well as in medical applications such as prosthetics and radiation therapy.

Cobalt is a critical enabler of the clean energy transition and modern manufacturing. It combines outstanding physical resilience with essential electrochemical performance, making it invaluable to both the energy and aerospace sectors. However, its extraction and trade present major ethical, environmental, and geopolitical challenges. The global focus is now shifting toward ensuring sustainable, traceable, and ethically sourced cobalt, alongside accelerating battery recycling and developing cobalt-free alternatives to support the growing demand for renewable energy and electric mobility.

Nickel (Ni)

Nickel (Ni) is critically recognized as a strong, silvery-white metal valued for its durability, versatility, and remarkable resistance to corrosion. It is one of the most indispensable industrial metals across various applications, including stainless steel production and high-performance alloys, and plays an increasingly vital role in energy storage technologies. Nickel's unique combination of mechanical strength, ductility, and chemical stability underpins its essential contribution to both traditional manufacturing processes and the burgeoning clean energy economy [66, 67].

The primary industrial application of nickel is in the form of stainless steel, where it contributes to about 70% of the global nickel demand [68, 69]. Nickel's incorporation into steel significantly enhances the material's corrosion resistance, toughness, and strength, making nickel-alloyed steel suitable for rigorous environments in diverse industries, such as construction, food processing, marine engineering, and chemical processing [68, 69]. The ability of nickel to improve the longevity and reliability of everyday products, ranging from cutlery to major structures like skyscrapers, further emphasizes its industrial significance [66].

In recent years, nickel's relevance has expanded notably due to the rapid growth of the electric vehicle (EV) market and the renewable energy sector, where it is a critical component in battery manufacturing. Nickel is predominantly utilized in the cathodes of lithium-ion batteries, especially in the chemistries of nickel-cobalt-aluminium (NCA) and nickel-manganese-cobalt (NMC). High nickel content in these formulations enhances energy density, enabling batteries to store energy more efficiently and extend the driving ranges of electric vehicles [66, 70]. Additionally, nickel improves battery stability and lifespan, which are crucial attributes for high-performance applications [66].

The global nickel market dynamics are influenced significantly by the types of geological deposits from which nickel is extracted. Nickel is mainly sourced from two types of deposits: sulphide and laterite [68, 70]. Sulphide reserves comprise a significant portion of global production, often yielding high-grade nickel concentrates suitable for the battery industry. In contrast, laterite deposits, which account for the majority of nickel reserves, are more complex and costly to process due to their mineralogical nature. Extraction from these deposits generally involves high-pressure acid leaching (HPAL) or pyrometallurgical techniques, which require substantial energy [68]. Major producers of laterite nickel include Indonesia and the Philippines, significantly impacting local economies and the global supply chain for battery materials [69].

Nickel production is categorized into "Class 1" and "Class 2" nickel. Class 1 nickel, boasting a purity level exceeding 99%, is primarily derived from sulphide ores and is essential for battery applications where impurity levels must be meticulously controlled [71]. In contrast, Class 2 nickel, which largely comes from laterite ores, is predominantly used in stainless steel and industrial alloys [68, 72]. The surging demand for Class 1 nickel, primarily driven by the electric vehicle battery sector, has sparked innovations related to nickel refining technologies [71].

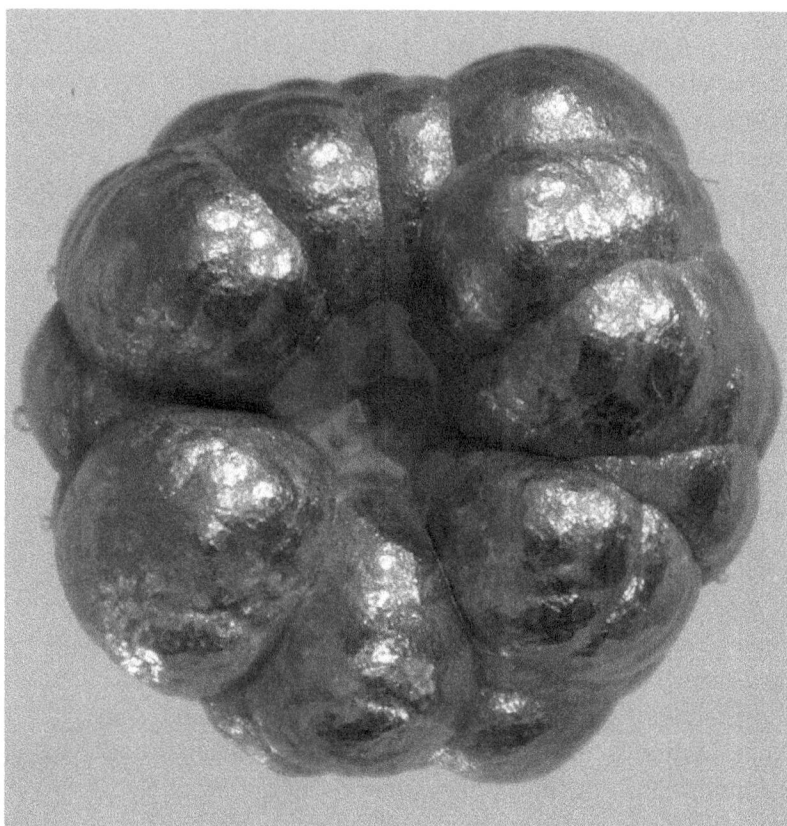

Figure 6: Pure Nickel, obtained by electrolysis of a rod. Jurii, CC BY 3.0, via Wikimedia Commons.

Graphite (C)

Graphite (C) is a naturally occurring crystalline form of carbon, characterized by its unique combination of physical and chemical properties, which include a distinctive soft, lightweight, opaque demeanour coupled with exceptional durability and thermal stability. The crystalline structure of graphite features layers of carbon atoms arranged in a hexagonal lattice formation. Within each layer, strong covalent bonds effectively hold the carbon atoms in place, while the layers themselves are interconnected by much weaker van der Waals forces. This unique layered configuration allows the sheets to easily slide over one another, contributing to graphite's lubricating properties and its recognizable slippery texture. Moreover, the arrangement of these layers facilitates the movement of electrons, rendering graphite an outstanding electrical and thermal conductor—attributes that align its performance with that of certain metals [73, 74].

The importance of graphite in the context of modern technology cannot be overstated, particularly in the burgeoning field of lithium-ion batteries (LIBs). In this application, graphite serves as a critical anode material, playing a vital role in the storage and release of lithium ions throughout the charging and discharging cycles of the battery. The inherent properties of graphite—its electrical conductivity, chemical stability, and layered structure—render it ideal for accommodating the frequent ion exchanges necessitated by these processes. For instance, a typical electric vehicle battery may contain anywhere from 50 to 100 kilograms of graphite, highlighting the increasing demand for this material, particularly in light of the global transition toward electrification and sustainable energy storage solutions [75, 76].

Figure 7: Graphite from Chardonnet in Haute-Savoie, France. Marie-Lan Taÿ Pamart, CC BY 4.0, via Wikimedia Commons.

In terms of geological formation, graphite exists in two primary forms: flake graphite and amorphous graphite, which differ significantly in both origin and application. Flake graphite is produced via metamorphic processes that expose carbon-rich sediments to elevated temperatures and pressures, resulting in a crystalline structure typically found in metamorphic rocks such as schist or gneiss. Such deposits are especially prized for industrial uses, including battery applications, due to their high purity and crystallinity. Conversely, amorphous graphite arises from hydrothermal processes or the metamorphism of coal seams, yielding fine-grained formations. Although these are generally characterized by lower purity and conductivity, amorphous graphite finds utility in applications that do not necessitate exceptionally high performance levels, such as lubricants, paints, and foundry facings [77, 78].

The primary global producers of natural graphite are China, Mozambique, and Madagascar, which collectively represent a sizeable share of the world supply. In particular, China leads the market in both natural and synthetic graphite production, while Mozambique's Balama mine and various deposits in Madagascar serve as significant sources of high-quality flake graphite. Additionally, emerging players such as Canada and Australia are actively focusing on the exploration and development of battery-grade graphite resources, a strategic move intended to diversify supply chains and mitigate dependence on Chinese sources [75, 79].

Beyond its applications in lithium-ion batteries, graphite's functionality extends across a broad spectrum of industrial and technological sectors. For instance, owing to its remarkable heat-resistant attributes, graphite is integral to the manufacturing of refractory materials employed in steel production and other high-temperature processes. Furthermore, its intrinsic lubricating properties make it essential for a variety of dry lubricants and greases in applications where liquid lubricants are impractical. In the nuclear industry, high-purity graphite functions as a neutron moderator in reactors, while its conductivity is leveraged in electronics for components such as conductive coatings and brushes for electric motors [80, 81].

Alongside natural graphite, synthetic graphite has carved a niche in industries that prioritize consistency, purity, and high performance. This synthetic variation is produced by the thermal treatment of petroleum coke or coal tar pitch at extremely high temperatures (exceeding 2,500°C), a method that aligns carbon atoms into a graphite-like structure. Although more costly to manufacture, synthetic graphite's reliability in terms of quality and its performance in critical applications—including battery anodes and electrodes—underscore its importance in high-demand sectors [82, 83].

Platinum Group Metals (PGMs)

Platinum Group Metals (PGMs) refer to a family of six closely related metallic elements—platinum (Pt), palladium (Pd), rhodium (Rh), iridium (Ir), ruthenium (Ru), and osmium (Os)—known for their unique physical and chemical properties. These metals exhibit high density, exceptional corrosion resistance, and remarkable stability under elevated temperatures, making them critical resources in various industrial applications [84, 85]. Their notable catalytic activity enables them to accelerate chemical reactions without undergoing consumption in the process, thereby making PGMs indispensable in a wide array of environmental and energy technologies [86].

Among the PGMs, platinum and palladium hold significant positions, particularly in the automotive sector, where they are essential components of catalytic converters. These devices play a pivotal role in transforming harmful emissions, such as carbon monoxide (CO), unburned hydrocarbons (HCs), and nitrogen oxides (NO_x), into less harmful substances like carbon dioxide (CO_2), water vapor, and nitrogen (N_2) [87]. Platinum is predominantly utilized in diesel engines due to its efficiency in oxidizing CO and HCs, while palladium is favoured for petrol engines [87]. The requirement for higher PGM content in catalytic converters relates to rising costs for manufacturers, thereby highlighting their economic importance and influence on global emission regulatory measures [87, 88].

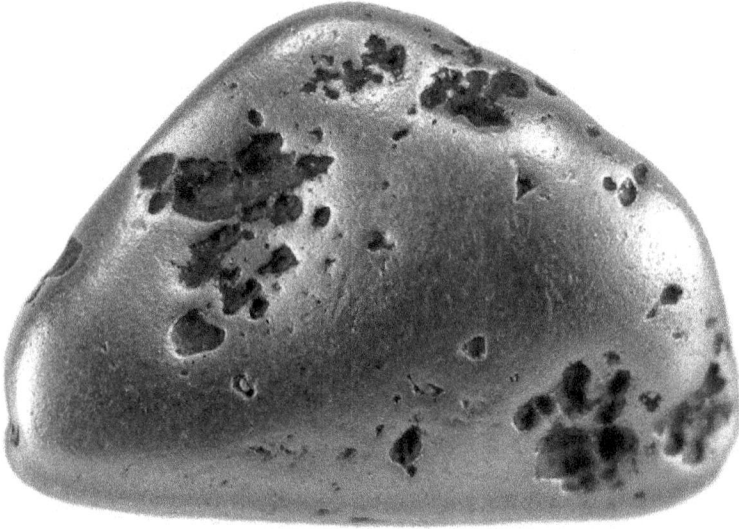

Figure 8: This is a lustrous, waterworn, metallic gray, platinum nugget. It is obviously from an alluvial deposit, and is a very attractive nugget with good heft for the size (7 grams). Rob Lavinsky, iRocks.com – CC-BY-SA-3.0, CC BY-SA 3.0, via Wikimedia Commons

Rhodium complements the role of platinum and palladium by enhancing the efficiency and selectivity of catalytic reactions, especially in reducing nitrogen oxide emissions, which are significant sources of pollution in vehicular exhaust [87]. Despite its limited availability and high cost, even small amounts of rhodium significantly improve the performance of catalytic converters, underscoring its value in emission reduction applications [87]. Beyond automotive uses, PGMs are gaining traction in hydrogen energy systems, with iridium and ruthenium serving critical roles in electrolysis processes and fuel cell technologies, facilitating the conversion of hydrogen into electricity [88, 89].

The rarity of PGMs on Earth contributes to their high value and strategic significance. These metals typically occur in very low concentrations within geological formations, predominantly as by-products of nickel and copper mining operations. Major PGM deposits are concentrated in regions such as the Bushveld Complex in South Africa, the Norilsk-Talnakh region in Russia, and the Great Dyke in Zimbabwe, with these three locations accounting for over 90% of global PGM supply [85, 90]. The geological processes that form these deposits include magmatic differentiation, which concentrates PGMs within specific sulphide layers, making their extraction technically challenging and geologically unique [85, 90].

Beyond their roles in catalysis and clean energy, PGMs are also utilized in jewellery, medical devices, chemical processing equipment, and electronic components due to their superior chemical stability and aesthetic appeal [88]. For instance, platinum's lustrous appearance and resistance to tarnishing make it a popular choice for high-end jewellery and dental materials [88]. The growing demand from the hydrogen economy and increasing application in green technologies have further intensified the strategic importance of PGMs, raising concerns about supply security, price fluctuations, and the environmental impacts associated with their extraction and refining [88, 90].

Other Emerging Critical Minerals

In recent years, an emerging consensus has recognized a broader array of elements as critical minerals beyond the traditional focus on rare earth elements and common battery materials such as lithium, cobalt, and nickel. This proliferation of critical minerals encompasses manganese (Mn), vanadium (V), niobium (Nb), gallium (Ga), and indium (In), as their roles in advanced technologies and the clean energy economy become increasingly vital.

Manganese, historically significant in steel production, has gained renewed importance with advancements in battery technologies. Its application in lithium-ion batteries—particularly in nickel-manganese-cobalt (NMC) and lithium manganese oxide (LMO) cathodes—enhances stability, safety, and energy capacity, rendering it a more abundant and cost-effective alternative to cobalt and nickel [91]. The global supply primarily originates from South Africa, Australia, and Gabon, where refining processes are geographically concentrated, heightening supply chain risks in response to escalating global battery demand [91, 92].

Vanadium is similarly pivotal in the transition to renewable energy, particularly through its role in vanadium redox flow batteries (VRFBs), which are designed for large-scale energy storage in renewable systems like solar and wind farms. These batteries are known for their durability and ability to undergo extensive charge-discharge cycles without degradation [93]. Moreover, vanadium's utility as a steel-strengthening additive underscores its importance across industrial applications [93]. Major producers include China, Russia, and South Africa, with its extraction often linked to by-products from iron and titanium mining, contributing to supply chain vulnerabilities [91].

Niobium enhances the performance of high-strength materials, particularly in aerospace and electronics, where it is used in superconductors that operate without resistance at low temperatures [91]. Brazil dominates global niobium production, holding over 80% of known reserves, thus making the geopolitical landscape of this mineral particularly sensitive [93].

Gallium and indium are critical components in the semiconductor and optoelectronics sectors, with gallium being essential for materials such as gallium arsenide (GaAs) and gallium nitride (GaN), prevalent in LEDs and communication technologies [92, 94]. Indium is primarily known through its association with indium tin oxide (ITO), vital for touchscreens and flat-panel displays [91, 93]. Both elements are predominantly recovered as by-products from the mining of aluminium, zinc, or copper, raising concerns about their supply stability due to dependencies on other metal industries [91, 92].

The classification of critical minerals is inherently dynamic, influenced by technological advancements and shifting market demands. As geopolitical circumstances evolve, nations such as the United States, Australia, Japan, and countries in the European Union continue to update their lists of critical minerals based on emerging industrial and strategic priorities [92, 95]. New discoveries and innovations reflect a growing recognition of the importance of these materials in supporting the clean energy transition and technological advancements essential for modern industries.

Overall, it is evident that the understanding of critical minerals encompasses a wider array of elements like manganese, vanadium, niobium, gallium, and indium. Their essential contributions to emerging technologies, renewable energy systems, and high-performance manufacturing highlight the increasing necessity for secure supply chains that align with modern industrial and environmental objectives.

Overview of the Global Supply Chain and Criticality Concept

The global supply chain for rare earth and critical minerals constitutes a multifaceted and interconnected network essential for modern technological systems, particularly in the realms of clean energy, advanced manufacturing, and digital infrastructure. This supply chain spans multiple stages including exploration, mining, processing, manufacturing, and recycling. The significance of this system is underscored by the increasing emphasis on transitioning to sustainable energy sources, raising the stakes for resource security amid geopolitical tensions.

At the outset of this supply chain, critical minerals are sourced from geographically concentrated areas with specific geological formations. For instance, Australia stands as a leader in lithium production, whereas the Democratic Republic of Congo (DRC) is notable for cobalt extraction. Other critical sources include Indonesia and the Philippines for nickel, along with South Africa and Russia, which dominate the market for platinum group metals [96]. The limited number of countries that can effectively yield these minerals complicates global supply dynamics, as described by Pawar and Ewing, who highlight that the rare earth element supply chain is significantly controlled by just a handful of nations [97].

The processing and refining of these minerals represent a crucial bottleneck within the supply chain. Specifically, it is estimated that China controls approximately 60–90% of the global refining capacity for key minerals like rare earths, graphite, and lithium compounds [98]. This concentration not only raises questions about overreliance on Chinese supply chains but also exposes international markets to considerable risks of disruption. The market for these minerals is influenced by geopolitical factors, drawing attention to strategic vulnerabilities, as noted by Nygaard, who indicates that a significant percentage of crucial minerals is sourced from politically unstable regions [98].

The manufacturing phase, wherein processed materials find application in high-value products such as electric vehicle batteries and renewable energy systems, is predominantly concentrated in developed economies including the U.S., Japan, and EU nations [96].

Rare Earth and Critical Mineral Operations and Processing

Industries are vulnerable to upstream supply risk due to their dependence on a narrow range of low-cost inputs from mineral powerhouse countries [99]. This phase of the supply chain highlights the dichotomy of technological advancement in economically stable nations, yet their vulnerability to potential supply disruptions rooted in upstream dependencies.

In addition to exploration and primary mining, recycling and urban mining have emerged as critical components of resource management in this supply chain. The potential to recover valuable minerals from electronic waste and spent batteries offers a path towards reducing dependency on primary mining operations [100]. Yet, challenges such as technical limitations and economic barriers hinder the widespread implementation of effective recycling practices [101]. Mining sectors can be potentially transformed by improving recovery techniques and advancing recycling technology [102], securing a secondary supply that aligns with principles of the circular economy and alleviates some environmental concerns associated with mining activities.

Assessing the criticality of different minerals is essential for determining which resources require strategic management to mitigate risks associated with supply disruptions. Critical minerals are defined by their economic importance and the high risk of supply interruption [103]. In this context, trade organizations and government agencies across the globe, including the U.S. Geological Survey and the European Union, regularly update their lists of critical minerals based on evolving technological needs and geopolitical contexts [104]. The transition to low-carbon energy systems has made materials like lithium and cobalt increasingly central to global supply discussions, deeming them essential for future sustainability while inherently complex regarding their supply chains [96].

Role of Australia and Other Major Producers

Australia plays a central and growing role in the global supply of rare earth and critical minerals, positioning itself as a secure, ethical, and technologically advanced supplier for the global energy transition. With abundant geological resources, world-class mining expertise, and strong regulatory and environmental frameworks, Australia has become a leading producer of several key minerals, including lithium, rare earth elements, nickel, cobalt, manganese, and graphite. These materials are critical for renewable energy technologies, electric vehicles (EVs), and advanced manufacturing systems.

Australia currently leads the world in lithium production, accounting for over half of global output. In 2022, Australia provided 52% of the global supply, and in 2023, it was the highest producer with 74,700 tonnes, accounting for 51% of the world's total, as reported by Natural Resources Canada [105]. The Greenbushes and Pilgangoora mines in Western Australia are among the largest and highest-grade spodumene deposits globally, supplying battery manufacturers across Asia, Europe, and North America. In addition, the Mount Weld mine, operated by Lynas Rare Earths, is one of the world's richest rare earth deposits outside China. The downstream processing of Mount Weld's concentrate in Malaysia and plans for expanded domestic refining at Kalgoorlie and other sites reflect Australia's broader strategy to move up the value chain—from raw material supplier to advanced materials processor.

As a case study example, Australia has a long and successful tradition of processing minerals and producing industrial chemicals, which has been a cornerstone of its economic development for decades. The sector not only provides essential inputs for agriculture, construction, and manufacturing but also underpins Australia's growing participation in high-value global markets for clean energy and advanced materials. By expanding mineral processing capacity, the nation aims to move beyond exporting unprocessed ores toward creating integrated value chains that convert its natural resources into refined products and specialised compounds for use in batteries, renewable technologies, and sustainable manufacturing.

A key example of this transformation is the work of Wesfarmers Chemicals, Energy and Fertilisers (WesCEF), a division of Wesfarmers Limited. WesCEF has long been involved in extracting, processing, and manufacturing essential commodities such as fertilisers, industrial chemicals, and energy products that support both Australian and international markets. The company exemplifies the shift from resource extraction to value-added mineral processing, where raw minerals are refined and converted into higher-value materials. This approach enhances economic returns, creates skilled jobs, and strengthens Australia's position in global supply chains [106].

One of WesCEF's most notable ventures is its joint project at Mt Holland in Western Australia, which focuses on lithium processing—a critical element in electric vehicle (EV) batteries and renewable energy storage systems. The Mt Holland project will use locally mined lithium as feedstock for an integrated refinery at Kwinana, where it will be converted into battery-grade lithium hydroxide. Once operational, the refinery is expected to produce approximately 50,000 tonnes per year of high-purity lithium hydroxide, positioning Australia as a key supplier to the global battery industry. By refining lithium domestically rather than exporting it as spodumene concentrate, this project demonstrates how Australia can capture more value within its borders while supporting the global shift to cleaner technologies [106].

To accelerate such developments, the Australian Government is actively supporting the establishment of onshore mineral and chemical processing operations through targeted investment and policy measures. One key initiative is the Critical Minerals Development Program, which provides funding to projects that expand domestic refining, processing, and battery material production. Under this program, the government has allocated $100 million to several strategic ventures, including [106]:

- Australian Energy Storage Solutions' battery precursor manufacturing operations, which will help create local capability for producing cathode and anode materials;

- Queensland Pacific Metals' energy chemicals refinery, which focuses on refining nickel and cobalt for battery applications using more sustainable processing technologies; and

- Alpha HPA's high-purity alumina (HPA) precursor production facilities, which support the manufacturing of advanced battery components and LED lighting materials.

These investments reflect a broader national strategy—outlined in the *Critical Minerals Strategy 2023–2030*—to position Australia as a trusted global supplier of not just raw materials but also refined, high-value products. By fostering domestic processing and manufacturing, Australia aims to reduce its reliance on foreign refining capacity (particularly in China), build resilient supply chains with like-minded trading partners, and contribute to the global transition toward net-zero emissions.

Australia also produces significant quantities of nickel, cobalt, and manganese, much of it extracted from lateritic deposits in Western Australia and Queensland. These materials are essential for battery cathode manufacturing. Projects such as BHP's Nickel West operation and emerging battery precursor plants in Kwinana are part of the nation's effort to establish an integrated battery materials industry, reducing dependence on offshore processing. Graphite exploration is also expanding rapidly in South Australia and the Northern Territory, while vanadium, rare earth, and hydrogen-linked projects are emerging as part of the country's broader critical minerals diversification strategy.

Australia's strategic positioning in the global marketplace extends beyond its abundant mineral resources. The country benefits from transparent governance and strong environmental standards that contribute to a stable investment atmosphere. These characteristics make Australia an appealing partner for nations seeking to diversify their supply chains, particularly in response to challenges posed by COVID-19, which highlighted vulnerabilities in chains heavily reliant on single sources like China. Businesses are increasingly looking to enhance resilience through diversified supply strategies [107, 108].

The *Australian Government's Critical Minerals Strategy (2023–2030)* [106] exemplifies this proactive approach, focusing on exploration, processing, recycling, and research related to critical minerals. This strategy has received significant financial support, including the establishment of the Critical Minerals Facility backed by an investment of A$4 billion. Strategic partnerships with allied nations such as the United States, Japan, South Korea, India, and the European Union further strengthen this initiative by providing diversified sources for essential materials [109, 110].

This framework aims not only to solidify Australia's position in the critical minerals supply chain but also to enhance resilience against global supply disruptions. The integration of international alliances with domestic policies underscores Australia's commitment to improving global supply chain resilience while promoting low-carbon industrial capabilities [111]. Understanding the implications of supply chain diversification in this context is crucial for ensuring the stability of critical resource supplies amid geopolitical tensions [112].

Moreover, these policies align with broader transformations in global trade influenced by recent geopolitical conflicts and export restrictions, particularly concerning critical materials essential for technological advancement. The diversification efforts serve as a response to immediate supply chain challenges and constitute a long-term strategic shift aimed at fostering sustainability and resilience in an increasingly interconnected global economy [113].

Globally, other major producers also play crucial and complementary roles. China remains the dominant force in the rare earth sector, controlling roughly 60–70% of mining and nearly 90% of processing and refining capacity. Its vertically integrated supply chain—from extraction to magnet manufacturing—gives it enormous strategic leverage. The Democratic Republic of Congo (DRC) supplies around 60% of the world's cobalt, although this dominance raises ethical and sustainability concerns related to artisanal mining and child labour. Indonesia and the Philippines are leading producers of nickel, with massive laterite deposits feeding Asia's stainless steel and battery industries. Russia, South Africa, and Zimbabwe are the primary suppliers of platinum group metals (PGMs), essential for catalytic converters, hydrogen fuel cells, and industrial catalysts.

Canada and the United States are also emerging as strategic players, investing in exploration and processing projects to reduce dependency on foreign supply chains. Canada's focus on battery metals (nickel, cobalt, lithium, and graphite) complements its strong environmental governance, while the United States is rebuilding domestic refining capacity and establishing alliances under frameworks such as the Minerals Security Partnership (MSP). Brazil, meanwhile, is a major source of niobium and hosts promising lithium and rare earth projects, adding to the global diversity of supply.

Key Terms and Concepts

Rare Earth Elements (REEs): A group of 17 metallic elements, including the fifteen lanthanides plus scandium and yttrium, valued for their unique magnetic, catalytic, and optical properties.

Critical Minerals: Minerals identified as essential for modern technologies and economic development but at risk of supply disruption due to limited availability or geopolitical factors.

Lanthanides: A series of fifteen elements from lanthanum (La) to lutetium (Lu) on the periodic table, forming the core of the rare earth family.

Strategic Materials: Raw materials vital to national security, defence, and advanced manufacturing sectors, often stockpiled or prioritised by governments.

Criticality Assessment: A framework used to evaluate minerals based on their economic importance and supply risk, guiding national and industrial resource strategies.

Clean Energy Technologies: Systems that generate power or mobility with low environmental impact, such as wind turbines, electric vehicles, and solar panels, all of which rely on critical minerals.

Supply Chain: The sequence of processes involved in producing and delivering a mineral product from exploration through extraction, processing, and end-use application.

Resource Security: The ability of a nation or organisation to ensure stable access to essential raw materials for industrial and strategic needs.

Geopolitical Influence: The impact of political, economic, and trade relationships on the control and distribution of mineral resources.

End-Use Applications: The final products and technologies—such as magnets, batteries, semiconductors, and defence systems—that depend on rare earth and critical minerals.

Economic Importance: The contribution of a mineral to national or industrial productivity, innovation, and competitiveness.

Supply Risk: The probability that access to a mineral resource may be disrupted due to scarcity, political instability, or market imbalance.

Key Terms and Concepts

Downstream Processing: Refining and value-added manufacturing stages that convert raw minerals into usable components or products.

Upstream Production: The early stages of the supply chain involving exploration, mining, and concentration of mineral ores.

Australia's Critical Minerals Industry: The network of exploration, mining, and processing enterprises positioning Australia as a global leader in sustainable supply of critical raw materials.

Chapter 1 Review Questions

Definitions (Short Answer)

1. Define rare earth elements (REEs).

2. Define critical minerals.

3. Explain how REEs differ from the broader category of critical minerals.

Classification & Identification

4. Classify each as LREE or HREE: Nd, Dy, Tb, La, Sm, Yb.

5. List four non-REE critical minerals central to modern batteries.

6. Match the mineral to its typical setting/use:

Lithium	(i) Brine salars & pegmatites
Cobalt	(ii) By-product of Cu/Ni; laterites & sulphides
Graphite	(iii) Anode material; flake in metamorphic rocks
PGMs	(iv) Catalysts; Bushveld/Norilsk magmatic systems

Applications (Short Answer)

7. Give one clean-energy application each for: (a) REEs, (b) lithium, (c) PGMs.

8. Name two electronics/telecom uses that rely on REEs.

9. State the primary use of nickel by volume and one growing use.

Deposits & Geology (Multiple Choice)

10. REEs are "rare" in practice mainly because:

    A. Geologically absent
    B. Always radioactive
    C. Rarely occur in economic concentrations & hard to separate
    D. Prohibited from mining

11. Deposit type most associated with HREEs and low-cost leaching:

    A. Carbonatite/alkaline hard rock
    B. Ion-adsorption clay regolith
    C. Placer only

D. Magmatic sulphides

12. Lithium occurs primarily in:

A. Laterites & placers
B. Pegmatites & brine salars
C. BIFs only
D. Kimberlites

Supply Chain & Criticality

13. List the main stages of the critical mineral supply chain in order (mine to market).

14. Name one common chokepoint in these supply chains and explain why.

15. Define the criticality concept and name two factors used in assessments.

16. Give two policy tools governments use to strengthen mineral security.

Geography & Producers

17. Identify the country most associated with cobalt supply concentration and ESG concerns.

18. Name two major nickel-producing countries and the deposit type most associated with each (laterite or sulphide).

19. Describe Australia's strengths in the critical minerals value chain and one area it is expanding into.

20. Contrast Australia's role in REEs with China's role.

Strategy, Risk & Sustainability

21. Provide two reasons recycling/urban mining matter for critical minerals.

22. Give two measures/metrics you would track to monitor graphite supply risk.

23. A battery maker reliant on NMC faces cobalt risk—suggest two technical and two supply-chain responses.

24. Briefly discuss how international cooperation can reduce supply risk without stifling trade.

Chapter 2

Geology, Occurrence, and Exploration

The transition to low-carbon and high-technology industries has made rare earth and other critical minerals indispensable to modern life. From electric-vehicle batteries and wind turbines to advanced electronics and defence systems, these resources underpin the global shift toward cleaner energy and digital infrastructure. Understanding where and how these minerals form is therefore essential for both geoscientists and policy makers.

This chapter introduces the geological foundations of rare earth and critical mineral deposits—how they originate, concentrate, and become economically viable. It explores the diverse geological systems responsible for their formation, identifies characteristic host rocks and ore minerals, and examines major deposit types found around the world. The discussion extends to the tools and techniques of mineral exploration—mapping, sampling, geophysical and geochemical analysis, drilling, and resource estimation—while considering the environmental and regulatory frameworks that guide responsible development. Together, these topics provide the knowledge base needed to interpret, evaluate, and sustainably manage critical mineral resources within the global supply chain.

Learning Outcomes	
This chapter aims to give you the ability to: 1. Explain the geological processes responsible for the formation and concentration of rare earth and critical mineral deposits, including magmatic, hydrothermal, sedimentary, and lateritic systems. 2. Identify the common host rocks and ore minerals associated with major critical elements, including bastnäsite, monazite, xenotime, spodumene, and other economically significant minerals. 3. Differentiate between key deposit types—such as carbonatite, pegmatite, placer, and laterite deposits—and recognise their characteristic mineral assemblages and formation environments.	

Learning Outcomes	

4. Describe major global examples of rare earth and critical mineral deposits and evaluate their significance in the global supply chain.
5. Outline and apply fundamental exploration methods, including geological mapping, geochemical sampling, geophysical surveys, drilling, and core logging, to locate and evaluate potential deposits.
6. Interpret basic assay results and geochemical data to assess the presence, quality, and potential value of critical minerals within an exploration area.
7. Discuss the concepts of ore grade, tonnage, and cut-off grade, and explain how these parameters influence project feasibility and economic decision-making in mineral exploration.
8. Recognise the importance of environmental, social, and regulatory considerations during exploration activities, particularly in sensitive or remote regions.

How Rare Earth and Critical Mineral Deposits Form

Rare earth and critical mineral deposits form through a variety of geological processes that concentrate elements dispersed throughout the Earth's crust into economically viable accumulations. These processes occur over millions of years under specific temperature, pressure, and chemical conditions that allow the metals to separate from surrounding rocks and fluids, creating zones rich enough to mine.

For rare earth elements (REEs), the main types of deposits include igneous, weathered (ion-adsorption), and sedimentary sources. Igneous processes, such as the crystallisation of carbonatite and alkaline magmas, form some of the world's largest rare earth deposits. These magmas are enriched in volatile components and incompatible elements, which facilitate the concentration of REEs as the magma cools. The Mount Weld deposit in Australia and Bayan Obo in China are classic examples formed this way. Over time, weathering and hydrothermal alteration can further enrich these deposits. In tropical and subtropical regions, prolonged weathering of REE-bearing rocks creates ion-adsorption clay deposits, where rare earth ions are weakly bound to clays and can be extracted through leaching — a key source of heavy rare earths (HREEs) in southern China.

Igneous processes are one of the main ways rare earth element (REE) deposits form. These processes begin deep within the Earth's mantle, where heat and pressure cause portions of the mantle rocks to partially melt. This partial melting produces magma that is enriched in certain elements, including rare earths, because these elements prefer to enter the liquid phase rather than remain in solid minerals. As this magma rises through the crust, it cools and changes composition—a process known as magmatic differentiation. During this stage, rare earth elements become concentrated in the residual melt, which can then crystallize to form REE-rich minerals.

Some of the most important REE sources are alkaline igneous rocks and carbonatites—unusual magmas that are rich in alkali elements (like sodium and potassium) or carbonates. These magmas are particularly effective at concentrating rare earths because they have low viscosity and high fluid content, allowing REEs to migrate and accumulate in late-stage mineral phases. Over time, as the magma cools, rare earth-bearing minerals such as bastnäsite, monazite, and xenotime can crystallize directly from the remaining melt or precipitate from hydrothermal fluids that circulate through fractures. In some cases, the remaining REE-rich melt can migrate into new intrusions, forming additional mineralized zones.

This process explains why major deposits like Mount Weld in Australia and Bayan Obo in China are closely linked to carbonatite intrusions. These igneous systems act as natural concentrators, gathering rare earth elements from deep mantle sources and locking them into solid rock as the magma evolves and cools.

Weathering and sedimentation are key surface processes that contribute to the formation of secondary rare earth and critical mineral deposits. These processes occur when existing rocks are exposed to the atmosphere, water, and biological activity, leading to the gradual breakdown, movement, and re-deposition of minerals. Unlike igneous deposits that form deep underground from molten material, weathering and sedimentation act at or near the Earth's surface, redistributing valuable elements into new, concentrated formations.

In tropical and subtropical climates, intense and prolonged weathering can produce laterite soils—thick, iron- and aluminium-rich layers that develop as soluble materials are leached away by rainfall, while less soluble elements such as iron, aluminium, and rare earth elements (REEs) remain behind. Over millions of years, this process can significantly enrich the upper soil layers in REEs, especially in areas underlain by REE-bearing rocks like carbonatites or granites. These ion-adsorption clay deposits—common in southern China—are among the most important global sources of heavy rare earth elements (HREEs), as the REE ions become weakly bound to clay minerals and can later be extracted through simple leaching.

Sedimentation plays another major role by concentrating resistant minerals into placer deposits. As weathering breaks down rocks containing REE-bearing minerals such as monazite and xenotime, these heavy minerals are released and transported by rivers and streams. Because of their high density, they tend to settle in riverbeds, beaches, or coastal sands, forming placer deposits rich in valuable materials. Such deposits are not only important for rare earths but also for other heavy minerals like titanium, zirconium, and tin.

For critical minerals such as lithium, cobalt, nickel, and graphite, different geological settings apply. Lithium forms in two main environments: hard-rock pegmatites, where it crystallises from late-stage granitic melts, and brine deposits, where lithium-bearing groundwater accumulates and evaporates in closed basins, forming concentrated saline lakes (e.g., Chile's Salar de Atacama). Cobalt and nickel are typically concentrated in magmatic sulphide deposits—formed from molten rock containing metal-rich fluids—or lateritic deposits, which develop through tropical weathering of ultramafic rocks that enrich cobalt and nickel near the surface. Graphite, on the other hand, forms through metamorphism of carbon-rich sediments, where heat and pressure realign carbon atoms into crystalline layers, creating the flake or vein graphite used in battery anodes.

The formation of critical minerals such as lithium, cobalt, nickel, and graphite occurs in distinctly different geological environments, each shaped by specific physical and chemical processes. These differences influence not only where deposits are found but also how they are mined, processed, and utilised in modern technologies like electric vehicles and renewable energy systems.

Lithium is found in two major types of deposits—hard-rock pegmatites and brine basins—each representing a unique geological setting. In hard-rock pegmatites, lithium crystallises from late-stage granitic melts that cool slowly beneath the Earth's surface. As these granitic magmas evolve, incompatible elements such as lithium, beryllium, and tantalum become concentrated in the final molten fraction, eventually forming minerals like spodumene, lepidolite, or petalite. These pegmatite deposits, found in regions such as Western Australia's Greenbushes and Pilgangoora, are mined through conventional open-pit or underground methods. In contrast, brine deposits develop in arid, closed basins where lithium-bearing groundwater accumulates and gradually evaporates, leaving behind highly saline lakes rich in lithium chloride and other salts. These so-called salars, such as Chile's Salar de Atacama or Bolivia's Salar de Uyuni, represent the world's most productive lithium sources due to the efficiency of solar evaporation in concentrating the dissolved lithium.

Cobalt and nickel typically form together in two main geological settings: magmatic sulphide deposits and lateritic deposits. Magmatic sulphide deposits originate deep within the Earth's crust, where molten rock rich in iron, magnesium, and sulphur cools and separates into metal-bearing layers. As the magma crystallises, dense sulphide liquids containing nickel, copper, and cobalt segregate from the silicate melt and settle to form ore bodies. Major examples include the Norilsk-Talnakh complex in Russia and Sudbury Basin in Canada. In contrast, lateritic deposits form closer to the surface under tropical climatic conditions. Here, intense weathering of ultramafic rocks—which are rich in magnesium and iron—causes mobile elements to leach away while nickel and cobalt become concentrated within iron and manganese oxide zones. Such lateritic deposits are widespread in Indonesia, the Philippines, and Australia, where they serve as key sources of battery-grade metals.

Graphite forms through an entirely different geological process known as metamorphism, which involves the transformation of existing rocks under heat and pressure. When carbon-rich sediments such as coal, shale, or limestone are buried deep within the Earth and subjected to metamorphic conditions, the carbon atoms rearrange into crystalline layers of graphite. This reorganisation creates the flake graphite commonly found in metamorphic rocks such as schist and gneiss. In some cases, graphite can also form as vein graphite, where carbon-bearing fluids migrate through cracks and solidify as pure graphite veins. Major graphite deposits occur in

Rare Earth and Critical Mineral Operations and Processing

China, Madagascar, Mozambique, and Canada, where both natural flake and vein types are extracted for use in battery anodes, refractories, and lubricants.

Together, these geological processes—igneous crystallisation, weathering and enrichment, and metamorphic transformation—illustrate the diverse origins of critical minerals. Each setting reflects a unique combination of temperature, pressure, fluid chemistry, and tectonic conditions that determine how economically viable concentrations of lithium, cobalt, nickel, and graphite form within the Earth's crust.

Other critical elements, including vanadium, niobium, gallium, and indium, have complex geological origins and are often recovered as by-products rather than from deposits mined specifically for them. Their occurrence is closely tied to the geochemical behaviour of more abundant metals, meaning they become concentrated within certain host rocks or mineral systems through subtle chemical processes. Although these elements are only present in small quantities, their unique properties—such as strength enhancement, superconductivity, and electronic conductivity—make them essential for advanced technologies in energy storage, electronics, and aerospace engineering.

Vanadium is most commonly found in titanium-bearing igneous rocks, iron ores, and shale formations. It tends to substitute for iron and titanium in crystal lattices during magma crystallisation, owing to its similar ionic size and charge. As a result, vanadium becomes concentrated in magmatic layered intrusions and mafic to ultramafic complexes, where minerals like magnetite and ilmenite host economically recoverable amounts. For instance, large vanadium-titanium magnetite deposits are found in South Africa's Bushveld Complex, Russia's Kachkanar deposits, and China's Panzhihua complex. Vanadium is also enriched in black shales and sandstone-hosted uranium deposits, where it occurs as a secondary mineral formed by the reduction of vanadium-bearing fluids. The metal's main industrial use is in vanadium redox flow batteries (VRFBs) for large-scale energy storage and in vanadium steel alloys, which greatly enhance strength, toughness, and corrosion resistance.

Niobium shares a close geological and chemical relationship with vanadium but is more often associated with carbonatite and alkaline igneous complexes. These unusual magmatic systems are rich in volatile elements such as carbon dioxide, fluorine, and phosphorus, which allow rare elements like niobium and tantalum to concentrate during the final stages of magma crystallisation. One of the world's most important niobium sources is the Araxá and Catalão carbonatite complexes in Brazil, which together supply more than 80% of global demand. Niobium also occurs in pyrochlore and columbite-tantalite minerals, often mined alongside rare earth elements. The element's ability to improve heat resistance and strength in superalloys makes it vital for jet engines, superconductors, and advanced electronics.

Gallium and indium differ from vanadium and niobium in that they are trace metals not concentrated enough to form their own ores. Instead, they are typically obtained as secondary products from the processing of more abundant base metals. Gallium is primarily recovered during the refining of bauxite (the main ore of aluminium) and, to a lesser extent, from zinc ores. It substitutes for aluminium in the crystal structure of bauxite minerals due to their chemical similarity. Once extracted, gallium is used to produce gallium arsenide (GaAs) and gallium nitride (GaN) semiconductors—key materials for high-speed electronics, LEDs, and solar cells. Indium, on the other hand, is mainly recovered from zinc and copper smelting residues. It tends to concentrate in sphalerite, a zinc sulphide mineral, where it substitutes for zinc in small amounts. Indium is best known for its use in indium tin oxide (ITO), a transparent conductive film essential for touchscreens, flat-panel displays, and photovoltaic cells.

Common Host Rocks and Ore Minerals (Bastnäsite, Monazite, Spodumene, Etc.)

Common host rocks and ore minerals play a crucial role in the concentration and economic extraction of rare earth elements (REEs) and other critical minerals. These minerals do not occur uniformly in the Earth's crust but are instead hosted within specific geological environments—known as host rocks—that provide the right chemical conditions for mineral formation and accumulation. Understanding the relationships between host rocks, ore minerals, and the geological processes that form them is essential for exploration, mining, and sustainable resource development.

Bastnäsite

Bastnäsite, a fluorocarbonate mineral expressed as $(Ce,La)(CO_3)F$, is one of the most significant rare earth minerals, serving as a primary source for vital light rare earth elements (LREEs) such as cerium, lanthanum, and neodymium. Its formation in carbonatite and alkaline igneous rock environments is crucial, as these geological formations are characterized by their rich carbonate and fluorine content. Studies indicate that rare earth elements, particularly LREEs, are often concentrated during the late stages of magma crystallization within these unique magmatic systems, facilitating their high-grade deposits [114-116].

Figure 9: Bastnäsite from Burundi. Kouame, CC BY-SA 3.0, via Wikimedia Commons.

The prominence of bastnäsite in the global rare earth mineral landscape is further supported by world-class deposits found in locations such as Mountain Pass in the USA, Bayan Obo in China, and Mount Weld in Australia. Each of these sites is recognized not only for its substantial reserves of bastnäsite but also as a crucial source of LREEs [117, 118]. Specifically, the Mountain Pass deposit has been noted for its complex carbonate ore mineralogy, which contributes to significant REE extraction [119]. Moreover, Bayan Obo is reported to account for a significant portion of global REE production and is closely linked to geological processes associated with carbonatite formations [120].

Importantly, while bastnäsite's relatively high content of REEs and its ease of processing compared to other minerals enhance its extractive value, there are notable environmental challenges associated with its mining and refining processes. This is primarily due to the mineral's association with radioactive thorium, necessitating careful management to mitigate ecological impacts [121, 122]. The extraction methods employed, particularly acid leaching, have been shown to effectively recover La and Ce from bastnäsite, but they also raise concerns regarding the management of thorium and other by-products during the processes [120].

Monazite

Monazite, a phosphate mineral with the chemical formula $(Ce,La,Nd,Th)PO_4$, plays a significant role as a source of rare earth elements (REEs), particularly the light rare earths, and is commonly found alongside bastnäsite in various geological settings. Monazite primarily forms in granitic pegmatites and metamorphic rocks such as gneiss and schist, where phosphate-rich fluids facilitate the crystallization of REE-bearing minerals [123, 124]. The presence of

41

monazite in such formations can be attributed to its genesis in environments characterized by high geological activity and alteration processes that promote the mineralization of REEs [125].

The high resistance of monazite to weathering significantly contributes to its accumulation in placers—secondary deposits created by the concentration of minerals via water flow in riverbeds or coastal settings [126, 127]. This property has made monazite a critical source of REEs, particularly in regions such as India, Brazil, and Australia, where these minerals are often mined from heavy mineral sands alongside bastnäsite and xenotime [128, 129]. The economic viability of extracting monazite and associated REE minerals is underscored by their prevalence in mining operations and the increasing demand for REEs in various high-tech and energy applications [130].

Figure 10: Monazite. Rob Lavinsky, iRocks.com – CC-BY-SA-3.0, via Wikimedia Commons.

However, the presence of thorium, a radioactive element in monazite, poses challenges for handling and storage due to necessary safety and environmental precautions [131]. Thorium concentration complicates the mining processes and necessitates careful assessment for environmental management to mitigate potential hazards associated with radioactivity [117]. As such, the extraction and processing of monazite are subjected to rigorous regulatory standards to ensure the safety and sustainability of REE production [132].

Rare Earth and Critical Mineral Operations and Processing

Xenotime

Xenotime, chemically represented as YPO_4, is a mineral characterized primarily by its high content of heavy rare earth elements (HREEs), especially yttrium, dysprosium, erbium, and ytterbium. The significance of xenotime arises from its composition, which is rich in these valuable HREEs that are rare and challenging to extract, rendering it economically important in the global market for rare earth elements [133, 134]. The mineral typically forms in granitic pegmatites, metamorphic rocks, and hydromorphic environments, often associated with minerals such as monazite, zircon, and quartz [135].

The geological formation process of xenotime generally occurs in environments rich in phosphorus and calcium, where it can precipitate under acidic conditions. This formation aligns with its occurrence alongside other heavy minerals and phosphates such as monazite and zircon [134]. Xenotime can also be concentrated into placer deposits through sedimentary processes, where its dense grains settle during the transport of sediments, contributing to the concentration of mineral sands in various regions [136, 137].

Regions noted for significant xenotime deposits include Malaysia, Indonesia, and Australia, where it often arises as a by-product of tin mining operations [137, 138]. This context not only emphasizes xenotime's economic relevance but also highlights the geological diversity of potential REE sources. In particular, Malaysia's rare earth element story, especially with reference to the "Amang," illustrates the presence of xenotime within heavy mineral concentrates and the technological approaches employed to recover it efficiently [138]. "Amang" is a local Malaysian term for tin tailings—the leftover sand and mineral residues from historical tin mining, particularly in areas like Perak and Selangor. These tailings often contain small amounts of valuable heavy minerals, including xenotime (YPO_4), monazite, zircon, and ilmenite.

Overall, the economic assessment of xenotime is underpinned by its higher proportions of heavy rare earth elements, making it a focal point for mining endeavours. Its dual existence in both hard rock formations, often associated with pegmatites and metamorphic landscapes, and in secondary placer deposits, marks its importance within both geological and economic narratives surrounding rare earth element extraction [139, 140].

Spodumene

Spodumene, chemically denoted as $LiAl(Si_2O_6)$, is acknowledged as one of the most significant lithium-bearing minerals, primarily occurring in granitic pegmatites—coarse-grained igneous rocks that crystallize from late-stage magma. This mineral is frequently found in conjunction with other lithium minerals such as lepidolite and petalite, underscoring its prominence in the geology of pegmatites [141, 142]. The concentration of lithium in spodumene can reach up to 8% Li_2O, affirming its status as a primary source for lithium extraction, vital for the production of lithium carbonate and lithium hydroxide utilized in lithium-ion batteries [143, 144].

Richard Skiba

Major deposits of spodumene are located in Western Australia, particularly at Greenbushes and Pilgangoora, as well as internationally in Tanco (Canada) and Bikita (Zimbabwe) [145, 146]. The persistence of mining activities at locations like Greenbushes since the late 19th century highlights the strategic importance and historical relevance of spodumene within the sector [147]. Large prismatic crystals of spodumene make it identifiable; however, to efficiently extract lithium, the mineral must undergo roasting followed by chemical processing. The roasting process is crucial as it converts α-spodumene into the more reactive β-spodumene required for subsequent extraction methods [148, 149].

The extraction of lithium from spodumene typically employs a sulfuric acid roasting process, which transforms the mineral into lithium sulphate, enabling further processing to yield battery-grade lithium compounds [150, 151]. The main methods for lithium extraction are the sulfuric acid method and the carbonate method, each utilizing high temperatures to facilitate the reaction [149, 152]. Moreover, advancements in extraction techniques have been noted, focusing on ways to enhance the efficiency and environmental sustainability of lithium recovery from spodumene ores [144, 153].

Other Important Ore Minerals

Several other ore minerals beyond rare earth and lithium-bearing species play crucial roles in the global critical mineral supply chain. These minerals are essential for producing key materials used in energy storage, electronics, steelmaking, and aerospace applications. Each mineral forms under specific geological conditions and is associated with characteristic host rocks that determine how and where they can be economically extracted.

Ilmenite ($FeTiO_3$) and Magnetite (Fe_3O_4)

Ilmenite and magnetite are iron-bearing oxide minerals that are major sources of titanium and vanadium, both of which are vital for high-strength alloys, aerospace materials, and energy storage technologies. These minerals are typically found in layered mafic and ultramafic intrusions, large igneous bodies that crystallised slowly from iron- and magnesium-rich magmas deep in the Earth's crust. As the magma cooled, dense minerals like magnetite and ilmenite settled out, forming distinct layers of mineral concentration known as magmatic layering.

One of the world's most famous examples of this type of deposit is South Africa's Bushveld Complex, which contains extensive layers rich in magnetite and ilmenite, along with valuable concentrations of vanadium, platinum group metals (PGMs), and chromium. Similar deposits occur in Russia's Kachkanar and Finland's Mustavaara complexes. Ilmenite serves as the principal ore of titanium, a metal prized for its strength-to-weight ratio and corrosion resistance—ideal for aerospace and renewable energy applications. Magnetite, in turn, often contains trace amounts of vanadium that can be recovered as a by-product, used to enhance the strength of steel and in vanadium redox flow batteries for large-scale energy storage.

44

Rare Earth and Critical Mineral Operations and Processing

Pyrochlore [(Na,Ca)$_2$Nb$_2$O$_6$(OH,F)]

Pyrochlore is the most important mineral source of niobium, a metal critical for producing superalloys, superconductors, and high-strength steel. It typically occurs in carbonatite and alkaline igneous complexes, which are rare, silica-poor magmatic systems rich in volatile components like fluorine, carbon dioxide, and phosphorus. These unusual magmas allow elements such as niobium, tantalum, and rare earths to concentrate during crystallisation.

The Araxá and Catalão carbonatite complexes in Brazil are the world's dominant sources of niobium, accounting for more than 80% of global production. Another notable deposit is the Niobec Mine in Canada, located within an alkaline intrusion associated with the Saguenay region of Quebec. Pyrochlore is typically mined through open-pit methods, and niobium is extracted using flotation and hydrometallurgical processes. Because niobium improves the strength, ductility, and corrosion resistance of steel, it is indispensable in pipeline construction, jet engines, and next-generation electronics that rely on superconductivity.

Graphite (C)

Graphite, a crystalline form of carbon, is one of the most versatile and technologically significant materials in the critical minerals category. It forms primarily in metamorphic rocks such as schist, gneiss, and marble, where organic carbon-rich sediments are transformed under high temperature and pressure into crystalline graphite through the process of metamorphism. This process realigns carbon atoms into a layered, hexagonal lattice structure, giving graphite its exceptional electrical conductivity, thermal stability, and lubricating properties.

Natural graphite occurs in several forms, including flake graphite, amorphous graphite, and vein graphite, each with distinct physical characteristics and industrial uses. Flake graphite—found in metamorphic deposits in China, Mozambique, and Madagascar—is particularly valuable for use in battery anodes, refractories, and lubricants. Vein graphite, formed from carbon-bearing fluids filling fractures, is rarer but highly pure, with significant production from Sri Lanka. As global demand for lithium-ion batteries continues to rise, graphite has become a key material in the clean energy transition, representing up to half the active material in electric vehicle (EV) batteries by weight.

Cobaltite (CoAsS) and Pentlandite ((Fe,Ni)$_9$S$_8$)

Cobaltite and pentlandite are the main ore minerals for cobalt and nickel, metals that are indispensable in battery manufacturing, superalloys, and catalysts. Both minerals are typically found in magmatic sulphide deposits, which form when molten magma rich in iron, nickel, and sulphur crystallises deep within ultramafic rocks. As the magma cools, immiscible sulphide liquids separate and sink, concentrating metals into distinct layers or zones.

ntandite, a nickel-iron sulphide, is the primary source of nickel globally and often contains significant amounts of cobalt as an associated element. Cobaltite, an arsenic-bearing cobalt sulphide, forms under similar magmatic or hydrothermal conditions and occurs in association with pyrite, chalcopyrite, and other sulphide minerals. Major deposits occur in Canada's Sudbury Basin, Russia's Norilsk region, and the Democratic Republic of Congo (DRC)—the latter being the world's leading cobalt producer, primarily from copper-cobalt oxide and sulphide ores. These minerals underpin the global battery supply chain, particularly for nickel-manganese-cobalt (NMC) and nickel-cobalt-aluminium (NCA) cathode chemistries used in electric vehicles.

Key Deposit Types and Global Examples

Igneous Deposits

Igneous deposits are some of the most important geological sources of rare earth elements (REEs), niobium, tantalum, and other critical minerals. They form when molten rock (magma) cools and solidifies, either beneath the Earth's surface as intrusive rocks (like granite or gabbro) or on the surface as extrusive rocks (like basalt or rhyolite). As magma cools, chemical differentiation occurs—meaning that different minerals crystallise at different temperatures and compositions. This natural "sorting" process can cause certain elements, including rare earths and other critical metals, to become concentrated in specific zones or rock types, eventually forming economically valuable mineral deposits.

Carbonatite deposits are one of the most important sources of rare earth elements globally. They are unusual igneous rocks composed predominantly of carbonate minerals (such as calcite and dolomite) rather than the silicate minerals typical of most igneous rocks. These magmas are enriched in volatile components like carbon dioxide, fluorine, and phosphorus, which promote the mobility and concentration of rare elements during crystallisation.

Because of their unique chemistry, carbonatite magmas can accumulate significant quantities of light rare earth elements (LREEs), along with niobium, fluorine, and phosphorus. Common REE-bearing minerals in these deposits include bastnäsite, monazite, and synchysite, while pyrochlore serves as the main ore mineral for niobium.

Two of the best-known examples of carbonatite deposits are:

- Mount Weld (Western Australia): One of the world's highest-grade rare earth deposits, containing large amounts of LREEs such as lanthanum, cerium, and neodymium. It is mined and processed by Lynas Rare Earths, one of the few major non-Chinese REE producers.

- Bayan Obo (China): The largest rare earth deposit in the world, producing a mix of LREEs, iron, and niobium. The deposit is hosted in a large carbonatite complex and supplies a substantial portion of global REE demand.

Rare Earth and Critical Mineral Operations and Processing

These deposits are formed when carbonatite magma intrudes into the crust and cools slowly, allowing REE-rich minerals to crystallise out of the melt. Later hydrothermal processes can further enrich these minerals, enhancing the deposit's economic potential.

Alkaline igneous complexes are another major source of rare earth elements, zirconium, hafnium, and sometimes tantalum and niobium. These rocks form from silica-poor, alkali-rich magmas that contain high levels of sodium and potassium. Because such magmas are volatile-rich, they retain elements like fluorine, chlorine, and carbon dioxide that inhibit early crystallisation of rare elements, allowing them to remain concentrated in the late-stage residual melt.

During this final stage of crystallisation, the magma becomes enriched in rare elements that do not fit easily into common mineral structures (these are called incompatible elements). As a result, minerals like zircon, eudialyte, allanite, and loparite can form, containing zirconium, REEs, and hafnium.

Important examples of these complexes include:

- Ilímaussaq Complex (Greenland): Hosts rare REE-bearing minerals like eudialyte and loparite.

- Khibina and Lovozero Massifs (Russia): Rich in REEs, zirconium, and niobium, these complexes are among the largest alkaline intrusions in the world.

The combination of silica undersaturation, slow cooling, and volatile retention makes alkaline complexes particularly effective at concentrating rare and exotic elements that are scarce in most other rock types.

Layered mafic intrusions are large, deep-seated igneous bodies formed from iron- and magnesium-rich magma. As these magmas cool slowly within the Earth's crust, they undergo a process called magmatic differentiation, where heavier minerals such as olivine, pyroxene, and magnetite crystallise first and settle to the bottom of the magma chamber. This process creates layered structures, each enriched in specific minerals and elements.

In certain layers, minerals such as ilmenite ($FeTiO_3$) and magnetite (Fe_3O_4) become concentrated, often containing valuable amounts of titanium and vanadium. Other layers may trap sulphide liquids, forming rich accumulations of nickel, copper, and platinum-group metals (PGMs).

The most famous example of this type of deposit is the Bushveld Complex in South Africa, which is the largest layered mafic intrusion on Earth. It hosts immense reserves of PGMs (platinum, palladium, rhodium), chromium, vanadium, and iron-titanium oxides. Other examples include the Stillwater Complex (USA) and Great Dyke (Zimbabwe), both important for PGM production.

These intrusions are significant not only for their metallic content but also for the insights they provide into magmatic processes deep within the Earth. Their layered structures act like geological records of magma evolution, showing how metals segregate, concentrate, and crystallise under different temperature and pressure conditions.

Recognising igneous deposits in the field or through exploration studies involves understanding their rock types, textures, mineral assemblages, and geochemical signatures. Because these deposits form from molten rock, they are typically associated with specific igneous environments, distinct mineral compositions, and diagnostic geological features. Each subtype—carbonatite, alkaline complex, and layered mafic intrusion—can be identified by a combination of field evidence, mineralogy, and analytical methods.

1. Field Identification and Geological Setting

Igneous deposits are most commonly associated with intrusive bodies such as dykes, sills, and plutons, or with volcanic centres that indicate ancient magma chambers. When mapping in the field, geologists look for:

- **Rock textures:** Igneous rocks have crystalline textures, often with visible interlocking crystals that indicate slow cooling underground (intrusive) or fine-grained, glassy textures from rapid cooling at the surface (extrusive).

- **Alteration zones:** Igneous deposits often exhibit hydrothermal alteration halos, where fluids have changed the colour or composition of the surrounding rocks.

- **Magnetic and radiometric anomalies:** Because many igneous rocks contain magnetic minerals (e.g., magnetite, ilmenite) or radioactive elements (e.g., thorium in monazite), they produce detectable signals in geophysical surveys.

- **Regional geology:** Igneous deposits typically occur in continental rift zones, alkaline provinces, and large igneous provinces, often associated with ancient tectonic or volcanic activity.

2. Recognising Carbonatite Deposits

Carbonatites can be recognised by their distinctive mineralogy and chemistry. Unlike most igneous rocks, which are silica-based, carbonatites are rich in carbonate minerals such as calcite or dolomite.

Field characteristics:

- They often appear as light-coloured intrusive bodies surrounded by more typical silicate rocks.

- Weathered surfaces may show rusty brown or grey colours due to iron oxidation.

- Carbonatites commonly occur in circular or pipe-like intrusions associated with rift zones or alkaline complexes.

Diagnostic minerals:

- Bastnäsite, monazite, and synchysite indicate REE enrichment.

- Pyrochlore signifies niobium potential.

- Accessory minerals such as fluorite, barite, or apatite are common, reflecting the volatile-rich chemistry of carbonatite magmas.

Analytical confirmation:

- Geochemical assays show high concentrations of REEs, niobium, phosphorus, and fluorine.

- Petrographic analysis under a microscope confirms carbonate dominance with minor silicate minerals.

- Magnetic and gravity surveys can help delineate dense carbonatite bodies beneath the surface.

Examples to recognise: Mount Weld (Australia) and Bayan Obo (China) both display circular intrusive patterns and high REE grades within carbonatite-hosted systems.

3. Recognising Alkaline Igneous Complexes

Alkaline igneous complexes are identifiable by their unusual chemistry—they are silica-poor but rich in sodium and potassium, leading to the crystallisation of rare, exotic minerals.

Field characteristics:

- These complexes form large ring-shaped intrusions or zoned plutons, often accompanied by carbonatites.

- They may show distinct layering or banding related to different magma pulses.

- Rocks are often fine- to medium-grained with a greasy or waxy lustre due to minerals like nepheline or aegirine.

Diagnostic minerals:

- Eudialyte, loparite, allanite, and zircon are key indicators of REE, zirconium, and hafnium enrichment.

- Sodalite and nepheline indicate silica undersaturation, a hallmark of alkaline magmas.

Analytical methods:

- X-ray fluorescence (XRF) and ICP-MS analyses reveal elevated levels of REEs, zirconium, niobium, and fluorine.

- Petrography and scanning electron microscopy (SEM) confirm exotic mineral assemblages that are absent in normal granitic systems.

Examples to recognise: The Ilímaussaq Complex (Greenland) and Lovozero Massif (Russia) show concentric zoning of alkali-rich rocks with distinct eudialyte-bearing zones, making them prime exploration models.

4. Recognising Layered Mafic Intrusions

Layered mafic intrusions are typically vast, sheet-like igneous bodies composed of iron- and magnesium-rich rocks (mafic rocks). They are easy to recognise by their distinct mineral layering and association with large-scale magmatic provinces.

Field characteristics:

- Outcrops often reveal visible layering, where different bands represent variations in mineral content (e.g., alternating pyroxene-olivine and magnetite layers).

- They are typically dark-coloured (black to greenish-grey), dense, and magnetic.

- Mafic intrusions are often associated with ancient continental cratons or rift zones, where deep magmas have intruded the crust.

Diagnostic minerals:

- Ilmenite ($FeTiO_3$) and magnetite (Fe_3O_4) are primary carriers of titanium and vanadium.

- Pentlandite, chalcopyrite, and pyrrhotite indicate nickel–copper–platinum-group mineralisation.

- PGMs occur as minute inclusions in sulphide minerals, often requiring microanalytical identification.

Analytical methods:

- Magnetic and gravity surveys can identify the dense, magnetite-rich layers.

- Drill core logging reveals distinct rhythmic layering of mineral-rich bands.

- Assay testing confirms elevated levels of Fe, Ti, V, Ni, Cu, and PGMs.

Examples to recognise: The Bushveld Complex (South Africa) shows extensive stratification, with clearly defined chromite, magnetite, and PGM layers; similar features are found in the Stillwater Complex (USA) and Great Dyke (Zimbabwe).

Pegmatite Deposits

Pegmatite deposits represent one of the most distinctive and economically important types of igneous formations, particularly in the context of critical minerals. These deposits form during the final stages of crystallisation of granitic magma, when the remaining melt becomes enriched in volatile components such as water, fluorine, boron, phosphorus, and lithium. Because these volatiles lower the magma's viscosity and melting temperature, they allow the residual melt to stay fluid for longer, enabling the transport and concentration of rare elements that do not fit into the crystal structures of common minerals. The result is a coarse-grained igneous rock, known as a *pegmatite*, often featuring exceptionally large crystals and a diverse mineral composition.

Pegmatites crystallise from silica-rich magmas, typically derived from granitic intrusions. As the magma cools, most of the common rock-forming minerals—such as quartz, feldspar, and mica—crystallise first, depleting the melt of major elements like silicon and aluminium. The remaining melt becomes enriched in incompatible elements, which are those that do not fit easily into the early-forming minerals' crystal lattices. This residual melt also traps water and other volatiles, creating a highly fluid, chemically enriched environment.

When this final portion of magma intrudes into fractures, shear zones, or cavities in the surrounding rock, it cools slowly and forms pegmatite veins or dykes. The slow cooling rate and volatile-rich nature of the fluid promote the growth of very large crystals—sometimes several metres long—and the formation of rare minerals that are seldom found in other rock types.

Pegmatites typically occur as:

- **Dykes or sills**, which are tabular bodies intruded into fractures in country rock.

- **Lenses or pods**, often concentrated near the margins of granitic intrusions.

- **Zoned bodies**, showing mineralogical variation from wall zones (feldspar- and quartz-rich) to core zones (containing lithium, tantalum, or beryllium minerals).

Pegmatites are economically important because they host several critical minerals essential to modern technology and clean energy industries. The most notable include:

- **Spodumene ($LiAl(SiO_3)_2$):** The primary ore of lithium, occurring as elongated prismatic crystals often green, pink, or white in colour. It is the dominant mineral in hard-rock lithium mining operations, used in lithium-ion batteries.

- **Lepidolite ($K(Li,Al)_3(Si,Al)_4O_{10}(F,OH)_2$):** A lithium-bearing mica that also contains rubidium and cesium. Lepidolite-rich pegmatites are mined both for lithium and for their rare alkali metal content.

- **Columbite–tantalite ($[(Fe,Mn)(Nb,Ta)_2O_6]$):** A family of niobium- and tantalum-rich minerals often collectively referred to as *coltan*. These are critical for producing capacitors, electronics, and high-strength alloys.

- **Beryl ($Be_3Al_2Si_6O_{18}$):** The main ore of beryllium, used in aerospace and nuclear applications. Varieties such as emerald and aquamarine are valued as gemstones.

- **Cassiterite (SnO_2)** and **tourmaline** may also occur as accessory minerals, depending on the pegmatite's chemistry.

These minerals typically form in the core zones of pegmatites, where the last, most volatile-rich fluids are concentrated. The outer zones usually contain more common minerals like feldspar, quartz, and muscovite.

Geologists recognise pegmatite deposits through a combination of field observations, mineralogy, and geochemistry. Key identifying features include:

- **Very coarse-grained texture:** Crystals often exceed several centimetres, and in some cases, metres, due to the slow crystallisation and volatile-rich environment.

- **Distinct zoning:** Pegmatites often display mineralogical layering or core–margin zonation, with rare minerals typically concentrated in the inner zones.

- **Association with granitic intrusions:** Pegmatites are spatially and genetically related to granitic magmas and are frequently found on the margins of granite batholiths or within metamorphic host rocks that were intruded by granitic melts.

- **Geochemical anomalies:** Elevated concentrations of lithium (Li), cesium (Cs), rubidium (Rb), niobium (Nb), tantalum (Ta), and beryllium (Be) are diagnostic.

- **Structural setting:** Pegmatite veins often exploit existing fractures or shear zones, suggesting emplacement under low-pressure, high-fluid conditions near the end of magmatic crystallisation.

Exploration geologists also use radiometric, geochemical, and remote sensing data to locate pegmatites. For example, lithium-bearing pegmatites often display distinct spectral signatures due to their alteration minerals.

Pegmatite deposits are found worldwide, but certain regions are especially renowned for their size, grade, and economic importance:

- **Greenbushes (Western Australia):** The world's largest and highest-grade lithium pegmatite deposit, producing spodumene concentrates used for battery-grade lithium hydroxide. The deposit also contains tantalum minerals.

- **Pilgangoora (Western Australia):** Another major lithium-tantalum pegmatite field, featuring multiple large spodumene-bearing dykes that supply the rapidly expanding electric vehicle battery market.

- **Tanco Pegmatite (Manitoba, Canada):** One of the most studied pegmatite deposits globally, it produces lithium (spodumene), tantalum (tantalite), and cesium (pollucite). The Tanco mine has served as a geological model for other rare-element pegmatite systems.

- Other significant pegmatites occur in Brazil, Zimbabwe, Mozambique, and China, reflecting the global distribution of granitic provinces where such magmas are generated.

Rare Earth and Critical Mineral Operations and Processing

Pegmatite deposits are increasingly vital in the transition to renewable energy and advanced technology manufacturing. Lithium, tantalum, and niobium from pegmatites are critical for batteries, superalloys, and capacitors. Pegmatite mining has advantages over brine-based extraction because it allows faster production cycles and often yields higher-purity concentrates, though it requires more intensive energy and chemical processing.

As demand for electric vehicles, grid-scale energy storage, and high-performance electronics continues to rise, pegmatite deposits have become strategic assets in global supply chains. Countries like Australia and Canada are positioning themselves as secure, sustainable suppliers of these essential minerals.

Pegmatite deposits can be recognised and classified through a combination of field mapping, mineralogical identification, structural observation, and geochemical analysis. Because pegmatites form during the final stages of granitic magma crystallisation, they display distinctive visual and geological features that make them relatively easy to identify once you know what to look for. However, determining whether a pegmatite is mineralised (economically valuable) requires more detailed investigation.

1. Field Recognition

In the field, pegmatites stand out for their coarse-grained texture and massive appearance. The individual crystals are often visible to the naked eye and may reach several centimetres or even metres in size. Pegmatites frequently form as dykes, sills, or lenses intruding into country rock—commonly metamorphic rocks such as schist, gneiss, or amphibolite.

Key features to observe in the field include:

- **Coarse crystals:** The rock typically consists of large interlocking grains of quartz, feldspar (often perthitic), and mica (muscovite or biotite).

- **Light colouration:** Most pegmatites are pale grey, pink, or white due to their high silica and feldspar content, although lithium-rich varieties may show green or purple hues.

- **Cross-cutting structures:** Pegmatite veins often cut across the foliation of host rocks, indicating their intrusive origin.

- **Contact zones:** The edges (margins) of pegmatite bodies may show finer-grained "chilled margins" where the magma cooled more rapidly against cooler surrounding rock.

When a pegmatite body is exposed, its **zoning** can often be seen clearly:

- **Wall zone:** Composed mainly of quartz, feldspar, and muscovite.

- **Intermediate zone:** May contain tourmaline, garnet, or topaz.

- **Core zone:** Hosts the most valuable minerals, such as spodumene, lepidolite, columbite–tantalite, or beryl.

This zonation reflects how volatile-rich fluids concentrated towards the centre of the cooling magma, forming rare-element minerals in the later stages.

2. Mineralogical Indicators

Identifying the specific minerals present is one of the most reliable ways to confirm that a rock body is a rare-element pegmatite rather than a simple granitic vein. Pegmatites rich in lithium, tantalum, niobium, or beryllium tend to contain certain diagnostic minerals:

- **Spodumene:** Long, prismatic crystals that can appear green, pink, or grey; often aligned parallel within the rock.

- **Lepidolite:** A lilac- to pink-coloured mica with a shimmering appearance, often occurring in fine sheets or masses.

- **Columbite–tantalite:** Black, heavy, metallic minerals often found in the pegmatite core; detectable with a magnet or by density testing.

- **Beryl:** Pale green, blue, or colourless hexagonal crystals, sometimes gem-quality (emerald, aquamarine).

- **Cassiterite and Tourmaline:** Secondary minerals that may indicate tin or boron enrichment, respectively.

If these minerals are observed, the pegmatite is likely a rare-element pegmatite, sometimes referred to as an "LCT pegmatite" (Lithium–Cesium–Tantalum type).

3. Geological and Structural Context

Pegmatite deposits are typically found in or around granitic intrusions, as they are genetically linked to the last stages of granitic magma crystallisation. In mapping studies, geologists look for:

- Proximity to granite plutons or batholiths, especially in craton margins and orogenic belts where magmatic activity was intense.

- **Structural controls:** Pegmatites often fill fractures, shear zones, or tension gashes, meaning they can appear in parallel swarms along structural trends.

- **Metamorphic host rocks:** The country rock surrounding pegmatites is often heavily altered by fluids, showing minerals like sericite, chlorite, or tourmaline along the contact zones.

These geological clues are valuable in exploration because pegmatite formation depends on the migration of volatile-rich fluids through structurally weak zones at the end of magmatic crystallisation.

4. Geochemical and Analytical Identification

Rare Earth and Critical Mineral Operations and Processing

In addition to field observations, laboratory analyses are essential for confirming and characterising pegmatite deposits:

- **Whole-rock geochemistry:** Elevated concentrations of Li, Cs, Rb, Nb, Ta, and Be are diagnostic.

- **Geochemical ratios:** High K/Rb and low Mg/Li ratios typically indicate pegmatitic evolution.

- **Mineral chemistry:** Electron microprobe or LA-ICP-MS analysis of spodumene, lepidolite, or columbite–tantalite can confirm ore-grade concentrations.

- **Spectral imaging and remote sensing:** Lithium pegmatites often exhibit distinctive reflectance spectra due to alteration minerals such as spodumene and lepidolite, allowing for satellite-based exploration.

Geophysical surveys (radiometric and resistivity methods) can also assist in identifying pegmatite zones beneath soil or weathered cover.

5. Economic Indicators

Not all pegmatites are mineralised; many are "barren," containing only common minerals like quartz and feldspar. Economic pegmatites are distinguished by:

- The presence of rare-element minerals (especially lithium, tantalum, or niobium-bearing phases).

- High concentration in the core zone rather than being scattered throughout the rock.

- Proximity to accessible infrastructure (important for mining feasibility).

- Zoned mineralogy, indicating strong fluid differentiation during crystallisation—often a good predictor of high-grade mineralisation.

Laterite and Regolith Deposits

Laterite and regolith deposits are secondary mineral accumulations that form near the Earth's surface through chemical weathering processes. They are among the most important sources of several critical minerals, including nickel, cobalt, aluminium, and rare earth elements (REEs). These deposits do not originate directly from magmatic or hydrothermal activity; instead, they result from the breakdown and alteration of pre-existing rocks over long geological timescales in warm, humid tropical climates.

The key to their formation lies in the interaction between rock, water, and atmosphere. Over time, intense tropical weathering causes soluble components (such as sodium, calcium, and silica) to be leached away by rainfall and groundwater, while insoluble elements (such as iron, aluminium, nickel, and cobalt) are left behind and gradually concentrated near the surface. The resulting material, known as *laterite*, forms thick, reddish-brown soil profiles rich in metal oxides and hydroxides.

Laterite and regolith deposits form through a combination of chemical leaching, oxidation, and residual enrichment. The process typically occurs in tropical and subtropical regions that experience high rainfall and good drainage, which encourages deep weathering over millions of years. The general sequence of formation includes:

Weathering of parent rock: The starting material is typically ultramafic, mafic, or granitic rock. Rainwater percolates through the rock, dissolving and carrying away mobile ions such as calcium (Ca), magnesium (Mg), and silica (Si).

Leaching and concentration: As soluble elements are removed, insoluble oxides of iron (Fe), aluminium (Al), and sometimes nickel (Ni) and cobalt (Co) accumulate in the weathering profile.

Profile development: The weathering process creates a layered structure, often several metres thick, consisting of:

- A soil or humic zone at the surface.
- A limonite (iron-rich) zone, containing hydrated iron oxides like goethite and hematite.
- A saprolite (silicate-rich) zone, preserving the structure of the original rock but chemically altered to clays and oxides.

Metal enrichment: In some cases, downward percolating fluids can remobilise metals like nickel and cobalt from upper zones and reprecipitate them in lower zones, forming high-grade horizons.

This weathering-driven process explains why laterite deposits are typically shallow, near-surface, and horizontally extensive, making them relatively easy to mine by open-cut methods.

Nickel and cobalt laterite deposits are the most economically significant type of lateritic ore system. They form through the weathering of ultramafic rocks, such as peridotite and dunite, which are rich in the minerals olivine and pyroxene. These rocks contain small but significant amounts of nickel and cobalt that become concentrated during weathering.

As the ultramafic rock breaks down, magnesium and silica are leached out, while iron oxides, nickel, and cobalt remain and are concentrated in the residual material. Over time, two principal ore zones develop:

- **Limonite Zone (upper layer):** Rich in iron oxides (goethite, hematite) and typically contains most of the cobalt.

- **Saprolite Zone (lower layer):** Rich in magnesium silicates (serpentine, garnierite) and generally contains most of the nickel.

Key characteristics and indicators:

- Red-brown to yellow soil colour due to oxidised iron.

Rare Earth and Critical Mineral Operations and Processing

- Often found in tropical climates with long-term geological stability (i.e., no recent glaciation or uplift).

- Commonly capped by an iron-rich crust or "duricrust."

Major examples include:

- **Indonesia and the Philippines:** The world's largest producers of lateritic nickel ores, where extensive tropical weathering has produced thick limonite–saprolite profiles.

- **New Caledonia:** Known for some of the highest-grade nickel laterite deposits globally, mined for more than a century.

- **Australia (Queensland and Western Australia):** Hosts large lateritic nickel–cobalt resources, including Murrin Murrin and Ravensthorpe.

These deposits are processed through hydrometallurgical methods such as High-Pressure Acid Leaching (HPAL) or Caron processes, which extract nickel and cobalt from oxide or silicate ores.

A particularly important subset of regolith-hosted mineralisation is the ion-adsorption clay (IAC) deposit, which is the dominant source of heavy rare earth elements (HREEs) such as dysprosium (Dy) and terbium (Tb). These deposits form through the deep weathering of REE-rich igneous rocks, such as granites and carbonatites, under warm and humid conditions.

During weathering, rare earth elements are released from the parent rock's minerals and are adsorbed onto the surfaces of clay minerals—particularly kaolinite and halloysite—through weak electrostatic bonding (ion exchange). Unlike traditional REE ores that require high-temperature smelting, these clays allow REEs to be extracted by simple chemical leaching, often using ammonium sulphate solutions.

Key characteristics of IAC deposits:

- Soft, fine-grained, clay-rich material.

- Low overall REE grade (typically 0.05–0.2% total REO) but high proportion of HREEs.

- Located in deeply weathered terrains, often on gentle hillslopes or plateaus.

- Environmentally less disruptive extraction methods compared to hard-rock mining.

Major examples include:

- **Southern China (Jiangxi, Guangdong, Guangxi provinces):** The world's leading source of heavy REEs. These deposits revolutionised REE supply because they can be mined through relatively low-cost, low-energy leaching operations.

- **Emerging potential in Myanmar, Madagascar, and Brazil:** Similar tropical weathering conditions have led to the identification of comparable ion-adsorption clays.

IAC deposits are especially strategic because HREEs are critical in the production of permanent magnets, wind turbines, and electric vehicle motors, and they are far less common than the light REEs (LREEs) found in carbonatite or monazite deposits.

Geologists recognise laterite and regolith-hosted mineral systems using a combination of field, mineralogical, and geochemical indicators:

Field Indicators:

- **Distinct colouration:** Reddish-brown to orange hues from iron oxides (hematite, goethite).

- **Layered profiles:** Presence of a hardened crust (duricrust) above soft clay or saprolite layers.

- **Topographic setting:** Commonly on stable plateaus or gently sloping terrains in tropical climates.

- **Weathering depth:** Profiles can exceed 10–30 metres in mature systems.

Mineralogical Indicators:

- **Limonite zone:** Dominated by iron oxides and hydroxides with adsorbed cobalt.

- **Saprolite zone:** Contains serpentine, garnierite, or other Mg–Ni silicates.

- **Clay zones (for REEs):** Rich in kaolinite and halloysite with adsorbed REE ions.

Geochemical Signatures:

- High concentrations of Fe, Al, Ni, Co, and Cr in laterites.

- Enrichment in HREEs, Y, and Sc in ion-adsorption clays.

- Low silica and high iron oxide content relative to parent rock.

Remote sensing and geophysical clues:

- Laterites have high reflectance in red spectral bands and can be detected by satellite imaging.

- Magnetic surveys often show subdued responses because iron oxides replace magnetite during weathering.

Laterite and regolith deposits play a critical role in the global energy and technology supply chain:

- Nickel and cobalt laterites provide essential materials for lithium-ion batteries, stainless steel, and aerospace alloys.

- Ion-adsorption clays are the main global source of heavy rare earths, essential for high-performance magnets in wind turbines, electric motors, and military technologies.

- Their near-surface nature makes them more accessible and generally cheaper to mine than deep hard-rock deposits.

However, environmental management is vital because these deposits often occur in biodiverse tropical regions, and their extraction can lead to soil erosion, deforestation, and water contamination if not properly managed.

Sedimentary and Placer Deposits

Sedimentary processes such as erosion, transport, and deposition play a critical role in forming some of the world's most accessible and economically valuable mineral deposits. These processes act as natural concentrators—breaking down rocks through weathering, moving particles via wind or water, and depositing heavier or chemically stable minerals in new locations. Among these are placer deposits, which are secondary accumulations of dense, resistant minerals formed by the mechanical action of water. In addition, certain chemical sedimentary environments also trap valuable elements such as manganese, vanadium, and rare earth elements (REEs) through geochemical precipitation.

The formation of placer and sedimentary-hosted deposits begins with the erosion of pre-existing rocks that contain valuable minerals. As these rocks—often granitic, metamorphic, or igneous in origin—are weathered by physical and chemical forces, minerals are liberated and carried away by streams, rivers, and coastal currents.

During transport, lighter and less durable minerals (such as quartz and feldspar) are broken down or carried further downstream, while heavier, more resistant minerals (like gold, zircon, monazite, and ilmenite) settle out earlier because of their higher specific gravity. Over time, these processes lead to the natural sorting and concentration of heavy minerals in riverbeds, deltas, beaches, and coastal dune systems, forming what are known as placer deposits.

Three main stages are involved:

1. **Erosion and liberation:** Weathering breaks down the source rock and releases resistant minerals.

2. **Transport and sorting:** Flowing water or wave action moves and segregates minerals by density and size.

3. **Deposition:** Heavy minerals accumulate in traps such as river bends, bars, or coastal strandlines where flow energy decreases.

These deposits are called secondary because the valuable minerals were originally formed in a different geological setting and later concentrated by sedimentary processes.

Placer deposits are among the most practical and historically significant sources of several critical and industrial minerals, particularly rare earth elements (REEs). The key REE-bearing minerals in placer systems are monazite and xenotime:

- **Monazite [(Ce,La,Nd,Th)PO$_4$]:** Rich in light rare earth elements (LREEs) such as cerium, lanthanum, and neodymium. It is often slightly radioactive due to trace thorium and occurs as small, dense, yellow to brown grains.

- **Xenotime [(Y,Yb,Er)PO$_4$]:** Enriched in heavy rare earth elements (HREEs) such as yttrium, dysprosium, and erbium. It typically occurs alongside monazite and zircon in fine-grained sands.

These minerals are highly resistant to both chemical weathering and mechanical abrasion, allowing them to survive multiple sedimentary cycles. They are often accompanied by other dense accessory minerals, including:

- **Zircon (ZrSiO$_4$):** A common accessory mineral in igneous rocks, concentrated in placers for use in ceramics and geochronology.

- **Ilmenite (FeTiO$_3$) and Rutile (TiO$_2$):** Major sources of titanium, frequently associated with REE-bearing heavy minerals in beach sands.

The presence of monazite or xenotime in a sedimentary environment generally indicates the erosion of REE-rich source rocks, such as granites, pegmatites, or metamorphic rocks containing allanite or bastnäsite.

Placer deposits can form in a range of sedimentary settings depending on the nature of transport and deposition:

- **Alluvial Placers:** Found in riverbeds and floodplains, where heavy minerals accumulate in riffles, meanders, and gravel bars. These are common sites for monazite, xenotime, and gold.

- **Beach Placers:** Formed by wave and tidal action along coastlines, where heavy minerals such as ilmenite, zircon, rutile, and monazite are concentrated in sand dunes and strandlines. Examples include deposits along the east coast of India and Western Australia.

- **Eolian Placers:** Wind-driven systems (less common) that can concentrate heavy minerals in arid or coastal dune environments.

- **Marine Placers:** Offshore accumulations formed by longshore currents and reworking of coastal sediments.

These depositional environments share one key characteristic: a natural mechanism for sorting and concentrating minerals by density and particle size.

Several countries have successfully developed placer mining industries that exploit REE and titanium-bearing minerals:

- **India:** The states of Kerala, Tamil Nadu, and Odisha are globally significant for coastal heavy mineral sands rich in monazite, ilmenite, rutile, and zircon. India's deposits have historically supplied monazite concentrates used for REE extraction.

Rare Earth and Critical Mineral Operations and Processing

- **Malaysia:** Known for "amang" deposits—residual tin tailings enriched in xenotime and monazite. These secondary deposits were originally derived from granitic rocks and have been processed to recover valuable REEs.

- **Australia:** The coastal regions of Western and South Australia contain large beach-sand placer systems (e.g., Eneabba, Capel–Bunbury) that host ilmenite, rutile, zircon, and monazite, making Australia a key supplier of titanium and zirconium minerals.

These deposits are typically mined by open-pit or dredge mining, followed by gravity and magnetic separation techniques to recover the heavy minerals.

In addition to mechanical concentration processes, some sedimentary environments also lead to chemical precipitation of metals and REEs. In these cases, dissolved ions in seawater or groundwater react under reducing (oxygen-poor) conditions, forming mineral-rich sediments.

Key examples include:

- **Manganese nodules and crusts:** Formed on the ocean floor through slow chemical precipitation, containing manganese, nickel, cobalt, and rare earth elements.

- **Vanadium-bearing shales:** Found in organic-rich marine or lacustrine environments where vanadium is trapped by organic matter under anoxic conditions.

- **Phosphorite deposits:** Marine phosphorite beds can host elevated concentrations of REEs, particularly in continental shelf environments where upwelling enriches phosphorus and associated trace metals.

These sedimentary-hosted deposits are particularly important because they form under low-temperature surface conditions, unlike magmatic or metamorphic systems. This makes them potentially easier to mine and process, though grades are often lower.

Geologists use a combination of field observations, sediment analysis, and geochemical indicators to identify potential placer and sedimentary-hosted deposits.

Field indicators:

- Concentrations of dark, heavy mineral sands along riverbeds, beaches, or dunes.

- Distinct layering or streaks of black sands rich in ilmenite, monazite, or zircon.

- Presence of REE-bearing accessory minerals in stream sediment samples.

- Proximity to granitic or metamorphic terrains, the likely source of the heavy minerals.

Geochemical and mineralogical indicators:

- Elevated levels of Ti, Zr, Th, Y, and REEs in sediment samples.

- Mineral assemblages dominated by ilmenite, rutile, zircon, monazite, and xenotime.

- Radiometric anomalies due to thorium in monazite or uranium in zircon.

Modern exploration often incorporates remote sensing, drone mapping, and magnetic surveys to trace heavy mineral accumulations along coastlines and river systems.

Sedimentary and placer deposits play a vital role in supplying critical and industrial minerals to global markets.

- REE-bearing monazite and xenotime placers support the production of rare earth oxides used in permanent magnets, catalysts, and electronics.

- Titanium-bearing ilmenite and rutile are key for producing titanium metal and white pigment (titanium dioxide).

- Zircon sands are used in ceramics, refractories, and nuclear applications.

- Phosphorite and manganese sediments contribute to fertiliser and battery precursor production.

Because placer deposits are near-surface, soft, and easily concentrated, they can be exploited with minimal drilling and complex processing, making them particularly attractive for developing countries. However, environmental management is critical—especially in coastal zones where mining can disrupt fragile dune and marine ecosystems.

Hydrothermal and Vein Deposits

Hydrothermal deposits are among the most diverse and economically significant mineral systems on Earth. They form when hot, mineral-rich fluids—known as *hydrothermal fluids*—move through cracks, fractures, and porous zones in the Earth's crust. As these fluids cool, react with surrounding rocks, or experience changes in pressure or chemistry, they precipitate minerals out of solution, leading to the formation of concentrated mineral veins, disseminations, or replacement bodies.

These deposits are responsible for the world's major concentrations of metals such as cobalt, nickel, copper, tungsten, tin, lead, and zinc, as well as for industrial minerals like graphite and fluorite. Because they can form from a variety of fluid sources and under different temperature–pressure conditions, hydrothermal systems are found across all geological environments—from volcanic arcs and mid-ocean ridges to stable continental crusts.

Hydrothermal deposits develop when circulating fluids dissolve elements from a source rock at high temperature and pressure, transport them in solution, and then precipitate them when conditions change. The main stages include:

Fluid generation:

- **Magmatic fluids:** Derived directly from crystallising magma, rich in metals, sulphur, chlorine, and volatiles such as H_2O and CO_2.
- **Metamorphic fluids:** Released during metamorphism when minerals dehydrate or decarbonate under heat and pressure.

- **Meteoric fluids:** Groundwater or seawater that percolates into the crust, becomes heated by magma or geothermal gradients, and interacts chemically with rocks. Often, hydrothermal systems involve mixing between these fluid types, enhancing metal transport and precipitation.

Fluid movement: The fluids migrate through fractures, faults, and permeable rock layers, sometimes driven by pressure from magmatic intrusions or tectonic deformation. These pathways provide the structural control for mineral deposition.

Precipitation of minerals: As the fluid cools or reacts with host rocks, dissolved elements form new mineral phases and precipitate along fractures or pore spaces. This process creates vein structures, breccia fillings, or disseminated ores, depending on the host rock and fluid chemistry.

Hydrothermal alteration often accompanies mineralisation, producing halos of new minerals (like chlorite, sericite, or quartz) that serve as exploration indicators.

Hydrothermal deposits form under a range of thermal regimes, typically grouped into three categories:

- High-temperature (magmatic-hydrothermal) systems: 300–700°C, close to magmatic intrusions. These systems produce deposits of copper, tungsten, and molybdenum.

- Medium-temperature systems: 200–400°C, often associated with volcanic and geothermal activity; responsible for lead–zinc–silver and gold–silver veins.

- Low-temperature (epithermal) systems: Below 200°C, forming near the surface; include precious metal and industrial mineral deposits.

This temperature variation explains why hydrothermal deposits are found from deep-seated granite intrusions to shallow volcanic settings.

Hydrothermal systems are efficient at transporting and concentrating metals, making them critical sources of many critical and base metals.

Cobalt, Nickel, and Copper: These metals are often enriched in magmatic-hydrothermal systems linked to mafic or ultramafic intrusions. The interaction of magmatic fluids with these rocks results in the formation of sulphide minerals:

- **Cobaltite (CoAsS):** A metallic, silver-grey mineral that forms in high-temperature hydrothermal veins and skarn deposits. It is an important cobalt ore, often found with nickel and arsenic minerals.

- **Chalcopyrite ($CuFeS_2$):** The most abundant copper sulfide mineral, forming in a range of hydrothermal systems from porphyry to vein-type deposits.

- **Pentlandite ($(Fe,Ni)_9S_8$):** The main nickel sulfide mineral, typically forming in association with magmatic intrusions, later remobilised and concentrated by hydrothermal fluids.

These minerals often occur together in massive sulphide veins or disseminated ores, commonly associated with alteration zones of quartz, calcite, and pyrite.

Examples of hydrothermal–magmatic systems rich in these metals include:

- **The Norilsk-Talnakh deposits (Russia):** Hosting nickel, copper, and cobalt formed from magmatic and hydrothermal processes.

- **Bou Azzer (Morocco):** A famous cobalt deposit where cobaltite and skutterudite occur in hydrothermal quartz–carbonate veins.

Tungsten (W) is another key element concentrated in hydrothermal systems, particularly in granitic and skarn environments.

- Scheelite ($CaWO_4$) and wolframite (($Fe,Mn)WO_4$) are the main tungsten minerals, precipitating from high-temperature fluids derived from granitic magmas.

- These typically form vein or greisen deposits, often accompanied by cassiterite (SnO_2) and molybdenite (MoS_2).

Example: The Mittersill deposit (Austria) and King Island (Australia) are classic examples of tungsten-bearing hydrothermal systems.

One of the less common but highly valuable hydrothermal deposit types is graphite veins. Unlike sedimentary or metamorphic graphite, which forms from carbon-rich organic material, hydrothermal graphite precipitates directly from carbon-bearing fluids—typically rich in CO_2, CH_4, or other hydrocarbons—that migrate through fractures in the crust.

As these fluids cool and pressure decreases, carbon is deposited in the form of pure crystalline graphite along fracture walls, forming veins that can be several centimetres to metres thick.

Key features of hydrothermal graphite veins:

- Exceptionally high carbon purity (often >95–99% C).

- Coarse, platy crystal structure ideal for industrial and battery-grade graphite applications.

- Strong association with metamorphic and granitic terrains, where carbon-bearing fluids are abundant.

Sri Lanka is the world's best-known example of this deposit type. Its high-purity vein graphite forms from fluids that migrated through high-grade metamorphic rocks during ancient tectonothermal events. The resulting veins are hand-mined for their exceptional quality and purity.

Geologists identify hydrothermal systems through a combination of field evidence, alteration mineralogy, and geochemical patterns.

Field indicators:

- **Vein structures:** Quartz, carbonate, or sulphide-filled fractures cutting across host rocks.

- **Alteration halos:** Zones of new mineral growth—chlorite, sericite, epidote, or silica replacement—surrounding veins.

- **Gossans:** Rusty, iron-stained surfaces formed from weathered sulphides (a common sign of buried hydrothermal mineralisation).

Mineralogical indicators:

- Sulphide minerals such as pyrite, chalcopyrite, galena, sphalerite, and arsenopyrite.

- Gangue minerals like quartz, calcite, and barite filling the vein matrix.

- Carbon-rich zones or platy graphite in the case of hydrothermal graphite.

Geochemical signatures:

- Elevated concentrations of elements such as Cu, Co, Ni, W, Sn, and As.

- Stable isotope ratios (O, H, C, S) that distinguish between magmatic, metamorphic, and meteoric fluid sources.

- High alteration index values reflecting fluid–rock interaction.

Hydrothermal deposits are vital to the global supply of critical and base metals, as well as high-purity industrial minerals.

- Cobalt, nickel, and copper are essential for electric vehicle batteries, renewable energy systems, and electrical wiring.

- Tungsten is used in high-strength alloys and electronics due to its extreme melting point and durability.

- Graphite from hydrothermal veins provides some of the highest-quality natural graphite for use in battery anodes and refractories.

Moreover, hydrothermal systems are structurally predictable—meaning they often follow fracture networks or faults—making them favourable targets for exploration. Modern geophysical methods, such as resistivity, induced polarisation (IP), and magnetotellurics, are used to map these hidden subsurface systems.

Metamorphic Deposits

Metamorphic deposits form when existing rocks are transformed by heat, pressure, and chemically active fluids deep within the Earth's crust. These forces do not melt the rock completely but instead cause it to recrystallise, changing its texture, structure, and mineral composition. During this process, valuable elements and minerals can be remobilised, concentrated, or recrystallised, leading to the formation of economically significant mineral deposits. Metamorphism therefore acts as both a modifier of pre-existing ores and a creator of new ones under the right physical and chemical conditions.

Metamorphic mineral deposits originate through a combination of recrystallisation, remobilisation, and chemical exchange within rocks that have been subjected to elevated temperature and pressure. These conditions typically occur during mountain-building (orogenic) events, deep burial, or contact metamorphism near intrusive magmas.

There are three key mechanisms:

1. **Recrystallisation:** Existing minerals change in size, shape, and arrangement without the rock melting. This can improve the purity or coarseness of certain minerals (e.g., graphite).

2. **Remobilisation:** Heat and pressure allow some elements or fluids to migrate over short distances, where they reprecipitate in new locations, forming concentrated veins or bands of valuable minerals.

3. **Metasomatism:** Chemically active fluids introduced during metamorphism can add or remove elements, altering the rock's composition and forming new minerals such as garnet, amphibole, or magnetite.

The end result is a rock that has undergone a complete transformation—both structurally and chemically—producing new ore minerals or enriching existing ones.

One of the best examples of a metamorphic mineral deposit is graphite, which forms from the metamorphism of carbon-rich sedimentary rocks such as coal, shale, or limestone containing organic material.

During metamorphism, as temperature and pressure rise:

- The original organic matter (kerogen or bituminous material) undergoes carbonisation, where hydrogen, oxygen, and nitrogen are expelled as gases.

- The remaining carbon atoms reorganise into a crystalline structure, forming graphite (pure carbon).

- The degree of crystallinity depends on the metamorphic grade—higher temperatures and pressures produce coarser, more ordered graphite crystals.

The process typically occurs in regional metamorphic belts, especially where sedimentary basins have been deeply buried and subjected to prolonged metamorphic conditions.

Key features of metamorphic graphite deposits:

- Occur in schists, gneisses, and marbles associated with high-grade metamorphism.

- Display flake or crystalline textures, which determine their commercial value.

- Often found in folded or sheared zones, where fluids and deformation have helped concentrate carbon.

Notable examples:

- **Sri Lanka:** Hosts world-famous vein-type graphite deposits, where high-purity carbon was mobilised by metamorphic fluids and redeposited in fractures.

- **India, Madagascar, and Canada:** Contain flake graphite deposits formed by regional metamorphism of carbonaceous sediments.

Certain rare earth element (REE)-bearing minerals and accessory phases—such as allanite, monazite, zircon, and xenotime—can survive metamorphism but often undergo recrystallisation or partial chemical alteration.

- **Allanite [(Ce,La,Ca)(Al$_2$Fe^{3+})(SiO$_4$)(Si$_2$O$_7$)O(OH)]:** A common REE-bearing silicate that becomes re-stabilised under metamorphic conditions. Fluids can remobilise some REEs, causing them to precipitate in micro-veins or metamorphic banding.

- **Zircon (ZrSiO$_4$):** Extremely resistant to both heat and chemical attack, zircon can partially dissolve and reprecipitate during metamorphism. This process often produces new zircon growth zones, which can trap REEs or uranium and are used in U-Pb geochronology to date metamorphic events.

- **Monazite [(Ce,La,Nd,Th)PO$_4$]:** Another REE phosphate that recrystallises during metamorphism, particularly in aluminous schists. It can either form new grains or grow as rims around older crystals, often enriching specific rock layers in light rare earths.

Through these processes, REEs and related elements are redistributed and locally concentrated, especially in high-grade metamorphic terranes where fluids are active.

Metamorphism can also upgrade or alter pre-existing mineral deposits, enhancing their economic potential.

- **Iron formations:** Banded Iron Formations (BIFs) subjected to metamorphism can recrystallise to form **magnetite-rich schists**, improving ore quality.

- **Talc, asbestos, and chlorite deposits:** Form through metamorphism of ultramafic rocks rich in olivine and pyroxene.

- **Manganese and tungsten deposits:** May form or be enriched through fluid-assisted metamorphism of sedimentary precursors.

These examples demonstrate that metamorphism not only creates new mineral assemblages but can also rejuvenate or upgrade older deposits, a process known as metamorphic remobilisation.

Metamorphic deposits can be recognised through their geological setting, texture, and mineralogy.

Field Indicators:

- Occurrence in regional metamorphic belts (gneiss, schist, or marble terranes).

- Banding or foliation caused by directed pressure.

- Presence of coarse recrystallised textures—minerals appear larger and more intergrown compared to their sedimentary or igneous precursors.

- Association with folds, faults, or shear zones where deformation has facilitated fluid flow and mineral concentration.

Mineralogical Indicators:

- Minerals stable under high temperature and pressure: garnet, sillimanite, kyanite, amphibole, and graphite.

- REE-bearing accessory minerals (allanite, zircon, monazite) showing metamorphic overgrowths or recrystallisation features.

- Carbon-rich rocks where organic matter has been converted to crystalline graphite.

Geochemical and Analytical Tools:

- Petrographic microscopy to identify recrystallised textures.

- Electron microprobe and LA-ICP-MS to detect REE distribution in metamorphic minerals.

- Isotopic studies (e.g., U-Pb dating) to distinguish metamorphic zircon or monazite growth from primary magmatic phases.

Metamorphic deposits, while not as voluminous as magmatic or hydrothermal systems, are highly valuable for their purity and stability.

- **Graphite:** Metamorphic graphite is a key material for battery anodes, lubricants, refractories, and nuclear applications. Vein graphite (as in Sri Lanka) commands the highest prices due to its exceptional purity (>99% carbon).

- **REE-bearing minerals:** Metamorphic recrystallisation of allanite, monazite, and zircon contributes to the redistribution of REEs within continental crusts and may form small but high-grade deposits.

- **Upgraded iron and manganese ores:** Metamorphism improves ore quality by coarsening grains and removing impurities.

From a strategic standpoint, metamorphic processes help preserve and enhance critical mineral potential in old continental terranes—especially those that have undergone multiple cycles of deformation and metamorphism, such as the Canadian Shield, Western Australia, and Southern India.

Exploration Methods: Mapping, Sampling, Drilling, and Assay Techniques

Exploration is the first and most crucial phase in discovering and developing rare earth and critical mineral deposits. It combines geological observation, scientific analysis, and technological tools to locate, evaluate, and quantify mineral resources before any mining begins. The main exploration methods—mapping, sampling, drilling, and assaying—work together to build a clear geological picture of a potential deposit's size, grade, and economic viability.

Geological Mapping

Geological mapping is the foundation of mineral exploration. It involves systematically recording the types of rocks, their structures, textures, and alteration features at the Earth's surface. The purpose is to identify geological formations that may host mineralisation and to understand the relationship between rock types, fault lines, and mineral occurrences.

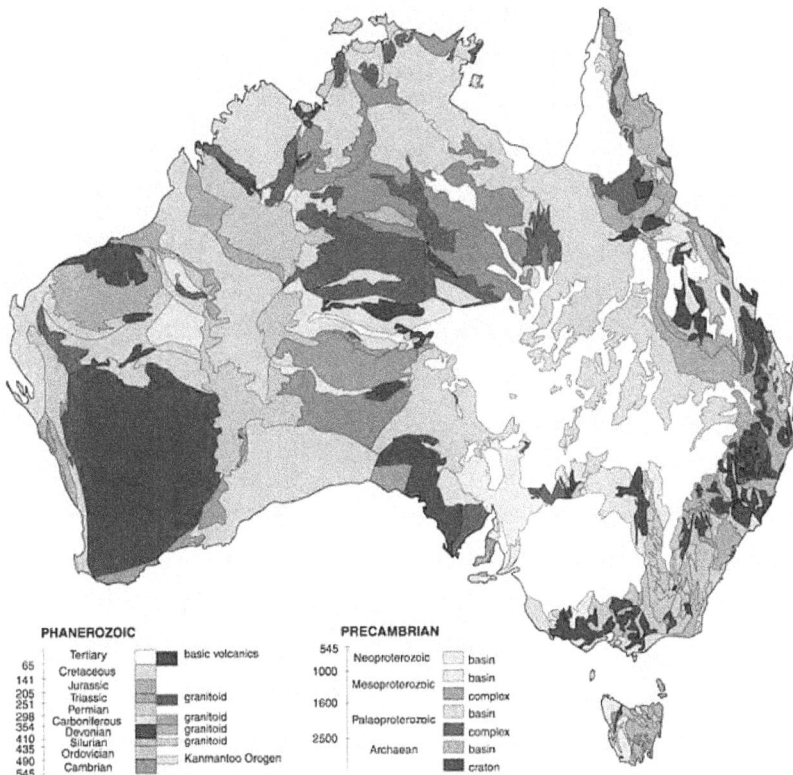

Figure 11: Basic geological units of Australia, after Addario et al., created in Arcinfo GIS from public domain geological mapping data, GFDL free use.

Field geologists use topographic maps, satellite imagery, and GPS to delineate rock boundaries and record mineral indicators such as veins, alteration zones, and gossans (rust-coloured outcrops caused by oxidised sulphides). In the context of rare earth and critical minerals, mapping often focuses on:

- Igneous complexes (e.g., carbonatites, pegmatites) where REEs, niobium, or lithium might occur.

- Weathered or lateritic zones for nickel, cobalt, and bauxite.

- Structural features such as faults, folds, and shear zones that act as fluid pathways for mineralising events.

Modern geological mapping integrates remote sensing, LiDAR, and geophysical surveys (magnetic, gravity, and radiometric data) to reveal subsurface structures that cannot be seen directly. This helps explorers prioritise targets for more detailed work.

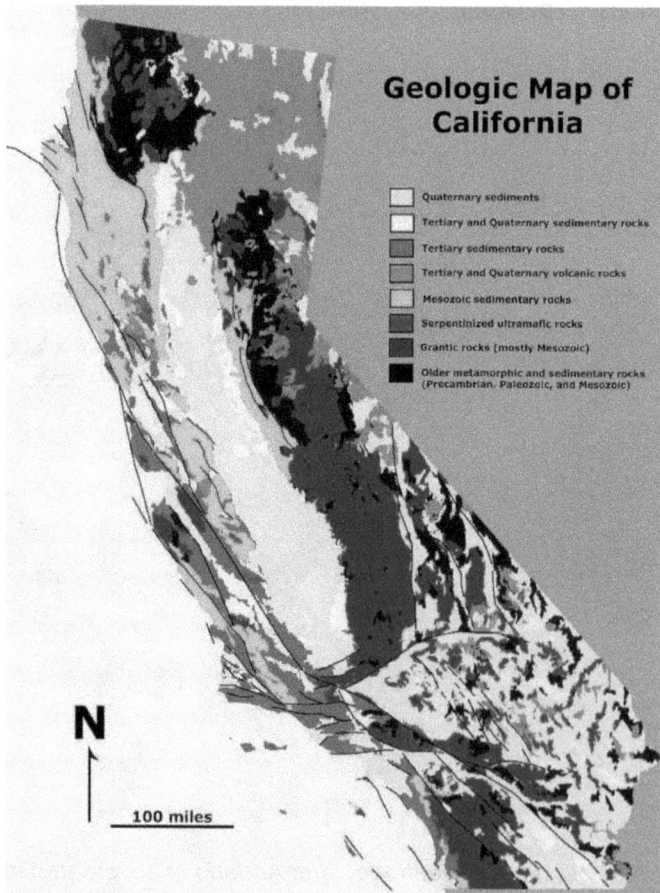

Figure 12: Geologic map California. U.S. Geological Survey, Public Domain, via Picryl.

Rare Earth and Critical Mineral Operations and Processing

Sampling

Sampling is the process of collecting representative materials—such as rock, soil, stream sediments, or drill core—for laboratory analysis. The goal is to measure the concentration of valuable elements and understand how these concentrations vary across an area.

Different types of sampling are used depending on the exploration stage:

- **Rock chip sampling:** Collects fragments directly from outcrops or trenches to test for mineral content.

- **Soil sampling:** Commonly used in areas with limited outcrop; helps identify geochemical anomalies by testing the fine fraction of soil.

- **Stream sediment sampling:** Detects mineral signatures carried downstream by erosion, useful for regional reconnaissance.

- **Channel sampling:** Involves cutting a continuous channel across a mineralised zone to obtain a representative cross-section of mineral content.

For example, in a potential lithium pegmatite field, explorers may collect rock chips of spodumene-bearing dykes and soil samples over a grid to trace lithium dispersion patterns. In REE exploration, sampling focuses on weathered profiles or clays to detect ion-adsorbed rare earths.

All samples are carefully logged, labelled, and stored under strict chain-of-custody protocols to ensure accuracy and traceability.

Drilling

Drilling is a critical stage in confirming whether the mineralisation identified at the surface continues at depth. It provides direct physical evidence of subsurface geology and helps define the geometry, grade, and continuity of a deposit.

Common drilling methods include:

- **Reverse Circulation (RC) Drilling:** Uses compressed air to bring rock chips to the surface. It is faster and cheaper than core drilling, ideal for initial resource definition.

Figure 13: RC drill rig. en:User:Blastcube, CC BY-SA 3.0, via Wikimedia Commons.

- **Diamond Core Drilling:** Extracts continuous cylindrical rock cores, allowing precise geological logging and structural analysis. It provides the most accurate samples for assay and metallurgical testing.

- **Auger or Aircore Drilling:** Used in soft sediments or weathered zones for early-stage exploration.

Drill cores are meticulously examined to identify mineral-bearing zones, alteration types, and geological structures. Each interval is logged, photographed, and sampled at measured intervals (commonly 1–2 metres). Geologists note visible minerals such as monazite, bastnäsite, spodumene, pyrochlore, or graphite, depending on the target commodity.

The resulting drill data form the basis for creating three-dimensional geological models and preliminary resource estimates.

Assay Techniques

Rare Earth and Critical Mineral Operations and Processing

Assaying is the laboratory process used to determine the exact chemical composition of rock, soil, or drill samples. It provides the quantitative data needed to evaluate the economic potential of a deposit—identifying how much of each valuable element (e.g., neodymium, lithium, cobalt, nickel) is present.

Assay techniques vary depending on the target mineral:

- **Inductively Coupled Plasma Mass Spectrometry (ICP-MS):** Highly sensitive and capable of detecting trace levels of rare earth elements, lithium, and other metals.

- **X-Ray Fluorescence (XRF):** A rapid, non-destructive method that provides semi-quantitative results for major and minor elements. Portable XRF devices are commonly used in the field for quick screening.

- **Fire Assay:** Traditional but highly accurate method used for precious metals such as gold and platinum group metals.

- **Atomic Absorption Spectroscopy (AAS):** Used for base metals like copper, nickel, and cobalt.

- **Loss on Ignition (LOI):** Determines volatile content such as water and carbonates in the sample, helping to interpret mineralogy.

Quality assurance and quality control (QA/QC) are essential parts of assaying. Laboratories insert blanks, duplicates, and certified reference materials into the sample stream to ensure results are precise and reproducible.

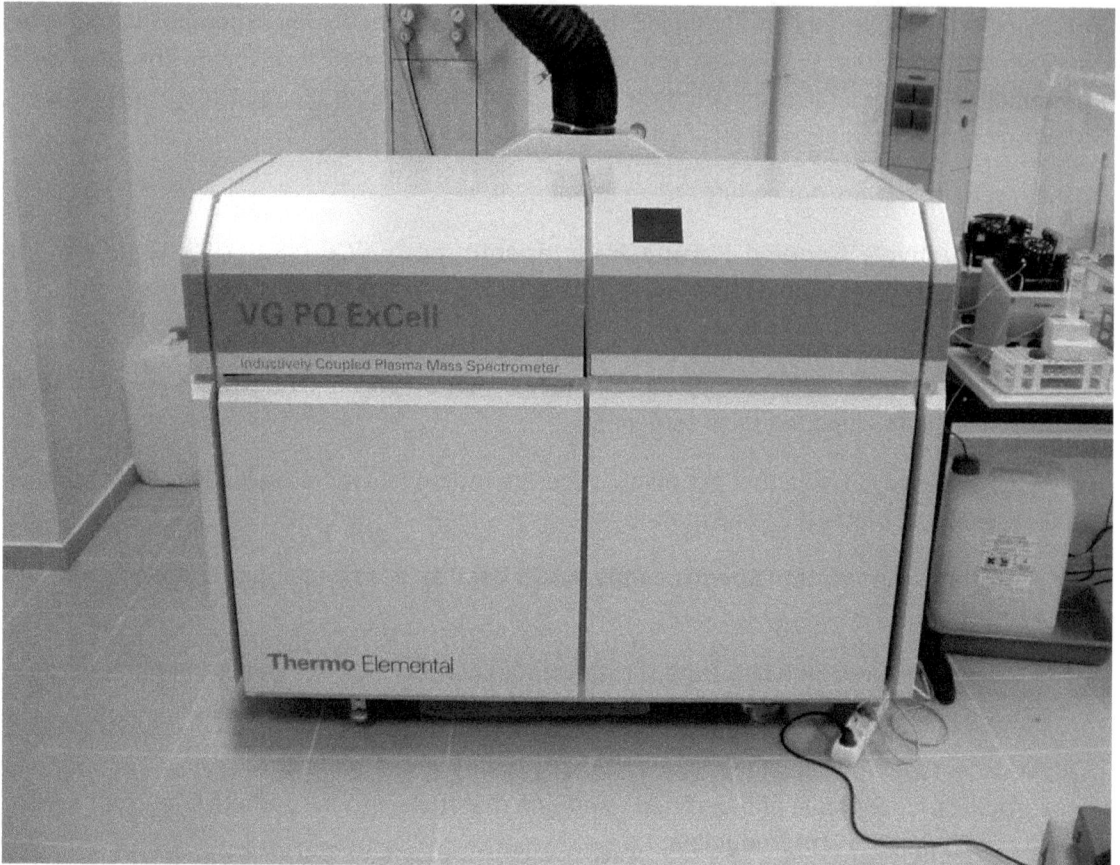

Figure 14: Inductively Coupled Plasma Mass Spectrometry (ICP-MS). Superchilum, CC BY-SA 3.0, via Wikimedia Commons.

Once assay results are received, geologists compile and interpret the data to map grade distribution and identify the richest zones for follow-up drilling or potential mining.

Integration and Interpretation

Modern mineral exploration relies on integrating data from all these methods—mapping, sampling, drilling, and assaying—into comprehensive geological and geochemical models. Advanced software allows 3D visualisation of ore bodies, estimation of resource tonnage, and assessment of economic feasibility.

For example:

- In a lithium pegmatite project, mapping identifies dykes, sampling confirms lithium anomalies, drilling defines ore zones, and assays quantify Li_2O grades.

- In a rare earth carbonatite system, geological mapping outlines the intrusive body, trench sampling reveals REE-bearing minerals, and drilling confirms vertical continuity of bastnäsite-rich layers.

The integration of these techniques transforms scattered field observations into a scientifically validated mineral resource model.

Ore Grade and Tonnage Considerations

In mineral exploration and resource evaluation, ore grade and tonnage are the two fundamental parameters that determine whether a deposit can be mined economically. Together, they define the size and quality of a mineral resource and play a decisive role in feasibility studies, investment decisions, and mine planning. Even if a deposit contains valuable elements such as rare earths, lithium, or cobalt, it can only be developed profitably if the concentration (grade) and total amount (tonnage) of extractable material meet specific thresholds.

Ore Grade

Ore grade is one of the most important measures used in mining and mineral exploration because it tells us how much of a valuable element is contained within a given mass or volume of rock. In simple terms, it indicates the quality or richness of the ore. A higher grade means that the rock contains more of the desired element, which can potentially make extraction more profitable. However, grade alone does not guarantee economic success — mining costs, recovery efficiency, and the physical and chemical characteristics of the minerals must also be considered.

Ore grade expresses the concentration of valuable elements in an ore body. Depending on the mineral and its market form, different units are used:

- **Percentage (%):** Used for elements present in relatively high concentrations. For example, a nickel ore grading 1.2% Ni means that each tonne of ore contains 12 kilograms of nickel metal.

- **Parts per million (ppm):** Used for elements found in trace amounts. For instance, 1 ppm equals 1 gram of metal per tonne of rock.

- **Grams per tonne (g/t):** Commonly used for precious metals such as gold or platinum. For example, a gold grade of **5 g/t** means 5 grams of gold per tonne of ore.

Different commodities require different grade reporting conventions because their abundance and economic value vary dramatically.

Each mineral commodity uses a specific way of expressing grade based on industry standards:

- **Lithium:** In hard-rock deposits (typically pegmatites), lithium grade is expressed as a percentage of lithium oxide (% Li_2O). A typical spodumene concentrate might average 1–2% Li_2O. This means that each tonne of ore contains between 10 and 20 kilograms of lithium oxide.

- **Rare Earth Elements (REEs):** These are usually reported as Total Rare Earth Oxides (TREO %), representing the combined weight of all rare earth oxides in the sample. A carbonatite deposit containing 3% TREO might hold 30 kilograms of REE oxides per tonne of rock.

- **Gold:** Because gold occurs in minute quantities, grades are expressed in grams per tonne (g/t). A gold grade of 10 g/t is considered high-grade, while 1 g/t can still be economically mined in large, low-cost open-pit operations.

- **Nickel and Cobalt:** In laterite and sulphide deposits, grades are reported as % Ni or % Co. A typical economic nickel laterite ore contains 0.8–2.5% Ni, and cobalt is often a by-product, present in concentrations of 0.05–0.2% Co.

While a higher grade generally indicates greater potential value, the relationship between grade and profitability is not direct or linear. This is because several other factors influence whether a deposit can be mined economically:

- **Mining and Processing Costs:** Lower-grade ore may require moving and processing more rock to extract the same amount of metal, increasing costs. However, if the deposit is near the surface and easily mined, it may still be profitable.

- **Recovery Rate:** This refers to how much of the valuable element can be successfully extracted during processing. If only 70% of the contained metal can be recovered, the effective grade is reduced.

- **Mineralogy and Metallurgy:** The form in which the element occurs (e.g., mineral type, chemical bonds) affects how easily it can be processed. Some rare earth minerals, for instance, are chemically complex and expensive to separate.

- **Market Prices:** The economic cut-off grade changes with fluctuations in commodity prices. Higher market demand for strategic minerals like lithium or neodymium can make lower-grade deposits viable.

Examples in practice:

- **Rare Earth Elements (REEs):** REE deposits often appear low-grade compared to base metals, with total rare earth oxide (TREO) concentrations ranging from 0.5–2%. However, they can still be extremely valuable because certain elements — notably neodymium (Nd), praseodymium (Pr), dysprosium (Dy), and terbium (Tb) — have very high prices due to their use in high-strength permanent magnets for electric vehicles and wind turbines. Even a small increase in the proportion of these critical REEs can dramatically boost the economic value of a deposit.

- **Lithium:** For lithium, grade and type of deposit are both key. In hard-rock pegmatite deposits like those in Australia, ore grades above 1% Li_2O are considered high-quality and suitable for producing battery-grade lithium hydroxide or carbonate. By contrast, lithium brine deposits — such as those found in South America's "Lithium Triangle" — have much lower grades (200–1,000 ppm, or 0.02–0.1% Li). Yet they remain profitable because the extraction process relies on solar evaporation in large salt flats (salars), which is far cheaper and less energy-intensive than hard-rock mining.

Ore grade directly influences mine design, processing plant capacity, and overall project economics. High-grade ores can justify smaller operations with lower capital investment, while low-grade ores require economies of scale and efficient processing technologies to remain competitive.

For instance:

- A lithium mine producing 1.5% Li_2O at high recovery rates can sustain profitability even if lithium prices fluctuate.

- A rare earth project with only 1% TREO can still succeed if the mineralogy is simple (easy to separate) and the deposit contains a high ratio of magnet elements like Nd and Dy.

Ultimately, ore grade defines how much metal can be obtained per tonne of ore and, together with tonnage, determines the total metal content (grade × tonnage). This total is what investors and governments use to assess whether a deposit is worth developing.

Tonnage

In mineral exploration and mining, tonnage refers to the total quantity of ore contained within a mineral deposit that can be economically mined under current technical and market conditions. It is a measure of the volume and mass of the ore body and is expressed in metric tonnes (t) or million tonnes (Mt). Tonnage is one of the key determinants of a deposit's potential value, alongside grade — the concentration of the valuable element or mineral within that ore.

While ore grade defines *quality*, tonnage defines *quantity*. The two must be considered together to assess whether a deposit can be developed profitably. A large deposit with low-grade material can still be worthwhile if it can be mined and processed inexpensively, whereas a smaller, high-grade deposit may yield strong profits if the contained material has high market value or can be extracted efficiently.

The total tonnage of a mineral deposit is not a random figure—it is determined by a combination of geological and physical characteristics that define the nature of the ore body. One of the most influential factors is the size, shape, and depth of the deposit. The three-dimensional geometry—its surface area, thickness, and orientation—controls how much material it can contain. Large, shallow, and flat-lying deposits, such as sedimentary iron formations or lithium brine basins, often have extremely high tonnages and are well suited to cost-effective open-pit or surface mining methods. In contrast, steep, narrow veins or deep-seated bodies, such as pegmatite or hydrothermal systems, typically hold smaller tonnages but may compensate with exceptionally high grades of valuable minerals.

Another critical factor is the density of the ore-bearing rock, which determines how much mass is contained within a given volume. Density is measured in tonnes per cubic metre (t/m^3). For instance, dense rocks like magnetite, with a density of about 5.0 t/m^3, contain significantly more mass—and potentially more metal—than lighter rocks such as granite, which averages around 2.6 t/m^3. Accurately measuring rock density through laboratory testing or downhole logging is vital for converting geological volume into tonnage and ensuring resource estimates are reliable.

A third key parameter influencing tonnage is the cut-off grade, which represents the minimum concentration of valuable mineral that can be mined profitably under current market and technological conditions. Material with grades below this threshold is classified as waste, while material above it contributes to the mineable ore tonnage. Adjusting the cut-off grade can have a major impact on total tonnage and economic viability. Lowering the cut-off grade increases tonnage by including more material, but it may reduce overall profitability because the added material has a lower metal content. Conversely, raising the cut-off grade decreases tonnage but focuses mining on higher-value ore zones.

These parameters—geometry, density, and cut-off grade—are combined to construct an ore block model, a three-dimensional digital representation of the deposit. This model forms the foundation for estimating both total tonnage and grade distribution, guiding mine design, production planning, and economic assessment.

The balance between grade and tonnage is central to determining a deposit's economic value. In general:

- High-grade, low-tonnage deposits can be profitable if the metal value per tonne of ore is high, mining costs are low, or the deposit contains critical or strategic minerals.

- Low-grade, large-tonnage deposits rely on economies of scale, efficient bulk mining methods, and low processing costs to achieve profitability.

For example:

- A carbonatite-hosted rare earth deposit like Mount Weld (Western Australia) contains about 20 million tonnes (Mt) of ore with grades of 8–10% total rare earth oxides (TREO). Despite its moderate size, the high grade and favourable mineralogy make it one of the richest REE deposits in the world.

- In contrast, lithium brine deposits such as Salar de Atacama (Chile) hold vast tonnages of brine containing only ~0.15% lithium (1,500 ppm). However, extraction costs are very low because lithium is recovered through solar evaporation, which requires little energy or infrastructure compared to hard-rock mining.

Rare Earth and Critical Mineral Operations and Processing

Thus, grade and tonnage are inversely related: high grades are usually found in smaller deposits, while large deposits tend to have lower grades. Yet both can coexist economically depending on extraction technology, infrastructure, and commodity prices.

Tonnage is not merely a geological measurement—it has significant implications for mine design, infrastructure development, and overall project economics. The size of a deposit directly influences how a mine is planned, how long it can operate, what mining methods are used, and how sensitive the operation will be to fluctuations in market conditions.

In terms of mine life, larger tonnage deposits generally support longer operational periods, which enhances financial stability and provides consistent returns over time. A long mine life also justifies the higher upfront investment required for large-scale processing plants, haul roads, and other infrastructure. These long-term projects often attract more investors and government support because they promise sustained employment and regional development.

The mining method chosen depends heavily on tonnage. Large-tonnage deposits are usually mined using bulk mining techniques, such as open-pit operations, which enable the extraction of massive volumes of ore efficiently and at relatively low cost per tonne. In contrast, smaller or deeper deposits with limited tonnage often necessitate underground mining, a more selective but significantly more expensive method that targets only the richest zones of ore.

Processing requirements also vary with tonnage. High-tonnage but low-grade deposits must rely on large, continuous processing plants to achieve economies of scale. While this approach can make lower-grade ores profitable, it also increases energy use, reagent consumption, and tailings generation, all of which influence both operating costs and environmental impact. Efficient recovery methods and sustainable waste management become critical factors in maintaining profitability and regulatory compliance.

Finally, market sensitivity differs between deposit types. Large-tonnage operations tend to be more resilient to market price swings because they can sustain production at high throughput and lower profit margins. High-grade, low-tonnage deposits, however, are far more sensitive to changes in commodity prices or operating costs. When prices fall or expenses rise, such deposits can quickly become uneconomic. Thus, understanding how tonnage interacts with mine design, cost structures, and market dynamics is essential for long-term project success.

Tonnage estimation is a critical step in evaluating the economic potential of a mineral deposit. It brings together information from geological mapping, drilling data, and three-dimensional (3D) modelling to calculate the total quantity of ore contained within defined geological and economic boundaries. This process transforms field data into a measurable estimate of how much material can potentially be mined.

The estimation process begins by defining the deposit's geometry, using data from drill holes, geological cross-sections, and surface mapping to outline the shape, size, and orientation of the ore body. Once the geometry is established, geologists calculate the total volume of ore-bearing rock within those boundaries. That volume is then multiplied by the rock's density—a measure of mass per cubic metre—to convert it into total tonnage. Finally, the cut-off grade is applied to exclude material that falls below the economic threshold, ensuring that only the mineable portion of the deposit is counted.

The outcome of this process is reported as a resource estimate, summarising both the tonnage and the average grade of the deposit. For example, a statement might read:

"The deposit contains 25 million tonnes (Mt) of ore grading 1.2% Li_2O, representing approximately 300,000 tonnes of contained lithium oxide."

These estimates are not static. As exploration continues and new drilling and assay data are collected, the resource model is refined, and the level of confidence in the estimate improves. Over time, what begins as an inferred resource can be upgraded to an indicated or measured resource, providing the foundation for mine planning, feasibility studies, and investment decisions.

The relationship between grade and tonnage is one of the most important economic considerations in the development of mineral deposits. These two factors often work in opposition—high-grade deposits tend to be smaller in size, while low-grade deposits are typically much larger. The balance between them determines not only the style of mining operation but also the role each deposit plays within the global critical minerals supply chain.

In the high-grade, moderate-tonnage category, deposits are smaller but contain a much higher concentration of valuable minerals. These are often mined for premium, high-value products that can sustain strong profit margins even with smaller volumes. For example, Mount Weld in Western Australia contains approximately 20 million tonnes (Mt) of ore grading 8–10% total rare earth oxides (TREO), making it one of the richest rare earth deposits in the world. Similarly, the Greenbushes lithium deposit, also in Western Australia, hosts around 60 Mt at 2.0% Li_2O, and is renowned for producing high-quality lithium concentrates for battery manufacturing.

Conversely, low-grade, large-tonnage deposits dominate the other end of the spectrum. These deposits are vast but contain lower concentrations of valuable elements, relying on economies of scale and efficient extraction methods to remain profitable. A prime example is the Salar de Atacama in Chile, which contains thousands of millions of tonnes of lithium brine averaging only 0.15% Li. Despite its low grade, the deposit is highly profitable due to low-cost, solar-based evaporation extraction. Similarly, Bayan Obo in China is one of the world's largest rare earth deposits, containing over 1,000 Mt at grades of 4–6% TREO, and serves as a cornerstone of global rare earth production.

Both deposit types play crucial roles in meeting global demand for critical minerals. High-grade, moderate-tonnage deposits supply specialised, high-purity products essential for advanced technologies, while large, low-grade deposits ensure long-term supply stability and help buffer the market against shortages. Together, they illustrate the economic trade-off between grade and tonnage that underpins modern resource development and supply chain resilience.

Rare Earth and Critical Mineral Operations and Processing

The Cut-Off Grade Concept

The cut-off grade is one of the most critical concepts in mineral resource estimation and mine planning. It represents the minimum concentration of a valuable element or mineral that can be mined, processed, and sold at a profit under current economic and technical conditions. In simple terms, it is the dividing line between ore and waste—only material above this threshold is considered economically viable for extraction, while material below it is discarded or stockpiled for potential future use.

In every mineral deposit, the grade of valuable material varies across different zones. Not all parts of a deposit contain the same concentration of metal, and mining all the rock indiscriminately would be uneconomic. The cut-off grade provides a rational basis for selecting which parts of the deposit to mine.

- Material above the cut-off grade is classified as ore, meaning it contains enough value to justify mining, processing, and sale.

- Material below the cut-off grade is classified as waste, as the costs of extracting and processing it would exceed its market value under current conditions.

This distinction allows mining companies to design efficient extraction plans, minimising waste and maximising profit.

The cut-off grade is not fixed; it fluctuates depending on a combination of economic, technical, and market factors, including:

- **Metal Prices:** When the price of the target element (e.g., neodymium, lithium, cobalt, nickel) increases, lower-grade material becomes more valuable, and the cut-off grade may be lowered to include additional ore. Conversely, a price drop can make previously economic material unviable.

- **Processing Costs:** If energy, reagents, or labour costs rise, the cut-off grade must be increased to maintain profitability. New, cost-saving processing technologies can reduce it.

- **Recovery Rates:** The proportion of metal that can be extracted during processing also affects the cut-off grade. Higher recovery efficiency means more metal can be obtained from lower-grade ore.

- **Technological Improvements:** Advancements in beneficiation, hydrometallurgy, and automation can make it feasible to profitably mine lower-grade or previously refractory material.

- **Environmental and Regulatory Constraints:** Stricter environmental regulations can raise costs and thus increase cut-off grades, while more efficient waste management or renewable energy use may have the opposite effect.

Because of these dynamic factors, the cut-off grade is regularly reviewed and updated during the life of a mining project to reflect changes in market conditions and operating costs.

The relationship between the cut-off grade and total mineable ore is inverse—as the cut-off grade decreases, tonnage increases, but average grade and profitability per tonne decrease. Conversely, increasing the cut-off grade reduces tonnage but focuses on the highest-value material.

For example:

- If the price of neodymium doubles due to rising demand for electric vehicle motors, a rare earth mine may lower its cut-off grade from 1.0% TREO to 0.6% TREO, effectively expanding its ore reserves.

- In a lithium brine operation, if processing technology improves and lowers extraction costs by 20%, the company may include brine with 200 ppm Li that was previously uneconomic, increasing total recoverable resources.

These adjustments can extend mine life, improve resource utilisation, and adapt operations to shifting market realities.

Geologists and mining engineers use advanced computer-based 3D geological models to apply cut-off grades across the deposit. Using drilling, sampling, and assay data, they divide the ore body into blocks—small 3D units, each assigned a grade value. By applying the cut-off grade threshold to these blocks, they can:

- Define the boundaries of economically mineable material.

- Calculate total ore tonnage and average grade.

- Estimate contained metal content for project evaluation.

This process forms the basis for resource and reserve classification, a critical step in reporting and investment decision-making.

To ensure consistency and transparency, mineral resource estimates are reported according to international standards such as:

- **JORC Code (Australia):** Developed by the Australasian Joint Ore Reserves Committee, this code classifies mineral resources as Inferred, Indicated, or Measured, based on geological confidence, and defines Ore Reserves as Probable or Proved once economic viability is demonstrated.

- **NI 43-101 (Canada):** A similar reporting framework that mandates detailed technical disclosure of exploration results, resource estimates, and mining feasibility.

Under both systems, the application of a realistic cut-off grade is a mandatory component of reserve reporting. It ensures that declared resources reflect material that can be mined profitably and responsibly under current or reasonably foreseeable conditions.

Rare Earth and Critical Mineral Operations and Processing

The cut-off grade also has broader implications beyond economics. Lowering it can maximise resource utilisation and reduce waste generation, aligning with sustainability goals and circular economy principles. However, it can also increase the amount of material mined and processed, requiring more energy and water, and generating more tailings. Thus, modern mine planning aims to find a balanced cut-off grade that optimises profitability, resource recovery, and environmental performance.

The cut-off grade is the cornerstone of resource estimation and mine design. It defines what portion of a deposit is viable for extraction, influences mine life and production strategy, and responds dynamically to changing market and technological conditions. By applying this threshold through sophisticated 3D modelling, geologists and engineers can accurately determine ore boundaries, tonnage, and grade, providing the foundation for reliable reporting under international standards such as JORC and NI 43-101.

In essence, the cut-off grade is where geology meets economics—a constantly evolving balance that determines whether a mineral deposit remains a scientific curiosity or becomes a profitable, sustainable mining operation.

Grade–Tonnage Relationship

The grade–tonnage relationship is a fundamental concept in economic geology and mine planning that describes how the concentration of valuable minerals (grade) and the total quantity of extractable material (tonnage) are interrelated. In most natural systems, these two variables have an inverse relationship—as one increases, the other tends to decrease. In other words, high-grade deposits are generally smaller and rarer, while low-grade deposits are usually larger but require more extensive mining and processing to extract the same amount of valuable material. Understanding this relationship is essential for determining whether a mineral deposit can be mined profitably and sustainably.

In geological terms, high-grade mineralisation occurs under specific conditions and tends to be concentrated in smaller, localized zones—such as veins, lenses, or pockets—where the right chemical and physical processes have acted together to enrich certain elements. Because these processes are relatively rare and restricted in scale, high-grade deposits are uncommon and limited in tonnage.

By contrast, low-grade deposits form under broader geological conditions and are more widespread. They typically consist of vast volumes of rock containing only small amounts of the desired mineral. Although these deposits are easier to find and mine in bulk, they require large-scale operations, significant energy input, and advanced processing to achieve economic returns.

For example:

- High-grade deposits (like Mount Weld in Australia, with 8–10% total rare earth oxides) yield high-value products from smaller ore bodies.

- Low-grade deposits (such as the Salar de Atacama lithium brine fields in Chile, with ~0.15% Li) cover enormous areas but depend on low-cost extraction to remain profitable.

This natural trade-off between grade and tonnage is one of the central challenges in modern mining economics.

The relationship between grade and tonnage is often visualised using a grade–tonnage curve, a graph that shows how total recoverable metal content changes as the cut-off grade (the economic threshold) varies.

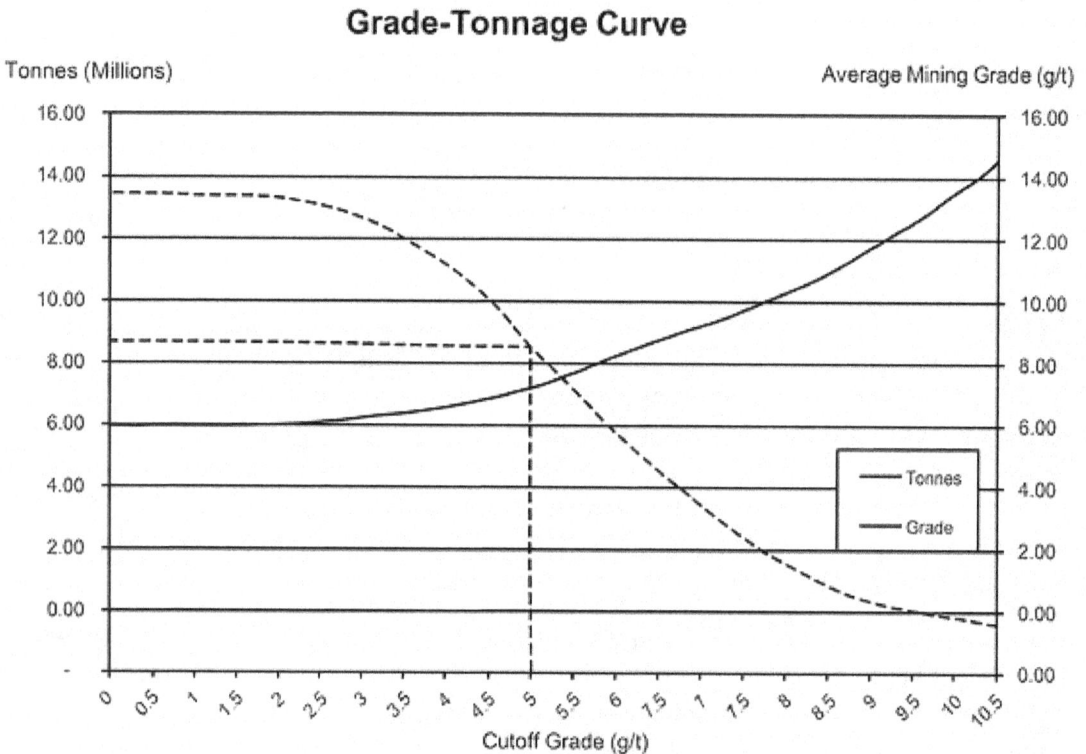

Grade-Tonnage Curve

Figure 15: Example of a grade–tonnage curve.

Here's how it works:

- As the cut-off grade increases, only the highest-grade zones are classified as ore. This reduces the total tonnage but increases the average grade and potential revenue per tonne.

Rare Earth and Critical Mineral Operations and Processing

- As the cut-off grade decreases, more material is included as ore, increasing total tonnage but lowering the average grade. This means more rock must be mined and processed, which raises costs and generates more waste.

The area under the curve represents the total contained metal, and the shape of the curve helps engineers determine the optimal operating range—the balance point where profits are maximised.

For instance, if a rare earth mine raises its cut-off grade from 1% to 2% TREO, tonnage might fall by half, but each tonne of ore mined will yield more rare earths, reducing processing costs per unit of metal. Conversely, lowering the cut-off to 0.5% TREO could double tonnage but make the operation less efficient and environmentally demanding.

Determining the optimal grade–tonnage balance is not purely a geological exercise—it requires integrating economic, technological, and environmental considerations. The goal is to find the sweet spot where revenue, costs, and sustainability align.

Several factors influence this balance:

- **Mining and processing costs:** Lower-grade ores require more energy, water, and reagents to produce the same output, which increases per-tonne costs.

- **Metal prices:** Higher market prices justify mining lower-grade material, while falling prices make only the richer portions of the deposit profitable.

- **Recovery rates:** Efficient processing technologies can make lower-grade deposits viable by improving metal recovery.

- **Environmental regulations:** The cost of waste disposal and land rehabilitation can discourage operations with excessive tonnage and low grades.

Modern mining companies use computer simulations and cost models to adjust these variables and determine the most profitable and sustainable operating grade.

Explorers and mining engineers use the grade–tonnage relationship to guide exploration strategies, feasibility studies, and mine design.

- During exploration, it helps assess whether a deposit's scale and grade justify continued investment.

- In mine planning, it determines which portions of the deposit should be mined first and how the cut-off grade should evolve over the mine's life.

- During production, companies may adjust cut-off grades dynamically in response to changing commodity prices or processing efficiency.

For example:

- In a lithium pegmatite project, if lithium prices rise, the company might lower the cut-off grade to include more tonnage, extending the mine life.

- In a nickel-cobalt laterite deposit, if energy or reagent costs increase, engineers might raise the cut-off grade to focus only on the highest-value ore zones.

These adjustments ensure that mining remains profitable even in fluctuating economic environments.

The grade–tonnage relationship has broader implications for national and global resource management.

- High-grade deposits are ideal for supplying premium, high-purity materials used in advanced technologies.

- Large, low-grade deposits provide the volume needed to stabilise long-term supply chains for essential commodities like lithium, nickel, and rare earths.

Balancing these two deposit types supports a resilient and diversified critical minerals industry—one capable of meeting both high-technology demand and mass-market energy transition needs.

In essence, the grade–tonnage relationship defines the economic heart of mining. High-grade deposits yield quality; low-grade deposits provide quantity. The optimal balance between the two ensures profitability, sustainability, and reliable resource supply.

By analysing grade–tonnage curves and adjusting cut-off grades, geologists and engineers can optimise resource extraction, adapt to market conditions, and ensure that both geological potential and economic reality align. This relationship remains a cornerstone of decision-making in the exploration, development, and operation of rare earth and critical mineral projects worldwide.

Recoveries and Ore Quality

While grade—the concentration of a valuable element within an ore—is a critical indicator of potential value, it does not by itself determine whether a deposit is economically viable. The true value of an ore depends just as much on its metallurgical recovery—the proportion of the contained metal that can actually be extracted and refined through processing. High-grade ore that is difficult to process may be less profitable than a lower-grade ore that yields high recoveries. Therefore, both grade and recovery must be evaluated together to assess a deposit's real economic potential.

Metallurgical recovery refers to the percentage of valuable metal or element recovered from the ore during extraction and processing. It measures how efficiently a processing plant can convert the ore's theoretical metal content into saleable product.

The formula is:

$$\text{Effective Grade} = \text{In-situ Grade} \times \text{Recovery Rate}$$

Rare Earth and Critical Mineral Operations and Processing

For example, if a lithium ore grades 1.5% Li_2O and has a recovery rate of 80%, the effective grade (or recoverable lithium) is 1.2% Li_2O. This means that only 80% of the lithium present in the ore is extracted as product, while the remaining 20% is lost during crushing, grinding, flotation, or leaching processes.

Similarly, a rare earth deposit might contain a high in-situ grade of 5% total rare earth oxides (TREO), but if its recovery rate is only 60%, the effective yield is reduced to 3% TREO. In practice, this can significantly lower profitability because processing costs per tonne remain constant while saleable output is reduced.

Mining companies ultimately make money not from the grade in the ground, but from the metal produced and sold. Even if an ore contains large amounts of valuable minerals, they must be liberated, concentrated, and refined efficiently to generate profit. Metallurgical recovery therefore plays a pivotal role in determining:

- Processing efficiency (how much of the metal can be extracted).

- Operating costs (energy, reagents, and time required per unit of recovered product).

- Environmental performance (amount of waste and tailings generated).

High recoveries mean more product from the same tonnage of ore, improving profitability. Low recoveries, on the other hand, increase the cost per unit of recovered metal and can make even rich deposits uneconomic.

The ease with which metals can be extracted depends heavily on mineralogy—the types of minerals that host the valuable elements and their physical and chemical characteristics. Some minerals release metals easily during processing, while others are complex, resistant, or chemically bound in forms that are difficult to treat.

- **Simple mineral forms:** Minerals like spodumene ($LiAl(SiO_3)_2$) for lithium or bastnäsite ($REECO_3F$) for rare earths are relatively straightforward to process. They respond well to conventional concentration methods such as flotation, gravity separation, or acid leaching, resulting in high recoveries and lower processing costs.

- **Complex mineral forms:** In contrast, minerals such as xenotime (YPO_4), eudialyte ($Na_4(Ca,Ce)_2Fe_2ZrSi_8O_{22}(OH,Cl)_2$), or monazite [$(Ce,La,Nd,Th)PO_4$] can be much more challenging. These minerals may contain multiple rare earths and trace radioactive elements (like thorium), requiring specialised chemical separation steps. This increases both processing costs and waste management complexity, while lowering overall recovery rates.

In essence, an ore's mineralogical complexity often dictates the choice of processing route— and thus the ultimate economic outcome.

Processing plants employ a combination of physical and chemical techniques—such as crushing, grinding, flotation, magnetic separation, roasting, and leaching—to liberate and extract metals. Each step involves inherent losses, and the aggregate efficiency of these processes determines the recovery rate.

Advances in technology can significantly improve recoveries:

- Hydrometallurgical innovations, such as ion-exchange and solvent extraction, have boosted recovery for rare earth elements from complex ores.

- Direct lithium extraction (DLE) technologies are increasing lithium recovery from brines beyond the traditional 50–60% achieved by solar evaporation.

- Bioleaching and nanofiltration techniques are emerging as low-impact alternatives for certain metals, improving both recovery and sustainability.

However, each technological improvement must balance efficiency with cost—recovering the last few percentage points of metal is often the most expensive.

Metallurgical recovery directly impacts the revenue, cost structure, and environmental footprint of a mining operation:

- Revenue: Higher recoveries mean more saleable product for the same mining cost.

- Costs: Lower recoveries increase the cost per tonne of recovered metal, as energy, labour, and reagents are spread over less output.

- Sustainability: Inefficient recovery generates more tailings and waste, requiring greater environmental management.

For instance, a rare earth project with a high-grade ore but only 60% recovery may underperform financially compared to a lower-grade operation with 90% recovery, simply because less of the contained metal is actually produced and sold.

When evaluating a deposit, geologists and engineers aim to find the optimal balance between grade, recovery, and cost. High recoveries are desirable, but they often require more energy, chemicals, or complex equipment, which can increase operational expenses. Similarly, chasing higher grades might mean mining deeper or narrower zones, raising extraction costs.

The most profitable operations achieve moderate-to-high recoveries at reasonable cost, using efficient processing methods matched to the ore's mineralogy. This balance is often refined through metallurgical test work and pilot plant studies before full-scale mining begins.

Metallurgical recovery is a key factor in determining the true economic value of a mineral deposit. While grade indicates how rich an ore is, recovery measures how much of that richness can actually be realised. Ores with simple mineralogy and high recovery rates are far more desirable than complex ores that yield only a fraction of their contained metals.

In critical mineral projects—such as those producing lithium, rare earth elements, and cobalt— understanding recovery behaviour is essential for accurate resource valuation, process design, and long-term profitability. Ultimately, successful mining depends not just on what is found in the ground, but on how efficiently and economically it can be brought to market.

The Role of Market and Technology

The concepts of ore grade and tonnage thresholds are not static; they change over time as technology, economics, and global demand evolve. A mineral deposit once considered uneconomic can become viable decades later due to improvements in extraction and processing technologies, shifts in market demand, or the emergence of new applications for the minerals it contains. Likewise, recycling and resource efficiency innovations can influence how we define what constitutes "ore." This dynamic nature of resource evaluation reflects the constant interplay between science, engineering, and economics in the mining industry.

Historically, the term *ore* referred strictly to material that could be mined and processed profitably with available technology. Anything below that threshold was called *waste rock*. However, as technology advances and market conditions shift, the boundaries between ore and waste often blur.

Several factors drive this evolution:

- **Technological innovation:** New extraction and processing techniques can recover metals from ores once deemed too low in grade or too complex in mineralogy.

- **Market price fluctuations:** Rising demand and higher prices for strategic minerals can suddenly make marginal deposits economically attractive.

- **Energy and environmental considerations:** Cleaner, more efficient methods reduce costs and environmental impacts, allowing the exploitation of lower-grade resources.

- **Recycling and secondary sources:** Recovering metals from waste or used products can complement or even replace traditional mining, reshaping what is considered viable ore.

Thus, what was considered sub-economic in one era may become highly valuable in another—especially for **critical and strategic minerals** central to renewable energy and advanced technologies.

Technological Advances Transforming Ore Viability

a. Hydrometallurgical Leaching and Ion-Adsorption Clays

A prime example of changing thresholds can be seen in the evolution of rare earth element (REE) extraction from ion-adsorption clays in southern China. These clays, formed through tropical weathering of REE-rich rocks, contain only 0.05–0.3% total rare earth oxides (TREO)—far lower than typical hard-rock deposits such as carbonatites or monazite sands.

For decades, these clays were considered too poor in grade to mine economically. However, the development of hydrometallurgical leaching technologies, particularly ammonium sulphate and weak acid leaching, revolutionised their viability. These methods allow rare earths—loosely bound to clay particles—to be selectively extracted under mild, low-cost conditions.

Today, these "low-grade" deposits are the world's primary source of heavy rare earth elements (HREEs) like dysprosium (Dy) and terbium (Tb), essential for high-performance magnets in electric vehicles and wind turbines. This transformation underscores how processing innovation can turn marginal geological formations into world-class resources.

b. Direct Lithium Extraction (DLE) from Brines

Another technological leap reshaping ore thresholds is Direct Lithium Extraction (DLE). Traditional lithium recovery from brine—such as in Chile's Salar de Atacama—relies on large evaporation ponds where lithium is concentrated over 12–18 months, achieving only 40–60% recovery. While this method is effective for high-grade brines, it struggles with lower-concentration or high-magnesium brines found in other regions.

DLE technology uses selective ion-exchange, adsorption, or membrane filtration to extract lithium directly from brine, achieving up to 80–90% recovery and drastically reducing water loss and environmental footprint. These advancements open the door for low-grade and previously uneconomic brines—including geothermal waters and oilfield brines—to become commercially viable.

As a result, countries like the United States, Argentina, and Australia are now reassessing lower-grade brine fields once dismissed as subeconomic, significantly expanding global lithium supply potential.

c. Demand-Driven Shifts in Rare Earth Economics

The surge in global demand for electric vehicles (EVs), wind turbines, and renewable energy technologies has altered the economic landscape for rare earth elements. Previously, deposits with moderate grades or challenging mineralogy were uneconomic due to limited demand and low prices.

However, as neodymium (Nd), praseodymium (Pr), dysprosium (Dy), and terbium (Tb) became critical components in high-efficiency permanent magnets, even moderate-grade deposits began to attract investment. The rising market value of these elements effectively lowers the economic threshold—meaning that deposits with lower grades or higher processing costs can now yield acceptable returns.

For example:

- A REE deposit with 2% TREO might have been uneconomic in the early 2000s but can now be developed profitably due to soaring magnet metal prices.

- Previously ignored bastnäsite or monazite-bearing sands are being re-evaluated as critical mineral sources in Australia, Africa, and North America.

In this sense, market-driven evolution complements technological advancement, making lower-grade deposits feasible and diversifying global supply chains.

Economic and Environmental Implications

Lowering grade and tonnage thresholds has profound implications for both the economics and sustainability of mining:

- **Economic expansion:** More deposits become mineable, extending mine life and supporting stable supply of critical minerals.

- **Resource efficiency:** Technologies that extract metals from lower-grade ores or waste materials reduce the need for new large-scale mining projects.

- **Environmental trade-offs:** Although more resources become accessible, processing lower-grade ores often requires more energy and reagents, increasing the carbon and waste footprint unless mitigated by new technologies.

This balance between expanded opportunity and environmental responsibility defines the modern mining industry's transition toward sustainable critical mineral production.

The Role of Recycling and Circular Economy Approaches

Beyond primary mining, recycling and urban mining are reshaping what counts as a "viable resource." Advanced metallurgical processes can now recover lithium, cobalt, nickel, and rare earths from spent batteries, magnets, and electronics, effectively creating secondary ore deposits.

As recycling efficiency improves, the global threshold for viable resource grade continues to fall—not because natural ore grades are rising, but because humanity is learning to extract value from ever-lower concentrations of material, whether in rock or waste.

Ore grade and tonnage thresholds evolve continuously as technology and market forces advance. Innovations such as hydrometallurgical leaching, direct lithium extraction, and improved rare earth processing have transformed the economics of previously marginal resources. At the same time, rising demand for green technologies is redefining what counts as economically viable ore.

In essence, today's waste rock may be tomorrow's strategic reserve. As extraction and recycling technologies improve, and as global markets increasingly prioritise supply security for critical minerals, the mining industry's definition of "ore" continues to expand—ensuring that resources once deemed too low-grade or too difficult to process become key contributors to the clean energy transition.

Key Terms and Concepts	

Ore Deposit: A naturally occurring concentration of minerals or metals that can be mined and processed economically under current conditions.

Mineralisation: The geological process by which valuable minerals accumulate and form ore deposits through magmatic, hydrothermal, sedimentary, or metamorphic activity.

Host Rock: The rock body that contains or surrounds an ore deposit, influencing mineral occurrence, mining methods, and processing requirements.

Bastnäsite: A carbonate-fluoride mineral that is one of the primary sources of light rare earth elements such as cerium, lanthanum, and neodymium.

Monazite: A phosphate mineral rich in rare earth elements and thorium, commonly found in placer deposits and heavy mineral sands.

Spodumene: A lithium-bearing pyroxene mineral mined primarily for lithium extraction used in batteries and ceramics.

Carbonatite Deposit: A rare igneous rock formation composed predominantly of carbonate minerals, often hosting rare earth elements and niobium.

Pegmatite: A very coarse-grained igneous rock that may contain economically valuable minerals such as lithium, tantalum, and beryllium.

Placer Deposit: A surface deposit formed by the mechanical concentration of heavy minerals through water or wind action, common for monazite and zircon.

Laterite Deposit: A weathered soil layer rich in iron and aluminium oxides, which can host nickel, cobalt, and rare earth elements in tropical regions.

Key Terms and Concepts

Geological Mapping: The process of systematically recording rock types, structures, and mineral occurrences on a map to aid exploration and resource evaluation.

Sampling: The process of collecting representative portions of rock, soil, or drill core for laboratory analysis to assess mineral content and grade.

Drilling: The mechanical process of penetrating the earth's surface to obtain subsurface rock or core samples for geological and geochemical assessment.

Assay: A laboratory test used to determine the concentration of valuable elements or minerals in a rock, soil, or concentrate sample.

Ore Grade: The concentration of a valuable element or mineral within an ore deposit, usually expressed as a percentage or parts per million (ppm).

Tonnage: The total quantity of ore or material available in a deposit, typically measured in tonnes, used to assess resource size and economic viability.

Cut-off Grade: The minimum grade at which a mineral deposit can be economically mined and processed under current market conditions.

Exploration Geophysics: The use of physical measurement techniques such as magnetics, gravity, or seismic surveys to detect subsurface mineralisation.

Chapter 2 Review Questions

Geological Processes

1. What are the main geological processes that form rare earth and critical mineral deposits?

2. How do magmatic processes such as carbonatite intrusion lead to REE concentration?

3. What role does weathering play in forming ion-adsorption clay deposits?

4. How are lithium brine deposits formed compared with hard-rock pegmatites?

Host Rocks and Ore Minerals

5. In which host rocks does bastnäsite commonly occur?

6. What distinguishes monazite from bastnäsite in terms of mineral composition and occurrence?

7. Why is xenotime important economically?

8. Where is spodumene typically found, and what is its industrial significance?

Key Deposit Types

9. Describe the key mineral assemblages in carbonatite deposits.

10. How can pegmatite deposits be identified in the field?

11. What processes form lateritic nickel–cobalt deposits?

12. What defines a placer deposit?

Global Examples

13. Name two globally significant carbonatite rare earth deposits.

14. Where are major lithium pegmatites found?

15. Which countries dominate ion-adsorption clay REE production?

16. Why is the Bushveld Complex significant to critical mineral supply?

Exploration Methods

17. What is the purpose of geological mapping in mineral exploration?

18. How do geochemical surveys assist in identifying mineralisation?

19. Describe the function of geophysical surveys in mineral exploration.

20. Why is core logging important after drilling?

Assay and Geochemical Data

Rare Earth and Critical Mineral Operations and Processing

21. What is an assay result used for in exploration?

22. How can geochemical data indicate ore potential?

23. Why are multiple samples analysed across drill cores?

Ore Grade, Tonnage, and Cut-Off Grade

24. Define ore grade and tonnage.

25. What is meant by "cut-off grade"?

26. How do grade and recovery interact to affect economic value?

27. Why can technological advances lower economic thresholds for mining?

Environmental, Social, and Regulatory Considerations

28. Why must exploration programs address environmental management early?

29. What social factors can influence exploration success?

30. How do regulations protect sensitive environments?

31. What are common environmental impacts of critical mineral exploration?

Richard Skiba

Chapter 3

Mineral Characterisation and Ore Testing

Understanding the physical and chemical properties of ores is the foundation of every successful mineral processing and metallurgical design. Before any beneficiation, extraction, or refining can occur, geologists and engineers must first know exactly what minerals are present, how they are distributed, and how they behave under mechanical, chemical, and thermal conditions. Chapter 3 builds this essential bridge between geology and metallurgy by examining the methods used to characterise ores and to test their response to processing.

The chapter begins by exploring the physical and chemical properties that define ore behaviour—from hardness, density, and magnetism to reactivity, solubility, and oxidation state. It then introduces the principal analytical techniques used in rare-earth and critical-mineral evaluation, including X-ray Diffraction (XRD), X-ray Fluorescence (XRF), Scanning Electron Microscopy (SEM), Mineral Liberation Analysis (MLA), and Inductively Coupled Plasma Mass Spectrometry (ICP-MS). Together, these tools reveal the mineralogical and elemental makeup of an ore, informing both exploration strategies and processing routes.

Finally, the chapter discusses bench-scale and pilot-scale testing—the critical transition between laboratory research and full-scale production. Through systematic testing of crushing, grinding, leaching, and separation steps, these programs validate metallurgical concepts, optimise process parameters, and reduce technical and financial risks. By the end of this chapter, you will understand how mineral characterisation and ore testing convert geological discovery into operational feasibility, forming the cornerstone of sustainable and economically sound rare-earth and critical-mineral projects.

Learning Outcomes

This chapter aims to give you the ability to:

96

Learning Outcomes	

1. Describe the key physical and chemical properties of ores that influence mineral processing performance, including hardness, density, magnetism, conductivity, and chemical composition.
2. Explain the importance of mineral characterisation in determining processing routes and metallurgical performance for rare earth and critical mineral ores.
3. Identify the main techniques used for mineral identification and liberation studies, and interpret their relevance to understanding ore texture, grain size, and mineral associations.
4. Operate or interpret results from key analytical tools—including X-ray diffraction (XRD), X-ray fluorescence (XRF), scanning electron microscopy (SEM), mineral liberation analysis (MLA), and inductively coupled plasma mass spectrometry (ICP-MS)—for accurate mineral and elemental analysis.
5. Differentiate between qualitative and quantitative analytical methods and evaluate their suitability for various ore types and project stages.
6. Discuss the purpose and procedures of bench-scale testing for assessing ore response to crushing, grinding, flotation, leaching, and other processing techniques.
7. Explain the role of pilot-scale testing in process design and demonstrate how data obtained from pilot plants inform scale-up, equipment selection, and economic evaluation.
8. Apply an understanding of mineralogical and analytical data to recommend appropriate process flowsheets and treatment methods for different rare earth and critical mineral ores.

Physical and Chemical Properties of Ores

Knowing the physical and chemical properties of ores isn't "nice-to-have geology"—it's what turns rock into revenue. In exploration, these properties shorten the path to discovery: density, magnetism, conductivity, and weathering behaviour guide geophysical/geochemical targeting and explain why REEs accumulate in specific trap sites (e.g., durable monazite/xenotime in placers versus ion-adsorption clays where bastnäsite has broken down). Geochemical pathfinders and ratios (La/Yb, Ce/Y, Nb/Ta, Zr/Hf, B–Li–Be) further discriminate carbonatite, alkaline, and pegmatite systems, saving metres of drill core and focusing drilling on the highest-potential targets.

Once a discovery is made, those same properties drive resource modelling and mine design. Grade only matters if it is recoverable: grain size, cleavage, and mineral intergrowth control liberation and therefore the "effective" grade the plant can actually produce. Hardness, texture, and density contrast steer open-pit versus underground methods, blasting strategies, and realistic cut-off grades. In metallurgy, mineralogy dictates the flowsheet: bastnäsite leans toward acid leach; monazite/xenotime typically need sulphuric roasting or alkaline cracking before REE extraction; ion-adsorption clays favour low-impact ammonium-salt ion exchange; spodumene requires high-temperature α→β conversion before leaching; and Nb–Ta systems hinge on fluoride complexation. Unit-operation selection (gravity, magnetic/electrostatic, flotation) and tight pH/redox control are set by these ore traits.

Product quality, marketability, and operating cost are equally governed by ore properties. Buyers pay for specifications—NdPr/Dy/Tb ratios, battery-grade LiOH quality, Nb with low Ta—and the ore's chemistry (solid solution, trace impurities) predicts penalty elements and blending needs. Hard, silica-rich ores elevate comminution power; refractory phosphates raise reagent and thermal budgets; carbonate ores consume acid; fluoride-rich circuits need HF capture. Conversely, clay-hosted REEs can reduce energy and reagents but require careful solution management. These choices shape C1 cost, water and reagent stewardship, emissions, residue storage, and the overall ESG footprint—often determining how smoothly a project permits.

Finally, ore properties underpin risk management, operations, and commercial strategy. Cut-off grades and pit shells can be re-optimised as prices for NdPr, Dy/Tb, or Li shift—if you know which mineral phases are actually recoverable. HREEs hosted in xenotime versus clays imply different geopolitical and processing dependencies. Storage and handling plans must anticipate surface oxidation (e.g., bastnäsite skin to CeO_2) that reduces leachability. In the plant, portable XRF, magnetic susceptibility, and particle-size tracking connect feed characteristics to real-time set-points, improving solvent-extraction staging and mass-balance closure. At the market interface, value-in-use models based on ore physics and chemistry forecast yields, reagent bills, and product slates, while by-product credits (Nb in carbonatites, Ta in LCT pegmatites, Sc in laterites) can materially lift project economics.

Although the name "rare earth elements" (REEs) implies scarcity, many of these elements are actually quite abundant in the Earth's crust, with distributions comparable to commonly mined metals such as copper, nickel, and zinc. Cerium (Ce), specifically, is the most abundant rare earth element and is found at approximately 66.5 parts per million (ppm), ranking as the 25th most common element in the Earth's crust, similar in abundance to copper and zinc [154-156]. Promethium (Pm), in stark contrast, is an exception due to its radioactivity and instability; it has no stable isotopes and a half-life of only 17.7 years, resulting in extremely low natural abundances.

The tendency for REEs to occur together is largely attributable to their similar chemical behaviours, resulting from their large ionic radii and high field strength. These characteristics classify them as incompatible elements; during partial melting processes, they preferentially remain in the melt phase rather than being incorporated into the crystal structures of primary minerals. Consequently, REEs tend to concentrate in distinct minerals such as bastnäsite and monazite, which are significant sources of these elements [154, 156, 157]. During Earth's early formation, denser materials migrated to the mantle, leading to a relative depletion of REEs in the crust compared to the mantle [133].

The REEs are generally categorized into light rare earth elements (LREEs) and heavy rare earth elements (HREEs) based on their atomic mass and ionic radii. This classification gives rise to the phenomenon known as lanthanide contraction, where ionic radii decrease from LREEs to HREEs. This lanthanide contraction leads to distinctive geochemical behaviours: LREEs like lanthanum and cerium are more mobile and can migrate into magmas, while HREEs such as dysprosium and yttrium are more likely to be found in residual minerals due to their compatibility with certain crystal structures [158, 159]. The development of enriched deposits is also affected by geological processes; for example, residual rocks enriched in HREEs reflect their confinement within mineral lattices, whereas larger crustal deposits of LREEs arise due to their greater mobility [133, 159, 160].

Magma generated through partial melting generally has higher concentrations of LREEs. This is why large ore bodies that contain heavy concentrations of LREEs, like cerium and neodymium, are typically discovered in economic deposits, while HREEs are rarer and often found in smaller deposits with significant challenges associated with their extraction [154, 156]. Major sources of HREEs today primarily derive from ion-adsorption clay deposits in southern China, where specific elemental distributions align with the Oddo–Harkins rule, demonstrating that even-numbered elements tend to be more abundant than their odd-numbered counterparts [154, 155].

Additionally, REEs are found in various minerals with LREE-rich sources, such as monazite and bastnäsite, while HREE-rich sources include xenotime and gadolinite. Significant commercial mining locations encompass vast resources such as Bayan Obo in China and Mountain Pass in the USA [157, 159]. These deposits typically arise from alkaline and carbonatite magmatism associated with volcanic activity, which can concentrate REEs through secondary geological processes such as weathering and sedimentation. In humid environments, the weathering of primary REE-bearing minerals leads to the formation of clay-rich deposits, making them accessible for mining [161, 162].

Table 1: Overview of Rare-Earth Metal Properties,

Z	Symbol	Name	Etymology	Selected Applications	Abundance (ppm)	Chemical and Physical Properties Summary
21	Sc	Scandium	From Latin *Scandia* (Scandinavia)	Light Al–Sc alloys for aerospace, metal-halide lamps, mercury-vapor lamps, tracer in oil refining	22	Silvery-white, lightweight transition metal; forms Sc^{3+} ions; density ≈ 3.0 g/cm^3; melts at 1541 °C; highly reactive with oxygen forming Sc_2O_3.

Z	Symbol	Name	Etymology	Selected Applications	Abundance (ppm)	Chemical and Physical Properties Summary
39	Y	Yttrium	After Ytterby, Sweden	YAG and YVO_4 lasers, YBCO superconductors, YSZ ceramics, alloys, LEDs, cancer treatment, optics	33	Soft, silver-metallic; forms Y^{3+}; density 4.47 g/cm^3; melts at 1526 °C; stable in air but oxidizes slowly; similar chemistry to HREEs.
57	La	Lanthanum	Greek *lanthanein* ("to be hidden")	Optical glass, camera lenses, hydrogen storage, NiMH batteries, FCC catalysts	39	Soft, ductile, silvery-white; oxidation +3; density 6.15 g/cm^3; melts at 920 °C; readily tarnishes in air; paramagnetic.
58	Ce	Cerium	After dwarf planet *Ceres*	Polishing powder, glass colouring, catalysts, lighter flints, coatings	66.5	Most abundant REE; oxidation +3/+4; density 6.77 g/cm^3; melts at 795 °C; easily oxidized; strong oxidant as Ce^{4+}.
59	Pr	Praseodymium	Greek *prasios didymos* ("leek-green twin")	Magnets, lasers, glass colorant, welding goggles, fibre-optic amplifiers	9.2	Soft, malleable, yellowish-white; oxidation +3; density 6.77 g/cm^3; melts at 931 °C; forms green salts; magnetic.
60	Nd	Neodymium	Greek *neos didymos* ("new twin")	Strong Nd–Fe–B magnets, lasers, capacitors, EV motors	41.5	Bright silvery metal; oxidation +3; density 7.01 g/cm^3; melts at 1024 °C; ferromagnetic; readily oxidized to pink/purple oxides.
61	Pm	Promethium	After Titan *Prometheus*	Nuclear batteries, luminous paint	1×10^{-15}	Radioactive; no stable isotopes; emits beta radiation; metallic appearance; density ≈ 7.3 g/cm^3 (estimated); half-life 17.7 yrs (Pm-145).
62	Sm	Samarium	After Vasili Samarsky-Bykhovets	Magnets, lasers, nuclear control rods	7.05	Hard, silvery metal; oxidation +2/+3; density 7.52 g/cm^3; melts at 1072 °C; forms SmCo

Rare Earth and Critical Mineral Operations and Processing

Z	Symbol	Name	Etymology	Selected Applications	Abundance (ppm)	Chemical and Physical Properties Summary
						magnets; good neutron absorber.
63	Eu	Europium	After Europe	Phosphors (red/blue), lasers, fluorescent lamps, NMR agents	2	Softest lanthanide; oxidation +2/+3; density 5.26 g/cm^3; melts at 826 °C; highly reactive; excellent luminescent activator.
64	Gd	Gadolinium	After Johan Gadolin	MRI contrast, neutron capture, alloys, magnets, refrigeration	6.2	Silvery-white, magnetic (paramagnetic→ferromagnetic < 20 °C); oxidation +3; density 7.90 g/cm^3; melts at 1313 °C; large magnetocaloric effect.
65	Tb	Terbium	After Ytterby, Sweden	Green phosphors, lasers, magnetostrictive alloys (Terfenol-D), fuel cells	1.2	Silvery-white, soft, ductile; oxidation +3/+4; density 8.23 g/cm^3; melts at 1356 °C; strong magnetostriction.
66	Dy	Dysprosium	Greek *dysprositos* ("hard to get")	Nd–Fe–B magnet additive, lasers, data storage	5.2	Metallic silver; oxidation +3; density 8.54 g/cm^3; melts at 1412 °C; highly magnetic; improves magnet heat resistance.
67	Ho	Holmium	From *Holmia* (Stockholm)	Lasers, spectrometer calibration, magnets	1.3	Soft, malleable silver metal; oxidation +3; density 8.79 g/cm^3; melts at 1474 °C; exhibits strongest magnetic moment of any element.
68	Er	Erbium	After Ytterby, Sweden	Fiber-optic amplifiers, infrared lasers, alloys	3.5	Soft, silvery-white; oxidation +3; density 9.07 g/cm^3; melts at 1529 °C; pink compounds used in glass tinting; infrared-emissive.
69	Tm	Thulium	Mythical *Thule* (northern land)	Portable X-ray sources, lamps, lasers	0.52	Silvery-gray; oxidation +3; density 9.32 g/cm^3; melts at 1545 °C; least abundant stable REE; emits soft X-rays when bombarded.

Z	Symbol	Name	Etymology	Selected Applications	Abundance (ppm)	Chemical and Physical Properties Summary
70	Yb	Ytterbium	After Ytterby, Sweden	Infrared lasers, reducing agent, decoy flares, sensors	3.2	Soft, ductile, silver-white; oxidation +2/+3; density 6.97 g/cm^3; melts at 824 °C; reacts slowly in air; luminescent in IR region.
71	Lu	Lutetium	After *Lutetia* (Paris)	PET scan detectors, high-index glass, catalysts, LEDs	0.8	Hardest, densest lanthanide; oxidation +3; density 9.84 g/cm^3; melts at 1663 °C; stable, corrosion-resistant, used in scintillators.

Across the lanthanide series, a gradual change occurs known as the lanthanide contraction. As you move from lanthanum (La) to lutetium (Lu), the atomic radius of each successive element decreases slightly. This happens because additional electrons are added to the inner 4f orbital, which poorly shields the outer electrons from the increasing nuclear charge. As a result, the electrons are pulled closer to the nucleus. This contraction leads to subtle but important effects: densities and melting points generally increase toward the heavier lanthanides, and their ionic radii become smaller. These variations also influence how the rare earths behave chemically and how easily they substitute for one another in minerals.

The oxidation state of rare earth elements (REEs) is typically +3, which dominates their chemistry in both natural and industrial settings. However, a few elements exhibit additional oxidation states under specific conditions. Cerium (Ce) can form Ce^{4+} in oxidizing environments, while europium (Eu) and ytterbium (Yb) can form stable +2 states under reducing conditions. Terbium (Tb) occasionally shows a +4 state as well. These variations in oxidation state explain why certain REEs behave differently during geochemical processes or in separation technologies.

In terms of physical characteristics, all rare earth metals are silvery-white and metallic in appearance. They are relatively soft and malleable, though they become harder and denser toward the end of the series. Most tarnish when exposed to air, forming a dull oxide layer. Many REEs are paramagnetic or ferromagnetic at low temperatures, giving them unique magnetic properties that make them useful in applications such as permanent magnets, data storage, and laser systems.

Chemically, rare earth elements are highly reactive, particularly with oxygen and water. When exposed to air, they readily oxidize to form REE oxides (RE_2O_3)—a process that underpins many industrial extraction and refining steps. In humid environments, they may also react with water to form hydroxides, further influencing their stability and processing behaviour. This reactivity must be carefully managed during mineral processing, refining, and storage, as it affects recovery efficiency, product purity, and safety in handling.

Physical Properties of Ores

The physical properties of ores are fundamental characteristics that determine how a mineral deposit can be identified, extracted, processed, and economically evaluated. Understanding these properties allows geologists, engineers, and metallurgists to predict the behaviour of the ore during mining and beneficiation. The following provides a detailed explanation of the key physical properties of ores and their practical significance.

1. Colour and Streak: The colour of an ore refers to its appearance in reflected light, which can vary depending on surface oxidation or impurities. For instance, hematite is typically steel-grey to reddish-brown, while chalcopyrite displays a brassy yellow colour. However, because colour can be misleading due to weathering or tarnish, geologists often use the streak test—the colour of a mineral's powdered form when rubbed against a porcelain plate. The streak provides a more reliable indicator of composition, such as the red-brown streak of hematite or the greenish-black streak of pyrite.

Rare earth and critical minerals show a wide variety of colours depending on composition and oxidation state.

- Bastnäsite is usually *honey-yellow to reddish-brown*, and its streak is *white*, making it easily recognisable in carbonatite ores.

- Monazite, another REE phosphate, often appears *brown, yellow, or greenish*, with a *white streak*.

- Spodumene, the key lithium ore in pegmatites, ranges from *colourless to light green or lilac*, depending on trace elements.

- Ilmenite, a titanium-iron oxide common in heavy-mineral sands, has a *black metallic sheen* with a *black streak*.

- Xenotime, which hosts heavy REEs and yttrium, shows *brownish-yellow to dark brown* hues.

Colour and streak together help distinguish between REE-bearing phosphates (usually lighter) and oxides (usually darker and metallic).

2. Lustre: Lustre describes how a mineral reflects light from its surface. Metallic ores typically have a metallic or submetallic lustre, while non-metallic minerals may appear vitreous (glassy), dull, or earthy. For example, galena has a bright metallic lustre, while cassiterite (tin ore) is submetallic. Lustre helps in field identification and provides insights into the mineral's surface chemistry and weathering behaviour.

The lustre of critical mineral ores varies from metallic to vitreous.

- Monazite and xenotime display a *resinous to vitreous lustre*, reflecting their non-metallic composition.

- Bastnäsite has a *vitreous to pearly lustre*, typical of carbonate minerals.

- Spodumene shows a *vitreous lustre* similar to quartz.

- In contrast, metallic critical minerals like ilmenite, cassiterite, and columbite-tantalite (coltan) exhibit *submetallic to metallic lustre*, often used as a diagnostic clue in placer deposits.

3. Hardness: The hardness of an ore refers to its resistance to scratching or abrasion. It is measured using the Mohs hardness scale (ranging from 1 for talc to 10 for diamond). For example, magnetite has a hardness of about 6, while graphite is very soft at 1–2. Hardness influences crushing and grinding requirements during mineral processing. Softer ores are easier to mill, while harder ores may require high-energy crushing equipment and wear-resistant liners.

Hardness controls how these ores are crushed and milled for beneficiation.

- Spodumene has a *Mohs hardness of 6.5–7*, similar to quartz, requiring strong crushing equipment.

- Bastnäsite is softer, *4–4.5*, allowing easier grinding.

- Monazite ranges from *5–5.5*, while xenotime is slightly harder (*4.5–5*).

- Metallic minerals like ilmenite (~5.5–6) and cassiterite (~6.5–7) resist abrasion and need energy-intensive comminution.

The hardness of these minerals influences liberation efficiency and equipment wear in processing plants.

4. Density (Specific Gravity): Density or specific gravity is one of the most important physical properties in mineral processing. It is defined as the ratio of the ore's mass to the mass of an equal volume of water. Ores of heavy metals such as galena (lead sulphide, ~7.5 g/cm^3) and cassiterite (tin oxide, ~6.8 g/cm^3) have much higher densities than gangue minerals such as quartz (~2.65 g/cm^3).

This contrast in density is exploited in gravity separation techniques like jigging, tabling, and spirals, which separate valuable minerals from waste based on differences in mass.

The density of REE and critical minerals is a key factor in gravity separation and ore classification.

- Monazite: 4.9–5.3 g/cm^3

- Xenotime: 4.4–5.1 g/cm^3

- Bastnäsite: 4.9–5.2 g/cm^3

- Spodumene (Li$_2$O host): 3.1–3.2 g/cm^3

- Cassiterite: 6.8–7.1 g/cm^3

- Columbite–tantalite: up to 8.0 g/cm^3

- Ilmenite: 4.7–5.0 g/cm^3

Because REE phosphates and oxides are denser than their host rocks, they can be recovered through gravity-based concentrators such as spirals, shaking tables, or centrifugal separators.

5. Magnetic and Electrical Properties: Some ores exhibit distinct magnetic or electrical conductivity properties that are critical in exploration and beneficiation:

- **Magnetic properties**: Minerals like magnetite (Fe$_3$O$_4$) and ilmenite (FeTiO$_3$) are strongly magnetic, allowing their separation using magnetic separators. Others, such as hematite (Fe$_2$O$_3$), are weakly magnetic and require high-intensity magnetic fields for recovery.

- **Electrical properties**: Conductive minerals such as chalcopyrite and galena can be separated from non-conductive gangue using electrostatic separation.

These properties also assist in geophysical exploration, where magnetic and electromagnetic surveys detect ore bodies beneath the surface.

Magnetic separation is crucial for differentiating REE-bearing minerals from gangue or associated oxides.

- Monazite and xenotime are *paramagnetic*, meaning they can be separated in high-intensity magnetic fields.

- Ilmenite is *weakly magnetic*, while magnetite (a common associate in carbonatite-hosted REE deposits) is *strongly magnetic*.

- Bastnäsite is *non-magnetic* and separated using electrostatic or gravity methods.

- Spodumene is *non-conductive* and often beneficiated using froth flotation.

These differences are exploited in both exploration (magnetic surveys) and beneficiation (dry magnetic/electrostatic separation).

6. Cleavage, Fracture, and Tenacity: Cleavage is the tendency of minerals to break along specific planes of atomic weakness. For instance, galena exhibits perfect cubic cleavage, breaking into small cubes.

- Fracture describes irregular breakage when cleavage planes are absent, such as the conchoidal (shell-like) fracture seen in quartz.

- Tenacity refers to a mineral's resistance to breaking, bending, or deforming. Malleable ores (like native gold) can be hammered into sheets, while brittle ores (like pyrite) shatter easily.

These properties affect the crushing, grinding, and liberation characteristics of ores during mineral processing.

- Spodumene shows *distinct cleavage* in two directions at nearly 90°, which helps identify it in pegmatite outcrops.
- Monazite and xenotime have *poor cleavage* but *conchoidal fracture*, producing sharp edges.
- Bastnäsite displays *perfect cleavage* in one direction, typical of carbonates.
- Metallic ores like ilmenite and coltan are *brittle* and break easily.

These structural characteristics determine how ores behave during crushing, with cleavage minerals often breaking along smooth planes while brittle ones produce irregular fragments.

7. Grain Size and Texture: The grain size, shape, and intergrowth texture of ore minerals greatly influence beneficiation efficiency. Fine-grained ores require more intensive grinding to liberate valuable minerals from the gangue, increasing energy costs. Coarse-grained ores are easier to separate.

Textures such as massive, disseminated, or vein-type indicate how minerals are distributed within the host rock and help in planning drilling, blasting, and concentration methods.

The grain size and texture of REE and lithium ores vary widely and strongly affect processing.

- Pegmatitic spodumene occurs as *coarse crystals* up to several metres long, ideal for mechanical separation.

- Carbonatite-hosted bastnäsite is often *fine-grained and disseminated*, requiring fine grinding.

- Monazite and xenotime in placer sands are *liberated grains* (<1 mm), easy to concentrate by gravity.

- Lateritic REE clays are extremely fine-grained, necessitating leaching rather than physical separation.

Understanding texture helps design appropriate comminution and beneficiation strategies.

8. Porosity and Permeability: Porosity refers to the proportion of void spaces in a rock, while permeability measures how easily fluids can pass through it. These properties are crucial in ores recovered by in-situ leaching or solution mining, such as uranium, copper, and rare earth elements in certain clay deposits. High porosity and permeability allow leaching solutions to move freely and dissolve target metals efficiently.

These properties are particularly relevant to clay-hosted REE deposits and lateritic systems.

- In ion-adsorption clays (southern China), high porosity and permeability allow weak acid leaching of REEs adsorbed on clay surfaces.

- In contrast, hard rock pegmatite ores like spodumene are dense and non-porous, requiring mechanical crushing and chemical conversion.

Thus, permeability directly determines whether hydrometallurgical extraction or physical beneficiation is more effective.

9. Optical and Reflective Properties: Under reflected light microscopy, metallic ores display distinct reflectance colours, bireflectance, and anisotropy, which aid in ore mineral identification. For example, chalcopyrite shows a yellow reflectance, while sphalerite is brownish-grey. These optical traits are particularly useful in polished section analysis and mineralogical studies for determining ore associations and paragenesis (the sequence of mineral formation).

Under reflected light microscopy, REE and critical minerals display unique optical signatures.

- Monazite and xenotime show high reflectivity and anisotropy, changing colour slightly as the stage is rotated under the microscope.

- Bastnäsite displays a moderate reflectance with pale yellow tones.

- Columbite-tantalite shows strong anisotropy with colour variations between brown and grey.

Such characteristics are essential in mineralogical and petrological studies, helping geologists trace REE mineralisation stages and metamorphic overprinting.

10. Brittleness and Malleability: Ores vary in how they deform under stress. Malleable minerals like gold and silver can be flattened or shaped without breaking, while brittle minerals like galena fracture easily. This property affects mechanical processing, as brittle ores tend to produce fine particles that may be lost during flotation or gravity separation, whereas malleable ores require careful milling to prevent smearing and coating effects.

Most REE-bearing minerals, such as monazite, xenotime, bastnäsite, and coltan, are brittle, fracturing easily during crushing. This brittleness aids in liberation but can generate excessive fines, complicating flotation recovery.

By contrast, native metals like tantalum or niobium (when metallic) are malleable, allowing reshaping without breaking—though such pure native forms are rare. Spodumene, though crystalline, can split cleanly along cleavage planes, giving predictable fragmentation during processing.

The physical properties of ores not only define their appearance and composition but also determine their economic value, exploration detectability, and processing behaviour. A thorough understanding of these characteristics allows geologists to correctly identify ore minerals in the field, assists engineers in designing efficient extraction and concentration systems, and helps metallurgists predict recovery performance in industrial operations.

Chemical Properties of Ores

The chemical properties of ores describe how minerals behave during chemical reactions, including how they interact with oxygen, water, acids, bases, and other reagents used in processing. Understanding these properties is critical in ore identification, beneficiation, extraction, and refining.

In the case of rare earth and critical minerals, chemical properties determine how elements are bonded, how easily they can be separated, and how they respond to leaching or reduction during processing. Below is a detailed explanation of the major chemical properties of ores, with examples from rare earth and critical mineral systems.

1. Chemical Composition and Elemental Substitution

Every ore has a distinct chemical formula that defines its composition, but natural variations occur due to elemental substitution within the mineral structure. Many elements of similar ionic size and charge can replace one another within a crystal lattice — a process known as isomorphous substitution.

- In rare earth minerals, the lanthanides commonly substitute for one another because they all form trivalent (+3) cations of similar size.

    o For example, monazite ($CePO_4$) can contain variable amounts of La, Nd, Sm, or Pr.

    o Bastnäsite [$(Ce,La)(CO_3)F$] may contain Ce^{3+}, La^{3+}, and Nd^{3+} in different proportions.

- Xenotime (YPO_4) can incorporate heavy rare earth elements (HREEs) such as Yb, Dy, and Er due to similar ionic radii.

- In spodumene ($LiAl(SiO_3)_2$), lithium may be partially replaced by sodium or magnesium in trace amounts.

This substitution behaviour affects chemical processing, since mixed-element minerals require selective leaching or separation to recover individual metals.

Rare earth and critical minerals often show extensive solid-solution series, where elements of similar ionic radii and charge substitute for one another.

- Monazite ($CePO_4$) — commonly contains variable proportions of La, Nd, Sm, and Pr due to trivalent substitution among light REEs.

- Bastnäsite [(Ce,La)(CO$_3$)F] — incorporates Ce^{3+}, La^{3+}, Nd^{3+}, and Pr^{3+}; the Ce/La ratio is a key exploration indicator.

- Xenotime (YPO$_4$) — a heavy REE (HREE) host where Y^{3+} is substituted by Er^{3+}, Yb^{3+}, Dy^{3+}, or Gd^{3+}.

- Columbite–tantalite [(Fe,Mn)(Nb,Ta)$_2$O$_6$] — shows continuous substitution between niobium and tantalum, which complicates refining.

- Spodumene (LiAl(SiO$_3$)$_2$) — may contain trace Na$^+$, Mg^{2+}, or Fe^{2+}, slightly altering melting point and reactivity.

- Ilmenite (FeTiO$_3$) — can incorporate Mn^{2+} or Mg^{2+} replacing Fe^{2+} and Cr^{3+} replacing Ti^{4+} in minor amounts.

This chemical flexibility is why REEs often co-occur, requiring selective leaching or solvent extraction to separate individual elements.

2. Oxidation and Reduction Behaviour

Oxidation-reduction (redox) reactions play a major role in the formation and processing of ores. These reactions determine mineral stability under different conditions and influence how easily elements can be extracted.

- Rare earth elements (REEs) mainly exist in the +3 oxidation state, but some (e.g., Ce^{4+}, Eu^{2+}, Yb^{2+}) can change valence depending on the environment.

  - Ce^{4+} in cerianite (CeO$_2$) forms in oxidizing conditions, while Ce^{3+} in bastnäsite or monazite forms under reducing conditions.

  - These redox states influence processing efficiency, as Ce^{4+} is harder to dissolve in acid than Ce^{3+}.

- Transition metal-bearing ores, like ilmenite (FeTiO$_3$) and magnetite (Fe$_3$O$_4$), also undergo redox changes during smelting or roasting, releasing oxygen and forming metallic phases.

- Graphite (C) is chemically stable but can act as a reducing agent in metallurgical reactions, helping extract metals from oxides.

Understanding redox chemistry is vital when designing roasting, leaching, and solvent extraction circuits, as it dictates which reagents and conditions will release the target metal efficiently.

The redox chemistry of REE and critical minerals governs their stability and extractability.

- Cerium (Ce) — the only lanthanide to form a stable +4 oxidation state; present as Ce^{4+} in cerianite (CeO_2) under oxidising conditions and Ce^{3+} in bastnäsite or monazite under reducing ones.

- Europium (Eu) — can exist as Eu^{2+} or Eu^{3+}; this valence flexibility is used in geochemical fingerprinting of REE sources.

- Ilmenite ($FeTiO_3$) — can be oxidised to hematite (Fe_2O_3) and rutile (TiO_2) during roasting, a step in titanium extraction.

- Magnetite (Fe_3O_4) — alternates between Fe^{2+} and Fe^{3+} states, making it both a magnetic and redox-active ore.

- Graphite (C) — inert under ambient conditions but serves as a reducing agent during metallurgical smelting, aiding metal recovery from oxides.

Understanding redox behaviour ensures that the right oxidation state is targeted during leaching or roasting processes.

3. Acid and Base Reactivity

Ores vary widely in their reactivity with acids and bases — a key property exploited in hydrometallurgical processing.

- Carbonate minerals such as bastnäsite and ancylite react vigorously with acids, releasing carbon dioxide (CO_2) and dissolving the REEs into solution:

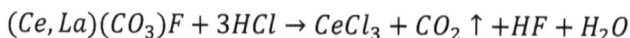

$$(Ce, La)(CO_3)F + 3HCl \rightarrow CeCl_3 + CO_2 \uparrow + HF + H_2O$$

This reaction forms soluble rare-earth chlorides, which can later be precipitated or extracted.

- Phosphate minerals like monazite and xenotime are more acid-resistant due to strong P–O bonds. They require high-temperature sulphuric acid digestion or alkaline cracking with sodium hydroxide to release REEs:

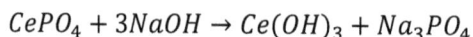

$$CePO_4 + 3NaOH \rightarrow Ce(OH)_3 + Na_3PO_4$$

- Spodumene ($LiAl(SiO_3)_2$) is resistant to both acids and bases in its natural (α) form, but after roasting to the β-phase at ~1000°C, it becomes reactive and can be leached with sulphuric acid to extract lithium.

- Ion-adsorption clays in China and Southeast Asia hold REEs weakly bound to clay minerals, which can be extracted using mild ammonium sulphate solutions without destroying the host material.

Thus, acid-base behaviour determines the leaching method and chemical selectivity of extraction.

Chemical reactivity with acids and bases determines the leaching route for REE and critical minerals.

- Bastnäsite [$(Ce,La)(CO_3)F$] — reacts readily with acids, releasing CO_2 and forming soluble chlorides or sulphates.

- Monazite ($CePO_4$) and Xenotime (YPO_4) — highly resistant to acids; require sulphuric acid roasting or alkaline NaOH cracking to liberate REEs.

- Spodumene ($LiAl(SiO_3)_2$) — unreactive until thermally converted to β-spodumene (~1000°C), after which it dissolves in sulphuric acid to release Li^+.

- Ancylite [$(Sr,Ce)(CO_3)_2 \cdot H_2O$] — carbonate REE mineral that dissolves easily in weak acids.

- Ion-adsorption clays — extractable with mild ammonium sulphate leach solutions, making them environmentally preferable to hard-rock REE ores.

These reactions dictate whether processing relies on acidic, alkaline, or neutral leach systems.

4. Solubility and Complex Formation

Solubility controls how easily metals dissolve during weathering, transport, or leaching. In REE systems, solubility is closely linked to complex formation — the ability of ions to form stable complexes with anions such as carbonate (CO_3^{2-}), fluoride (F^-), chloride (Cl^-), or phosphate (PO_4^{3-}).

- Light rare earth elements (LREEs) form stable carbonate and fluoride complexes, explaining their enrichment in bastnäsite [$(Ce,La)(CO_3)F$].

- Heavy REEs (HREEs) tend to form phosphate complexes, stabilising them in xenotime (YPO_4) or monazite.

- During acid leaching, REE^{3+} ions can form soluble chloride or sulphate complexes, improving extraction efficiency.

- In contrast, niobium and tantalum form highly stable fluoride complexes (e.g., $[NbF_6]^-$, $[TaF_7]^{2-}$), which complicates separation during refining.

- Lithium in spodumene dissolves as Li^+, forming soluble salts like Li_2CO_3 or LiOH after precipitation.

Understanding solubility and complexation is crucial for designing selective solvent extraction and ion exchange systems used in modern REE refineries.

Solubility and complexation define REE mobility in fluids and extraction efficiency.

- Bastnäsite forms carbonate (CO_3^{2-}) and fluoride (F^-) complexes, stabilising LREEs in carbonatite environments.

- Monazite and Xenotime favour phosphate (PO_4^{3-}) complexes, binding HREEs more strongly and reducing solubility.

- Columbite–tantalite forms fluoro-complexes ($[NbF_6]^-$, $[TaF_7]^{2-}$) during hydrofluoric acid digestion, allowing separation of Ta and Nb.

- Spodumene-derived Li^+ ions form soluble salts like Li_2CO_3, $LiCl$, or $LiOH$, which are easy to recover.

- In hydrothermal fluids, REEs travel as chloride and sulphate complexes, precipitating when pH or temperature changes.

Complexation behaviour directly influences solvent extraction selectivity and REE partitioning in natural systems.

5. Chemical Stability and Weathering Resistance

The stability of ores under surface weathering conditions determines whether they accumulate as secondary deposits (placers, laterites) or dissolve easily.

- Monazite and xenotime are chemically stable and resist weathering, which is why they accumulate in placer deposits along with zircon and ilmenite.

- Bastnäsite, being a carbonate mineral, breaks down easily in acidic surface waters, releasing REEs that can migrate or form secondary minerals.

- Ilmenite and magnetite are moderately stable and often persist in heavy-mineral sands.

- Spodumene, being a silicate, is relatively resistant to chemical alteration, but in tropical climates, it can weather into clay minerals, potentially forming lithium-bearing laterites.

This property influences exploration targets (residual vs. transported deposits) and guides environmental management during mining.

Weathering resistance controls whether a mineral remains intact or forms secondary deposits.

- Monazite and Xenotime — chemically robust, persist in placer deposits after long-term erosion.

- Bastnäsite — decomposes easily under acidic weathering, releasing REEs into soil and groundwater, forming ion-adsorption clays.

- Ilmenite and Rutile (TiO_2) — resist weathering, concentrating in beach sands with zircon and magnetite.

- Spodumene — stable under most conditions but can alter to smectite or kaolinite clays, potentially forming secondary lithium-bearing laterites.

- Gadolinite and Eudialyte — moderately weatherable, releasing REEs and Y into soil during hydrothermal alteration.

Stability differences explain why LREEs (e.g., from bastnäsite) dominate primary deposits, while HREEs (e.g., from xenotime or clays) occur in secondary ones.

6. Ion Exchange Capacity

Certain ore types, particularly ion-adsorption clays, exhibit significant ion exchange capacity (IEC)—the ability to adsorb and release REE ions through chemical exchange with other cations in solution.

- In these deposits, REE^{3+} ions are weakly held on the surface of kaolinite or halloysite clays and can be exchanged with ammonium (NH_4^+) or magnesium (Mg^{2+}) ions:

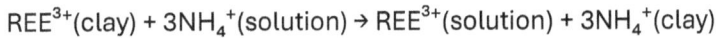

$$REE^{3+}(clay) + 3NH_4^+(solution) \rightarrow REE^{3+}(solution) + 3NH_4^+(clay)$$

- This reversible process allows extraction using weak salt solutions without aggressive acids,   reducing   environmental   impact.
  This property is unique to weathered REE deposits and forms the basis of eco-friendly extraction methods used in southern China and Southeast Asia.

Ion exchange governs extraction from REE-bearing clays, especially in tropical environments.

- Ion-adsorption clays (southern China, Myanmar, Vietnam) contain REEs weakly adsorbed to kaolinite, halloysite, and montmorillonite surfaces.

- Lanthanum (La^{3+}), Cerium (Ce^{3+}), and Yttrium (Y^{3+}) ions exchange with ammonium (NH_4^+) or magnesium (Mg^{2+}) ions during mild leaching.

- $REE^{3+} + 3NH_4^+ \rightarrow REE^{3+}(solution) + 3NH_4^+(clay)$

- This reversible process extracts REEs without breaking down the clay lattice, producing low-impact, eco-friendly extraction systems.

This behaviour distinguishes ion-adsorption REE clays from hard-rock ores, which require aggressive acids.

7. Reactivity with Water and Air

Exposure to air and moisture can lead to oxidation, hydration, or hydrolysis, altering the composition and processing behaviour of ores.

- Bastnäsite can oxidize to form cerium oxide (CeO_2) on exposure to air, changing its colour and reactivity.

- Monazite may slowly alter to form secondary phosphates like rhabdophane ($REEPO_4 \cdot H_2O$) in tropical environments.

- Spodumene and petalite are stable, but finely ground lithium minerals may absorb moisture and react with CO_2 to form lithium carbonate (Li_2CO_3) films.

Such transformations influence both storage stability and processing chemistry, especially for finely milled concentrates.

Exposure to water or air alters many critical minerals chemically and physically.

- Bastnäsite oxidises to CeO_2 (cerianite) on exposure, changing its surface chemistry and colour.

- Monazite weathers slowly to form rhabdophane ($REEPO_4 \cdot H_2O$) or churchite ($YPO_4 \cdot 2H_2O$).

- Spodumene and Petalite remain stable, but ground Li-minerals absorb moisture, reacting with CO_2 to form Li_2CO_3 films.

- Graphite remains inert but can oxidise to CO_2 at high temperatures in air.

- Ilmenite can oxidise to hematite + rutile at elevated temperatures, important during roasting.

These reactions influence ore storage stability, pre-processing handling, and surface reactivity during leaching.

8. Acid-Base Buffering and pH Dependence

The pH of the surrounding solution affects how readily metals are released from ores.

- In acidic conditions, REEs dissolve easily from bastnäsite or lateritic clays.

- In neutral to alkaline conditions, REEs may precipitate as carbonates or phosphates, limiting mobility.

- During processing, pH control is essential: slightly acidic solutions (pH 4–5) maximise recovery while minimising gangue dissolution.

This sensitivity to pH underpins the chemistry of leaching circuits, solvent extraction, and wastewater treatment in mineral processing.

Rare Earth and Critical Mineral Operations and Processing

REE and critical mineral solubility is strongly pH-dependent.

- At low pH (<4), LREE carbonates like bastnäsite dissolve rapidly.

- In neutral to slightly alkaline conditions, REEs precipitate as $REE(OH)_3$ or $REE_2(CO_3)_3$, forming secondary minerals.

- Lateritic clays release REEs at pH 4–6, optimal for ammonium-salt leaching.

- Phosphate minerals (monazite, xenotime) buffer leach solutions, maintaining pH balance during digestion.

pH control is essential to maximise REE recovery while avoiding unwanted dissolution of impurities such as Fe, Al, or Si.

9. Association with Fluorine, Carbonates, and Phosphates

Many REE and critical minerals are chemically associated with fluorine (F), carbonate (CO_3), or phosphate (PO_4) groups.

- Fluorine stabilises minerals like bastnäsite and fluocerite, affecting their solubility and thermal behaviour.

- Carbonate groups enhance REE mobility in hydrothermal fluids, leading to carbonate-hosted REE deposits.

- Phosphate bonds in monazite and xenotime make them resistant to acid attack, necessitating roasting or alkaline digestion before leaching.

These chemical associations define the processing route, determining whether the ore is treated via acidic, alkaline, or ion-exchange pathways.

REE and critical minerals often contain anion complexes that define their chemistry and processing.

- Fluorine (F) — stabilises bastnäsite [$(Ce,La)(CO_3)F$] and fluocerite (CeF_3), lowering melting points and aiding concentration in carbonatites.

- Carbonate (CO_3) — essential in bastnäsite, ancylite, and parisite, facilitating REE mobility in hydrothermal systems.

- Phosphate (PO_4) — strengthens monazite and xenotime structures, making them acid-resistant but requiring alkaline cracking.

- Niobium and tantalum are associated with fluoride complexes, explaining their occurrence in fluorine-rich pegmatites.

These associations directly influence processing pathways (acidic, alkaline, or ion-exchange).

10. Chemical Indicators in Exploration

Certain chemical properties act as geochemical indicators in exploration.

- REE-rich rocks often show high ratios of La/Yb, Ce/Y, or Th/Sc, reflecting enrichment in LREE or HREE.

- Carbonatite-hosted systems display high CO_2, F, and P content, while pegmatite systems show B, Li, and Be anomalies.

These chemical fingerprints guide geologists in identifying prospective REE and critical mineral deposits.

Geochemical ratios and anomalies act as exploration guides for REE and critical minerals.

- High La/Yb and Ce/Y ratios — indicate LREE-enriched carbonatites (e.g., Mount Weld, Bayan Obo).

- Elevated Y/Ho and Dy/Yb — signal HREE-rich xenotime-bearing systems.

- Nb/Ta and Zr/Hf ratios — distinguish between NYF-type and LCT-type pegmatites (used in lithium and niobium exploration).

- B, Li, and Be anomalies — highlight pegmatitic environments (e.g., Greenbushes, Pilgangoora).

- High F, P, and CO_2 contents — point to carbonatite intrusions with REE potential.

These chemical fingerprints are essential tools in geochemical surveys, soil sampling, and exploration geochemistry.

Mineral Identification and Liberation Studies

Mineral identification and liberation studies are essential steps in understanding how an ore will behave during processing and how effectively valuable minerals can be separated from waste material. These studies combine mineralogical, chemical, and textural analyses to determine what minerals are present, in what proportions, and how they are physically intergrown within the rock. In the rare earth and critical mineral industry, accurate mineral identification is particularly important because economically valuable elements (such as neodymium, dysprosium, or lithium) may occur within complex mineral assemblages like monazite, bastnäsite, xenotime, or spodumene. Each of these minerals responds differently to crushing, grinding, and leaching, so proper identification underpins the entire process design.

Rare Earth and Critical Mineral Operations and Processing

The identification phase typically begins with techniques such as reflected and transmitted light microscopy, scanning electron microscopy (SEM), and energy-dispersive X-ray spectroscopy (EDS). These allow mineralogists to visually recognise mineral textures and determine chemical compositions at microscopic scales. X-ray diffraction (XRD) is then used to confirm crystalline structures, while electron microprobe and laser ablation inductively coupled plasma mass spectrometry (LA-ICP-MS) provide quantitative element data. The combination of these methods helps to build a detailed mineral map showing which minerals contain the target elements, how they are associated with gangue, and how fine-grained or interlocked the ore is.

Identification Phase

The identification phase of mineral characterization involves a sequence of analytical techniques that progressively increase in precision—from visual and morphological observations to detailed chemical and structural analysis. This phase is crucial for determining the mineral species, chemical composition, and textural relationships that define how an ore will respond to beneficiation and extraction. Below is a detailed explanation of how each method contributes to the process:

Reflected and Transmitted Light Microscopy: This is the starting point for most mineral identification work.

- Reflected light microscopy is used for opaque minerals such as sulphides, oxides, and rare earth phosphates (e.g., monazite, xenotime, and bastnäsite). Under reflected light, minerals are examined for their colour, reflectivity, anisotropy, and internal reflections, which provide clues to mineral identity and alteration states.

- Transmitted light microscopy is used for transparent or translucent minerals such as silicates (e.g., spodumene, quartz, feldspar) and carbonates. By passing light through thin sections (30 µm thick), mineralogists can observe optical properties such as birefringence, pleochroism, interference colours, and refractive index.

These methods establish the basic mineralogical framework, identifying key ore and gangue minerals and revealing textures such as intergrowths, inclusions, and alteration zones.

Scanning Electron Microscopy (SEM): After optical microscopy, samples are examined using SEM, which provides high-resolution images of mineral surfaces at magnifications from 50× to over 100,000×.

- SEM operates by scanning a focused electron beam across the polished sample surface. The interaction between the electrons and the atoms of the sample produces secondary and backscattered electrons, which are used to generate detailed images of the surface morphology and composition.

- SEM is particularly valuable for identifying fine-grained intergrowths and distinguishing minerals that appear similar under optical microscopes.

This method also reveals grain boundaries, cleavage, fractures, and textural associations, all of which influence liberation during grinding and processing.

Energy-Dispersive X-ray Spectroscopy (EDS): Most SEM instruments are fitted with EDS detectors, which allow for rapid semi-quantitative chemical analysis of individual mineral grains.

- When the electron beam interacts with the sample, atoms emit characteristic X-rays corresponding to specific elements. The EDS system detects and measures these X-rays, providing a chemical "fingerprint" of the mineral.

- By analysing spectra peaks, mineralogists can determine which elements are present and in what approximate proportions. For example, a bastnäsite grain may show peaks for Ce, La, F, and C, while a monazite grain may show Ce, P, and Nd.

EDS helps confirm mineral identity and detect minor or trace substitutions (e.g., Nd in Ce-dominant minerals or Ta substituting for Nb in columbite–tantalite).

X-ray Diffraction (XRD): To confirm the crystal structure and phase identity, powdered samples are analysed by XRD.

- In this technique, an X-ray beam is directed at a finely powdered sample, and the diffracted rays produce a pattern of peaks that correspond to specific atomic lattice spacings.

- Each mineral has a unique diffraction pattern, allowing precise phase identification even when chemical compositions overlap.

XRD is critical for differentiating structurally similar minerals—for example, distinguishing between bastnäsite and parisite (both REE carbonates) or between spodumene and petalite (both Li-bearing silicates).

Electron Microprobe Analysis (EMPA): The electron microprobe provides quantitative elemental data at micrometre-scale resolution.

- Like SEM, it uses an electron beam but is equipped with wavelength-dispersive spectrometers (WDS) that measure characteristic X-rays with higher precision and accuracy.

- EMPA is used to determine exact elemental compositions, detect zoning patterns, and quantify trace substitutions (e.g., Y in xenotime or Fe/Mn ratios in columbite–tantalite).

The data are reported as oxide weight percentages, which can be recalculated to determine mineral formulae and assess chemical variability within ore phases.

Laser Ablation Inductively Coupled Plasma Mass Spectrometry (LA-ICP-MS): For trace-element and isotopic analysis, LA-ICP-MS offers unparalleled sensitivity.

- A focused laser beam ablates a microscopic portion of the mineral surface, turning it into an aerosol. This aerosol is carried into a high-temperature plasma, where the atoms are ionised and analysed by a mass spectrometer.

- The resulting spectra reveal concentrations of trace elements (often at parts-per-million or parts-per-billion levels), such as the full suite of lanthanides in a single monazite grain.

This technique helps determine REE distribution patterns, isotope ratios, and geochemical signatures that can indicate the deposit's origin or enrichment process.

Integration into a Mineral Map: The data from these techniques are integrated to build a quantitative mineralogical model or "mineral map."

- Automated mineralogy platforms (e.g., QEMSCAN or Mineral Liberation Analyzer) combine SEM imaging with EDS data to map mineral phases across polished sections.

- The output includes mineral abundance, grain size, associations, and liberation statistics.

This integrated model shows which minerals host the target elements, how they are distributed through the ore body, and how readily they can be liberated for processing—information essential for designing efficient extraction and beneficiation circuits.

Liberation Studies

Liberation studies focus on how easily the valuable minerals can be physically separated from surrounding gangue minerals during comminution (crushing and grinding). Liberation describes the degree to which mineral grains are freed from one another, a factor that strongly affects recovery in flotation, magnetic, or gravity separation circuits. Automated mineralogy systems, such as QEMSCAN or MLA (Mineral Liberation Analyzer), are often used to quantify liberation by measuring grain size distributions, intergrowth textures, and mineral associations. For instance, monazite grains may be easily liberated in coarse fractions, whereas fine-grained bastnäsite within a carbonatite matrix may require more intensive grinding.

Liberation analysis begins with the careful collection of representative ore samples from drill core, bulk samples, or mine feeds.

- Samples are crushed and split into manageable portions to maintain a consistent size distribution.

- Each portion is ground to various particle size fractions—commonly 150 μm, 75 μm, and 38 μm—to simulate different stages of comminution.

- Subsamples from each size fraction are embedded in epoxy resin, polished to a mirror finish, and carbon-coated to make them conductive for electron-beam analysis.

This preparation allows scientists to study how particle size affects the degree of mineral liberation.

Before automated analysis, mineralogists often perform a preliminary optical examination under reflected light or scanning electron microscopy (SEM).

- This step identifies the dominant ore and gangue minerals and observes textures such as inclusions, intergrowths, or exsolution lamellae.

- Features such as fractures, cleavage planes, and alteration zones indicate how easily minerals might break apart during grinding.

For example, coarse monazite grains in quartz-rich matrix may appear distinct and separable, whereas bastnäsite intergrown with calcite may appear more finely disseminated and harder to free.

The most powerful tools for liberation analysis are automated mineralogy platforms—specifically, QEMSCAN (Quantitative Evaluation of Minerals by Scanning Electron Microscopy) and the Mineral Liberation Analyzer (MLA). These systems combine SEM imaging with energy-dispersive X-ray spectroscopy (EDS) to analyse thousands of particles automatically.

- How it works:
    - The polished sample is scanned with an electron beam in a grid pattern.
    - At each pixel (often 1–5 μm spacing), an X-ray spectrum is collected and compared to a reference mineral database.
    - Each pixel is classified by mineral type, generating a false-colour mineral map where every grain or phase is identified.
    - The software reconstructs particle boundaries and calculates the proportion of each mineral, grain size, shape, and degree of liberation.

- Key data outputs:
    - Grain size distribution: Identifies the range of particle sizes for each mineral.

- o   Mineral association: Shows which minerals are in contact, e.g., monazite intergrown with quartz or bastnäsite embedded in calcite.

- o   Liberation percentage: Quantifies how much of a mineral is exposed versus locked within other minerals.

- o   Locking textures: Classifies grains as "liberated," "middlings" (partly locked), or "locked."

These outputs create a quantitative understanding of how comminution affects recovery efficiency.

Liberation is typically expressed as a percentage—the proportion of a mineral's surface area that is free of contact with gangue minerals.

- A fully liberated grain has >90% of its boundary surface composed of the target mineral.

- Partially liberated (middling) grains have 50–90% exposure, while locked grains have <50%.

By comparing liberation data across different grind sizes, engineers can determine the optimal particle size that maximises liberation while minimising energy costs.

For example:

- Monazite in a granitic host might reach 90% liberation at 75 µm.

- Bastnäsite in a carbonatite might only achieve 60% liberation at the same size, requiring finer grinding (e.g., 45 µm) to improve recovery in flotation.

Liberation data are integrated with bench-scale beneficiation tests to evaluate processing performance.

- **Gravity separation:** Works best when high-density REE minerals (e.g., monazite, xenotime) are well-liberated from lighter gangue like quartz or feldspar.

- **Magnetic separation:** Relies on clear textural boundaries between magnetic and non-magnetic minerals (e.g., xenotime vs quartz).

- **Flotation:** Requires adequate surface exposure of minerals like bastnäsite to reagents; locked grains reduce flotation recovery.

By correlating liberation percentages with recovery data, metallurgists refine the comminution circuit design, adjusting grinding time, mill type (ball vs. HPGR), and classification parameters.

For complex or fine-grained ores, 3D tomography and micro-CT scanning may be used to visualise mineral relationships in three dimensions.

- These methods reveal hidden inclusions or internal porosity that 2D imaging can miss.

- Combining 3D and 2D data improves the accuracy of liberation models, especially for ores with anisotropic textures or fibrous minerals.

In a carbonatite-hosted REE deposit, QEMSCAN data might reveal that bastnäsite grains average 15 µm in size and are commonly locked in calcite. Liberation tests show that at 80 µm grind size, only 55% of bastnäsite is free; increasing grinding to 45 µm raises liberation to 85% but doubles energy consumption. Metallurgists then design a two-stage grinding circuit with coarse regrind and selective flotation to optimise cost and recovery.

Outcomes

The outcomes of mineral identification and liberation studies directly influence flowsheet development and processing efficiency. Knowing which minerals contain the economic elements determines which separation techniques—magnetic, gravity, flotation, or hydrometallurgical—will be most effective. Understanding grain size and textural relationships ensures that energy input during grinding is optimised for maximum liberation without generating excessive fines. In short, these studies transform geological knowledge into metallurgical strategy, ensuring that the physical and chemical characteristics of the ore are matched to the most efficient, cost-effective, and environmentally responsible processing route.

Analytical Tools: XRD, XRF, SEM, MLA, and ICP-MS

X-Ray Diffraction (XRD) – Identifying Crystal Structures

XRD identifies the *crystalline structure and mineral phases* in powdered or solid samples. It is essential for distinguishing between polymorphs (e.g., α- and β-spodumene) and for confirming the presence of REE minerals such as monazite, bastnäsite, and xenotime.

X-Ray Diffraction (XRD) is a powerful analytical technique used to determine the crystalline structure, phase composition, and atomic arrangement of minerals and materials. In geology and mineral processing—especially in the study of rare earth and critical minerals—XRD plays a vital role in confirming which minerals are present and how they are structurally organised. It provides insights into mineral identity, purity, and crystal structure, supporting exploration, processing, and quality control efforts across the value chain.

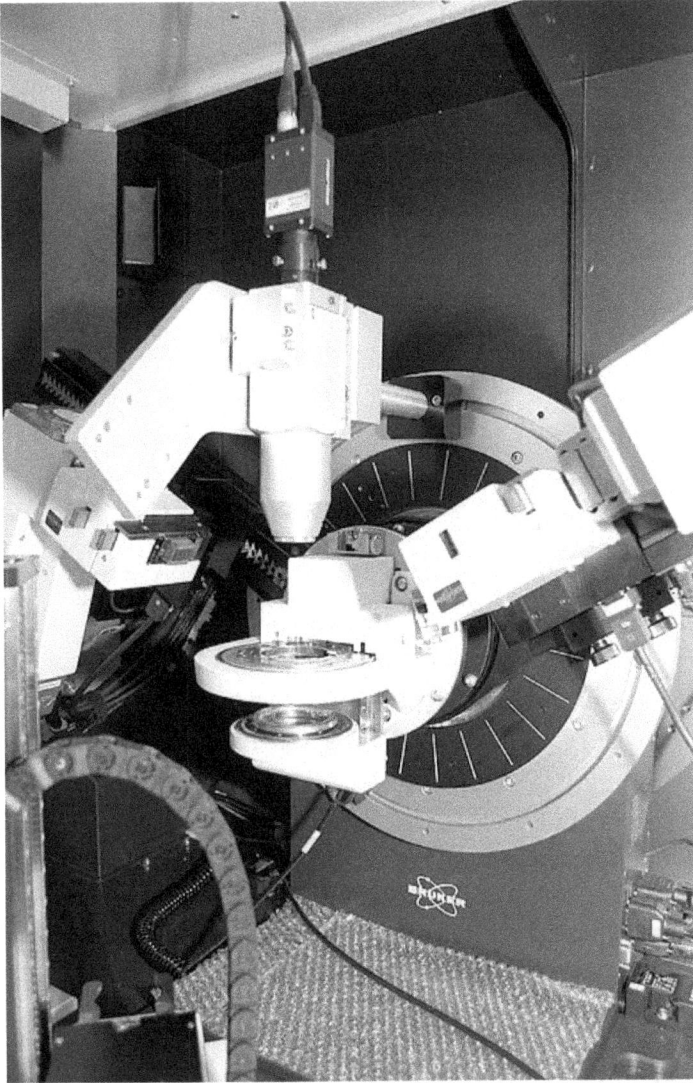

Figure 16: X-Ray diffractometer (XRD). Mirolka, CC BY-SA 3.0, via Wikimedia Commons.

At the atomic level, every crystalline material consists of atoms arranged in a repeating three-dimensional lattice. When a beam of X-rays—high-energy electromagnetic radiation with very short wavelengths (typically 0.5–2.5 Å)—strikes this lattice, the X-rays are scattered by the electrons surrounding the atoms. These scattered rays can interfere constructively or destructively depending on the geometric arrangement of the atoms. Constructive interference occurs when the scattered waves are in phase, producing a detectable diffraction peak, whereas destructive interference occurs when they cancel out. This relationship is expressed by Bragg's Law ($n\lambda = 2d \sin\theta$), which defines how the wavelength (λ), lattice spacing (d), and diffraction angle (θ) interact to produce measurable diffraction peaks. When Bragg's condition is satisfied for a particular atomic plane spacing, the diffracted X-rays reinforce one another, forming the characteristic peaks of an XRD pattern.

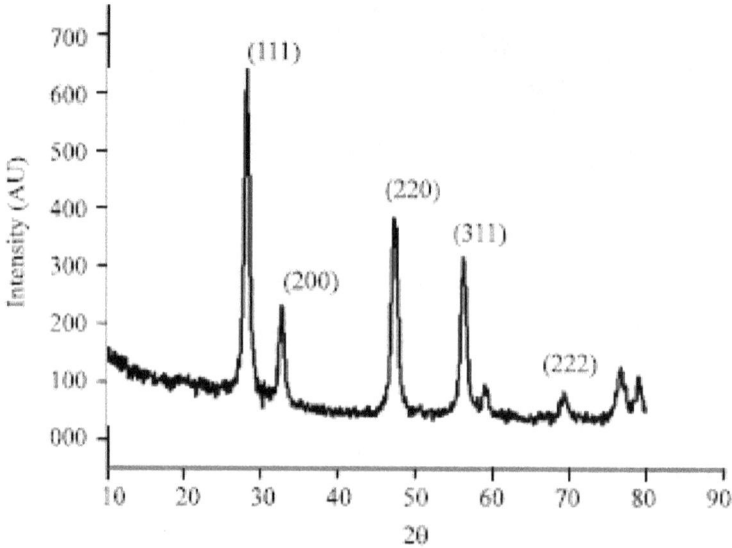

Figure 17: X-ray analysis of cerium oxide. Esteteius, CC BY-SA 4.0, via Wikimedia Commons.

A modern XRD instrument, known as a diffractometer, typically consists of four main components: an X-ray source, a sample holder, a goniometer, and a detector. The X-ray source generates radiation, often using a copper target that emits Cu Kα radiation (λ = 1.5418 Å). The beam is directed at the sample mounted on a stage, usually as a fine powder (<75 μm) to ensure that all possible crystal orientations are represented. The goniometer precisely controls the angles of the X-ray tube, sample, and detector, allowing the detector to sweep through a range of 2θ angles as it records diffracted X-rays. The detector measures the intensity of the diffracted beams and produces an XRD pattern—a plot of intensity versus diffraction angle (2θ)—which forms the fundamental dataset for analysis.

Each crystalline mineral produces a unique diffraction pattern, effectively serving as its structural fingerprint. The position of the peaks reveals the interplanar spacings, while their intensity reflects the abundance and atomic type of the diffracting planes. The width of the peaks provides clues about crystallite size and structural disorder—broader peaks often indicate smaller or more imperfect crystals. For example, quartz displays strong peaks near 20.8°, 26.6°, and 50.1° 2θ, while monazite ($CePO_4$) and xenotime (YPO_4) exhibit distinctive peak sets that allow clear differentiation. Similarly, α- and β-spodumene show different XRD patterns, which are used in lithium processing to verify successful phase transformation during roasting at approximately 1000 °C.

Once a diffraction pattern is recorded, it is compared against reference databases such as the ICDD PDF-4+ to identify the mineral phases present. Software automatically matches the observed peaks with known mineral entries, enabling both phase identification and quantitative phase analysis. The latter is achieved using the Rietveld refinement method, which fits theoretical patterns to the observed data to estimate the relative proportions of each phase. For example, XRD analysis might determine that a sample contains 65% bastnäsite, 20% calcite, 10% barite, and 5% quartz.

In the context of rare earth and critical minerals, XRD has a wide range of applications. During exploration, it helps confirm the presence of REE-bearing minerals such as monazite, xenotime, and bastnäsite. In processing, it is used to verify mineral phase transformations, such as the conversion of α- to β-spodumene before lithium leaching. XRD also supports environmental monitoring by identifying secondary minerals in tailings (e.g., cerium oxides or iron hydroxides) and contributes to quality control by confirming the purity of concentrates and detecting unwanted crystalline phases in refined products.

<u>Step-by-Step Procedure</u>

a. Sample Preparation

1. Crushing and Grinding: Grind a representative portion of the ore sample using an agate mortar or mill to a particle size of <75 μm.

2. Homogenization: Mix the powder thoroughly to ensure uniformity.

3. Mounting:

- Pack the sample into an XRD sample holder or smear it onto a glass slide.
- Ensure a smooth, flat surface to avoid diffraction errors.

4. Preferred Orientation Check: If minerals have plate-like or fibrous habits (e.g., clays or micas), randomize particle orientation by back-loading or spray-drying.

b. Instrument Setup

1. Power On and Calibration:

- Turn on the X-ray generator and allow the tube to stabilise.
- Calibrate using a standard such as silicon (Si) or corundum (Al_2O_3).

2. Set Parameters:

- Target: Cu Kα radiation ($\lambda = 1.5418$ Å) is most common.
- Voltage/Current: ~40 kV and 40 mA.
- 2θ Range: 5°–70° for general mineral scans.
- Step size: 0.02° 2θ, count time: 0.5–1 s per step.

c. Data Acquisition

- Run the scan using the chosen parameters.

- The instrument records the intensity of diffracted X-rays versus the angle (2θ), producing a diffraction pattern (series of peaks).

d. Data Processing

1. Peak Identification: Use pattern-matching software (e.g., PANalytical HighScore, Bruker EVA) to compare measured peaks with databases such as ICDD PDF-4+.

2. Phase Quantification: Apply the Rietveld refinement method for quantitative phase analysis if multiple minerals are present.

3. Interpretation: Identify major and minor phases; note if REE minerals co-occur (e.g., bastnäsite with barite or calcite).

In summary, X-Ray Diffraction works by measuring how X-rays are diffracted by the atomic planes within a crystal. By analysing the resulting diffraction pattern, scientists can accurately determine the mineralogical composition, structural order, and processing state of materials—making XRD an indispensable tool in both geological research and the commercial exploitation of rare earth and critical mineral resources.

X-Ray Fluorescence (XRF) – Determining Bulk Chemical Composition

XRF measures *elemental concentrations* (major, minor, and trace elements) in solid or powdered samples. It provides bulk chemistry used in ore classification, grade control, and geochemical fingerprinting of REE and critical minerals.

X-Ray Fluorescence (XRF) is a widely used analytical technique that determines the elemental composition of a material. It is particularly valuable in geology, mining, and mineral processing for rapid, non-destructive chemical analysis of ores, rocks, soils, and concentrates. In rare earth and critical mineral studies, XRF is used to measure the abundance of elements such as cerium (Ce), neodymium (Nd), lanthanum (La), lithium (Li), niobium (Nb), and tantalum (Ta), providing essential data for exploration, resource estimation, and process control.

The XRF process is based on the interaction between high-energy X-rays and the atoms in a sample. When a beam of primary X-rays (produced by an X-ray tube) strikes a material, it excites the atoms by ejecting inner-shell (core) electrons—usually from the K or L electron shells. This creates a vacancy in the electron structure, making the atom unstable.

Rare Earth and Critical Mineral Operations and Processing

To regain stability, an electron from a higher energy level (for example, the L or M shell) drops down to fill the vacancy. During this transition, the atom releases the excess energy as secondary X-rays—known as fluorescent X-rays.

The key point is that the energy (or wavelength) of these emitted X-rays is unique to each element, because it depends on the difference between the atom's discrete electron shell energy levels. This means that by measuring the energy of the emitted X-rays, we can determine which elements are present. Similarly, by measuring the intensity of the emitted X-rays, we can determine how much of each element is present.

In wavelength-dispersive XRF (WDXRF), the detection system uses a crystal and the principle of Bragg's Law ($n\lambda = 2d \sin\theta$) to separate fluorescent X-rays based on their wavelength.

- Different elements emit X-rays of different wavelengths.

- By adjusting the angle (θ), the detector measures X-rays that satisfy the diffraction condition for each wavelength.

In energy-dispersive XRF (EDXRF), however, the system measures energy directly using a semiconductor detector, converting X-ray photons into electrical signals. EDXRF is faster and ideal for field or handheld instruments, while WDXRF is more precise and typically used in laboratory settings.

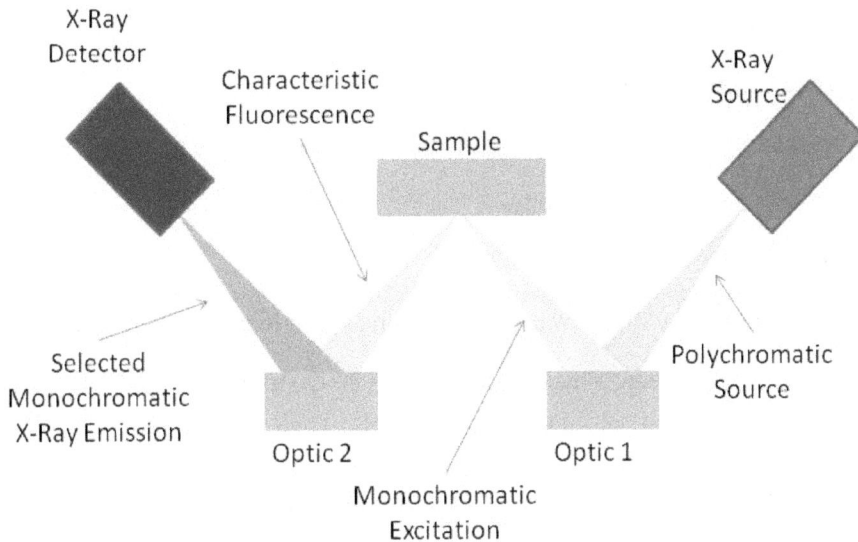

Figure 18: Schematic of Monochromatic Wavelength Dispersive X Ray Fluorescence (MWD XRF). UteForLife, CC BY-SA 3.0, via Wikimedia Commons.

An XRF spectrometer generally consists of four major components:

X-Ray Tube (Source): Generates primary X-rays by bombarding a metal target—commonly rhodium (Rh), molybdenum (Mo), or tungsten (W)—with high-energy electrons. These X-rays excite atoms in the sample.

Sample Stage/Holder: Holds a solid, pellet, or fused bead sample in place. The X-ray beam passes through a collimator and strikes a small spot on the sample surface.

- Powders and pressed pellets are often used for quantitative analysis.

- Fused glass disks provide the most accurate results for silicate materials by eliminating particle-size effects.

Detector: Measures the energies and intensities of the fluorescent X-rays emitted by the sample.

- **EDXRF detectors** (silicon drift detectors or proportional counters) measure photon energy directly.

- **WDXRF detectors** use crystals to separate X-rays by wavelength.

Computer and Software System: Converts detector signals into spectra—graphs of intensity versus energy or wavelength—and processes the data to calculate elemental concentrations.

Step-by-Step Procedure

a. Sample Preparation

1. Pulverise the sample to <75 µm using a tungsten-carbide or agate mill.

2. Mix thoroughly to ensure homogeneity.

3. Pressed Pellet Method (for routine analysis):

- Mix ~4 g of powder with a few drops of binding agent (e.g., polyvinyl alcohol).

- Press into a pellet using a hydraulic press at ~20–30 tonnes.

4. Fused Bead Method (for precise analysis):

- Mix ~1 g of sample with 9 g of lithium tetraborate flux.

- Melt at ~1000 °C and pour into a platinum mould to create a glass disk.

b. Instrument Setup

1. Turn on the spectrometer and stabilise the X-ray tube.

2. Select measurement program (major or trace element mode).

3. Calibrate using certified reference materials (CRMs) similar to the matrix of your samples (e.g., geological standards such as USGS BIR-1 or NIST 2711a).

c. Data Acquisition

- Insert the sample pellet or bead into the sample chamber.

- The instrument bombards the sample with primary X-rays, causing each element to emit characteristic secondary X-rays.

- Detector measures X-ray intensities for specific wavelengths.

d. Data Processing

- Software converts intensities to elemental concentrations using calibration curves.

- Data output includes oxide percentages (e.g., SiO_2, Al_2O_3, Fe_2O_3, RE_2O_3).

- Check for matrix effects and apply corrections if necessary.

e. Interpretation

- Identify REE enrichment trends (e.g., CeO_2, Nd_2O_3 levels).

- Compare ratios like Nb/Ta, Zr/Hf, or Th/Y to classify deposit types (carbonatite vs. pegmatite).

The resulting XRF spectrum is a plot of X-ray intensity versus energy (or wavelength). Each peak corresponds to an element, and its height reflects abundance. The software identifies and quantifies all detectable elements within the calibration range.

For example: A sample spectrum from a carbonatite REE ore might show high peaks for Ce, La, Nd, Pr, and Ca, indicating bastnäsite or monazite presence.

A lithium pegmatite might show strong Li (indirectly measured via stoichiometry), Al, Si, and K peaks, confirming spodumene or petalite.

In essence, X-Ray Fluorescence works by bombarding a sample with primary X-rays, which excite atoms and cause them to emit characteristic secondary (fluorescent) X-rays. By measuring the energy and intensity of this emitted radiation, XRF determines the types and quantities of elements present.

For rare earth and critical mineral applications, XRF provides a rapid and reliable way to measure bulk chemistry, monitor process performance, and classify ore types—making it a cornerstone analytical tool in both field exploration and laboratory-based mineral analysis.

Scanning Electron Microscopy (SEM) – Imaging and Microanalysis

SEM provides high-resolution images of mineral textures and microstructures, and when coupled with Energy Dispersive Spectroscopy (EDS), identifies elemental composition at micro-scales.

Scanning Electron Microscopy (SEM) is an advanced imaging technique that uses a focused beam of electrons, rather than light, to produce highly detailed images of a sample's surface. It is one of the most important analytical tools in geology, materials science, and mineral processing because it allows researchers to observe mineral textures, grain boundaries, and microstructures at magnifications far greater than those possible with optical microscopes. In rare earth and critical mineral studies, SEM is essential for identifying mineral phases, assessing liberation, and understanding ore mineral associations at the micron to nanometre scale.

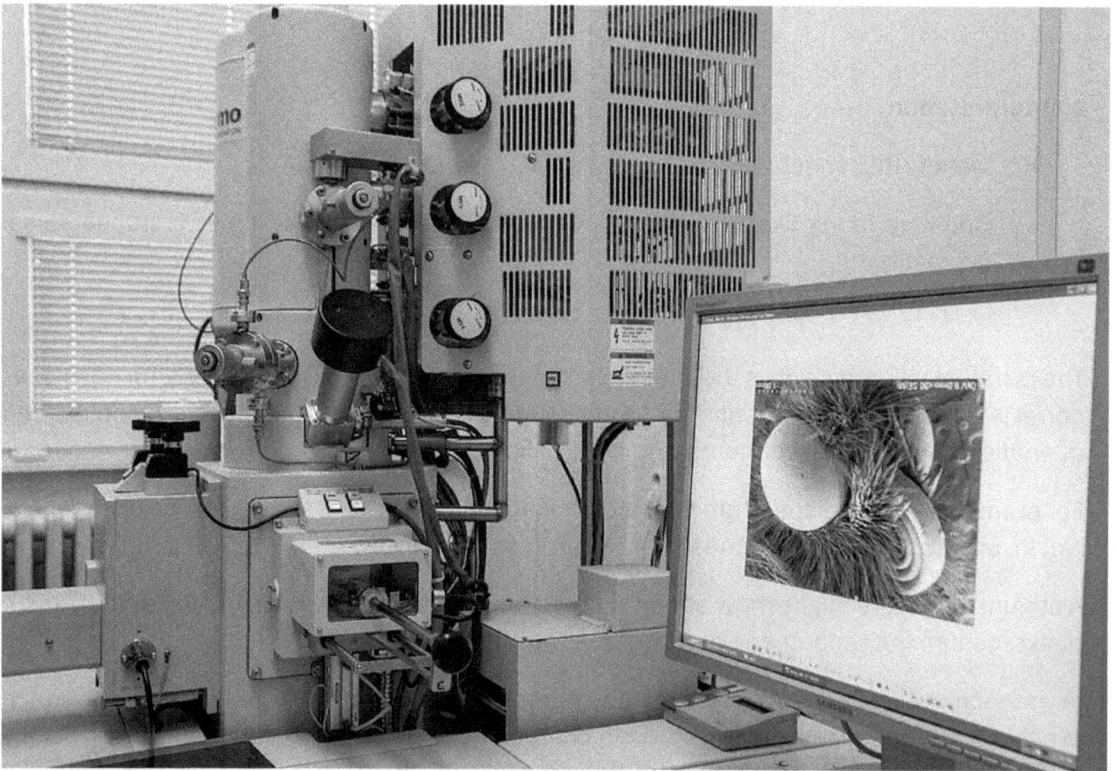

Figure 19: Scanning electron microscope. Tadeáš Bednarz, CC BY-SA 4.0, via Wikimedia Commons.

SEM works by directing a focused beam of high-energy electrons onto the surface of a solid sample. When these electrons interact with the atoms in the sample, they generate several types of signals—mainly secondary electrons, backscattered electrons, and characteristic X-rays. Each signal provides different information about the material's surface or composition.

- Secondary electrons (SEs) are produced when the incident beam knocks loose low-energy electrons from the outer shells of atoms. They give information about the surface topography—the fine texture and shape of the sample's surface.

- Backscattered electrons (BSEs) are high-energy electrons that are reflected back from the sample after colliding with atomic nuclei. Their intensity depends on atomic number, so BSE imaging highlights compositional contrast—heavier elements appear brighter than lighter ones.

- Characteristic X-rays are generated when inner-shell electrons in the sample are displaced and replaced by higher-energy electrons, emitting X-rays with energies unique to specific elements. These are detected using Energy-Dispersive X-ray Spectroscopy (EDS) for chemical analysis.

A typical SEM system consists of several key components working together to generate and detect electron signals:

- Electron Gun (Source): Produces a focused beam of electrons. Modern SEMs often use tungsten filaments, lanthanum hexaboride (LaB_6) cathodes, or field emission guns (FEGs) for high brightness and resolution.

- Electromagnetic Lenses and Coils: Shape and focus the electron beam into a fine spot (1–10 nm) using magnetic fields, allowing precise scanning across the sample.

- Scanning System: Deflects the beam in a raster (grid-like) pattern across the sample's surface. The signal intensity is synchronised with beam position to build the image pixel by pixel.

- Detectors: Capture the various signals emitted by the sample. Secondary electron detectors provide topographic images, while backscatter detectors give compositional contrast.

- Vacuum System: Maintains a high vacuum inside the chamber (10^{-4} to 10^{-6} torr) to prevent electron scattering by air molecules. Some modern SEMs use variable pressure or environmental modes to analyse non-conductive or moist samples without full vacuum.

- Display and Control System: Converts the detector signals into images and spectra, allowing the user to adjust magnification, contrast, and brightness in real time.

Proper sample preparation is critical for obtaining clear SEM images. Samples must be dry, stable, and conductive to avoid charging under the electron beam.

- Conductive samples (such as metal ores) can be mounted directly on a stub using carbon tape.

- Non-conductive samples (like silicates or clays) are coated with a thin conductive layer—usually gold, platinum, or carbon—using a sputter coater to prevent surface charging.

- The sample is then placed inside the vacuum chamber. The electron beam is focused on the surface, and the microscope scans in a raster pattern.

- Detectors collect the emitted signals and convert them into images that reveal surface features, texture, and composition with magnifications ranging from 10× to over 300,000×.

Figure 20: Spherical Tungster Carbide under Scanning Electron Microscope. Gabriela P., CC BY 4.0, via Wikimedia Commons.

When SEM is coupled with EDS, it becomes a powerful imaging and compositional analysis tool. The EDS detector identifies the energy of emitted X-rays from the sample, allowing elemental mapping and point analysis. This is especially useful for determining which minerals contain rare earth or critical elements and in what proportions. For example:

- In a REE-bearing carbonatite, EDS can distinguish bastnäsite $(Ce,La)(CO_3)F$ from barite $(BaSO_4)$ or calcite $(CaCO_3)$.

- In a lithium pegmatite, it can confirm spodumene $(LiAl(SiO_3)_2)$ or lepidolite $(K(Li,Al)_3(Si,Al)_4O_{10}(F,OH)_2)$ by their elemental composition.

- EDS elemental maps show the spatial distribution of REEs, Nb, Ta, and Ti, helping metallurgists design efficient separation processes.

SEM is used across multiple stages of exploration and mineral processing:

- **Exploration:** Identifies ore minerals, textures, and grain boundaries in drill core samples.

- **Liberation Studies:** Measures grain size and intergrowths to evaluate how easily minerals can be separated from gangue during crushing and grinding.

- **Process Control:** Monitors concentrate quality and detects unwanted phases in flotation or magnetic separation products.

- **Failure and Tailings Analysis:** Examines residues or precipitates to identify lost or unrecovered REE minerals.

For example, SEM analysis can reveal whether monazite grains are fully liberated or locked within quartz or feldspar, directly impacting recovery rates. It can also detect nanoscale weathering rims or oxidation products that affect leaching efficiency.

Step-by-Step Procedure

a. Sample Preparation

1. Mounting: Embed a representative sample in epoxy resin and polish to a mirror finish using diamond paste (down to 0.25 μm grit).

2. Cleaning: Ultrasonically clean the sample to remove polishing residue.

3. Coating: Apply a carbon coating (~20 nm) using a vacuum coater to ensure electrical conductivity.

b. Instrument Setup

1. Vacuum System: Load the sample into the SEM chamber and pump down to high vacuum ($\sim10^{-5}$ mbar).

2. Operating Conditions:

- Accelerating Voltage: 15–20 kV for geological samples.

- Beam Current: ~1 nA for imaging, higher for EDS.

- Working Distance: ~10 mm.

3. Detector Selection: Use Secondary Electron (SE) detector for topography and Backscattered Electron (BSE) detector for compositional contrast.

c. Data Acquisition

- Focus and align the beam on a region of interest.

- Capture BSE images: bright areas indicate high atomic number minerals (e.g., monazite, xenotime); dark areas indicate lighter elements (e.g., quartz, feldspar).

- Use EDS to collect X-ray spectra and determine elemental composition of selected points or areas.

d. Interpretation

- Identify REE-bearing phases by comparing EDS spectra (Ce, La, Nd peaks indicate monazite or bastnäsite).

- Map element distributions using EDS mapping to visualise associations or zoning.

- Record textural observations (grain size, intergrowths, alteration).

In simple terms, Scanning Electron Microscopy works by scanning a focused beam of electrons across a sample's surface and detecting the resulting emitted signals to create high-resolution images and elemental data. It bridges the gap between imaging and compositional analysis, making it a cornerstone of modern mineral characterization. In rare earth and critical mineral research, SEM provides critical insights into mineral textures, associations, and liberation characteristics that underpin exploration success, process optimization, and metallurgical recovery.

Mineral Liberation Analyzer (MLA) – Quantifying Mineral Associations and Liberation

MLA combines SEM imaging with automated EDS analysis to quantify mineralogical composition, grain size distribution, and liberation characteristics—vital for process optimisation.

Rare Earth and Critical Mineral Operations and Processing

A Mineral Liberation Analyzer (MLA) is an advanced automated mineralogy system that combines scanning electron microscopy (SEM) with energy-dispersive X-ray spectroscopy (EDS) to quantitatively determine the mineral composition, texture, and liberation characteristics of ore samples. It is a critical tool in modern mineral processing and metallurgy because it allows engineers to understand how minerals are distributed within a rock, how well they are liberated during grinding, and how this affects recovery in flotation, magnetic, or gravity circuits.

In rare earth and critical mineral studies, MLA is especially valuable for characterising complex ores like bastnäsite, monazite, xenotime, and columbite–tantalite, where multiple minerals can be finely intergrown.

The MLA works by scanning a polished sample surface (such as a resin-mounted grain section or process concentrate) using an electron beam, just like a scanning electron microscope (SEM).
As the beam interacts with the sample, it generates backscattered electron (BSE) images and characteristic X-rays that are unique to each mineral's composition.

The MLA system automates this process — it:

1. Scans the entire sample surface in a grid pattern.

2. Captures BSE images at each point to identify textural and compositional contrast.

3. Collects EDS spectra at selected points or pixels to determine the elemental composition.

4. Matches spectra to a database of known minerals to classify each grain or pixel.

5. Generates quantitative mineral maps showing phase distribution, grain size, and liberation data.

A typical MLA setup integrates several components:

- **Scanning Electron Microscope (SEM):** Provides the electron beam and imaging capabilities. The MLA operates as a software-controlled automation layer over the SEM.

- **Backscattered Electron (BSE) Detector:** Measures signal intensity variations caused by atomic number differences — higher atomic number minerals (e.g., monazite with Ce, La, Th) appear brighter than lighter ones (e.g., quartz or feldspar).

- **Energy-Dispersive X-ray Spectrometers (EDS):** Detect characteristic X-rays emitted from the sample and measure their energy, identifying the elements present at each point.

- **MLA Software System:** Controls the SEM, processes image and spectral data, identifies minerals, and compiles textural and compositional statistics automatically.

a. Sample Preparation

1. Prepare polished epoxy mounts (30 mm diameter) from representative ore samples.

2. Polish to <0.25 μm finish and apply a thin carbon coating for conductivity.

b. Instrument Setup

1. Load sample into the MLA chamber.

2. Set analysis mode:

- **Particle Mode:** for liberation and association studies.

- **BSE Mode:** for high-resolution mineral identification.

The MLA can operate in several modes depending on the study objectives:

- **Particle X-ray Mapping (PXM):** Captures full EDS spectra for each pixel — highly accurate but time-consuming.

- **Sparse Phase Mapping (SPM):** Collects EDS data only from particles of interest or at specific intervals — faster, used for large datasets.

- **BSE Gray Level Classification:** Classifies minerals based solely on grayscale intensity (atomic number contrast) — useful for rapid screening when composition differences are large.

- **Line Scan or Point Mode:** Used for specific compositional analyses or verification.

3. Define pixel resolution (1–5 μm typical for liberation work).

c. Data Acquisition

- MLA scans the entire mount using SEM imaging.

- At each pixel, an EDS spectrum is acquired and compared to a reference mineral library.

- The software automatically identifies each mineral and constructs a mineral map.

d. Data Processing

- The MLA software calculates:

    o Modal mineralogy (% area or weight)

    o Grain size distribution

    o Mineral association matrices

    o Liberation index (% of exposed mineral surface area)

    o Locking texture classification (free, middlings, locked)

The MLA produces a range of quantitative outputs, including:

- **Modal mineralogy:** e.g., 35% bastnäsite, 15% monazite, 10% barite, 40% gangue.

- **Grain size distributions:** showing how minerals are distributed across particle size fractions.

- **Liberation curves:** indicating how mineral exposure changes with grinding fineness.

- **Association tables:** identifying which minerals are locked together (important for separation strategy).

These data are exported as tables, images, and reports that guide metallurgical testwork, plant design, and recovery optimisation.

e. Interpretation

- Identify which minerals host target elements (e.g., Nd, Dy in monazite; Li in spodumene).

- Quantify liberation vs. particle size—essential for designing grinding circuits.

- Correlate liberation data with flotation or magnetic recovery performance.

The Mineral Liberation Analyzer (MLA) combines the imaging power of SEM with the chemical precision of EDS to deliver automated, quantitative mineralogical analysis. It reveals how minerals are distributed, how well they are liberated, and how they associate with other phases — all crucial information for designing and optimising rare earth and critical mineral processing circuits.

By transforming microstructural data into numerical process insights, MLA bridges the gap between geology and metallurgy, making it a cornerstone tool in modern ore characterisation and mineral processing research.

Richard Skiba

Inductively Coupled Plasma Mass Spectrometry (ICP-MS) – Trace Element Analysis

ICP-MS detects and quantifies trace elements and isotopes at parts-per-billion (ppb) levels. It is the gold standard for REE analysis, providing accurate distribution patterns across the lanthanide series.

Inductively Coupled Plasma Mass Spectrometry (ICP-MS) is one of the most sensitive and precise analytical techniques for determining trace and ultra-trace element concentrations in geological, environmental, and industrial materials. In rare earth and critical mineral analysis, ICP-MS is indispensable because it can detect all the rare earth elements (REEs) — often down to parts per trillion (ppt) — and provide accurate elemental ratios such as La/Yb or Ce/Y, which are vital for exploration geochemistry, mineral processing, and environmental monitoring.

ICP-MS works by ionising a liquid sample in a very hot argon plasma and then using a mass spectrometer to separate and measure those ions according to their mass-to-charge ratio (m/z).

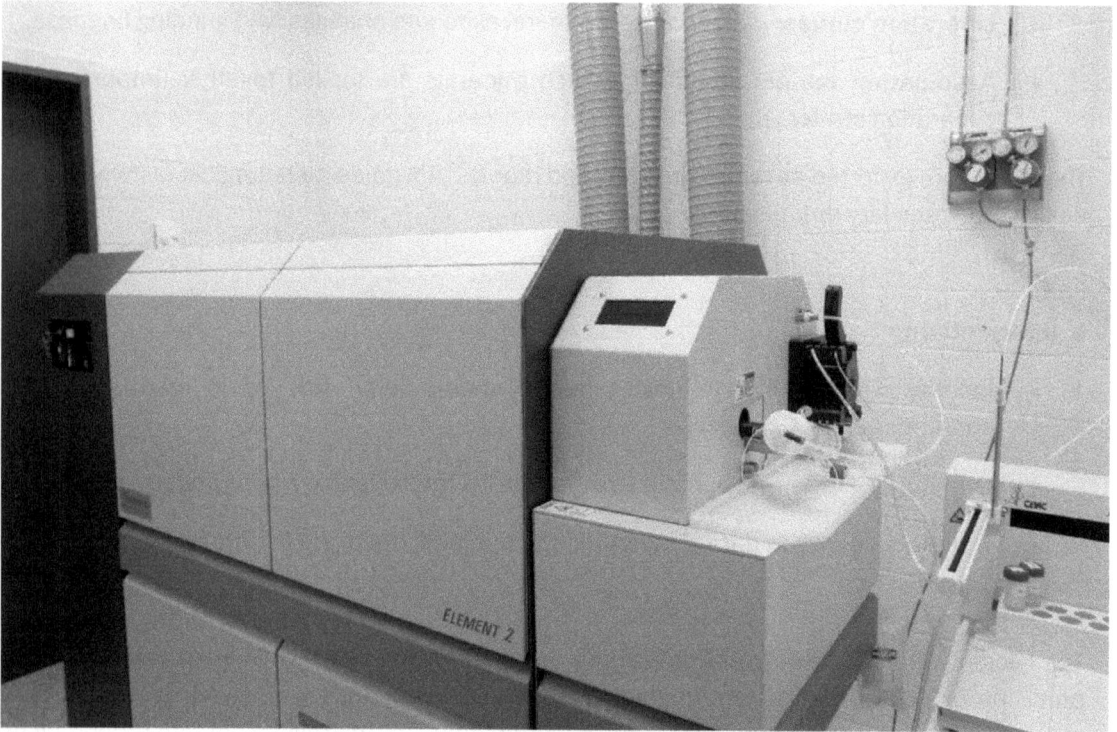

Figure 21: HR-ICP-MS, high-resolution inductively coupled plasma ionization mass spectrometer used for multi-element analysis, in CAFIA laboratory, Czech Republic. Sarka Na kopci, CC BY-SA 4.0, via Wikimedia Commons.

In simpler terms:

Rare Earth and Critical Mineral Operations and Processing

1. The sample (usually dissolved in an acid solution) is converted into ions in a plasma that's hotter than the surface of the Sun (~10,000 K).

2. These ions are then drawn into a mass spectrometer, where they're sorted by mass and counted.

3. The result is a highly accurate measurement of each element's concentration, even if only a few atoms are present.

Major components and how they work:

(a) Nebuliser and Spray Chamber – Turning Liquid into an Aerosol

- The liquid sample (usually an acid digest of rock, ore, or soil) is introduced through a nebuliser, which uses a stream of argon gas to break it into a fine aerosol mist.

- The spray chamber filters out large droplets, allowing only the finest mist (1–10 µm) to enter the plasma.

(b) Inductively Coupled Plasma (ICP) Torch – Creating the Plasma

- The heart of the system is the plasma torch, which contains flowing argon gas surrounded by a radio-frequency (RF) induction coil.

- The coil produces an alternating electromagnetic field (typically 27–40 MHz).

- When a spark is introduced, the argon gas becomes ionised, creating a stable plasma at 6,000–10,000 °C.

- The aerosol droplets from the nebuliser enter this plasma and are sequentially dried, vaporised, atomised, and finally ionised:

$$M \rightarrow M^+ + e^-$$

where M is an atom or element in the sample.

This plasma is so energetic that nearly all elements in the periodic table are converted into singly charged ions (M^+), which makes the system ideal for multi-element analysis.

(c) Interface Region – Extracting Ions from the Plasma

- The plasma operates at atmospheric pressure, but the mass spectrometer requires a high vacuum.

- To bridge this, an interface with two small cones — the sampler and skimmer cones — allows ions to pass from the plasma into the vacuum system.

- Neutral particles and photons are largely excluded, while ions continue through to the ion optics.

(d) Ion Optics – Focusing and Steering the Ion Beam

- Electrostatic lenses guide and focus the positively charged ions into a narrow beam and remove neutral or interfering species.

- The goal is to maximise transmission efficiency and minimise noise before the ions reach the mass analyser.

(e) Mass Analyser – Separating Ions by Mass

- The ion beam enters a mass spectrometer, most commonly a quadrupole (though sector-field and time-of-flight types are also used).

- The quadrupole consists of four parallel rods with alternating electrical fields that act as a mass filter:

  - Only ions with a specific *mass-to-charge ratio (m/z)* can pass through at any given moment.

  - By rapidly scanning across m/z values, the system detects all elements sequentially.

- This step distinguishes, for example, $^{140}Ce^{+}$ from $^{142}Ce^{+}$ or $^{146}Nd^{+}$ from $^{147}Sm^{+}$.

(f) Detector – Counting Ions

- The separated ions strike an electron multiplier detector, which counts them individually as electrical pulses.

- The number of pulses per second corresponds directly to the concentration of that element in the sample.

To translate ion counts into concentrations, the instrument must be calibrated using standards of known composition (multi-element solutions).

- A series of standards (e.g., 0, 1, 10, 100 ppb) are analysed to generate a calibration curve for each element.

- Internal standards (such as indium or rhodium) are added to correct for signal drift, matrix effects, and plasma fluctuations.

- The unknown sample's signal intensity is then compared to the calibration curve to determine precise concentrations.

ICP-MS routinely achieves detection limits in the parts-per-trillion (ppt) to parts-per-billion (ppb) range for most elements.

ICP-MS is widely used across the exploration, processing, and environmental stages of rare earth and critical mineral projects:

- **Geochemical Exploration:** Detects subtle REE anomalies (e.g., high La/Yb, Ce/Y, or Eu anomalies) in soils, sediments, and rocks to identify REE-bearing systems such as carbonatites or pegmatites.

- **Ore Characterisation:** Quantifies the total REE content in mineral concentrates or bulk ores, distinguishing between light (LREE) and heavy (HREE) enrichment patterns.

- **Process Control:** Monitors REE recovery during leaching, solvent extraction, and precipitation by analysing solution chemistry.

- **Environmental Monitoring:** Measures trace REE or critical metal concentrations in tailings leachates, wastewater, and soils for regulatory compliance.

For example:

- A sample from a monazite-rich placer may show Ce, La, and Nd dominating the REE profile.

- A xenotime-bearing sample may reveal high Y, Dy, and Er content, indicating heavy REE enrichment.

Step-by-Step Procedure

a. Sample Preparation

1. Digestion:

- Use acid mixtures to dissolve minerals:

    - **REE minerals:** HNO_3 + HCl + HF (carefully controlled).

    - **Silicates:** Fusion with $LiBO_2$ followed by acid dissolution.

- Heat in Teflon beakers or microwave digestion systems.

2. Dilution: Dilute the solution to an appropriate concentration (e.g., 1:1000) with ultrapure water.

3. Filtration: Filter to remove insoluble residues before analysis.

b. Instrument Setup

1. Plasma Ignition: Start the argon plasma (~10,000 K) using the instrument software.

2. Tuning and Calibration:

- Tune for optimal sensitivity using a standard solution (e.g., 1 ppb Be, Co, In, U).

- Calibrate with multi-element REE standards covering expected concentration ranges.

3. Sample Introduction: Feed the liquid sample through a nebulizer and spray chamber into the plasma torch.

c. Data Acquisition

- The plasma ionises atoms, and the mass spectrometer separates ions by their mass-to-charge ratio (m/z).

- Record intensities for REEs (La–Lu), Y, Sc, Nb, Ta, Li, etc.

- Include internal standards (e.g., Rh, Re) to correct for drift.

d. Data Processing and Interpretation

- Convert signal intensities to concentrations using calibration curves.

- Normalise REE concentrations to chondritic or shale standards to interpret enrichment trends (LREE vs. HREE patterns).

- Identify fractionation signatures that indicate deposit type (e.g., carbonatite vs. granitic pegmatite).

Overall, ICP-MS works by transforming a liquid sample into ions using an extremely hot argon plasma and then measuring those ions' masses to determine elemental composition.

It is the benchmark technique for quantitative multi-element and isotopic analysis, particularly in the rare earth and critical minerals sector, where precise detection of trace REEs and associated elements (Nb, Ta, Li, Sc) is crucial for exploration success, process design, and product quality assurance.

Bench-Scale and Pilot Testing for Process Design

Bench-scale and pilot testing are essential transitional stages between laboratory research and full-scale mineral processing operations. They provide the critical link that validates metallurgical concepts, determines processing parameters, and reduces the technical and financial risk of scaling up to production. In rare earth and critical mineral projects—where each deposit can have unique mineralogy and chemistry—bench and pilot testing form the foundation for designing efficient, economical, and environmentally compliant extraction flowsheets.

Rare Earth and Critical Mineral Operations and Processing

The main goal of bench-scale and pilot testing is to verify the feasibility and optimise the design of the proposed processing route before large capital investments are made. Laboratory tests might demonstrate that a mineral can be leached or separated, but only controlled, scaled experiments can determine whether that process works efficiently, continuously, and economically at industrial scale.

These tests provide essential data on:

- Metallurgical recoveries (how much of the valuable element is extracted).

- Reagent consumption and costs.

- Solid–liquid separation behaviour (filtration, thickening).

- Tailings characteristics and environmental impacts.

- Process stability and control requirements.

- Equipment sizing and power demands.

Without these data, a process design would be largely theoretical, exposing a project to major operational and financial risks.

Bench-scale testing represents the first step beyond laboratory assays. It involves conducting small-scale experiments—usually in batch mode—under controlled conditions that replicate anticipated processing steps such as flotation, leaching, precipitation, or solvent extraction. Typical sample masses range from a few hundred grams to several kilograms.

Key aspects of bench-scale testing encompass several interrelated steps that collectively define how an ore behaves under controlled processing conditions. These tests provide essential data that inform flowsheet design, recovery efficiency, and process optimisation before moving to pilot-scale trials.

The first step is mineral characterisation, which focuses on understanding the ore's mineralogical composition and texture before any processing trials begin. Various analytical tools are employed for this purpose. X-ray Diffraction (XRD) is used to identify mineral phases and determine the crystalline structure of each component in the ore. Scanning Electron Microscopy with Energy Dispersive Spectroscopy (SEM-EDS) provides detailed images and elemental maps, revealing mineral textures and grain boundaries. Automated mineralogy systems such as QEMSCAN or MLA (Mineral Liberation Analyser) quantify mineral liberation, intergrowths, and associations. Together, these techniques help identify which minerals host the target elements—such as monazite, bastnäsite, or xenotime—and predict how they will respond during comminution, beneficiation, and chemical treatment.

The second stage involves physical separation tests, which simulate mechanical concentration processes. These include gravity separation methods (using Wilfley tables, spirals, or jigs), magnetic separation at both low and high intensity, and electrostatic separation for minerals like rare earth phosphates and carbonates. Flotation tests, performed in laboratory-scale Denver or micro-flotation cells, assess how surface chemistry and reagents influence recovery. These tests determine the ore's amenability to beneficiation by identifying the particle size range, reagent types, and operational conditions that yield the best recovery and grade.

Following physical beneficiation, hydrometallurgical tests evaluate the chemical processing routes. These experiments study leaching under various conditions—acidic, alkaline, or ionic—to determine optimal temperature, pH, and reagent concentrations for dissolving target elements. Solvent extraction and ion exchange trials are then used to selectively separate and purify individual rare earth elements, such as neodymium, praseodymium, dysprosium, or terbium. Finally, precipitation and crystallisation tests produce refined compounds such as rare earth oxides (REOs) or lithium hydroxide, which are market-ready products. Each step's yield, reagent consumption, and residue composition are carefully measured to construct a preliminary mass balance and estimate reagent costs for process design.

The final phase of bench-scale work is metallurgical modelling, where experimental data are integrated to develop preliminary process models. These models predict key operational parameters such as throughput, recovery efficiency, energy requirements, and reagent usage. The outputs from this stage form the technical and economic foundation for pilot-scale validation, where the proposed process is tested under semi-continuous conditions to confirm its scalability and robustness.

Pilot testing scales up the process to a semi-continuous or continuous flow system, usually processing several tonnes of ore. The purpose is to replicate the entire flowsheet under conditions closely resembling those of a full-scale plant.

Pilot-scale facilities include:

- Small grinding circuits (ball or rod mills).

- Continuous flotation cells or columns.

- Leach reactors with continuous reagent feed.

- Counter-current decantation (CCD) thickeners.

- Solvent extraction and ion exchange columns.

- Filtration and product recovery units.

Typical pilot plants operate at 1/100 to 1/1000 of the full-scale throughput but maintain realistic hydrodynamics, temperature control, and material handling.

Key outcomes of pilot testing include:

a. Process Validation

Confirms that the laboratory-proposed flowsheet works as intended under realistic conditions. Any issues with scaling, mixing, or kinetics become apparent here.

Rare Earth and Critical Mineral Operations and Processing

b. Kinetic and Scale-Up Data

Determines rate constants for reactions (leaching, precipitation, extraction) and equipment performance factors (e.g., residence time, agitation efficiency).

c. Material and Energy Balances

Provides real-world data for developing a mass and energy balance, critical for designing and sizing full-scale equipment and estimating utility requirements (electricity, water, steam).

d. Product Quality and Environmental Testing

Evaluates the purity of the recovered concentrate or oxide and characterises waste streams for environmental compliance—particularly important for radioactive by-products (e.g., thorium in monazite).

In rare earth and critical mineral projects, each deposit's unique mineralogy dictates the processing challenges and determines the specific combination of physical, thermal, and chemical steps required to achieve efficient recovery. The mineralogical variability of these ores means that no single extraction method fits all deposits—each must be tested and tailored through bench-scale and pilot-scale experimentation.

For example, bastnäsite ores, which are carbonate-fluoride minerals, require controlled acid leaching to dissolve the rare earth elements (REEs) while carefully managing CO_2 evolution during decomposition. This ensures safe and efficient processing while preventing excessive gas formation that can disrupt leach kinetics and reactor operation.

In contrast, monazite and xenotime—both phosphate minerals—are chemically resistant and require high-temperature cracking to break down their phosphate matrix before leaching. This is usually achieved using concentrated sulphuric acid roasting or alkaline caustic digestion with sodium hydroxide (NaOH). Only after these steps can the REEs be effectively dissolved and recovered through hydrometallurgical techniques.

Ion-adsorption clays, which host REEs weakly bound to clay minerals such as kaolinite and halloysite, demand a much gentler approach. These deposits are processed using mild ammonium-salt leaching solutions, such as ammonium sulphate, to exchange REE ions without dissolving the clay structure. This method minimises contamination and environmental impact, making it one of the most sustainable extraction routes in the REE industry.

For spodumene, the principal lithium ore, a thermal phase transformation is essential before chemical leaching can occur. Spodumene naturally exists in the α-phase, which is dense and resistant to acid attack. Roasting the ore at around 1000°C converts it to the more open β-phase, which is far more reactive and amenable to sulphuric acid leaching.

Bench-scale and pilot testing are used to determine which combination of thermal, mechanical, and chemical processes achieves the highest recovery rates at the lowest operating cost and environmental footprint. These tests provide the data necessary to optimise process parameters—such as temperature, pH, and reagent concentration—ensuring that the final process flowsheet is both economically viable and environmentally responsible.

Integration into Process Design

The results derived from bench-scale and pilot testing are essential in the strategic development of rare earth and critical mineral projects. These results serve as empirical evidence that supports critical engineering, financial, and environmental decisions necessary for transitioning projects from concept to viable operational mines and processing facilities. Numerous studies emphasize the importance of empirical data in the feasibility assessment and design of such projects.

The primary application of test results is in the preparation of Definitive Feasibility Studies (DFS) or Bankable Feasibility Studies (BFS). These studies include thorough technical and economic evaluations that determine a project's viability for construction and financing. Bench and pilot test data are crucial for validating proposed processing routes, which encompass recovery rates, reagent consumption, energy requirements, and waste characteristics, fostering greater confidence among investors and regulators [163]. For instance, a pilot test demonstrating consistent 90% recovery of rare earth elements (REEs) over extended runs is fundamental in securing funding and regulatory approvals [164].

Equally important, the outcomes of these tests directly inform equipment selection and the development of process flow diagrams (PFDs). Each processing stage—from crushing and grinding to leaching and solvent extraction—must be aligned with appropriately sized equipment based on real-world performance data obtained from pilot trials [165]. Such trials clarify how variations in reagents, temperature, and other conditions affect throughput and recovery rates, enabling engineers to design scalable and efficient processing circuits [166]. For example, performance metrics from pilot leach tests dictate the specifications for critical equipment, such as leach tanks and filtration systems [167].

Additionally, pilot and bench testing are vital for estimating capital and operating costs (CAPEX and OPEX). Data on reagent consumption, energy expenditure, and waste generation at the pilot scale can be extrapolated to predict overall production costs, providing an accurate financial outlook for the project [168]. This accuracy is crucial for evaluating process options aimed at maximizing returns on investment. For instance, a new leaching method that uses 20% less reagent than traditional approaches can notably reduce operating costs and associated transport and disposal expenses [169].

Environmental Impact Assessments (EIA) also significantly benefit from bench and pilot test data. These assessments require a comprehensive quantification of the process's environmental footprint, including emissions and waste by-products, which can be identified early through strategic testing [170, 171]. This enables the incorporation of risk mitigation strategies into the project design, ensuring compliance with environmental regulations and enhancing sustainability metrics.

In essence, integrating bench and pilot testing results into project design yields a comprehensive understanding of operational dynamics, allowing feasibility studies, equipment selections, financial modelling, and environmental evaluations to be based on validated performance data rather than assumptions. This empirical foundation not only improves economic efficiency and environmental sustainability but also enhances the project's attractiveness to investors seeking reliable, bankable ventures [172].

Case Study Scenario: Bench-Scale and Pilot Testing for a Rare Earth Project

Project Background

A mining company, *Southern Elements Ltd.*, discovers a rare earth deposit in Western Australia containing bastnäsite, monazite, and minor xenotime. Initial laboratory assays confirm high concentrations of cerium (Ce), neodymium (Nd), and dysprosium (Dy), but the mineralogical complexity of the ore poses uncertainty regarding its processability. To move beyond laboratory proof-of-concept results and de-risk the investment, the company initiates a structured bench-scale and pilot-scale testing program before developing a full-scale processing plant.

Bench-Scale Testing Phase

The bench-scale program begins with comprehensive mineral characterisation using X-ray Diffraction (XRD), Scanning Electron Microscopy (SEM-EDS), and Mineral Liberation Analyser (MLA) techniques. Results reveal that REEs are mainly hosted in bastnäsite (70%) and monazite (25%), with traces in xenotime. Liberation analysis shows that bastnäsite grains are coarse and easily separable, while monazite is fine-grained and locked with quartz.

Next, physical separation tests are conducted to concentrate REE minerals. Gravity and magnetic separation trials achieve 60% recovery, while flotation tests using fatty acid collectors raise recovery to 82% with a concentrate grade of 45% REO. However, residual thorium in the monazite fraction signals potential radiological management challenges.

During hydrometallurgical testing, acidic leaching of bastnäsite using hydrochloric acid yields 90% REE recovery, while monazite requires high-temperature caustic cracking to achieve 85%. Solvent extraction tests successfully separate light and heavy REEs, and final precipitation produces high-purity rare earth oxides suitable for magnet production. Data from these trials establish reagent consumption rates, leaching kinetics, and residue composition.

Pilot Testing and Process Validation

Following successful bench trials, a pilot plant is constructed to operate at 1/500 of full scale, processing 5 tonnes of ore per day. The integrated circuit includes crushing, grinding, flotation, leaching, solvent extraction, and precipitation stages. Over a three-month continuous run, the pilot plant confirms that REE recovery exceeds 88%, acid consumption is 18% lower than predicted, and the system maintains stable pH and temperature control. Importantly, tailings are found to meet environmental safety limits for thorium and fluorine, supporting environmental compliance.

Integration into Feasibility and Design

The pilot results feed directly into the project's Definitive Feasibility Study (DFS) and Environmental Impact Assessment (EIA). Engineers use real process data to design the full-scale Process Flow Diagram (PFD) and select equipment suited to continuous operation. Cost analysts integrate reagent use, energy requirements, and recovery data into CAPEX and OPEX models, confirming a 12% reduction in expected operating costs.

By basing design and economics on pilot-derived evidence, *Southern Elements Ltd.* eliminates major scale-up risks. For example, pilot results demonstrate that bastnäsite concentrate achieves 90% REE recovery with 20% lower acid use, prompting the adoption of this optimized leach circuit in the final design. The validated data reassure investors, accelerate environmental approvals, and support project financing.

Outcome

The project advances to construction with a verified, environmentally responsible, and economically optimized processing route. The case illustrates how bench-scale and pilot testing bridge the gap between laboratory innovation and industrial reality, ensuring that mineral processing flowsheets are both technically sound and commercially viable before large capital investments are made.

Key Terms and Concepts

Mineral Characterisation: The detailed analysis of a mineral's physical, chemical, and structural properties to determine its composition, associations, and processing behaviour.

Key Terms and Concepts

Ore Texture: The physical appearance and arrangement of minerals within an ore, including grain size, shape, and intergrowths, which influence liberation and processing.

Liberation: The process by which valuable minerals are freed from surrounding gangue material during crushing and grinding, enabling effective separation.

Gangue: The non-valuable minerals that occur with ore minerals and must be separated during processing.

X-ray Diffraction (XRD): An analytical technique used to determine the crystalline structure and mineral phases present in an ore sample.

X-ray Fluorescence (XRF): A rapid analytical method that measures the elemental composition of a sample by detecting secondary X-rays emitted from its surface.

Scanning Electron Microscopy (SEM): A high-resolution imaging technique that provides detailed information about mineral morphology and microstructure.

Mineral Liberation Analysis (MLA): A computer-assisted SEM-based technique that quantifies mineral composition, grain size, and liberation characteristics in ores.

Inductively Coupled Plasma Mass Spectrometry (ICP-MS): A highly sensitive analytical technique used for trace and ultra-trace elemental analysis of ores and solutions.

Bench-Scale Testing: Small-scale laboratory experiments conducted to evaluate ore performance under controlled processing conditions before pilot testing.

Pilot Testing: Intermediate-scale process trials used to validate laboratory results and gather data for full-scale plant design and operation.

Chemical Assay: A laboratory analysis used to determine the precise concentration of metals or elements in an ore or solution.

Quantitative Analysis: The measurement of the amount or concentration of specific elements or compounds within a sample.

Key Terms and Concepts

Qualitative Analysis: The identification of which elements or minerals are present in a sample, without determining their exact concentration.

Process Flowsheet: A schematic diagram showing the sequence of operations and equipment used to process an ore into a final product.

Metallurgical Testing: The evaluation of an ore's response to physical and chemical processing methods to determine recovery efficiency and product quality.

Chapter 3 Review Questions

Physical and Chemical Properties of Ores

1. What are the key physical properties of ores that influence mineral processing performance?

2. How does ore hardness affect crushing and grinding efficiency?

3. Why is density contrast important in gravity separation?

4. How do magnetic and electrical properties assist in mineral beneficiation and exploration?

5. What is the significance of chemical composition and elemental substitution in REE minerals?

6. How does oxidation state variability (e.g., Ce^{3+}/Ce^{4+}) influence rare earth element processing?

7. Why is understanding acid and base reactivity critical in selecting hydrometallurgical processes?

Mineral Characterisation and Processing Routes

8. What is mineral characterisation, and why is it crucial in determining the appropriate processing route for rare earth and critical minerals?

9. How does grain size and mineral intergrowth affect liberation and recovery?

10. Why must mineralogical data be integrated into resource modelling and process design?

Mineral Identification and Liberation Studies

11. What are the main analytical techniques used in mineral identification?

12. How does reflected versus transmitted light microscopy contribute to ore characterisation?

13. What information can scanning electron microscopy (SEM) combined with energy dispersive spectroscopy (EDS) provide?

14. How does the Mineral Liberation Analyzer (MLA) assist in quantifying liberation?

15. Why is liberation analysis performed at different particle size fractions?

Analytical Tools for Mineral and Elemental Analysis

16. What information does X-ray diffraction (XRD) provide, and how is it used in REE mineral studies?

17. How does X-ray fluorescence (XRF) differ from XRD, and what type of data does it produce?

18. What are the advantages of using inductively coupled plasma mass spectrometry (ICP-MS) in rare earth element analysis?

19. What is the difference between qualitative and quantitative analytical methods, and when is each appropriate?

20. How can multiple analytical tools be combined to produce a complete mineralogical model?

Bench-Scale Testing

21. What is the purpose of bench-scale testing in mineral processing?

22. How do bench-scale flotation and leaching tests differ in the data they provide?

23. What types of data are obtained from bench-scale experiments that inform process optimisation?

24. Why is mineralogical variability an important consideration during bench-scale testing?

Pilot-Scale Testing and Process Design

25. What is the role of pilot-scale testing in process validation?

26. How does pilot-scale testing inform equipment selection and process flow diagrams (PFDs)?

27. In what way do pilot plant results influence capital and operating cost estimates?

28. How are environmental impact assessments (EIAs) supported by pilot testing data?

Application to Process Flowsheets

29. How can mineralogical and analytical data be used to recommend suitable processing methods for bastnäsite, monazite, and xenotime?

30. What are the main differences in processing between hard-rock and ion-adsorption clay REE deposits?

31. How can integrating mineralogical, chemical, and pilot-scale data reduce project risk?

Chapter 4

Comminution, Classification, and Particle Preparation

In this chapter, you will explore how efficient comminution and classification form the foundation of modern mineral processing. Before you can separate valuable minerals from their surrounding gangue, you must first liberate them through controlled crushing and grinding. You will learn about the key principles, equipment, and operational strategies used to achieve effective particle size reduction—from primary crushers and grinding mills to advanced classifiers that control particle size distribution. As you progress, you will see how energy efficiency, liberation, and particle size control directly influence the performance of downstream processes such as flotation, magnetic separation, and leaching. By the end of this chapter, you will understand how to select and apply the right comminution and classification techniques to optimise recovery, reduce energy consumption, and prepare ores for successful beneficiation in rare earth and critical mineral processing.

Learning Outcomes	
This chapter aims to give you the ability to: 1. Explain the fundamental principles of comminution, including the mechanisms of crushing, grinding, and particle size reduction, as applied to mineral processing. 2. Identify and describe the major types of comminution equipment, including jaw crushers, cone crushers, ball mills, rod mills, and high-pressure grinding rolls (HPGR), and evaluate their appropriate applications for different ore types. 3. Discuss the purpose and operation of classification equipment, such as screens, hydrocyclones, and classifiers, in controlling particle size distribution within a processing circuit. 4. Analyse the relationship between energy input, particle size reduction, and grind efficiency, and apply energy efficiency concepts such as Bond's Work Index in process planning.	

Learning Outcomes	

5. Recognise the importance of achieving optimal mineral liberation prior to downstream separation processes, and assess how over- or under-grinding can affect recovery and concentrate quality.
6. Interpret particle size distribution data and use it to evaluate comminution circuit performance and efficiency.
7. Demonstrate an understanding of how ore hardness, texture, and mineral associations influence comminution strategy and equipment selection.
8. Apply safety, maintenance, and operational best practices when working with crushing, grinding, and classification equipment in mineral processing environments.

Fundamentals of Crushing and Grinding

Crushing and grinding are the first stages in the comminution process — the mechanical size reduction of mined ore to liberate valuable minerals from the surrounding gangue. Together, they form the foundation of mineral processing, as the degree of size reduction directly affects mineral liberation, recovery efficiency, and energy consumption in downstream separation stages such as flotation, magnetic, or gravity processing.

The primary goal of comminution is to reduce the ore to a size small enough to liberate the target minerals from the host rock. In rare earth and critical mineral projects, this means freeing minerals like bastnäsite, monazite, xenotime, or spodumene so they can be physically or chemically separated from gangue. Properly controlled size reduction improves both processing efficiency and economic recovery by maximising mineral exposure to reagents during leaching or flotation.

Crushing is the first stage of comminution and typically reduces large run-of-mine (ROM) ore lumps (up to 1 metre) down to 10–20 millimetres. It is mainly a dry, mechanical process that uses compressive forces to break the rock.

- **Primary Crushers:** Jaw crushers and gyratory crushers handle coarse, hard rock. They operate by compression between two rigid surfaces.

- **Secondary Crushers:** Cone crushers and impact crushers further reduce the material to finer sizes.

- **Key Parameters:** Feed size, reduction ratio (size in/size out), and energy efficiency determine crushing performance.

The objective of crushing is not to achieve liberation, but to prepare the ore for grinding. Efficient crushing minimises excessive fines, which can complicate grinding and separation.

Figure 22: Cone Crushers for 100 to 300 TPH Aggregate / Stone Crushing plants. Mvrajeshalind, CC BY-SA 4.0, via Wikimedia Commons.

Grinding follows crushing and involves reducing the ore to fine particles (often below 100 μm) where mineral grains become fully liberated. It is usually performed in wet conditions using rotating mills filled with grinding media.

There are several types of mills used in the grinding stage of mineral processing, each designed for a specific particle size range and ore type. Rod mills use long steel rods as the grinding medium and are primarily employed for coarser grinding applications, where the goal is to reduce particle size without producing excessive fines. They are particularly effective in preparing feed for ball mills or in circuits requiring uniform particle distribution.

Ball mills, in contrast, use spherical steel balls to achieve finer grinding and are among the most common mill types used in rare earth and critical mineral processing circuits. The impact and abrasion between the balls and the ore reduce the material to the desired size, making ball mills ideal for achieving high degrees of mineral liberation.

Figure 23: Ball mill with steel ball is milling iron oxide and potassium nitrate. GOKLuLe 盧樂, CC BY-SA 3.0, via Wikimedia Commons.

Autogenous (AG) and semi-autogenous (SAG) mills represent more advanced milling systems, where the ore itself serves as the primary grinding medium. In AG mills, no additional media are used, while SAG mills incorporate a small amount of steel balls to enhance grinding efficiency. These mills are suitable for large-scale operations handling high tonnages, such as in the initial stages of ore processing.

For fine and ultrafine grinding—often below 20 micrometres—vertical and stirred mills are used. These mills employ a vertical chamber with rapidly rotating impellers that agitate small grinding media, producing intense shear and attrition forces. They are particularly effective for complex or refractory ores, where fine liberation of minerals like monazite or bastnäsite is essential for efficient downstream processing.

Grinding performance depends on mill speed, charge load, slurry density, and the ore's hardness. The finer the grind, the greater the liberation, but excessive grinding can create slimes that reduce recovery in flotation or leaching.

Crushing and grinding are the most energy-intensive stages in mineral processing—accounting for up to 50–70% of total plant energy use. Therefore, energy efficiency and process optimisation are critical. The Bond Work Index (Wi) is often used to quantify the ore's resistance to grinding and to design energy-efficient circuits. Technologies such as high-pressure grinding rolls (HPGR) and advanced process control systems help reduce energy consumption while maintaining throughput.

The ultimate purpose of comminution is mineral liberation. Partial liberation leads to lower recoveries, while overgrinding increases processing costs and loss of fine particles. Liberation studies using tools like the Mineral Liberation Analyzer (MLA) or QEMSCAN quantify how mineral grains separate as particle size decreases, guiding adjustments in grind size targets and mill operation.

In rare earth and critical mineral processing, crushing and grinding play a decisive role in determining how efficiently valuable elements can be extracted from complex ores. These early comminution stages are not just about reducing particle size—they are about creating the right conditions for mineral liberation, chemical reactivity, and energy efficiency throughout the beneficiation and extraction process.

During crushing, large chunks of run-of-mine ore are reduced to smaller fragments that can be handled by downstream mills. In rare earth deposits—such as bastnäsite-bearing carbonatites or monazite-rich sandstones—the goal is to break the host rock (often carbonates, silicates, or phosphates) without generating excessive fines that may cause material loss or slurry handling problems. Primary crushers, such as jaw or gyratory units, handle the coarse feed, while secondary cone or impact crushers produce the 10–20 mm feed typically required for grinding circuits. Proper control of crushing parameters, including feed size and reduction ratio, ensures that valuable minerals remain intact and recoverable.

The grinding stage then takes over, reducing the crushed ore to fine particles—often less than 100 micrometres—to achieve mineral liberation. Liberation is crucial because rare earth and critical minerals like monazite, xenotime, and bastnäsite often occur as fine inclusions within gangue minerals such as quartz, feldspar, or barite. Grinding is commonly done in wet circuits using rod, ball, autogenous, or stirred mills. Each type of mill serves a particular function: rod mills produce uniform coarse particles ideal for intermediate stages; ball mills deliver finer liberation suitable for flotation or leaching; AG and SAG mills handle large-scale feed efficiently; and vertical or stirred mills generate ultrafine product (<20 µm) needed for refractory or complex ores.

For instance, bastnäsite-bearing ores typically undergo fine grinding before flotation to ensure the mineral is adequately liberated from carbonates. In monazite and xenotime processing, controlled grinding prevents overproduction of slimes, which can trap valuable particles and reduce separation efficiency. Spodumene ores used for lithium extraction require grinding after roasting to convert α-spodumene to the β-phase, which is more reactive to acid leaching. In niobium-tantalum minerals such as columbite-tantalite, precise grinding allows gravity or magnetic separation to work effectively.

Because comminution consumes 50–70% of total plant energy, rare earth and critical mineral operations place strong emphasis on energy-efficient grinding. Techniques such as high-pressure grinding rolls (HPGR) and variable-speed mill drives improve performance while lowering costs. Liberation studies using automated tools like the Mineral Liberation Analyzer

(MLA) or QEMSCAN quantify how minerals separate with size, helping engineers determine optimal grind targets that balance recovery and power consumption.

Ultimately, in rare earth and critical mineral processing, effective crushing and grinding transform geological material into a processable product. These stages set the foundation for all subsequent beneficiation and hydrometallurgical operations—where even minor improvements in liberation or energy efficiency can lead to significant gains in recovery, product purity, and project profitability.

Equipment Types: Crushers, Mills, and Classifiers

In mineral processing—especially in rare earth and critical mineral operations—crushers, mills, and classifiers form the core of the comminution circuit, which prepares mined ore for concentration and extraction. Each equipment type serves a specific function in breaking down, grinding, and separating material to achieve the required particle size and mineral liberation before downstream processing.

Crushers are the first step in the comminution process. Their purpose is to reduce large run-of-mine (ROM) ore chunks, sometimes over a metre in diameter, into smaller fragments (typically 10–20 mm) suitable for grinding. Crushing is mainly a dry process relying on compressive and impact forces.

- **Jaw Crushers:** Commonly used in primary crushing, they compress ore between a fixed and a moving jaw. They are ideal for hard, coarse materials found in carbonatite-hosted rare earth deposits or spodumene pegmatites.

- **Gyratory Crushers:** Handle higher capacities and continuous feed, often used in large-scale operations such as iron ore or niobium-tantalum mining.

- **Cone Crushers:** Used in secondary or tertiary stages, cone crushers apply compressive force between a rotating cone and a stationary shell, producing finer material for milling.

- **Impact Crushers:** Apply dynamic impact forces to break softer or brittle ores and are suitable for producing consistent feed for downstream grinding.

Efficient crushing minimises the generation of fines and provides uniform feed size to improve grinding performance and reduce energy consumption.

A jaw crusher is a fundamental piece of equipment used as a primary crusher in mining and mineral processing operations. Its main purpose is to reduce large chunks of rock or ore into smaller, manageable sizes suitable for further processing. In rare earth and critical mineral projects, jaw crushers are crucial at the initial stage of comminution, breaking run-of-mine (ROM) ore into smaller fragments that can then be processed by secondary crushers or grinding mills.

Rare Earth and Critical Mineral Operations and Processing

The jaw crusher operates on the principle of compression. It crushes material by squeezing it between two jaws—one fixed and one that moves back and forth. As the movable jaw oscillates, it applies compressive force to the material, causing it to fracture. The crushed material then passes through the discharge opening at the bottom, known as the "set" or "closed side setting," once it reaches the desired size.

Several key components enable this process. The fixed jaw provides a stationary surface for compression, while the movable jaw—driven by an eccentric shaft and flywheel—creates the crushing motion. The flywheel stores mechanical energy, ensuring smooth operation and consistent force. The toggle plate acts as both a safety device and a motion transfer mechanism, breaking under extreme stress to protect the crusher. Jaw liners, made of hardened manganese steel, cover both jaws and can be reversed when worn to extend service life.

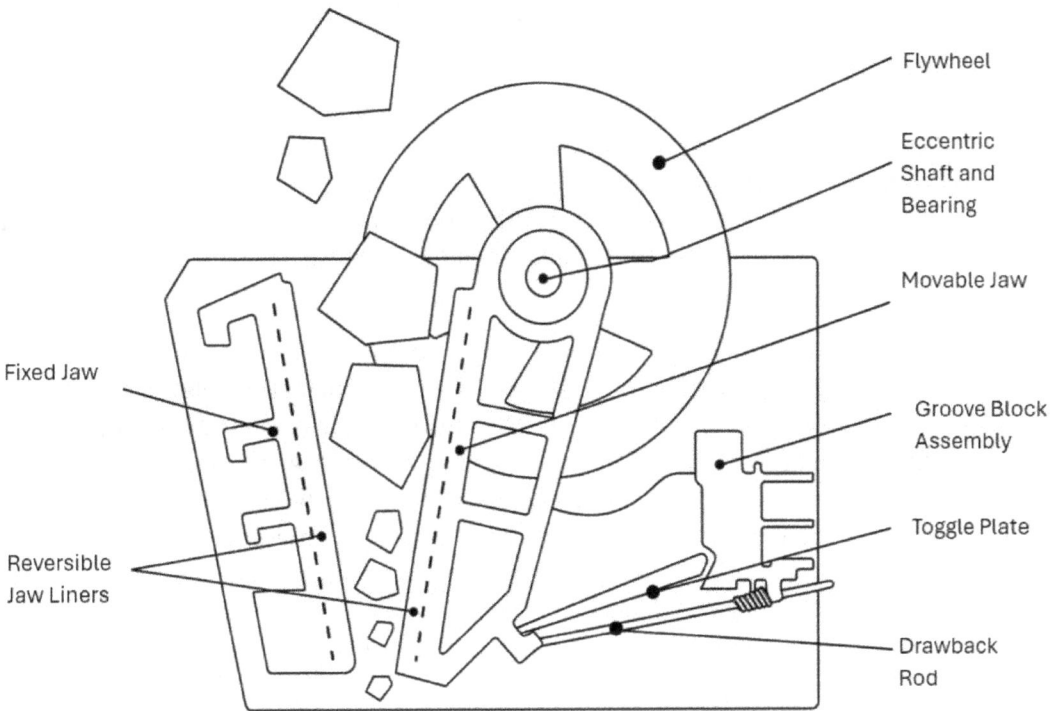

Figure 24: Illustration of a jaw crusher chamber showing the replaceable jaw plates.

Crushing occurs in four stages. During the feed stage, large ore pieces are introduced through the top opening. In the compression stage, the movable jaw moves toward the fixed jaw, crushing the material between them. The discharge stage follows, where the crushed fragments fall through the bottom gap once small enough. Finally, the crusher operates continuously, repeating this cycle hundreds of times per minute to achieve a steady flow of material.

There are two main types of jaw crushers: single-toggle and double-toggle. The single-toggle type, more common today, uses one toggle plate and a top-mounted eccentric shaft, making it simpler and easier to maintain. The double-toggle version uses two toggles and two shafts, delivering stronger crushing force but requiring more maintenance.

In rare earth and critical mineral operations, jaw crushers prepare ores such as bastnäsite, monazite, xenotime, and spodumene for downstream processing. They are valued for their reliability, ability to handle hard materials, and adjustable output sizes. However, they also have limitations—producing more fines than some other crushers and struggling with sticky or clay-rich ores.

Overall, the jaw crusher remains an essential tool in mineral processing. It is simple yet robust design enables efficient size reduction, paving the way for subsequent separation, beneficiation, and recovery of valuable rare earth and critical minerals.

A gyratory crusher is a large, robust machine used as a primary crusher in major mining and mineral processing operations. Its main function is to reduce large rocks into smaller, more manageable fragments for subsequent processing. While it performs the same fundamental task as a jaw crusher, it operates continuously and at much higher capacity. Gyratory crushers are particularly suited for handling hard, abrasive ores and are often employed in the early stages of processing rare earth and critical minerals, such as bastnäsite, monazite, and spodumene-bearing pegmatites.

The gyratory crusher operates on the principle of compression, much like a jaw crusher. However, instead of two flat jaws, it uses a gyrating mantle (a conical spindle) that moves within a concave outer shell. The mantle is attached to a main shaft suspended from a spider at the top and supported at the bottom by a hydraulic bearing or piston. As the main shaft rotates, the mantle moves in an eccentric, conical motion against the stationary concave surface. This cyclical movement continuously compresses and releases the material trapped between the mantle and concave, gradually breaking the rock into smaller fragments until it is fine enough to pass through the discharge opening.

The spider and top shell form the upper structure of the crusher and support the main shaft while also providing the feed opening for ore to enter the crushing chamber. The mantle is the moving, cone-shaped surface attached to the main shaft, which performs the gyrating motion that crushes the ore. Surrounding it is the concave liner, the stationary surface that forms the outer wall of the crushing chamber. The gap between the mantle and concave—known as the closed-side setting (CSS)—determines the size of the crushed product.

The main shaft and eccentric assembly generate the gyratory motion, with the eccentric mechanism causing the mantle to move in a circular orbit that produces zones of compression and release. A hydraulic support or piston keeps the main shaft vertically aligned and provides overload protection by allowing limited upward movement if excessive pressure occurs. The drive system, consisting of a motor and gear assembly, powers the eccentric rotation that drives the entire crushing process.

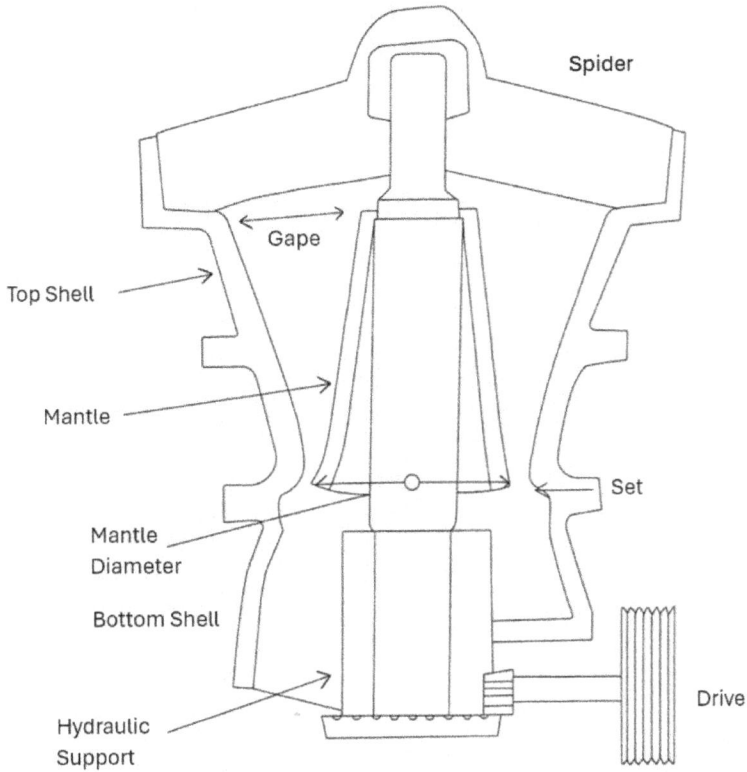

Figure 25: Diagram of a Gyratory Crusher.

Crushing begins as large rocks or ore lumps are fed through the top opening (spider) into the crushing chamber. As the mantle gyrates toward the concave, the ore is compressed and fractured under high pressure. When the mantle moves away, the smaller particles discharge through the bottom opening. This process is continuous, unlike the intermittent motion of a jaw crusher, allowing for high throughput and consistent product flow.

Figure 26: Gyratory crusher used as a primary crusher in the granite quarry at Glensanda, Scotland. Ludowingischer at de.wikipedia, CC BY-SA 2.0 DE, via Wikimedia Commons.

Gyratory crushers are classified by their gape (the feed opening) and mantle diameter. The product size is controlled by adjusting the closed-side setting (CSS), which can be modified hydraulically to maintain consistent particle size and throughput. This flexibility allows operators to optimise performance according to ore hardness, feed size, and downstream requirements.

In rare earth and critical mineral processing, gyratory crushers are particularly valuable for treating hard, abrasive ores found in carbonatite and pegmatite deposits. They are typically used as the first stage of comminution before secondary crushing or grinding. Their main advantages include high capacity, continuous operation, uniform product size, and low power consumption per tonne compared to other primary crushers.

However, they also have limitations. Gyratory crushers are large, heavy machines that require strong foundations and regular maintenance. They are not well suited for sticky or clay-rich materials, which can clog the crushing chamber.

Rare Earth and Critical Mineral Operations and Processing

A cone crusher is a type of secondary or tertiary crusher used to reduce rock and ore into smaller, more uniform sizes after primary crushing, which is usually performed by a jaw or gyratory crusher. It is an essential machine in mining, quarrying, and mineral processing operations—particularly in rare earth and critical mineral industries—where precise control of particle size is crucial for efficient grinding, beneficiation, and chemical processing.

A cone crusher operates on the principle of compression, much like a gyratory crusher, but features a smaller feed opening and shorter spindle. The machine consists of two main components: a mantle (the moving inner cone) mounted on a main shaft, and a concave (the stationary outer surface). As the main shaft rotates eccentrically, the mantle moves toward and away from the concave in a continuous circular motion. When ore or rock enters through the feed opening at the top, it becomes trapped between these two surfaces. The rock is crushed when the mantle moves toward the concave and released when it moves away. This repeated cycle of compression and release creates a continuous crushing action that produces fine, uniform material with a high reduction ratio.

The major components of a cone crusher each serve specific functions. The feed opening and cone feed plate distribute incoming material evenly into the crushing chamber. The mantle, attached to the main shaft, performs the gyrating motion responsible for crushing the rock. The concave, also known as the liner, forms the stationary surface against which the material is compressed, with the gap between the mantle and concave determining the final particle size. The main shaft, powered by an eccentric bushing and crown gear, generates the mantle's circular motion. A hydraulic tramp release cylinder acts as a safety feature, allowing the mantle to lift automatically if uncrushable material enters, preventing damage. The drive system, consisting of a motor and gears, provides the mechanical energy for operation, while the main frame supports the entire assembly and absorbs the operational loads.

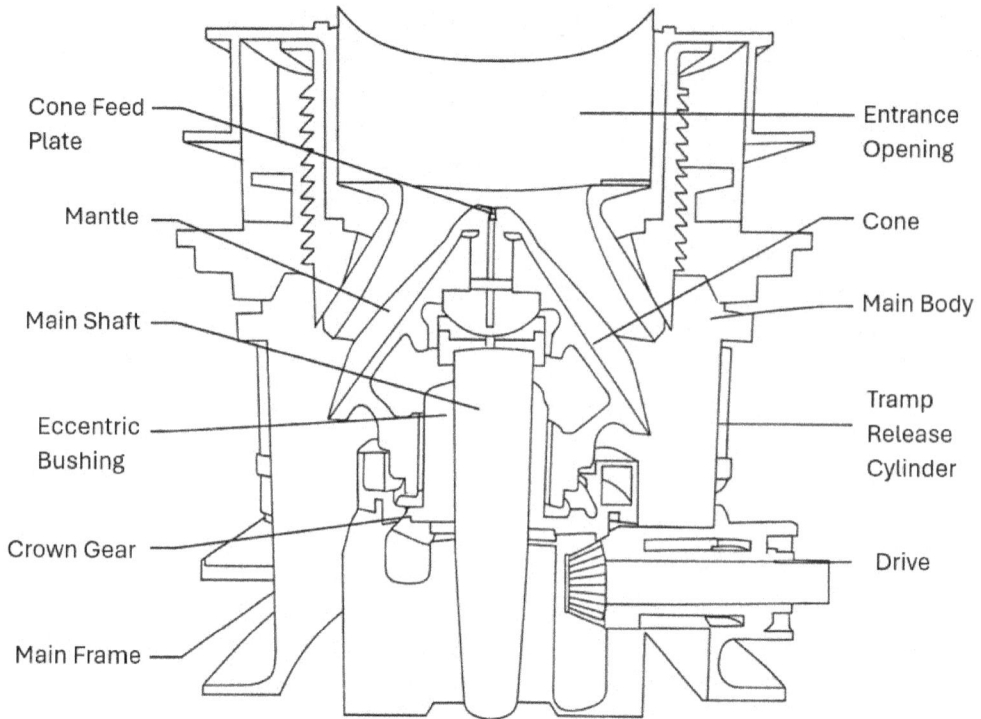

Figure 27: Cone Crusher major components.

The crushing process within a cone crusher is continuous. Material enters through the top opening and descends into the crushing chamber under gravity. As the mantle rotates eccentrically, the gap between the mantle and concave narrows, compressing the ore and breaking it into smaller particles. When the particles are small enough to pass through the narrowest gap (the closed side setting, or CSS), they exit at the bottom of the crusher. This consistent motion ensures a steady flow of product with uniform particle size.

Cone crushers are available in several configurations, each suited to specific applications. The standard (coarse) cone crusher is typically used for secondary crushing, producing larger product sizes. The short-head cone crusher is used for tertiary or fine crushing, with a steeper chamber that delivers finer output. Hydraulic cone crushers use advanced hydraulic systems for automatic adjustment, tramp release, and overload protection—making them ideal for modern, automated processing plants. Single-cylinder and multi-cylinder hydraulic cone crushers offer enhanced precision, capacity, and ease of maintenance compared to older mechanical types.

In rare earth and critical mineral processing, cone crushers play a vital role in achieving the desired liberation of valuable minerals. For example, in spodumene ($LiAlSi_2O_6$) processing, cone crushers are used after roasting to prepare material for grinding and leaching. In bastnäsite and monazite beneficiation, they produce consistent feed sizes for downstream gravity, magnetic, or electrostatic separation. Their ability to deliver fine, uniform particles with minimal overgrinding ensures high recovery rates and optimised energy efficiency throughout the circuit.

Cone crushers offer several advantages, including a high reduction ratio, continuous operation with high throughput, adjustable settings for variable product sizes, suitability for hard and abrasive materials, and built-in hydraulic protection against tramp materials. However, they also have limitations: they are more complex and costly than jaw crushers, can be sensitive to feed size and moisture (which may cause clogging), and require regular liner replacement due to wear.

The cone crusher is a precise, high-capacity machine designed for controlled size reduction. Its efficient compression mechanism, combined with modern hydraulic systems and adjustable settings, makes it indispensable in the processing of rare earth and critical minerals, where optimal liberation and consistent product quality are essential for economic and environmental performance.

An impact crusher is a machine that breaks rocks and ores using impact energy rather than compression. It is widely used across mining, quarrying, recycling, and mineral processing industries, particularly for medium-hard to soft materials. In rare earth and critical mineral operations, impact crushers are often employed as secondary or tertiary crushers to produce fine, uniform particle sizes that are ideal for subsequent grinding, flotation, or chemical processing stages.

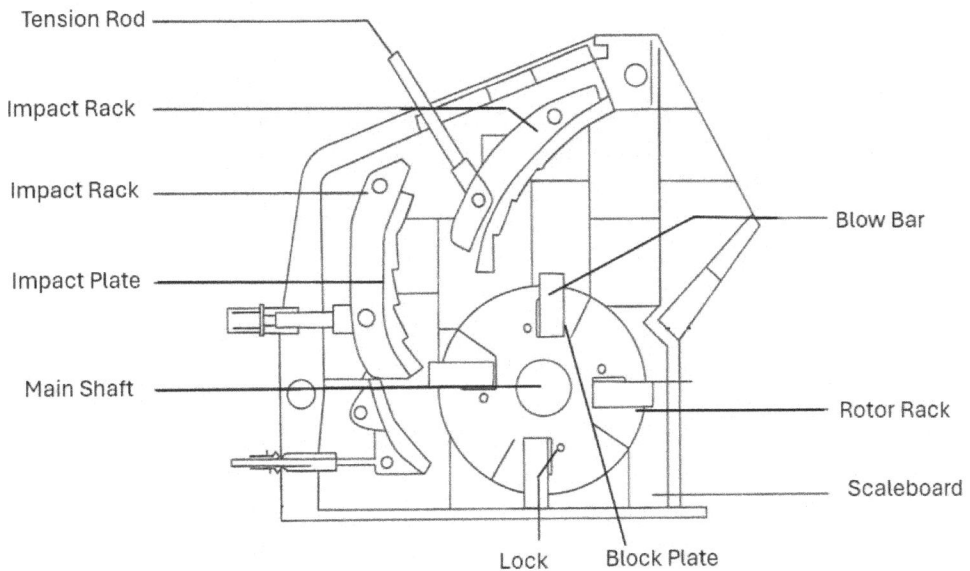

Figure 28: Impact Crusher design and components.

The working principle of an impact crusher is based on the rapid transfer of kinetic energy. Unlike jaw or cone crushers that apply steady compressive pressure, an impact crusher propels material at high velocity against hard surfaces known as impact plates (breaker plates). Feed material enters through the top and is struck by a fast-spinning rotor equipped with blow bars or hammers. The rotor, rotating between 500 and 2,000 revolutions per minute, hurls the rock outward, causing it to collide violently with the breaker plates lining the chamber. These collisions fracture the rock along natural grain boundaries and weak zones. The shattered fragments are then thrown back into the rotor's path, where they undergo repeated impacts until small enough to exit through the discharge opening at the base. This multi-impact process produces fine, cubic-shaped particles, which are ideal for high-quality aggregates or mineral feedstocks.

Several key components work together to achieve this efficient crushing action. The feed inlet directs material into the chamber, with feed size determining whether the crusher is used for primary, secondary, or tertiary reduction. The rotor, the heart of the crusher, carries blow bars that deliver the initial impact. These blow bars—made from wear-resistant alloys such as manganese steel or chromium iron—are replaceable and designed for high durability. The impact plates are fixed surfaces that absorb the energy of the rock's impact, helping to fracture and refine the material. The gap between the rotor and impact plates determines final product size. The casing encloses the crushing chamber, providing structure and containment, while the discharge opening allows appropriately sized material to exit. Power is provided by a drive mechanism, typically an electric motor connected to the rotor via belts or direct coupling.

There are two main types of impact crushers, each suited to specific applications. Horizontal Shaft Impact (HSI) crushers feature a horizontally mounted rotor and are used for medium-hard materials such as limestone, dolomite, or bastnäsite-bearing carbonatites. They are commonly installed in secondary or tertiary circuits and produce a well-shaped, uniform product. Vertical Shaft Impact (VSI) crushers, on the other hand, have a vertically oriented rotor that propels material outward at very high speed against anvils or rock shelves. They are ideal for fine crushing and particle shaping, such as in sand production or fine mineral processing, and are particularly effective at producing cubical or uniformly sized particles.

The crushing process in an impact crusher follows a continuous sequence. Material enters through the feed inlet and falls onto the spinning rotor, where it is struck by the blow bars and accelerated toward the impact plates. This constitutes the primary impact. The fragmented material rebounds off the plates and either strikes the rotor again or collides with other particles, creating secondary impacts that further reduce particle size. Once particles become small enough to pass through the gap between the impact plates and rotor path, they are discharged from the crusher. This high-speed cycle of impact and rebound continues until all material reaches the target size.

In rare earth and critical mineral processing, impact crushers are particularly valuable due to their ability to produce fine, clean, and well-shaped particles. They are used in bastnäsite and monazite beneficiation circuits to achieve finer liberation before flotation or magnetic separation. In spodumene (lithium ore) processing, impact crushers prepare feed material for grinding and roasting. They are also effective in graphite and fluorite operations, where fine crushing is needed without generating excessive slimes. Their energy-efficient, high-velocity impact action enables precise size reduction while maintaining mineral integrity and reducing contamination.

The advantages of impact crushers include a high reduction ratio, excellent product shape, lower energy consumption compared to compression crushers, and ease of maintenance and liner replacement. They are particularly effective for soft to medium-hard materials and allow flexible adjustment of product size. However, like other crushers, they also have limitations. Impact crushers are not suitable for extremely hard or abrasive ores, as these cause rapid wear on blow bars and liners. They tend to generate more dust and fine material, are sensitive to uneven feed, and require regular monitoring and replacement of wear parts to maintain consistent performance.

After crushing, mills take over to achieve fine particle sizes (usually below 100 μm), liberating valuable minerals from gangue. Grinding is typically carried out in wet conditions using mechanical motion and grinding media.

- **Rod Mills:** Use long steel rods to produce coarse, uniform particles, often as an intermediate stage before fine grinding.

- **Ball Mills:** Employ spherical steel balls for finer grinding, widely used in rare earth and lithium processing circuits to achieve complete liberation of minerals such as bastnäsite, monazite, and spodumene.

- **Autogenous (AG) and Semi-Autogenous (SAG) Mills:** In AG mills, the ore itself acts as grinding media, while SAG mills use a mix of ore and steel balls. These are suitable for large-scale operations with variable feed sizes.

- **Vertical and Stirred Mills:** Designed for fine and ultrafine grinding (<20 μm), they use intense agitation of small grinding beads. These mills are critical in processing refractory or fine-grained ores where minerals are interlocked at microscopic scales, such as heavy rare earth-bearing xenotime or ion-adsorption clays.

Grinding parameters such as mill speed, media load, and slurry density are adjusted to balance energy use with mineral liberation efficiency.

A ball mill and a rod mill are both types of grinding equipment used in mineral processing to reduce ore size for further processing or liberation of valuable minerals. They are part of the comminution process, which follows crushing and aims to grind the material to fine particles suitable for beneficiation techniques like flotation, leaching, or magnetic separation.

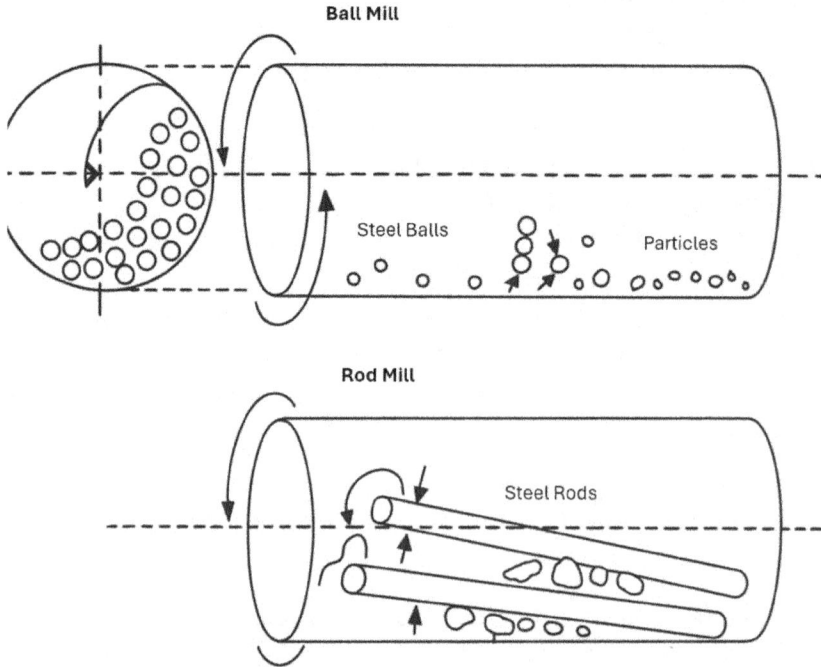

Figure 29: Ball and Rod Mill operations.

A ball mill uses steel balls as the grinding media inside a rotating cylindrical shell. The mill is partially filled with both the ore and the balls. As the cylinder rotates around its horizontal axis, the balls are lifted along the inner wall of the mill by centrifugal force. When they reach a certain height, gravity causes them to fall onto the material below. This repeated process of impact (when balls drop) and attrition (when they roll and slide) breaks down the ore into finer particles.

The efficiency of the ball mill depends on several factors:

- **Speed of rotation:** Too slow and the balls merely roll; too fast and they cling to the wall (centrifuging), reducing grinding action.

- **Ball size distribution:** Larger balls break coarse material; smaller balls create finer grinding.

- **Feed size and moisture:** Finer feed and controlled moisture improve efficiency.

In rare earth and critical mineral processing, ball mills are used to finely grind materials like bastnäsite, monazite, xenotime, or spodumene before flotation or leaching. They are essential in liberating fine-grained minerals from gangue material and achieving high recovery rates.

A rod mill operates on a similar principle but uses long steel rods instead of balls as the grinding medium. The rods are slightly shorter than the mill's length and lie parallel to the axis. When the mill rotates, the rods cascade in a linear motion, grinding the ore primarily through line contact rather than point contact. This produces a coarser product with fewer fines compared to a ball mill.

Because of this, rod mills are often used for primary grinding or when a controlled particle size distribution is required before feeding into a ball mill. They are ideal for materials like spodumene pegmatites, lithium ores, or soft carbonatite-hosted rare earth ores, where excessive fines can reduce downstream separation efficiency.

Ball mills are preferred for fine grinding and when the objective is to achieve full mineral liberation for flotation or leaching. Rod mills are used for coarser grinding, producing a product suitable for further reduction in a ball mill or classification circuit. Both types can operate in wet or dry modes, though wet grinding is more common in mineral processing because it reduces dust and enhances efficiency.

Autogenous (AG) and Semi-Autogenous (SAG) mills are large, rotating grinding machines used in mineral processing to reduce ore size through impact and attrition. Unlike conventional grinding mills that depend entirely on steel media such as balls or rods, AG and SAG mills use the ore itself—either partially or completely—as the grinding medium. These mills are used during the early stages of mineral processing and are capable of replacing multiple crushing and grinding steps. They are particularly effective for large-scale operations dealing with hard, competent ores, including those containing rare earth elements (REEs), lithium, or base metals.

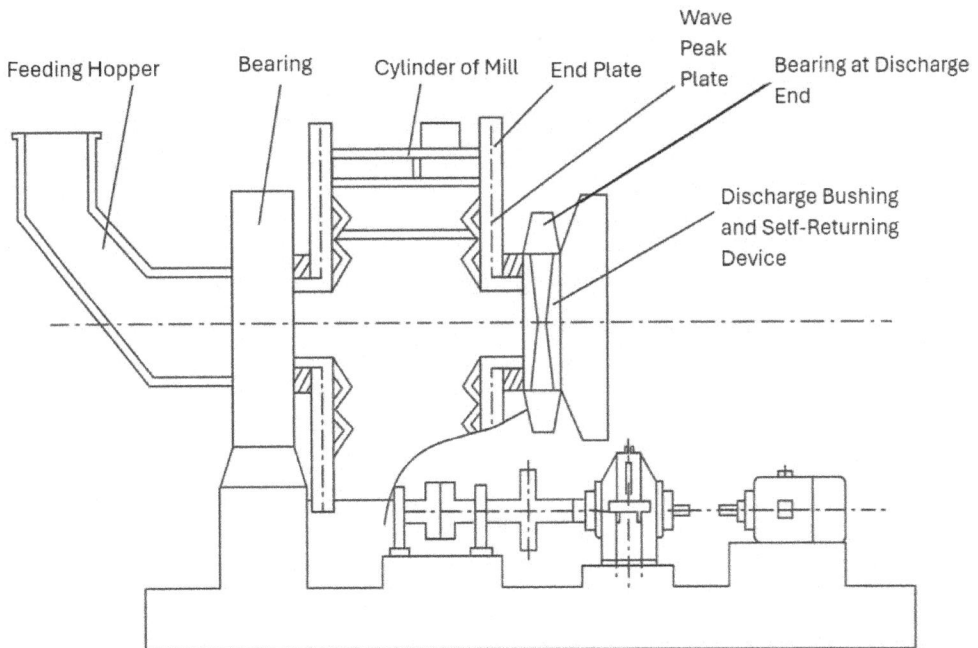

Figure 30: Autogenous Mill diagram.

169

Both AG and SAG mills operate as rotating cylindrical shells that tumble the ore and any grinding media inside. As the mill rotates, friction lifts the ore up along the inner wall until gravity causes it to fall. This tumbling action creates two key grinding forces: impact and attrition. Impact occurs when large rocks fall from near the top of the mill and strike other rocks or the mill bottom, fracturing the material. Attrition takes place when smaller particles are trapped between larger fragments or grinding media and are abraded into finer sizes. Ore is continuously fed into one end of the mill and moves toward the discharge end as it is ground to the required size. Depending on the design, material may exit through grates or overflow sections.

In an autogenous (AG) mill, the ore itself acts as the only grinding medium. Large, competent pieces of ore collide and grind against each other to achieve particle size reduction. The efficiency of an AG mill depends largely on the hardness, size distribution, and internal structure of the ore. Softer or friable ores may not generate enough coarse material to serve as grinding media, resulting in lower throughput. AG mills typically accept feed sizes of 150–300 mm and produce material around 2–3 mm in size, depending on downstream processing needs. They have the key advantage of lower operating costs, as they do not require steel grinding media. AG mills are best suited for coarse grinding applications involving hard, competent ores, such as certain bastnäsite or carbonatite-hosted rare earth deposits.

A semi-autogenous (SAG) mill works in a similar way but includes a small proportion (typically 5–15%) of steel balls or rods to supplement the grinding process. These additional media improve grinding efficiency, particularly when the ore is not hard enough to grind itself effectively. SAG mills handle larger feed sizes—up to about 350 mm—and produce finer outputs ranging from 0.1 to 1.0 mm. Their combination of ore-to-ore and ore-to-ball impact provides greater throughput, more stable performance, and finer control over product size. SAG mills are commonly used in processing circuits for lithium (spodumene), nickel, copper, and rare earth elements, where they prepare material for subsequent fine grinding or flotation.

The performance of AG and SAG mills depends on several operational parameters. Mill speed—typically operated at 70–80% of the critical speed—affects the balance between impact and cascading motion. Ore hardness and competency determine whether additional grinding media are required, while a consistent feed rate and size distribution ensure stable operation. Liner design protects the mill shell and influences the internal motion of the ore. AG and SAG mills also draw significant power, often making them the most energy-intensive equipment in a mineral processing plant.

In rare earth and critical mineral processing, AG and SAG mills play a crucial role ahead of fine grinding stages. For example, in spodumene (lithium ore) processing, SAG mills prepare the material for thermal conversion and leaching. In bastnäsite and monazite ores, AG/SAG mills facilitate mineral liberation before flotation or magnetic separation. Similarly, in nickel, cobalt, and graphite operations, these mills handle high-throughput requirements while achieving coarse liberation essential for subsequent beneficiation.

Rare Earth and Critical Mineral Operations and Processing

Vertical and stirred mills are fine and ultrafine grinding machines used in mineral processing when very small particle sizes are required—typically below 50 micrometres (μm) and, in some cases, down to just a few microns. These mills are especially important in rare earth and critical mineral operations, where fine liberation of valuable minerals such as monazite, bastnäsite, xenotime, spodumene, or graphite is essential for effective downstream processing such as flotation, leaching, or magnetic separation. Their efficiency and precision make them a vital component in achieving high recovery rates and purity in complex ore systems.

The fundamental working principle of vertical and stirred mills is based on intense agitation and attrition rather than impact. Unlike traditional tumbling mills such as ball, rod, or SAG mills, these mills use a vertically oriented chamber filled with fine grinding media—typically small steel, ceramic, or high-density alloy beads ranging from 1–10 mm in diameter. A central rotating agitator or impeller stirs the media at high speed, creating turbulence and shear forces within the slurry (a mixture of water and finely ground ore). As the impeller rotates, the grinding media move rapidly, colliding and rubbing against ore particles suspended in the slurry. The resulting attrition and abrasion gradually reduce particle size through thousands of micro-collisions. The vertical configuration helps gravity keep the media and slurry in constant motion, improving energy efficiency and achieving a uniform grind.

A vertical or stirred mill consists of several key components. The grinding chamber is a tall, cylindrical or conical vessel where grinding takes place; it is lined with wear-resistant materials to handle continuous contact with the slurry and media. The agitator shaft and impellers, mounted vertically, rotate to generate high-intensity mixing and grinding. The grinding media—small beads made from steel, ceramic, or zirconia—are critical to the process; their small size increases the contact surface area, enabling finer and faster grinding. The feed and discharge system allows slurry to enter at the bottom and exit at the top once the target particle size is reached. This design helps classify the particles naturally: finer material exits first, while coarser particles remain in the chamber longer. Finally, the motor and drive assembly provides the high torque required to move dense slurry and media efficiently, often through a gearbox that controls speed and torque.

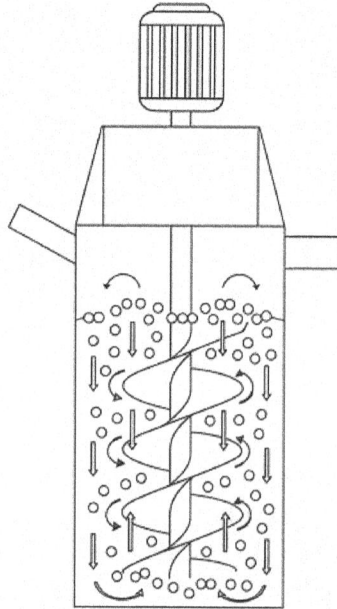

Figure 31: Stirred Mill diagram.

The grinding mechanism in these mills is primarily based on attrition or frictional grinding. The media move randomly within the stirred zone, creating localized zones of high stress. Ore particles trapped between moving beads experience compressive and shear forces that fracture them into finer particles. Because of this mechanism, stirred mills are far more energy-efficient for fine and ultrafine grinding compared to traditional ball mills. Typical operating parameters include media sizes of 1–10 mm, feed sizes below 1 mm, and product sizes between 5–50 μm. Energy consumption can be up to 50% lower than that of an equivalent ball mill producing the same fineness.

There are several types of vertical and stirred mills used in mineral processing. Tower mills are early designs that use simple screw-type agitators and rely on gravity-assisted grinding for coarse-to-medium fine applications. Vertimills (developed by Metso Outotec) are modern vertical mills with screw agitators and a stationary shell, widely used for efficient fine grinding in base metal and rare earth circuits. IsaMills, developed by Glencore Technology, feature high-intensity horizontal chambers that use fine ceramic media for ultrafine grinding, often to below 10 μm, and are commonly used in regrind or polishing stages. Stirred Media Detritors (SMDs) are vertical mills that operate at high tip speeds, capable of achieving extremely fine product sizes, and are particularly useful for liberating complex mineral intergrowths in fine regrind applications.

In rare earth and critical mineral processing, vertical and stirred mills play a vital role in achieving the fine or ultrafine liberation necessary for effective separation. In rare earth carbonatite ores (such as bastnäsite or monazite), fine grinding ensures that REE minerals are fully liberated from gangue minerals like calcite, barite, or quartz before flotation or leaching. In spodumene (lithium) ores, fine grinding enhances lithium recovery during flotation or chemical conversion. Graphite and cobalt ores benefit from improved purity and liberation through fine grinding before magnetic separation or refining, while nickel laterites and oxide ores require fine feed material for leaching, where increased surface area accelerates dissolution rates.

Vertical and stirred mills offer several advantages: they are highly energy-efficient for fine and ultrafine grinding, produce narrow particle size distributions, occupy less floor space than conventional ball mills, and operate with low noise and reduced environmental impact. They are also capable of continuous, high-throughput operation, making them ideal for large-scale processing plants. However, they also have some limitations. Their capital cost is higher than simpler tumbling mills, and maintenance can be more complex due to wear on impellers and liners. They are not suitable for very coarse feed, which must first be pre-ground, and their performance can be sensitive to changes in slurry density or viscosity.

Classifiers are used to separate ground material based on particle size, ensuring that only adequately fine material proceeds to the next processing stage. Oversized particles are returned to the mill for further grinding, creating a closed circuit that maintains consistent product quality.

- **Spiral Classifiers:** Use gravity and water flow to separate coarse particles from fine ones. Commonly found downstream of ball mills in traditional circuits.

- **Hydrocyclones:** Employ centrifugal force to classify particles rapidly in slurry form. They are compact, efficient, and widely used in modern rare earth and lithium plants.

- **Air Classifiers:** Used in dry grinding systems to separate fine powders based on aerodynamic properties, useful for high-purity rare earth oxide production.

Effective classification is essential for maintaining optimal particle size distribution—too coarse reduces liberation and recovery, while too fine leads to losses in flotation or leaching circuits.

Together, crushers, mills, and classifiers form an integrated system that transforms mined rock into a processable feedstock. In rare earth and critical mineral operations, the choice and configuration of this equipment determine not only liberation efficiency and recovery rates, but also energy consumption, reagent use, and overall processing cost. Properly designed comminution and classification circuits ensure that minerals like bastnäsite, monazite, xenotime, and spodumene reach the required fineness for efficient beneficiation and extraction, forming the essential first link in the critical mineral value chain.

A spiral classifier is a mechanical separation device used in mineral processing to sort or classify particles according to size and density after grinding. It serves an essential role in closed-circuit grinding systems by separating fine material, which is ready for the next processing stage, from coarse material that must be returned to the mill for further reduction.

Spiral classifiers are typically paired with ball mills, rod mills, or other grinding units in base metal, rare earth, and lithium ore beneficiation circuits.

The spiral classifier operates on the principle of settling velocity—the rate at which particles of different sizes and densities settle in a liquid, usually water. When the slurry, a mixture of finely ground ore and water, enters the inclined tank of the classifier, heavier and coarser particles settle more rapidly to the bottom, while finer and lighter particles remain suspended and move toward the overflow end. Inside the tank, a rotating spiral screw mechanism stirs the slurry and pushes the settled coarse material upward along the inclined trough to the discharge end. This movement lifts the coarse "sand" particles out of the slurry, returning them to the mill for additional grinding. Meanwhile, the fine "slime" fraction overflows at the top and proceeds to the next processing stage, such as flotation or leaching. This continuous cycle prevents overgrinding of fines and maintains efficient throughput in the grinding circuit.

The classifier comprises several key components. The feed inlet is where the slurry from the grinding mill enters the tank. The tank itself is an inclined, elongated trough partially filled with water, where separation takes place. The spiral screw or helix rotates slowly, lifting the coarse material to the discharge point, and is powered by a drive system consisting of a motor, gearbox, and coupling. The overflow weir allows fine particles and water to exit, while the underflow discharge at the lower end returns coarse material to the grinding circuit.

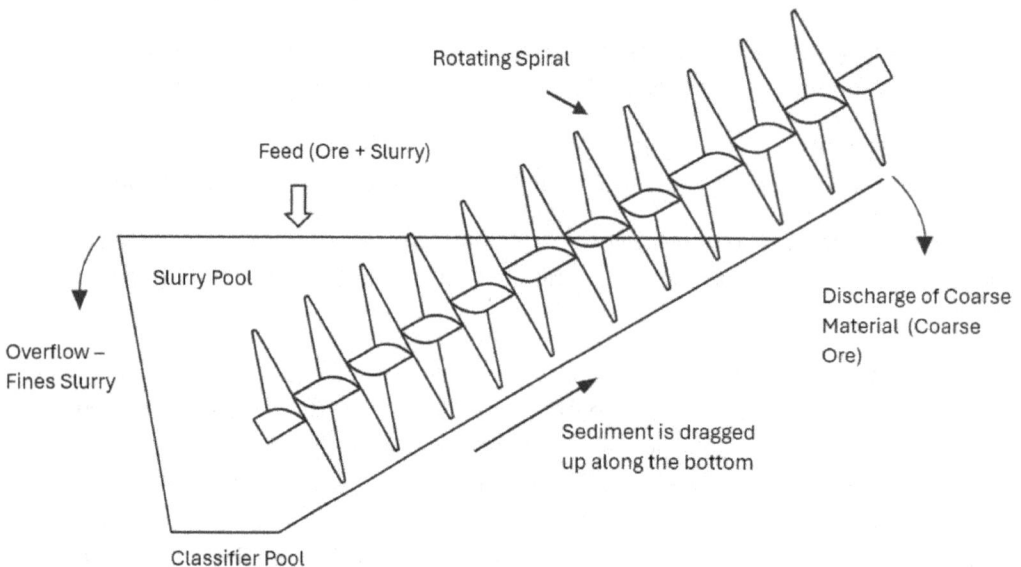

Figure 32: Spiral Classifier diagram.

Rare Earth and Critical Mineral Operations and Processing

There are two main types of spiral classifiers: the high weir type, which positions the spiral axis high and the overflow weir near the tank top, making it suitable for coarse particle classification (0.15–0.3 mm); and the submerged type, in which most of the spiral is immersed in slurry, improving fine particle recovery (0.007–0.15 mm) and classification stability.

The classification process involves several steps. First, the slurry is fed into the tank and begins to settle under gravity. Coarse particles sink faster while fine ones stay suspended. The spiral then conveys the coarse settled particles toward the discharge end, while the fine particles overflow at the top. The coarse "sand" is returned to the mill for further grinding, completing the closed circuit.

In rare earth and critical mineral processing, spiral classifiers are commonly used after grinding to prepare fine material for separation techniques such as flotation or magnetic separation. For example, in bastnäsite and monazite ores, classifiers ensure that rare earth mineral particles are properly sized for flotation. In spodumene (lithium ore) circuits, they regulate feed size entering leaching or flotation stages. In graphite or cobalt operations, classifiers promote fine liberation while preventing overgrinding that could damage flake integrity or particle structure.

Spiral classifiers offer several advantages: they are simple and robust in design, easy to operate, and capable of continuous and automatic classification. The spiral speed and overflow weir height can be adjusted to achieve precise control. They also consume less power than hydrocyclones of similar capacity and efficiently handle large slurry volumes. However, they have some limitations: they occupy significant floor space due to their length and inclination, provide lower precision than hydrocyclones, and are less efficient for particles finer than 50 micrometres. Additionally, regular maintenance is required for spiral blades and bearings due to wear.

A hydrocyclone is a highly efficient classification and separation device used in mineral processing to separate particles in a slurry according to their size, shape, and density. Unlike traditional mechanical classifiers, it has no moving parts, which makes it simple, compact, and low-maintenance. Hydrocyclones are widely applied in grinding circuits, dewatering systems, and desliming operations for materials such as base metals, rare earth elements (REEs), lithium, graphite, and other critical minerals.

The hydrocyclone operates on the principle of centrifugal separation. Slurry is fed tangentially under pressure into the cylindrical upper section of the cyclone, creating a high-speed, swirling flow inside the conical body. This motion generates centrifugal forces that are many times greater than gravity, acting on each particle within the slurry. The heavier and coarser particles, which experience stronger outward forces, are thrown toward the wall of the cyclone and spiral downward along the cone to exit through the underflow (apex) at the bottom. Meanwhile, finer and lighter particles remain closer to the centre of the vortex, where the pressure is lower, and are drawn upward through the vortex finder (overflow) at the top. This continuous motion results in efficient particle classification: coarse, dense solids are discharged as underflow, while fine, light solids and liquid are carried upward and exit as overflow.

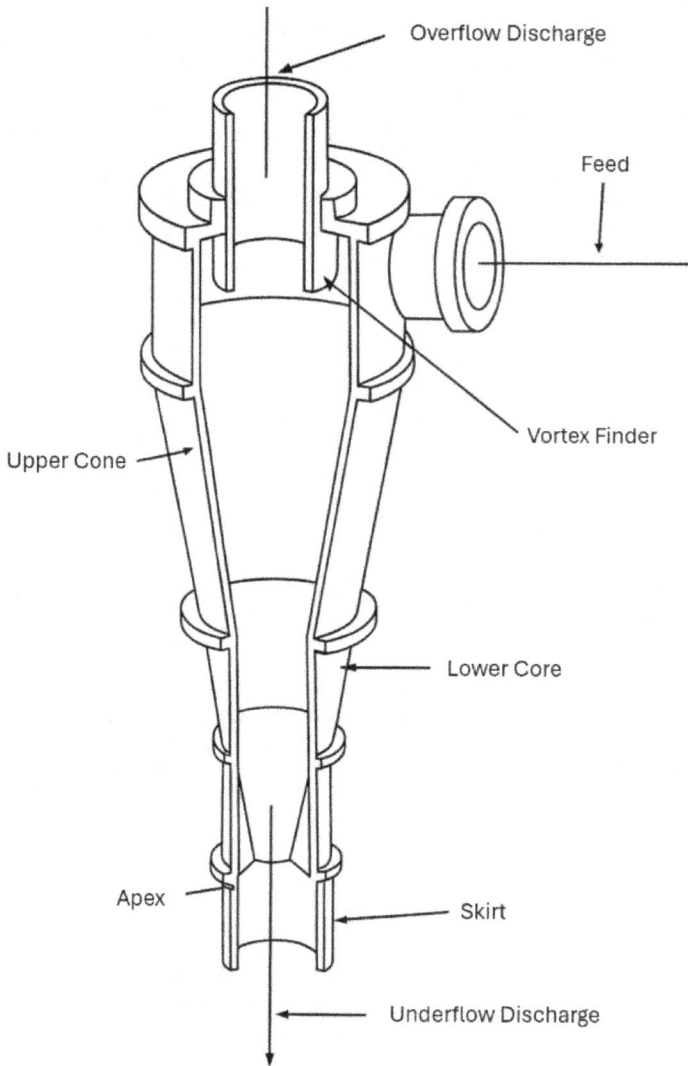

Figure 33: Hydrocyclone Working Principle.

A hydrocyclone consists of several key components. The feed inlet introduces the slurry tangentially, initiating the rotational flow that drives separation. The cylindrical section forms the upper body of the cyclone where the vortex develops and centrifugal forces begin to act. The conical section narrows downward, concentrating the heavier solids and directing them toward the underflow outlet. The vortex finder, or overflow pipe, is a central tube that allows fine particles and liquid to exit through the top. Finally, the spigot or apex is the narrow opening at the bottom through which the coarse, dense particles are discharged. The size of this opening determines the flow split between underflow and overflow streams.

Rare Earth and Critical Mineral Operations and Processing

The separation process inside a hydrocyclone occurs in several stages. As the slurry enters tangentially, it produces a high-speed rotational motion. Two distinct vortices form: the outer vortex, which spirals downward and carries coarse particles, and the inner vortex, which spirals upward through the centre and carries fine particles. The centrifugal and drag forces acting on the particles cause larger or denser ones to move outward and downward, while finer or lighter particles remain suspended and follow the upward flow. Eventually, the coarse solids exit through the underflow, and the finer material leaves through the overflow outlet.

Figure 34: Hydrocyclone configuration. Peter Craven, CC BY 2.0, via Wikimedia Commons.

Several factors influence hydrocyclone performance. Feed pressure is critical—higher pressure increases centrifugal force and improves separation but can accelerate wear. Cone angle and body size affect capacity and classification efficiency, with smaller cone angles favouring finer separations. The apex and vortex finder dimensions determine the ratio of coarse to fine discharge, while slurry density and viscosity influence settling behaviour and overall classification precision.

In rare earth and critical mineral processing, hydrocyclones serve multiple functions, including classification, desliming, and thickening. For instance, in bastnäsite and monazite processing, they remove ultra-fine slimes before flotation, improving recovery and selectivity. In spodumene (lithium ore) circuits, hydrocyclones separate fine gangue from coarse spodumene before flotation or roasting. In graphite beneficiation, they help eliminate clays and silicates while preserving valuable flake structure. Similarly, in nickel and cobalt laterite operations, they are used to manage particle size and pulp density in hydrometallurgical circuits.

Hydrocyclones offer several advantages. They have no moving parts, which simplifies operation and minimizes maintenance. They are compact, handle large slurry volumes, and provide efficient separation over a wide particle size range. Their adjustable design allows operational flexibility, and they are both cost-effective and energy-efficient compared to mechanical classifiers. However, they also have limitations: efficiency drops when handling very fine particles (below 10 micrometres), they are subject to wear at the apex and feed inlet due to high velocity, and they require stable feed conditions for optimal performance. Additionally, they provide less precise control over cut size than specialized precision separators.

An air classifier is a dry separation device used in mineral processing to classify or separate particles according to their size, shape, and density using a controlled stream of air rather than water. It operates based on aerodynamic principles, where particles are suspended and sorted in either upward or horizontal airflows depending on their terminal velocities. Air classifiers are particularly valuable in fine and ultrafine classification for dry processing circuits, especially in cases where water usage is restricted or undesirable. They are commonly employed in rare earth, lithium, graphite, and industrial mineral operations to achieve precise particle size control while maintaining product purity.

The working principle of an air classifier depends on the interaction of gravitational, centrifugal, and drag forces acting on particles suspended in an airflow. When a feed containing mixed fine and coarse particles is introduced into the classifier, a high-velocity stream of air carries the finer and lighter particles upward while the heavier and coarser ones fall due to gravity or are directed toward the coarse discharge by centrifugal action. The airflow velocity is adjusted carefully so that only particles below a specific "cut size" are carried to the fine product outlet. This ability to precisely control the separation point allows air classifiers to produce consistent particle size distributions without the need for water, making them vital in dry processing environments.

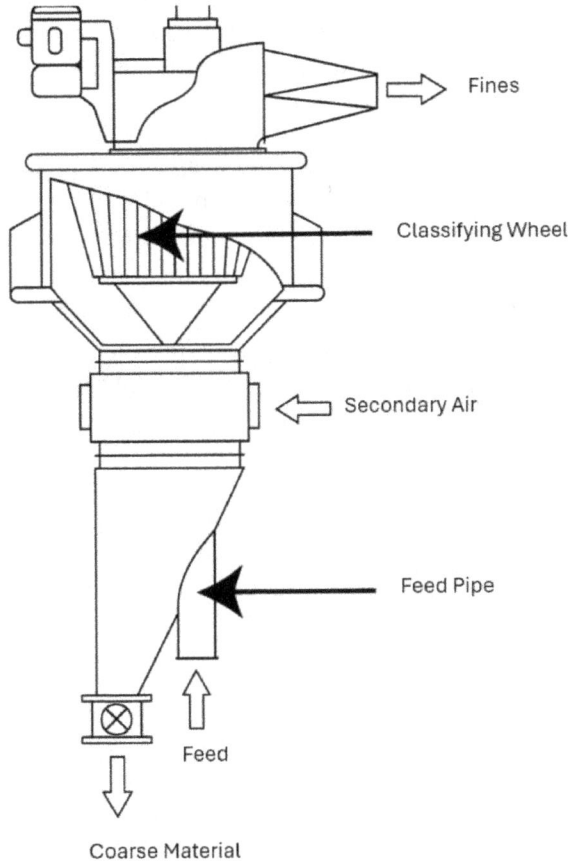

Figure 35: Air Classifier diagram.

Air classifiers are composed of several key components, each serving a specific function. The feed inlet is where the pre-dried material enters, often conveyed pneumatically or mechanically. Inside the classifying chamber, separation occurs as air velocity and direction determine the efficiency of classification. In dynamic classifiers, a rotor or cage wheel spins at high speed to generate a centrifugal field that forces coarse particles outward while allowing finer ones to pass through. The airflow system, consisting of fans or blowers, maintains the desired velocity and can be configured for either vertical or tangential flow. Separated fine particles are collected in a cyclone or dust collector, while the coarse and fine discharge outlets remove the respective fractions—coarse material is typically returned for further grinding, and fine product is collected for downstream processing.

The classification process begins with feeding the mixed material into the classifier, where it encounters an air stream of controlled velocity. As the air accelerates the particles, finer ones are easily suspended and entrained, while larger, denser particles resist motion. The combined action of gravity, drag, and centrifugal forces dictates each particle's path—fine particles are carried with the airflow toward the fine outlet, and coarse particles settle or are pushed outward and removed separately. The fine fraction, carried with the air, is typically recovered in a cyclone or dust collector, and the coarse fraction exits through a lower outlet for recirculation or waste disposal.

There are several types of air classifiers, each suited to specific applications. Static classifiers (also known as gravity or zigzag classifiers) rely solely on airflow and gravity and are best suited for coarse separations. Dynamic classifiers, sometimes referred to as centrifugal or rotor classifiers, use a rotating cage or wheel to generate additional centrifugal force, improving precision and control for ultrafine separations often required in rare earth or lithium processing. Multi-stage or high-efficiency classifiers combine multiple classification stages in one unit, achieving sharper size separations and higher recovery rates of fine particles.

Figure 36: The Whirlwind-type / Sturtevant Inc. (also marketed by N.N. Zoubov Engineers SMCE) centrifugal air classifier / air separator is a dry particle-size-classification machine designed to separate fine particles from coarse ones in a feed stream using controlled air flows and centrifugal action. Zoubov Engineers - Sturtevant Mill Company of Europe (SMCE), CC BY-SA 4.0, via Wikimedia Commons.

In rare earth and critical mineral processing, air classifiers are vital in dry beneficiation circuits where water scarcity or strict purity requirements make wet classification unsuitable. For instance, in monazite and bastnäsite ore processing, air classifiers separate fine rare earth minerals from gangue following dry grinding, ensuring the feed is well-prepared for magnetic or electrostatic separation. In graphite processing, air classification helps preserve flake integrity by avoiding the degradation caused by wet grinding. Similarly, in spodumene (lithium ore) processing, air classifiers are used to size roasted material prior to leaching or flotation. They are also used in refining high-purity silica, zircon, and titanium oxides, where precise particle size control is required for final product quality.

The advantages of air classifiers include their ability to operate entirely without water, making them ideal for dry or arid regions and for materials sensitive to moisture. They allow precise particle size control by adjusting air velocity and rotor speed and can operate continuously with compact, energy-efficient designs. They integrate easily into dry grinding or pneumatic conveying systems, producing minimal contamination and requiring low maintenance. However, air classifiers also have limitations. They are less efficient for very fine particles (below about 5 μm) or for sticky materials prone to agglomeration. Stable airflow and controlled particle loading are necessary to maintain performance, and they often require dust collection systems to manage airborne fines. Additionally, they are less accurate than wet classifiers for extremely fine or submicron separations.

Table 2 below summarises the preferred crushing, grinding, and classification pathways for a range of critical and energy-related minerals, showing how each ore type requires a distinct approach to achieve optimal liberation and downstream recovery. In essence, the table links mineralogy to process selection—it demonstrates that comminution and sizing strategies are not generic but are carefully tailored to the physical properties, texture, and target liberation size of each deposit type.

The table highlights clear process differentiation by mineral type. Hard, blocky ores such as bastnäsite, spodumene, and niobium-bearing carbonatites require robust primary crushers, such as jaw or gyratory units, followed by multi-stage reduction using cone or impact crushers to handle competent feed. In contrast, soft or friable materials like graphite and ion-adsorption clays demand minimal or gentle crushing to preserve valuable textures and avoid generating excessive fines.

Each mineral exhibits a distinct grind size target based on liberation requirements. Ball mills dominate fine-grinding duties for minerals such as rare-earth element (REE) phosphates and coltan. Rod mills or other gentle circuits are used where preserving flake or crystal morphology is important, as in graphite or spodumene processing. Semi-autogenous (SAG) and autogenous (AG) mills are selected where throughput and coarse liberation are priorities, such as in lithium or nickel–cobalt operations. Stirred or vertical mills, by contrast, are used for ultrafine grinding (typically below 50 μm) when fine liberation is essential, such as in xenotime and complex REE concentrates.

Classification acts as a critical control stage in these circuits. The use of hydrocyclones, spirals, and air classifiers maintains tight particle-size distribution (PSD) control, which is vital for efficient flotation, leaching, and magnetic or electrostatic separation. Fine desliming, particularly in REE circuits, enhances selectivity by removing slimes that would otherwise impair recovery.

The "Notes / Liberation Target" column in Table 2 directly connects comminution outcomes with downstream beneficiation strategies. Flotation is the principal method for REEs; roasting and leaching are typical for lithium; gravity and magnetic separation dominate for niobium–tantalum and titanium–zircon minerals; and chemical purification is critical for graphite. This relationship underscores that comminution is not an isolated step but a foundational component of overall process efficiency.

Finally, the table reveals extremes in processing requirements. At one end, niobium–iron carbonatites and nickel–cobalt laterites require large-scale, abrasive, high-energy circuits. At the other, ion-adsorption REE clays need almost no comminution, relying instead on desliming and direct leaching. Together, these examples demonstrate the full spectrum of mechanical energy inputs across mineral types.

Overall, the table conveys that comminution and classification design must be mineral-specific, liberation-driven, and recovery-oriented. It encapsulates the engineering logic behind critical-mineral processing—balancing energy efficiency, product control, and mineral integrity to ensure that each ore type achieves its optimum separation performance in downstream circuits.

Table 2: Comminution and Classification Matrix – Critical and Energy Minerals.

Mineral / Ore Type	Primary Crushing	Secondary / Tertiary Crushing	Milling / Grinding	Classification & Sizing	Notes / Liberation Target
Bastnäsite / Monazite (REE carbonatites)	Jaw or gyratory – handles hard, blocky ROM	Cone – 10–12 mm product	Ball mill (P80 75–150 μm)	Hydrocyclone deslime or air classify	Flotation; stirred regrind for complex gangue
Monazite (LREE phosphate)	Jaw	Cone	Ball mill	Cyclone deslime	Magnetic/e-static separation or direct cracking
Xenotime (HREE phosphate)	Jaw	Cone	Ball or stirred mill (<50 μm)	Fine hydrocyclone or air classifier	High-intensity magnetic/electrostatic separation
Spodumene ($LiAlSi_2O_6$)	Jaw or gyratory (hard pegmatite)	Cone or impact (post-roast trim)	SAG → Ball → Stirred mill	Hydrocyclone / spiral / air class	Roast α→β, then flotation or acid leach; tight PSD

Mineral / Ore Type	Primary Crushing	Secondary / Tertiary Crushing	Milling / Grinding	Classification & Sizing	Notes / Liberation Target
Columbite–Tantalite (Nb–Ta)	Jaw	Cone	Ball mill or stirred regrind	Spiral or hydrocyclone	Gravity + magnetic separation; avoid over-sliming
Graphite (flake)	Gentle jaw crush	— or light cone / HSI	Rod or ball mill (line-contact)	Hydrocyclone or air classifier	Preserve flake integrity; deslime to remove clays
Ilmenite / Rutile / Zircon (hard-rock HMS)	Jaw	Cone	Ball mill	Hydrocyclone / air classify	Magnetic + electrostatic separation
Niobium / Pyrochlore / Iron-rich complexes	Gyratory (high throughput)	Cone	SAG → Ball	Hydrocyclone	Abrasive, hard ores; feed to magnetic or flotation circuits
Nickel / Cobalt sulphide or laterite hubs	Gyratory	Cone	AG/SAG	Hydrocyclone	Coarse liberation before leach or float; PSD control critical
Ion-adsorption REE clays	Minimal crushing	—	—	Deslime only	Direct leach; comminution avoided to preserve recovery

Key:

- *PSD* = Particle Size Distribution

- *P80* = 80 % passing size

- Equipment selection reflects ore competency, tonnage, and liberation target.

- Fine control (<50 μm) often achieved using stirred or vertical mills for rare-earth phosphates and graphite purification.

Energy Efficiency and Particle Size Control

Rare Earth and Critical Mineral Operations and Processing

Energy efficiency and particle size control represent critical aspects of comminution and classification in mineral processing, significantly impacting downstream processes and overall economic viability. The interplay between these factors is particularly significant in complex mineral ores, where efficient processing methods can lead to substantial reductions in energy use and improvements in recovery efficiencies.

Energy Efficiency in Comminution

Comminution, encompassing both crushing and grinding processes, is responsible for a large portion of energy consumption in mineral processing—often estimated to account for 50% to 70% of total energy usage [173]. This high energy demand underscores the necessity for improvements in energy efficiency to minimize operational costs and reduce environmental impacts associated with mineral extraction and processing. Technologies designed to enhance efficiency include High-Pressure Grinding Rolls (HPGR) and stirred mills, which are recognized for their potential to achieve energy savings while also maintaining or enhancing the liberation of valuable minerals [174, 175].

HPGR, unlike traditional mills that rely on impact or abrasion, utilizes interparticle breakage, enabling effective size reduction with reduced energy consumption [176]. Additionally, operational strategies such as circuit optimization—including pre-classification and the recycling of oversize material—are vital for further reducing energy wastage by ensuring that only unliberated particles are subjected to additional grinding [177]. By focusing on these energy-efficient comminution techniques, plants can achieve cost reductions and lower their carbon footprints significantly.

Particle Size Control and Distribution

Particle size distribution (PSD) is essential for efficient mineral liberation and separation. An inappropriate PSD may hinder recovery by either locking valuable minerals within waste or generating excessive fines that complicate subsequent processing stages [178]. Effective classification systems, such as hydrocyclones and vibrating screens, play a crucial role in producing a targeted and uniform PSD, thereby facilitating optimal downstream processes such as flotation and leaching [179].

Different beneficiation methods impose varying PSD requirements: flotation typically operates best within an optimal range of 75 to 150 µm, while magnetic and electrostatic separations may require tighter control, often below 100 µm [173, 180]. In processes involving high-value minerals like spodumene or bastnäsite, precise control over PSD is vital not just for selectivity but also for effective reagent performance and product quality [181, 182]. Thus, employing robust classification techniques directly supports improved recovery and operational efficiency.

Balancing Energy and Size Control

A considerable challenge for mineral processing engineers is balancing energy input with particle size control. An excessively coarse grind may lead to poor mineral liberation, while a grind that is too fine can waste energy and complicate separation processes [183, 184]. Advanced technologies, including real-time particle size sensors and closed-loop control systems, provide innovative solutions for dynamically optimizing these variables to ensure that both energy efficiency and ideal PSD are maintained concurrently [185].

Ultimately, integrating strategies that foster energy efficiency while controlling particle size represents a holistic approach to mineral processing. This optimization is essential not only for immediate operational concerns but also from a sustainability perspective, as it promotes the reduction of waste and enhances the overall environmental impact of mineral extraction and processing.

Importance of Liberation Before Separation

Liberation is the foundation of all mineral processing operations. It refers to the degree to which valuable minerals have been freed from the surrounding gangue material after crushing and grinding. In simple terms, liberation determines whether separation can actually occur—because physical and chemical separation techniques, such as flotation, gravity concentration, magnetic, or electrostatic separation, rely on differences in particle properties that can only be expressed when minerals exist as discrete particles rather than locked composites.

When an ore is mined, valuable minerals are typically intergrown with unwanted gangue minerals in complex textures. The goal of comminution (crushing and grinding) is to reduce particle size until most of the valuable minerals are liberated from the host rock. If the particles remain locked, separation methods cannot effectively distinguish between valuable and waste components, leading to poor recovery and low-grade concentrate.

For example, in rare-earth element (REE) ores such as bastnäsite or monazite, liberation at a fine particle size (P80 around 75–150 μm) is necessary before flotation or magnetic separation can efficiently recover REE minerals. In spodumene (lithium) ores, liberation is essential before roasting and flotation or leaching; otherwise, unliberated grains will not convert properly to the reactive β-phase, reducing lithium yield. Similarly, in graphite processing, overgrinding damages flakes and decreases product value, illustrating that the target is not maximum breakage but *selective liberation*.

Liberation depends on the textural relationships and grain size of minerals within the ore. Coarse-grained ores may liberate at relatively large sizes, whereas fine-grained or interlocked ores require much finer grinding. However, as particle size decreases, the surface area and energy demand increase exponentially, and the risk of generating slimes (ultrafine particles that behave poorly in separation circuits) rises. Therefore, liberation must be optimised rather than maximised—achieving the smallest size necessary for effective separation, but no smaller.

The effectiveness of any downstream separation process is strongly influenced by the degree of liberation:

- Flotation relies on differences in surface chemistry; if valuable and gangue minerals remain locked, reagent selectivity is lost and recovery drops.

- Gravity separation depends on density contrast; composite particles with mixed density behave unpredictably.

- Magnetic and electrostatic separation require clean mineral surfaces for distinct response to magnetic fields or electrical charge.

- Leaching and hydrometallurgy depend on exposed surface area; locked minerals may remain unreacted, reducing extraction yield.

In every case, insufficient liberation directly translates to lower recovery, poorer concentrate quality, and higher losses to tailings.

Achieving the right level of liberation also has strong economic and environmental impacts. Overgrinding wastes energy and increases processing costs, while undergrinding reduces recovery and increases waste volumes. A well-liberated feed maximises separation efficiency, reduces reagent use, and improves overall plant performance. Moreover, by enhancing recovery, it reduces the quantity of tailings per unit of product, lowering environmental footprint and improving sustainability.

Liberation before separation is essential because it establishes the physical and chemical independence of mineral particles—the condition upon which all subsequent beneficiation depends. Without adequate liberation, even the most advanced separation technologies cannot produce high recoveries or grades. Therefore, the design and control of comminution circuits must always aim to deliver optimal liberation: sufficient to expose valuable minerals, minimal enough to conserve energy and preserve particle integrity, and precise enough to enable efficient downstream processing and cleaner production outcomes.

Key Terms and Concepts

Comminution: The mechanical process of reducing solid materials from one average particle size to a smaller size through crushing and grinding.

Crushing: The initial stage of comminution in which large ore fragments are reduced in size using mechanical force, typically with jaw or cone crushers.

Grinding: The secondary stage of comminution where smaller particles are further reduced in size using mills such as ball, rod, or SAG (semi-autogenous grinding) mills.

Particle Size Distribution (PSD): A measure of the range and frequency of particle sizes within a sample, used to assess grinding efficiency and product quality.

Liberation: The process of freeing valuable minerals from the surrounding gangue to make them accessible for separation and concentration.

Bond's Work Index: A standard measure of ore hardness that estimates the energy required for size reduction during grinding.

Specific Energy Consumption: The amount of energy used to grind a specific quantity of ore, expressed in kilowatt-hours per tonne (kWh/t).

Screening: The process of separating particles based on size using vibrating or static screens with various aperture sizes.

Classification: The process of separating fine and coarse particles using equipment such as hydrocyclones or classifiers, typically based on settling velocity.

Hydrocyclone: A cone-shaped device that separates particles in a fluid stream by centrifugal force according to their size and density.

Closed-Circuit Grinding: A grinding system in which oversized material is continuously returned to the mill for further reduction until the desired size is achieved.

Open-Circuit Grinding: A grinding system without recirculation of coarse material, typically used for coarse grinding applications.

Key Terms and Concepts

Size Reduction Ratio: The ratio of the feed particle size to the product particle size achieved in a crushing or grinding process.

Overgrinding: The excessive reduction of particle size, leading to energy inefficiency and potential losses during subsequent separation stages.

Undergrinding: Inadequate size reduction that results in incomplete liberation of valuable minerals, reducing recovery efficiency.

Milling Circuit: A system of interconnected equipment—such as crushers, mills, classifiers, and conveyors—used to control and optimise particle size.

Chapter 4 Review Questions

Fundamental Principles of Comminution

1. Short Answer: Define comminution and explain its primary purpose in mineral processing.

2. Multiple Choice:

   Which statement best describes the goal of comminution?

   a) To change mineral chemistry

   b) To reduce particle size for mineral liberation

   c) To increase ore density

   d) To mix reagents uniformly

3. True/False: Crushing and grinding together account for most of the energy consumed in a mineral processing plant.

4. Applied: Explain how the mechanisms of impact, compression, and attrition differ in the way they reduce particle size.

Major Types of Comminution Equipment

5. Matching:

   Match the equipment type to its primary use.

   Jaw crusher

   Cone crusher

   Ball mill

   Rod mill

   HPGR

   Options:

   a) Fine grinding to achieve liberation

   b) Coarse primary crushing of hard ore

   c) Interparticle compression grinding for energy efficiency

   d) Coarse, uniform grinding with minimal fines

   e) Secondary or tertiary crushing producing 10–20 mm product

6. Short Answer: Describe the operating principle of a jaw crusher.

Rare Earth and Critical Mineral Operations and Processing

7. Multiple Choice:

    \Which mill is best suited for ultrafine grinding (< 20 μm) in rare-earth processing?

    a) SAG mill

    b) Rod mill

    c) Vertical or stirred mill

    d) Cone crusher

8. Applied: For a spodumene ore requiring roasting and leaching, explain why both cone crushers and ball mills are typically used in sequence.

Classification Equipment

9. Short Answer: What is the main function of classification in a comminution circuit?

10. Multiple Choice: Which classifier operates using centrifugal force rather than gravity?

    a) Spiral classifier

    b) Screen

    c) Hydrocyclone

    d) Grizzly feeder

11. True/False: Air classifiers are used when water is limited or product purity requirements demand dry processing.

12. Applied: Explain how a hydrocyclone separates coarse and fine particles within a slurry.

Relationship between Energy Input, Particle Size, and Grind Efficiency

13. Short Answer: What does Bond's Work Index measure, and how is it used in process design?

14. Multiple Choice: Energy efficiency in comminution can be improved by:

    a) Increasing mill speed above critical speed

    b) Using pre-classification and recycling of oversize material

    c) Using smaller feed sizes regardless of ore hardness

    d) Increasing slurry density beyond optimal range

15. True/False: A finer grind always increases mineral recovery and should therefore be the target in all operations.

16. Applied: Explain the trade-off between energy consumption and particle size reduction when planning grind targets for flotation feed.

Importance of Liberation before Separation

17. Short Answer: Why is mineral liberation essential before separation?

18. Multiple Choice: Which of the following best describes the effect of under-grinding?

   a) Loss of valuable minerals to tailings due to incomplete liberation

   b) Increased reagent efficiency in flotation

   c) Higher concentrate grades

   d) Reduced energy consumption without affecting recovery

19. True/False: Overgrinding can damage graphite flakes and reduce product value.

20. Applied: Using examples, explain how the optimal grind size differs between bastnäsite flotation and spodumene leaching.

Particle Size distribution (PSD) Data

21. Short Answer: What information does the term "P80" represent in a grinding circuit?

22. Multiple Choice: In a flotation circuit, which PSD range is typically optimal?

   a) 10–30 µm

   b) 40–60 µm

   c) 75–150 µm

   d) > 300 µm

23. True/False: A narrow particle size distribution generally improves downstream separation performance.

24. Applied: How would you use hydrocyclone overflow and underflow PSD data to judge if a grinding circuit is over- or under-performing?

Ore Hardness and Texture

25. Short Answer: How does ore hardness influence the choice between AG/SAG and ball milling?

26. Multiple Choice: Which ore characteristic most affects liberation behaviour?

   a) Mineral grain size and texture

   b) Colour

   c) Deposit depth

   d) Moisture content only

27. True/False: Coarse-grained ores liberate at smaller grind sizes than fine-grained, interlocked ores.

28. Applied: Given two ores—one coarse-grained bastnäsite carbonatite and one fine-grained xenotime greisen—recommend appropriate grinding and classification strategies for each.

Safety, Maintenance, and Operational Best Practices

29. Short Answer: Name two safety precautions that should be followed before performing maintenance on a jaw crusher.

30. Multiple Choice: What is the purpose of the toggle plate in a jaw crusher?

    a) Increase crushing speed

    b) Act as a safety release under overload

    c) Control feed rate

    d) Reduce liner wear

31. True/False: All rotating mills must be locked out and tagged out before internal inspection.

32. Applied: Describe how regular liner inspections and lubrication contribute to safe and efficient comminution operations.

Chapter 5

Physical Separation Processes

In mineral processing, physical separation methods form the foundation of ore beneficiation, transforming complex natural mixtures into concentrated products ready for refining. These techniques exploit measurable physical property differences—such as density, magnetic susceptibility, and electrical conductivity—to distinguish valuable minerals from gangue. Before any chemical processing can occur, effective physical separation ensures that target minerals are liberated, concentrated, and recovered efficiently, reducing downstream energy consumption and reagent demand.

This chapter introduces the core physical separation processes used across rare earth element (REE), critical mineral, and industrial mineral operations. You will explore how gravity, magnetic, and electrostatic principles are applied in a range of modern equipment—from jigs, spirals, and shaking tables to high-intensity magnetic and electrostatic separators—and how process design integrates these stages for optimal recovery. The discussion also extends to flotation fundamentals, where surface chemistry is manipulated to achieve selective separation through collectors, frothers, and modifiers. By the end of this chapter, you will understand how these interconnected processes form the backbone of beneficiation flowsheets for REE, lithium, and other strategic resources, and how operational parameters, reagent selection, and particle characteristics influence recovery, selectivity, and product quality.

Learning Outcomes	
This chapter aims to give you the ability to: 1. Explain the fundamental principles of physical separation processes, including gravity, magnetic, and electrostatic methods used in the beneficiation of rare earth and critical mineral ores. 2. Identify and describe the main types of gravity separation equipment—such as jigs, spirals, shaking tables, and dense media separators—and evaluate their suitability for various ore types.	

Learning Outcomes	

3. Differentiate between low-intensity, high-intensity, and high-gradient magnetic separation techniques and apply knowledge of magnetic properties to appropriate mineral separation strategies.
4. Describe the principles and applications of electrostatic separation and assess its role in upgrading rare earth and heavy mineral concentrates.
5. Explain the fundamentals of froth flotation, including the chemistry of surface interactions, the role of collectors, frothers, modifiers, and depressants, and the formation and stabilisation of froth.
6. Analyse the factors influencing recovery and selectivity in physical separation processes, such as particle size, density, surface chemistry, and fluid dynamics.
7. Interpret basic flowsheets for processing rare earth and lithium ores and relate each stage of separation to its purpose and outcome within the beneficiation circuit.
8. Recognise how comminution, classification, and liberation influence the efficiency of physical separation methods.
9. Apply safety and environmental considerations in the operation and optimisation of physical separation equipment.

Gravity, Magnetic, and Electrostatic Separation Principles

Separation processes in mineral processing are designed to exploit physical property differences between valuable minerals and gangue. Once minerals are adequately liberated by comminution, separation methods such as gravity, magnetic, and electrostatic separation can be applied to concentrate the valuable components. Each of these methods relies on distinct physical principles and equipment configurations, and their selection depends on the nature of the ore, particle size, and required product purity.

Gravity Separation

Gravity separation operates on the difference in specific gravity (density) between valuable minerals and gangue. When particles are subjected to a gravitational or centrifugal field, the denser particles settle faster than lighter ones. The separation efficiency depends on particle size, shape, density contrast, and the fluid medium (air or water).

In a fluid medium, denser particles experience greater gravitational force and settle toward the bottom, while lighter particles remain suspended or are carried by the fluid flow. The separation process may occur under the influence of simple gravity (as in sluices and shaking tables) or enhanced gravity using centrifugal force (as in spiral concentrators and Knelson concentrators).

Typical equipment includes:

- **Jigs:** Pulsating water currents separate particles by density and size.

- **Shaking Tables:** Produce stratification based on density differences along a sloped, vibrating surface.

- **Spiral Concentrators:** Helical channels use flowing water and centrifugal action to separate heavy from light minerals.

- **Centrifugal Concentrators (e.g., Knelson, Falcon):** Generate high centrifugal fields to recover fine, dense particles such as gold or columbite-tantalite.

Used for heavy mineral sands (ilmenite, rutile, zircon), tin, tungsten, gold, niobium–tantalum, and occasionally for coarse rare earth or lithium minerals before flotation.

A jig is a gravity separation device that separates minerals based on their differences in specific gravity (density) within a pulsating fluid—usually water. Jigs are among the oldest and most widely used gravity concentrators and are valued for their simplicity, efficiency, and ability to process relatively coarse particle sizes compared with other gravity devices such as shaking tables or spirals.

The fundamental principle of jigging is the differential settling of particles in a pulsating water medium. When a mixture of particles of different densities and sizes is subjected to alternating upward and downward water currents, each particle experiences a balance of gravitational, buoyant, and drag forces. Heavier or denser particles tend to penetrate downward through the jig bed, while lighter or less dense particles are lifted upward and carried away by the fluid flow. This process creates a stratified bed in which the heaviest material accumulates at the bottom and the lightest material remains near the top.

A jig consists of several key components that work together to achieve this separation. The feed system introduces the ore slurry (a mixture of water and particles) onto the top of the jig screen or bed. The screen or sieve plate supports the jig bed and allows fine particles to pass through while retaining coarser material. The jig bed, also known as the hutch bed, contains coarse heavy material known as *ragging* (such as steel shot or gravel) that enhances separation and stabilises the bed. A pulsator or diaphragm generates alternating pulses of water that move upward and downward through the bed, either mechanically or pneumatically. Beneath the screen lies the hutch chamber, where heavy particles accumulate as concentrate before being discharged. The tailings outlet removes the lighter gangue material that remains on the upper layers of the bed.

Jig operation consists of three main stages: pulsation, suction, and stratification with discharge. During the upward stroke, water is forced upward through the jig bed, loosening and suspending the particles so they can rearrange according to density and size. As the water flow reverses during the downward stroke, heavier particles settle more quickly and move downward through the spaces in the bed, while lighter particles remain above. Repeated pulsation and suction cycles create a stable, density-stratified bed in which the heavier mineral-rich layer moves to the hutch compartment, and lighter gangue material is washed away through the tailings outlet.

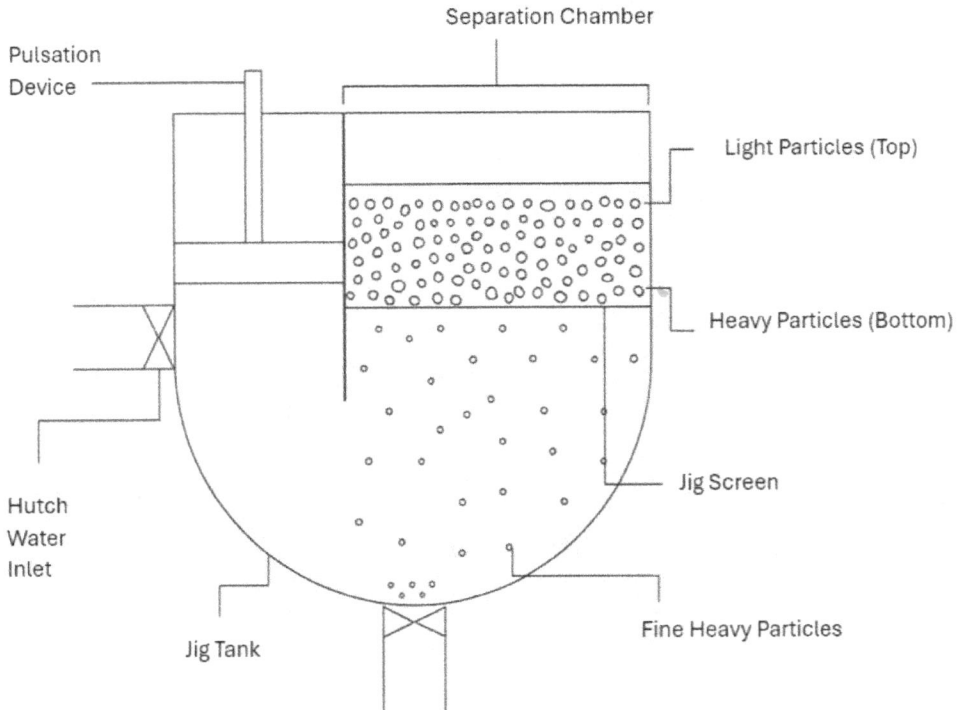

Figure 37: Jigging process.

There are several types of jigs used in mineral processing. Diaphragm jigs use a flexible diaphragm driven mechanically or hydraulically to create pulsations, while pneumatic jigs use compressed air to generate pulsations without mechanical movement. Moving-bed jigs, such as Harz or Denver types, have a movable screen or bed that enhances stratification. Large industrial systems such as Baum and Batac jigs are used in coal and heavy mineral processing and operate continuously with automatic discharge control.

Jigs are particularly effective for coarse and medium-sized particles, typically in the range of +2 mm to 0.1 mm, although modern fine jigs can handle smaller sizes. They are commonly applied in the recovery of gold and alluvial minerals, the beneficiation of tin, tungsten, and tantalum ores (such as cassiterite, wolframite, and coltan), the processing of iron and manganese ores, coal cleaning and upgrading, and the pre-concentration of heavy mineral sands including ilmenite, zircon, and rutile.

The main advantages of jigs include high throughput with low energy consumption, simple design and robust operation, and the ability to handle a wide feed size range while producing a relatively high-grade concentrate with minimal water use. However, jigs are less effective for very fine particles (below 0.1 mm) and can be sensitive to variations in feed consistency and particle size distribution. They also require steady pulsation control and regular cleaning of the bed to maintain performance.

Overall, jigs work by exploiting differences in particle density through pulsating fluid motion, which induces stratification within a bed of material. The heavier particles settle to the bottom and are collected as concentrate, while lighter particles are removed as tailings. Because of their simplicity, reliability, and efficiency, jigs remain an important and cost-effective tool in modern gravity separation circuits.

A shaking table is a gravity separation device used to separate minerals based on differences in density, particle size, and shape. It combines mechanical vibration with fluid flow to stratify particles into layers and move them across a gently inclined deck. Shaking tables are particularly effective for fine to moderately coarse particles, typically between 3 mm and 0.037 mm, and are widely applied in the concentration of precious metals, heavy minerals, and rare earth elements.

Figure 38: Geevor tin mine shaking tables. Rich257, CC BY-SA 3.0, via Wikimedia Commons.

The operating principle of a shaking table is based on differential motion and density stratification in a thin film of flowing water. When a mixture of minerals is fed onto the table deck, the combined action of vibration and water flow causes particles to separate according to their density and size. As the deck oscillates back and forth, heavier and denser particles settle through the flowing film of water and move toward the concentrate end, while lighter particles remain suspended longer and are carried toward the tailings side by the water flow. This process produces a fan-shaped distribution of materials along the deck, with heavy minerals forming a narrow band and lighter materials spreading more widely.

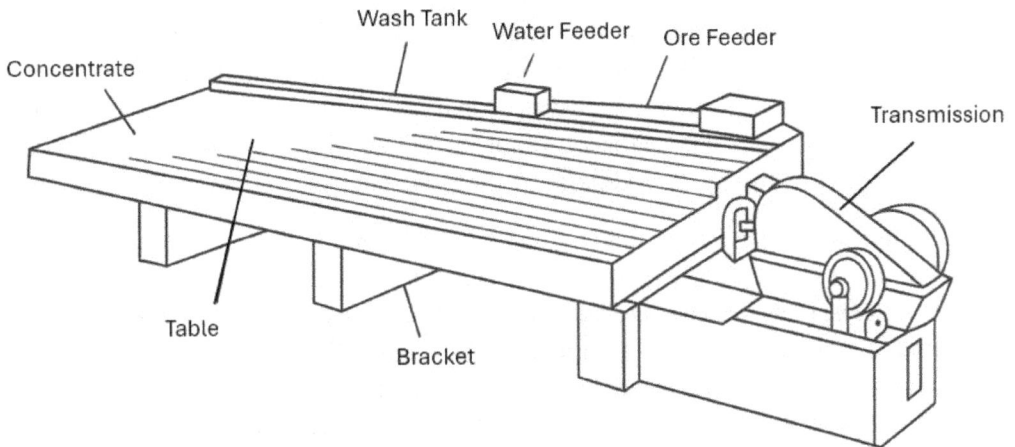

Figure 39: Shaking Table diagram.

A shaking table is made up of several key components that work together to achieve efficient separation. The deck or table surface is a slightly inclined rectangular platform, often made of fiberglass, wood, or metal, with riffles running perpendicular to the shaking motion. These riffles trap and convey the heavy particles. The feed box introduces the slurry (a mixture of water and ground ore) evenly across the deck, while the water supply system delivers a controlled flow of wash water to assist in the removal of lighter particles. The drive mechanism produces the table's reciprocating motion through a crankshaft or eccentric drive that imparts a slow forward stroke and a rapid return stroke. At the discharge end, launders collect the separated products—concentrate, middlings, and tailings.

During operation, the process occurs in several stages. First, the feed slurry is distributed evenly across the upper edge of the deck, where it mixes with wash water flowing down the slope. As the table vibrates, stratification takes place: denser particles settle into the riffle grooves, while lighter ones remain near the surface. This stratified layer is then transported by the asymmetric shaking motion, which propels heavier particles toward the concentrate end and lighter ones toward the tailings side. Intermediate-density particles form a middle zone known as the middlings. Finally, the separated materials are collected in distinct zones at the discharge end. Operators can fine-tune performance by adjusting the table tilt, stroke length, and water flow rate.

There are several types of shaking tables used in mineral processing. The Wilfley table is the most common design and is widely used for gold, tin, tungsten, and rare earth element recovery. Holman and Deister tables are industrial-scale versions that provide precise control and high recovery rates, while Gemini or laboratory tables are compact units suited for small-scale testing and pilot plants.

Shaking tables are used across a range of applications, including gold and precious metal recovery from both alluvial and hard-rock ores, concentration of tin, tungsten, and tantalum minerals such as cassiterite, wolframite, and coltan, and recovery of heavy mineral sands including ilmenite, zircon, and rutile. They are also employed in the separation of rare earth minerals like monazite and xenotime, as well as in coal cleaning and laboratory-scale mineral testing. Because of their high precision, shaking tables are particularly effective as final cleaning devices after other gravity separation methods such as jigs or spirals.

The advantages of shaking tables include their high separation accuracy, excellent recovery of fine heavy minerals, and simple operation with low power requirements. The visual nature of the process allows easy monitoring and adjustment during operation, and the table produces well-defined concentrate, middling, and tailings streams. However, shaking tables have limitations: they have limited throughput and are best suited for small-scale or finishing operations, are less effective for extremely fine particles (below 0.037 mm), and are sensitive to variations in feed density, particle size, and water flow. Consistent performance requires careful setup and maintenance.

A spiral concentrator is a gravity separation device that uses the combined forces of gravity, centrifugal action, and fluid dynamics to separate mineral particles according to their density and size. It provides a simple, efficient, and low-cost method for concentrating heavy minerals and is particularly effective for fine to medium-sized particles, typically between 2 mm and 0.037 mm. Spiral concentrators are widely used in the processing of heavy mineral sands, chromite, iron ore, gold, and rare earth element ores, where they deliver consistent and energy-efficient separation performance.

The operation of a spiral concentrator is based on the differential settling of particles within a helically shaped trough through which water and slurry flow under gravity. As the slurry flows down the spiral, the combined effects of gravity and centrifugal force cause particles to separate according to their density, size, and shape. Heavier and coarser particles move toward the inner part of the spiral trough, where the flow velocity is slower, while lighter and finer particles are carried outward by the faster-moving water. This continuous movement and layering create a natural stratification, forming distinct bands of material along the spiral's cross-section. The concentrate, consisting of the heaviest particles, is collected near the inner edge, while the middlings occupy the middle zone and the tailings, containing the lightest particles, flow along the outer edge.

Rare Earth and Critical Mineral Operations and Processing

A spiral concentrator consists of several key components that work together to achieve efficient separation. The spiral trough is a helical channel or launder made of fiberglass or polyurethane that allows slurry to flow downward under gravity. The feed distributor ensures even distribution of slurry into the top of the spiral for uniform flow and consistent separation. A wash water system introduces a controlled flow of water near the top or along the spiral to aid in the removal of lighter particles. Splitter assemblies at the base of the spiral act as adjustable dividers to separate the different product streams—concentrate, middlings, and tailings. Finally, collection launders gather and direct the separated products to storage or further processing.

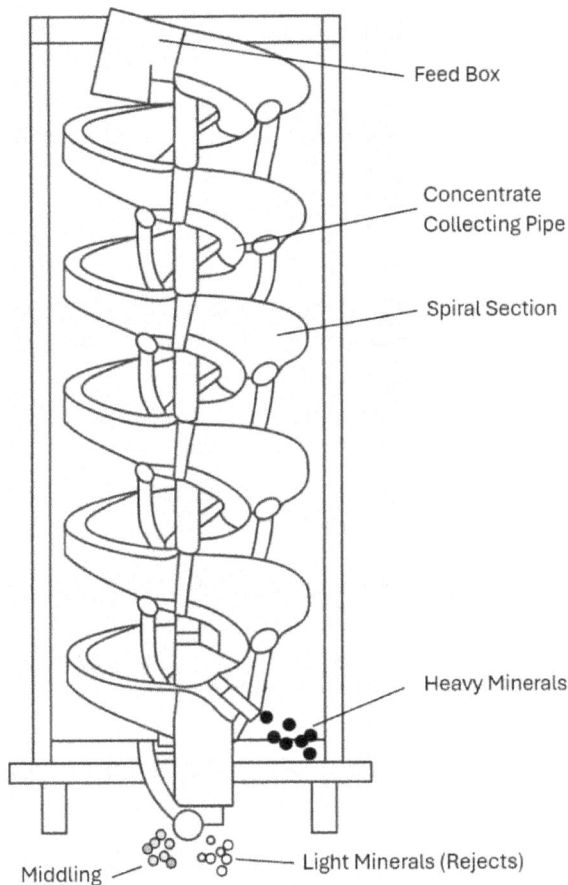

Figure 40: Spiral Concentrator diagram.

The operation of a spiral concentrator can be understood in four main stages. First, the feed introduction stage involves evenly distributing the slurry into the spiral through a feed distributor. During flow and stratification, as the slurry moves downward in a thin film, gravity and centrifugal forces cause denser particles to migrate inward toward the spiral's core while lighter particles move outward. In the separation zone, continuous flow results in layering of the different density fractions along the spiral's width. Finally, in the product recovery stage, adjustable splitters at the discharge end divide the flow into multiple product streams, typically heavy mineral concentrate, middlings, and tailings.

There are several types of spiral concentrators designed for different stages of processing. Rougher spirals are used in the initial stage to recover the bulk of heavy minerals, while cleaner spirals refine the concentrate from rougher stages to improve purity. Recleaner or scavenger spirals recover additional valuable minerals from tailings to maximise overall yield. Multistart spirals, which incorporate multiple spiral starts on a single frame, are employed to increase plant throughput without requiring additional floor space.

Spiral concentrators have a wide range of applications across the mineral processing industry. They are extensively used in the beneficiation of heavy mineral sands such as ilmenite, rutile, zircon, and monazite, as well as in the concentration of chromite and iron ores. They also play an important role in gold and tin recovery from alluvial and placer deposits, coal cleaning and desliming, and the concentration of rare earth elements from monazite- or xenotime-bearing sands.

The main advantages of spiral concentrators include low operating costs, minimal power requirements, and high reliability due to the absence of moving parts. They can handle large volumes of material continuously, are effective for fine to moderately sized particles, and produce distinct product streams that allow for flexible circuit design. However, there are also some limitations. Spirals are less effective for very fine particles (below 0.037 mm) or very coarse material, and they require a consistent feed rate and pulp density for stable operation. Efficiency can be reduced by excessive slimes or variations in mineral density, and precise adjustments to water flow and splitter settings are often necessary for optimal results.

A centrifugal concentrator is a gravity separation device that uses the combined effects of centrifugal force, fluid flow, and particle density differences to separate valuable heavy minerals from lighter gangue materials. These units enhance the efficiency of traditional gravity separation by replacing the weak gravitational field (1 g) with an artificially generated high-gravity field, typically ranging from 50 g to 300 g. The result is faster and more precise recovery of fine and ultrafine heavy minerals such as gold, platinum, tin, tungsten, tantalum, and rare earth elements.

The operation of a centrifugal concentrator is based on enhanced gravitational settling. When a slurry containing both heavy and light minerals is introduced into a rapidly rotating bowl, the rotation generates a strong centrifugal field. This force pushes all particles outward against the bowl wall; however, heavier particles experience greater force and migrate more rapidly toward the wall, while lighter particles remain suspended in the slurry and are carried away with the overflow. At the same time, a controlled stream of fluidisation water is introduced through small holes in the bowl wall. This water counteracts the outward drag on lighter particles, keeps the bed of material in gentle motion, prevents compaction, and allows continuous stratification. The outcome is the formation of a thin, dense layer of heavy mineral concentrate near the bowl wall, while lighter gangue particles are washed away through the top or sides as tailings.

Rare Earth and Critical Mineral Operations and Processing

A centrifugal concentrator includes several key components that work together to achieve efficient separation. The rotating bowl, also known as the concentrating cone or ring, is the primary separation zone where heavy minerals accumulate along the inner surface. The feed system delivers the slurry into the centre or base of the bowl for even distribution, while the fluidisation water system injects water through fine openings to control bed density and improve stratification. A drive mechanism powers the high-speed rotation necessary to create strong centrifugal forces. Finally, the tailings and concentrate discharge systems manage the removal of lighter gangue material and the collection of the heavy concentrate, either continuously or in cycles depending on the model.

The separation process occurs in several stages. First, during feed introduction, slurry enters the rotating bowl and is accelerated outward by centrifugal force. In the stratification and concentration stage, heavy particles move outward and settle along the bowl wall, while lighter particles remain suspended and are carried upward with the water flow. The countercurrent fluidisation water maintains a stable separation layer by preventing the buildup of material. In the product recovery stage, heavy minerals accumulate as a thin, high-density layer along the bowl's inner surface. Depending on the design, the concentrate is either continuously removed or periodically flushed out after each collection cycle. Finally, tailings discharge occurs as lighter particles are carried away by the wash water through designated outlets or over the bowl lip.

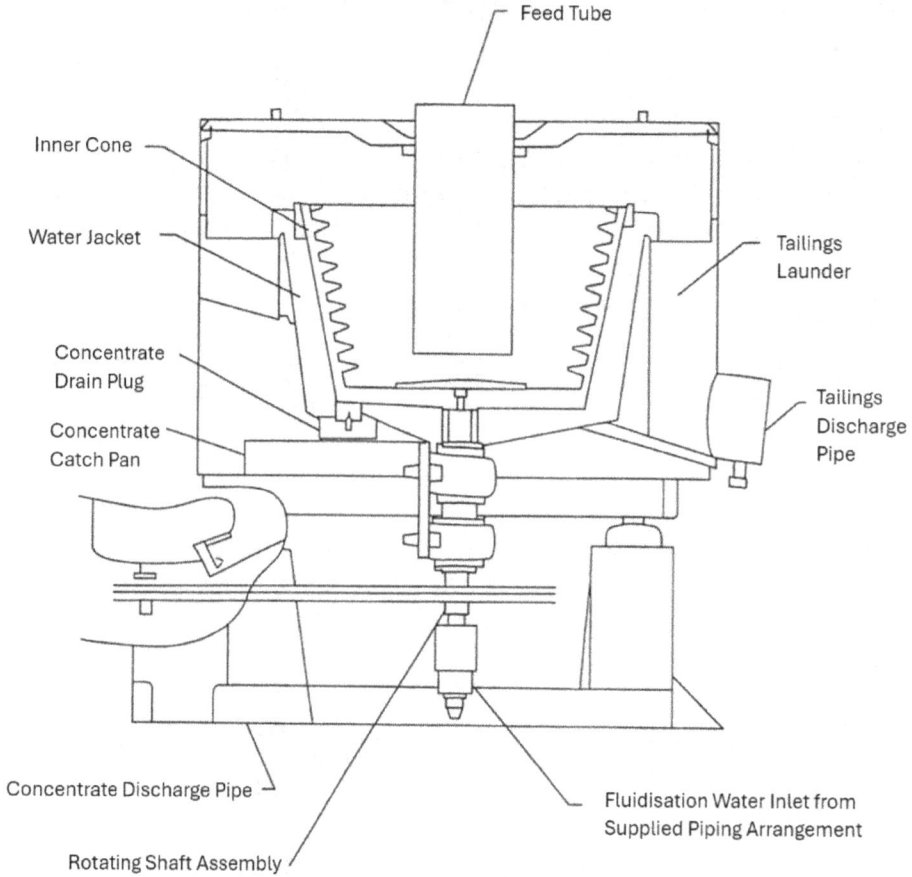

Figure 41: Knelson Concentrator diagram.

There are several types of centrifugal concentrators used in mineral processing. Batch concentrators, such as the Knelson and Falcon SB models, operate in timed cycles where concentrate is collected and then flushed out during a brief pause. These are particularly effective for gold recovery. Continuous concentrators, such as the Falcon C or Multi-Gravity Separator, enable simultaneous collection of concentrate and discharge of tailings, making them suitable for high-capacity applications. Enhanced gravity separators combine centrifugal force with shaking or oscillation to improve recovery of ultrafine particles in complex ores.

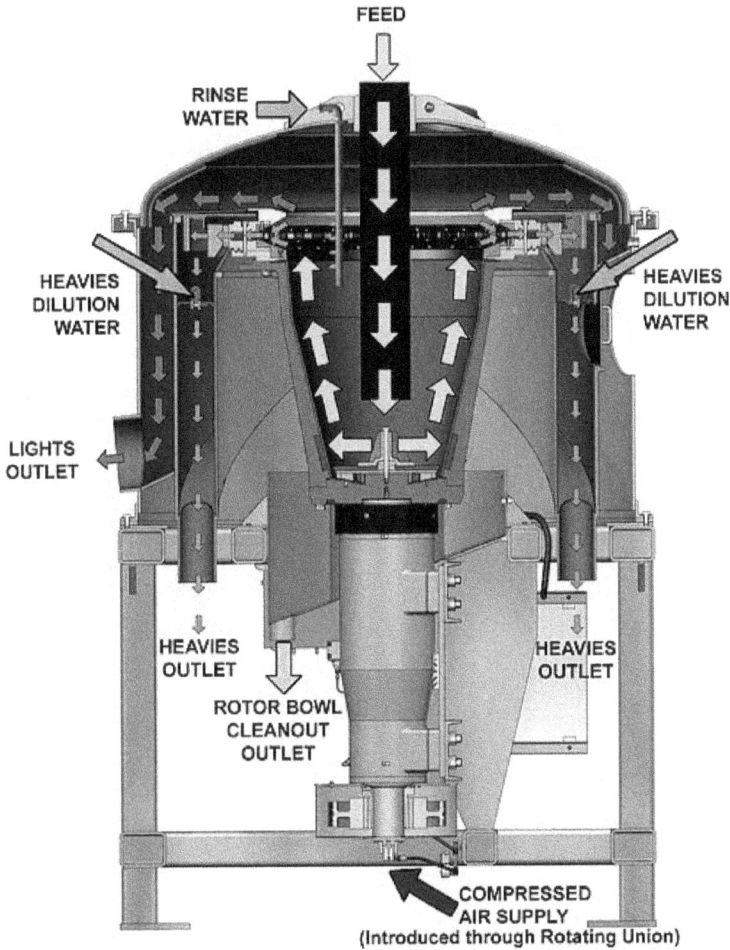

Figure 42: A cross section view of the Falcon Continuous (C) Concentrator. Sepro Mineral Systems, CC BY-SA 3.0, via Wikimedia Commons.

Centrifugal concentrators are widely used for recovering fine and ultrafine minerals that are difficult to capture with conventional gravity methods. They are essential in the recovery of gold and other precious metals from alluvial, placer, or milled ores, and in the processing of tin, tungsten, and tantalum ores. They are also applied in the concentration of rare earth elements and heavy minerals, the pre-concentration of sulphide minerals before flotation, and, in some cases, fine coal cleaning. In modern circuits, they are often used as final concentrators after jigs or spirals, or as scavenger units to recover fine particles from flotation tailings.

Figure 43: A Falcon C2000 Continuous Concentrator. Sepro Mineral Systems, CC BY-SA 3.0, via Wikimedia Commons.

The main advantages of centrifugal concentrators include extremely high separation efficiency for fine and ultrafine particles, compact design with high throughput, and low maintenance requirements. They are easy to operate, require no chemical reagents, and have a low environmental impact. Their operational parameters, such as bowl speed, fluidisation water flow, and cycle time, can be adjusted to suit specific ores and feed conditions. However, they also have limitations. Feed material must be relatively free of coarse particles, as large grains can disrupt separation. Excessive slimes or fine clays may block fluidisation holes and reduce efficiency. Periodic cleaning and maintenance are necessary to remove buildup inside the bowl, and batch systems require downtime for concentrate flushing. Additionally, power consumption is generally higher than that of passive gravity devices like spirals or shaking tables.

Rare Earth and Critical Mineral Operations and Processing

Centrifugal concentrators operate by generating high-speed rotational motion that creates powerful centrifugal forces, accelerating the separation of minerals based on density differences. Heavy particles are driven outward and retained in the bowl, while lighter gangue is washed away by fluidisation water. The process yields a clean, high-grade concentrate with excellent recovery of fine and ultrafine particles. Due to their efficiency, small footprint, and adaptability, centrifugal concentrators have become a vital component of modern mineral processing—especially in gold, tin, tungsten, and rare earth element recovery, where fine-particle capture is critical.

Magnetic Separation

Magnetic separation relies on differences in magnetic susceptibility—the degree to which materials are attracted to a magnetic field. When a mixture of particles passes through a magnetic field, magnetic minerals are attracted and retained, while non-magnetic minerals pass through unaffected.

As the feed passes over or through a magnetic separator, particles experience magnetic force proportional to their susceptibility and the field gradient. High-susceptibility minerals (e.g., magnetite) are easily separated in low-intensity fields, while weakly magnetic minerals (e.g., monazite, hematite) require high-intensity or high-gradient fields.

Typical equipment includes:

- **Low-Intensity Magnetic Separators (LIMS):** Drum or roll separators for strongly magnetic materials like magnetite and ilmenite.

- **High-Intensity Magnetic Separators (HIMS):** Induced-roll or rare-earth drum types for weakly magnetic minerals such as hematite or garnet.

- **High-Gradient Magnetic Separators (HGMS):** Matrix-type separators with very fine wire meshes that produce strong field gradients, suitable for ultrafine particles and rare earth minerals (e.g., monazite, xenotime).

Used for separating ferromagnetic and paramagnetic minerals in REE, iron, and heavy-mineral sands operations—particularly for cleaning bastnäsite, monazite, and ilmenite concentrates prior to electrostatic or flotation processing.

A Low-Intensity Magnetic Separator (LIMS) is a magnetic separation device designed to recover strongly magnetic minerals such as *magnetite, pyrrhotite,* and *ilmenite* from non-magnetic or weakly magnetic gangue materials. These separators operate at relatively low magnetic field strengths—typically below 0.2 Tesla (2,000 Gauss)—which are sufficient to attract ferromagnetic particles while allowing non-magnetic minerals to pass unaffected. LIMS units are commonly used in iron ore beneficiation, taconite processing, and other operations where efficient recovery of coarse, strongly magnetic minerals is required.

The principle of operation is based on differences in magnetic susceptibility between minerals. When a mineral mixture passes through a magnetic field, magnetic particles are drawn toward the magnetic source, while non-magnetic particles continue their normal trajectory. In LIMS, this field is generated by permanent magnets or electromagnets housed inside a rotating drum or beneath a conveyor. As the ore—either in slurry or dry form—passes through the magnetic zone, the magnetic particles are attracted to and held against the drum's surface. They are carried out of the non-magnetic stream until they move beyond the influence of the magnetic field, where they are released as a magnetic concentrate. The remaining non-magnetic materials are collected separately as tailings.

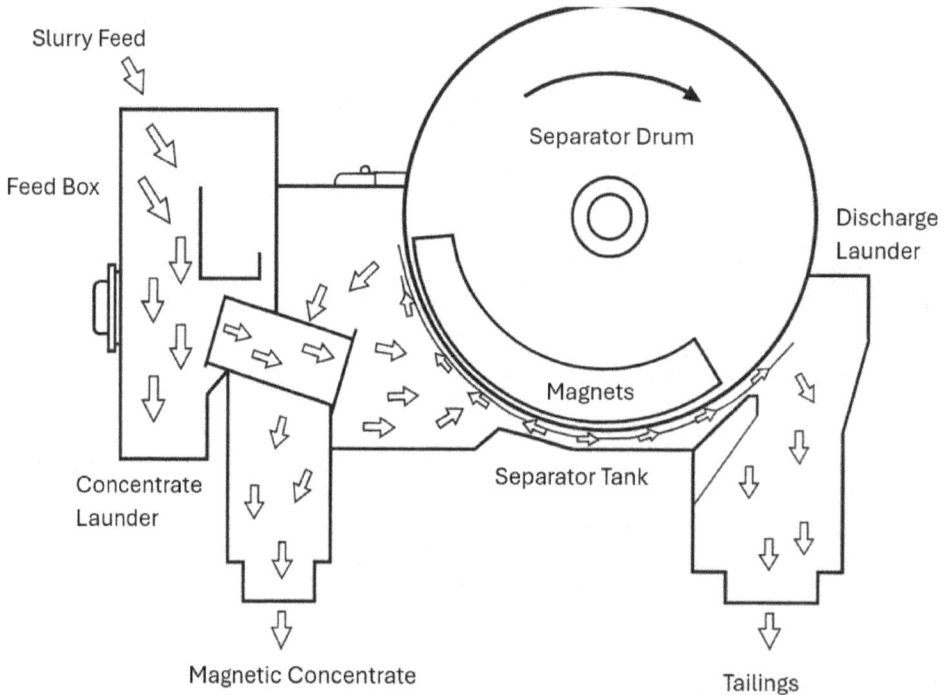

Figure 44: Magnetic Drum Separator.

A typical LIMS consists of several key components. The magnetic drum or roll is a rotating cylindrical shell that contains internal magnets generating the magnetic field. Inside the drum, the magnet assembly—made of permanent magnets or electromagnets—produces a uniform magnetic field along the drum's surface. The feed system introduces ore onto the drum in a thin, even layer, ensuring efficient separation. The tank or housing encloses the drum and regulates slurry flow to maintain consistent contact with the magnetic field. Finally, discharge chutes collect and separate the magnetic concentrate from non-magnetic tailings.

The separation process follows several stages. During feed introduction, either wet or dry feed is delivered to the drum surface. In wet systems, the drum is partially submerged in a slurry tank, where separation takes place under water. In the magnetic capture stage, the rotating drum attracts magnetic particles to its surface while non-magnetic particles remain in suspension. The transport and separation stage follows, where captured magnetic particles are carried upward on the drum's surface until they exit the magnetic field. They are then released by gravity or removed by a scraper and collected as concentrate. Meanwhile, non-magnetic particles flow away from the drum and are discharged as tailings through a separate outlet.

There are several LIMS configurations suited to different applications. Wet drum separators are used for slurries of fine particles, typically below 3 mm. The drum's partial immersion in water helps reduce entrainment and improves separation efficiency. Dry drum separators handle coarser particles, generally above 3 mm, or materials that cannot be processed in water. These rely on belt or vibratory feeders to control the flow across the drum. The direction of slurry flow relative to drum rotation—counter-current, concurrent, or counter-rotation—also affects performance, allowing operators to optimise recovery or concentrate grade for specific ore characteristics.

LIMS units are used extensively across various applications. In iron ore beneficiation, they recover magnetite from taconite and banded iron formations. In heavy mineral sands, they separate magnetic ilmenite from non-magnetic rutile and zircon. In coal preparation, they remove magnetic pyrite and iron-bearing impurities. They are also used in industrial mineral processing to clean and upgrade materials like silica sand and feldspar contaminated with magnetic particles.

The main advantages of LIMS include a simple and durable design, low operating and maintenance costs, and continuous, automatic operation suited to large-scale plants. They offer high recovery rates for strongly magnetic minerals, can process both wet and dry feeds, and operate efficiently due to their low magnetic field energy requirements. However, there are limitations. LIMS are ineffective for weakly magnetic or paramagnetic minerals such as hematite, garnet, or monazite. They are best suited for coarse to medium-sized particles, typically above 20 microns. Performance can decline when the feed contains excessive fine slimes or variable particle sizes, and optimal results depend on maintaining a steady feed rate and slurry density.

A High-Intensity Magnetic Separator (HIMS) is a specialized magnetic separation device designed to recover weakly magnetic minerals that cannot be efficiently separated using low-intensity methods. While Low-Intensity Magnetic Separators (LIMS) are effective for strongly magnetic minerals such as magnetite or pyrrhotite, HIMS units are used for minerals with lower magnetic susceptibility, including *hematite, ilmenite, chromite, garnet, monazite, xenotime,* and *wolframite*. These machines operate with magnetic field strengths typically ranging from 0.8 to 2.0 Tesla (8,000–20,000 Gauss) and play a vital role in the beneficiation of paramagnetic and fine-grained ores commonly found in rare earth, tungsten, and titanium processing.

The operation of a HIMS is based on the differential attraction between magnetic and non-magnetic particles within a high-intensity magnetic field. When a mixture of minerals passes through the field, weakly magnetic particles experience sufficient magnetic force to be deflected or captured, while non-magnetic particles continue along their natural path and are discharged as tailings. The magnetic field is generated using electromagnets or rare-earth permanent magnets, which are often arranged around a matrix of ferromagnetic materials such as steel wool, grooved plates, or rods. These matrix elements create localized high-gradient zones that amplify the magnetic field and enhance particle capture. As the ore—either as slurry or dry feed—flows through the magnetic zone, weakly magnetic particles are drawn to these concentrated field areas and retained, while non-magnetic materials pass through or are washed away.

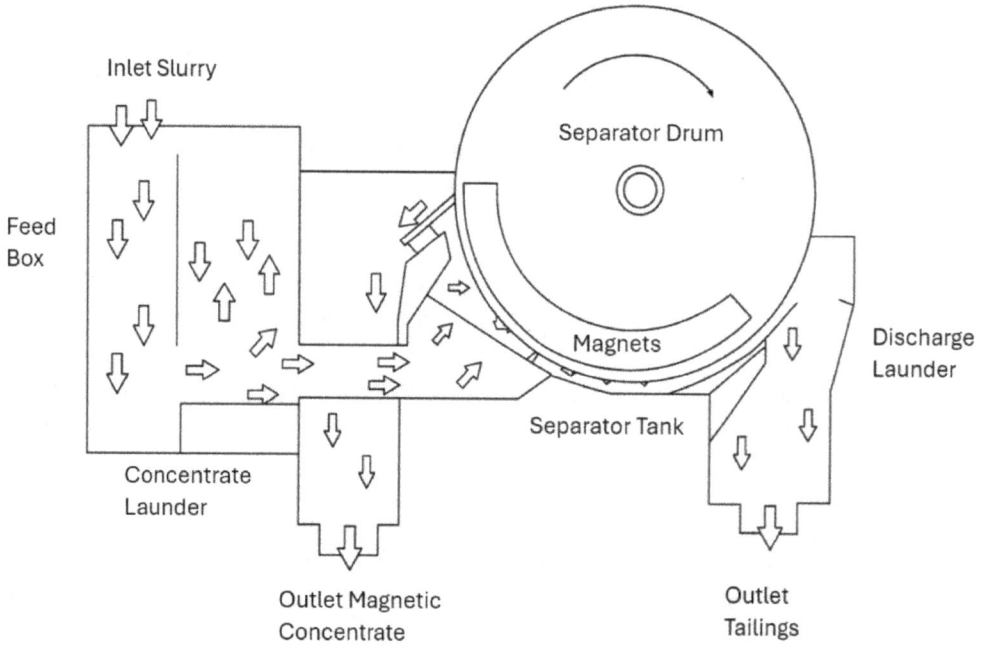

Figure 45: Wet High Intensity Magnetic Separation.

A typical High-Intensity Magnetic Separator includes several essential components that work together to achieve efficient separation. The magnetic field generator produces a strong, high-gradient field using electromagnets or rare-earth magnets. The matrix or capture zone—made from ferromagnetic wires, rods, or steel wool—intensifies the field locally and traps weakly magnetic particles. The feed system introduces the ore, either wet or dry, in a controlled and uniform manner to ensure consistent separation. Discharge systems are used to separate and collect the magnetic concentrate and non-magnetic tailings, while cooling and electrical systems maintain stable operation by dissipating the heat generated during high-current electromagnet operation.

The separation process occurs in several stages. During feed introduction, the ore is delivered into the separator—usually as a slurry in wet systems, or as a dry flow using vibratory or belt feeders. In the magnetic capture stage, weakly magnetic particles are attracted to the ferromagnetic matrix elements, which act as microscopic traps that hold even fine paramagnetic minerals. The retention and transport stage follows, where captured particles adhere to the matrix while non-magnetic material continues through unaffected. As the matrix moves or rotates, the magnetic particles are carried out of the field's influence. Finally, during cleaning and discharge, the trapped particles are released from the matrix—either by mechanical agitation, water backwashing, or demagnetization—and collected as the magnetic concentrate. The remaining non-magnetic material exits as tailings.

High-Intensity Magnetic Separators come in several configurations, each suited to specific applications. The Wet High-Intensity Magnetic Separator (WHIMS) is the most common design for processing fine paramagnetic minerals in slurry form. Water assists in particle transport and reduces entrainment losses. Examples include the Jones and SLon separators. The Dry High-Intensity Magnetic Separator (DHIMS) is used where water is unavailable or unsuitable, such as in arid regions or with moisture-sensitive materials. These systems use vibrating feeders and strong magnetic gradients to treat fine dry particles. The High-Gradient Magnetic Separator (HGMS) is a specialized variant of HIMS that employs an extremely high magnetic field gradient—up to 20,000 Gauss—within a dense matrix of fine steel wool or wire mesh. It is particularly effective for ultrafine and complex ores, including rare earth element recovery.

HIMS technology is widely applied in the recovery and purification of weakly magnetic minerals and non-metallic materials. It is used in iron ore processing for the recovery of hematite and goethite from non-magnetic gangue, and in tungsten and tantalum ore processing for separating wolframite and coltan (columbite–tantalite). In titanium mineral processing, HIMS assists in upgrading ilmenite and related alteration products such as leucoxene. It also plays a vital role in rare earth element beneficiation, concentrating monazite and xenotime, and in industrial mineral purification, removing iron impurities from silica sand, feldspar, and kaolin.

The main advantages of High-Intensity Magnetic Separators include their ability to recover weakly magnetic and fine-grained minerals, produce high-grade concentrates with minimal losses, and operate flexibly in either wet or dry modes. The magnetic field intensity can be adjusted to suit specific minerals, allowing fine-tuned optimization for different ores. Their compact and continuous operation also makes them well-suited to modern processing plants. However, HIMS systems have certain limitations. They consume more power than low-intensity systems, require regular maintenance to prevent matrix clogging or buildup, and are less efficient for very coarse or non-magnetic materials. Wet systems also demand careful control of pulp density and viscosity, while equipment costs and energy requirements are relatively high.

In summary, High-Intensity Magnetic Separators (HIMS) use strong magnetic fields and high-gradient matrices to capture weakly magnetic minerals from ore mixtures. By intensifying the magnetic field within ferromagnetic materials, they can separate fine paramagnetic particles that would otherwise escape recovery. HIMS units are indispensable in processing hematite, ilmenite, tungsten, and rare earth ores, forming a critical part of modern mineral beneficiation circuits focused on fine-particle and high-value mineral recovery.

A High-Gradient Magnetic Separator (HGMS) is an advanced magnetic separation system that uses extremely strong magnetic fields and very high field gradients to recover weakly magnetic and fine-grained minerals. While similar in principle to High-Intensity Magnetic Separators (HIMS), HGMS units are engineered to achieve even higher magnetic capture efficiency—especially for ultrafine particles and minerals with low magnetic susceptibility such as *hematite, ilmenite, wolframite, monazite, xenotime,* and various rare earth elements (REEs). These separators typically generate magnetic fields of up to 2 Tesla (20,000 Gauss), with localized gradients exceeding 1,000 Tesla per meter, allowing them to recover valuable minerals that would otherwise be lost in conventional processes.

Figure 46: High Gradient Cyclic Kaolin Magnetic Separator SALA. Antonín Ryska, CC0, via Wikimedia Commons.

The principle of operation of an HGMS is based on the concentration of magnetic force through a fine ferromagnetic matrix. When a mineral slurry flows through the separator, weakly magnetic particles experience magnetic attraction due to the high field gradient produced by a network of ferromagnetic wires, rods, or meshes inside the magnetic field. The magnetic force acting on a particle depends on both the magnetic field strength and the gradient; thus, even a moderate field can create a strong pulling force when the gradient is extremely high. As the slurry passes through the matrix, weakly magnetic particles are drawn to the surfaces of the wires or fibres—where the field is most intense—while non-magnetic particles continue flowing through and exit as tailings.

Figure 47: High-gradient magnetic separator SKODA with classic (copper) winding. Antonín Ryska, CC0, via Wikimedia Commons.

A typical High-Gradient Magnetic Separator is composed of several main components. The superconducting or electromagnetic coil generates a strong, uniform magnetic field that surrounds the separation chamber. The magnetic matrix, usually made of stainless steel wool, expanded metal, or steel rods, amplifies the field gradient and provides trapping sites for magnetic particles. The feed system introduces the slurry evenly into the separation zone to ensure that all particles are exposed to the field. The pulp chamber or canister is the central area where the matrix and field interact to separate particles. Discharge systems collect magnetic concentrates and non-magnetic tailings separately, while cooling and power systems maintain stable temperature and electrical performance in the magnetic coils.

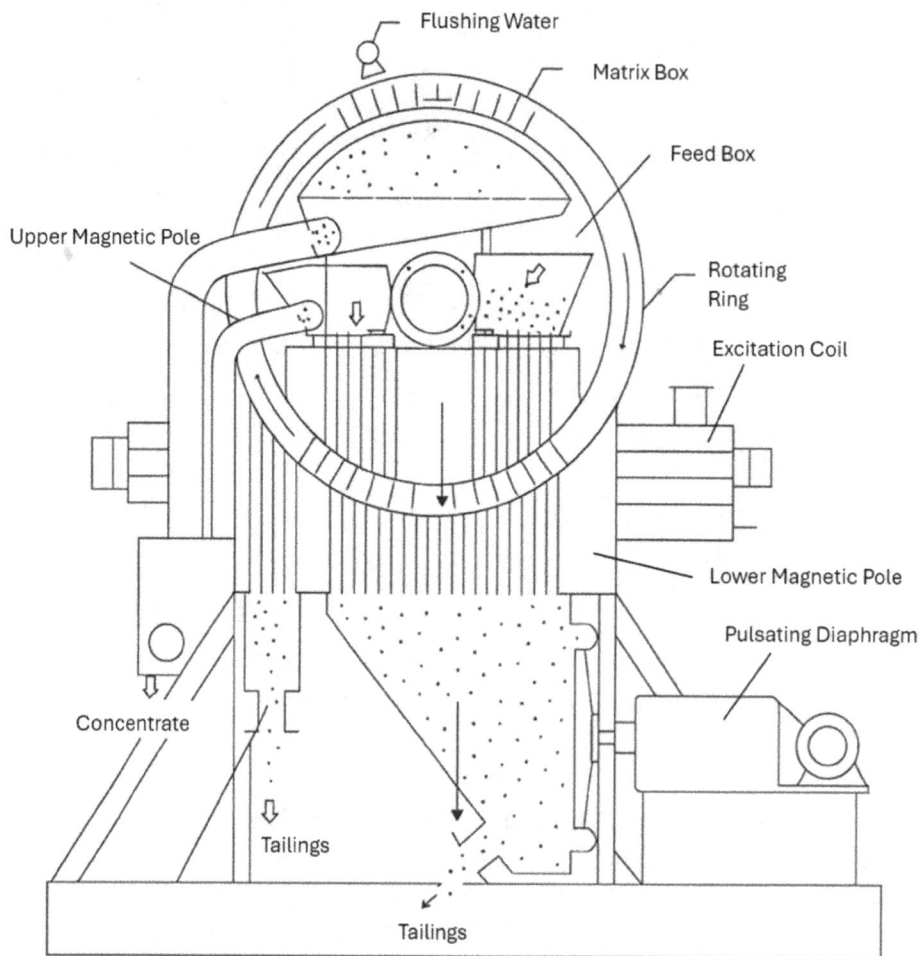

Figure 48: Vertical Ring Pulsating High Gradient Magnetic Separator.

The HGMS operates through several distinct stages. During feed introduction, the ore slurry containing both magnetic and non-magnetic particles is pumped into the separation chamber, flowing through the magnetic matrix. In the magnetic capture stage, weakly magnetic particles are attracted to and trapped on the surfaces of the ferromagnetic matrix elements, where the local field gradient is strongest. As this process continues, the particles accumulate and form a magnetic layer in the retention stage, while non-magnetic materials pass through unaffected. When the matrix becomes saturated, the system enters the rinsing or cleaning cycle: the magnetic field is reduced or turned off, and water or vibration dislodges the trapped particles, flushing them out of the matrix. Finally, during concentrate and tailings discharge, the recovered magnetic particles are collected as concentrate, while the remaining non-magnetic fraction exits as tailings. The separator is then re-magnetized for the next cycle.

Rare Earth and Critical Mineral Operations and Processing

There are several types of HGMS configurations, each designed for different operational requirements. Periodic (batch) HGMS systems operate in cycles, alternating between magnetic capture and demagnetization with rinsing—ideal for small-scale or high-purity operations. Continuous HGMS designs use rotating canisters or moving matrices that transport captured materials out of the magnetic field for cleaning, allowing uninterrupted processing. Superconducting HGMS systems employ superconducting coils to achieve ultra-high magnetic fields—up to 5 Tesla—with reduced energy consumption. These are used in high-value or difficult-to-process ores, including rare earth and kaolin applications.

HGMS technology is widely used in both mineral and industrial processing. In iron ore beneficiation, it recovers fine hematite and goethite from tailings and slimes. In tungsten, tantalum, and tin operations, it separates wolframite, coltan, and cassiterite mixtures. In rare earth element (REE) processing, it concentrates monazite, xenotime, and bastnäsite-bearing minerals. HGMS is also valuable for industrial mineral purification, where it removes iron and other trace contaminants from quartz, feldspar, kaolin, and phosphates. Beyond mining, it finds environmental applications such as removing iron oxides and heavy metals from wastewater or contaminated soils.

The advantages of HGMS are significant. It offers extremely high recovery rates for weakly magnetic and ultrafine particles, produces high-purity concentrates with minimal loss, and adapts to a wide range of feed materials and particle sizes. Superconducting designs lower energy consumption while maintaining high magnetic strength, and systems can operate either continuously or intermittently, depending on processing needs. However, HGMS systems also have limitations. They require high capital and operational investment, and performance depends on careful control of slurry viscosity, feed rate, and field strength. The magnetic matrix can become clogged with fine particles, demanding regular cleaning, while the cooling and power requirements are substantial for large-scale electromagnetic systems. HGMS is also less effective for coarse-grained or entirely non-magnetic materials.

High-Gradient Magnetic Separators use powerful magnetic fields and extremely high field gradients to capture fine and weakly magnetic minerals from complex mixtures. As slurry flows through a ferromagnetic matrix within a strong magnetic field, the fine magnetic particles are trapped and later released during cleaning cycles, producing a high-grade concentrate. Thanks to their precision, high recovery rates, and ability to process ultrafine materials, HGMS units are indispensable in iron ore beneficiation, rare earth recovery, tungsten and tantalum concentration, and industrial mineral purification, making them one of the most advanced technologies in modern mineral processing.

Electrostatic Separation

Electrostatic separation exploits differences in electrical conductivity and surface charge between minerals. When particles are exposed to an electric field, conductive minerals lose their charge quickly and are deflected by electrostatic forces, whereas non-conductive minerals retain charge longer and follow a different trajectory.

Dry, free-flowing particles are fed onto a grounded rotating drum or belt that passes through a high-voltage electric field. Conductive particles discharge rapidly and are thrown off the drum, while non-conductors remain attached until they are mechanically released. Effective separation requires precise control of particle size, humidity, and surface cleanliness.

Typical equipment includes:

- **High-Tension Roll Separators:** Conductive and non-conductive particles are separated by differential attraction to the electrode.

- **Plate and Screen Electrostatic Separators:** Used for finer or more delicate separations.

- **Corona Discharge Systems:** Enhance particle charging efficiency, improving selectivity for complex mixtures.

Commonly used in rare earth and heavy mineral processing—for separating conductive ilmenite from non-conductive rutile, zircon, and monazite, and for refining bastnäsite or xenotime concentrates after magnetic separation. Also used in graphite purification and certain recycling operations.

A High-Tension Roll Separator (HTR)—also known as an Electrostatic Separator or Electrostatic High-Tension Separator—is a dry separation device that uses electrical conductivity and electrostatic forces to separate minerals based on their electrical properties. This technique is essential for processing heavy mineral sands, rare earth element (REE) ores, and industrial minerals, where minerals vary significantly in their ability to conduct electricity. Common applications include the separation of ilmenite, rutile, zircon, monazite, xenotime, and cassiterite, as well as the removal of conductive minerals like graphite or sulphides from non-conductive materials.

The principle of operation relies on differences in the way minerals acquire and lose electrical charge when exposed to a strong electric field. When dry, free-flowing mineral particles are fed onto a rotating grounded roll positioned beneath a high-voltage electrode, they are charged through a process known as corona charging. Conductive minerals—such as ilmenite, hematite, or graphite—quickly lose their surface charge to the grounded roll, which causes them to be attracted to and held tightly against the roll's surface. In contrast, non-conductive minerals like zircon, quartz, and monazite retain their charge longer. The resulting electrostatic repulsion pushes them away from the roll, causing them to be deflected outward into separate product streams. This difference in charging and discharge behavior enables the efficient separation of minerals based on electrical conductivity rather than magnetic or density properties.

A High-Tension Roll Separator is composed of several critical components. The high-voltage electrode (often a corona wire or electrode bar) generates a strong electric field—typically between 20 and 40 kilovolts (kV)—which ionizes air molecules and charges the particles. The grounded rotating roll beneath the electrode serves as both the collection and transport surface for conductive particles. The feed system, which may include a vibrating feeder or a controlled chute, introduces a steady and uniform flow of dry material onto the roll. The electrostatic field zone between the electrode and roll is where particle charging and separation occur. Splitter blades and collection hoppers at the discharge end direct the separated conductive, semi-conductive, and non-conductive fractions into their respective bins. A control system manages key operational parameters such as voltage, roll speed, and feed rate to ensure optimal efficiency and recovery.

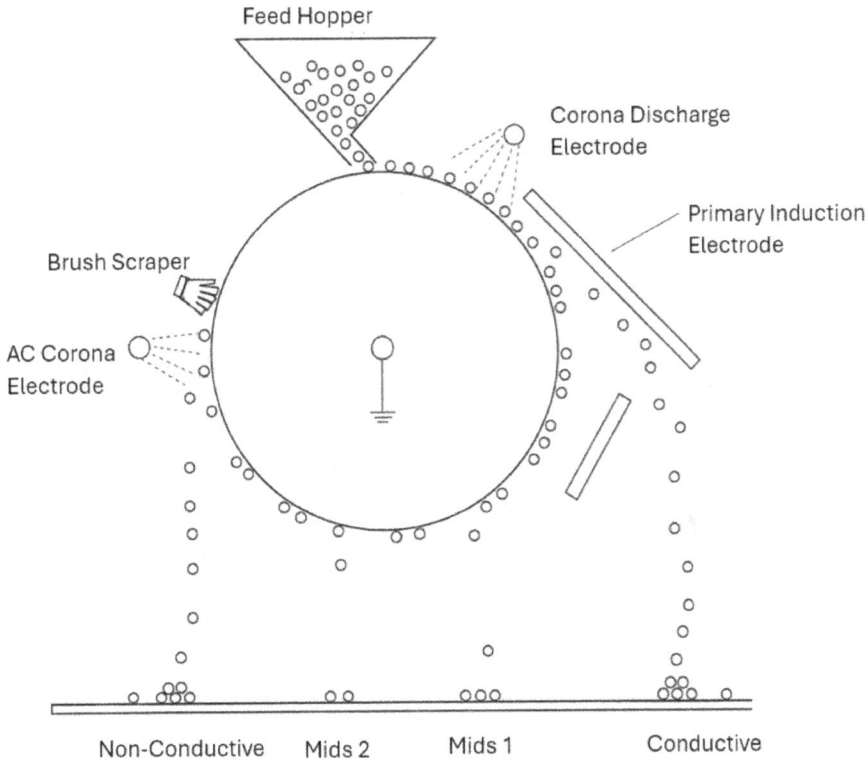

Figure 49: Electrostatic Separation.

The separation process occurs in several distinct stages. In the feed introduction stage, the dry mineral mixture—ideally within the size range of 75 µm to 2 mm—is evenly fed onto the rotating grounded roll. During corona charging, particles are exposed to the electric field, acquiring an electrical charge according to their conductivity. As the particles move across the roll, conductive minerals discharge rapidly through the roll and adhere to its surface, while non-conductive minerals retain their charge and are repelled outward by the electric field. During the product discharge stage, brushes or scrapers remove conductive particles adhering to the roll and deposit them into a separate container. Non-conductive minerals, which are repelled earlier, fall into a different collection area, while semi-conductive materials may be directed to an intermediate stream. Operators then fine-tune voltage levels, roll speed, and splitter blade positioning to optimize separation performance for different mineral compositions.

Several types of electrostatic separators are used in mineral processing, each suited to specific feed types and operational requirements. The High-Tension Roll Separator (HTR) is the most common, ideal for heavy mineral sand and rare earth applications. Plate or screen-type separators use stationary electrodes and grounded plates for coarser materials. Free-fall electrostatic separators work by allowing fine particles to fall between electrodes under gravity, while drum-type separators use a continuous rotating drum instead of a roll for high-capacity operations.

High-Tension Roll Separators have a broad range of applications. In heavy mineral sands, they separate conductive minerals such as ilmenite and leucoxene from non-conductive zircon and rutile. In rare earth processing, they are used to concentrate conductive monazite and xenotime from quartz-rich gangue. In tin, tantalum, and tungsten circuits, they separate conductive cassiterite or coltan from silicate minerals. They are also applied in industrial mineral purification, such as cleaning quartz, feldspar, and phosphate minerals by removing conductive contaminants, and in recycling industries for separating metals from plastics and other non-conductive materials.

The advantages of High-Tension Roll Separators include their ability to operate with dry feed, eliminating the need for water or chemicals. They offer high selectivity for fine-particle separation based on conductivity, low operating costs once installed, and continuous operation with adjustable parameters to optimize grade and recovery. These features make them ideal for both heavy mineral and industrial mineral processing. However, they also have limitations. Feed material must be completely dry and free-flowing, as moisture reduces charging efficiency. Fine slimes or dust can disrupt the electric field and reduce performance, while very fine particles below 75 µm are difficult to separate effectively. The process requires precise control of feed rate, roll speed, and voltage, and buildup of non-conductive dust on electrodes or rolls must be prevented through regular cleaning.

A Plate and Screen Electrostatic Separator is a dry mineral separation device that uses electrostatic forces to distinguish minerals based on their differences in electrical conductivity. Operating under the same basic principle as High-Tension Roll Separators (HTR), these separators differ in configuration by using stationary or inclined conductive plates, screens, or grids instead of a rotating roll. The design makes them particularly effective for coarse-grained or bulk mineral separations, such as those encountered in heavy mineral sands, rare earth element (REE) ores, and industrial mineral purification. They are widely used to separate minerals like ilmenite, rutile, zircon, monazite, and quartz, and to remove conductive impurities such as graphite or iron oxides from non-conductive materials.

The principle of operation is based on how minerals acquire and lose electrical charge when exposed to a high-voltage electric field. When a mixture of dry, free-flowing particles is introduced into the separator, each particle becomes electrically charged through either corona charging—where ionized air transfers charge to the particles—or triboelectric charging, which occurs through frictional contact between particles and the separator surfaces. Conductive minerals such as ilmenite, hematite, or graphite discharge their charge almost immediately upon contact with grounded metal plates or screens and are attracted toward the conductive surfaces. In contrast, non-conductive minerals like zircon, monazite, or quartz retain their charge for a longer period and are repelled by the electric field. This difference in discharge behaviour results in the separation of conductive and non-conductive fractions, which can then be collected separately.

Unlike roll-type separators, where particles are carried on a rotating drum, plate and screen systems achieve separation along stationary inclined surfaces, with gravity assisting particle flow. This design allows for efficient separation of coarse and heavy minerals, particularly where mechanical motion might disturb the electrostatic forces.

A typical Plate or Screen Electrostatic Separator includes several essential components that work together to achieve efficient mineral separation. The high-voltage electrode, usually a corona wire or discharge bar, generates an electric field between 20 and 50 kilovolts (kV), ionizing the surrounding air and imparting charge to the particles. Beneath this, the grounded conductive plates or screens act as collection surfaces where conductive particles discharge and adhere due to electrostatic attraction. The feed system, typically a vibrating feeder or chute, ensures that the dry mineral feed is distributed evenly across the separator. The separation zone lies between the electrode and grounded plate or screen, where charging and separation take place under controlled flow conditions. Splitter plates and collection bins located at the base of the separator divide the conductive, semi-conductive, and non-conductive fractions. Finally, a power supply and control system maintains voltage stability and allows operators to adjust variables such as electrode spacing, voltage level, and feed rate to optimize performance.

The operation of a Plate and Screen Separator occurs in several key stages. During feed introduction, the dry mineral mixture—typically within a particle size range of 100 μm to 2 mm—is evenly distributed across the top of the inclined plate or screen. In the charging phase, the particles pass through the corona discharge field or make contact with charged surfaces, acquiring an electrostatic charge. Next, during particle separation, conductive particles quickly lose their charge upon contacting the grounded plate and adhere to it, while non-conductive particles retain their charge and are repelled by the field, moving farther down the separator under gravity. Semi-conductive particles exhibit intermediate behaviour and form a distinct middle stream. As the particles move downward, splitter plates direct each stream—conductive, semi-conductive, and non-conductive—into separate collection hoppers. Finally, during parameter control, operators fine-tune the voltage, electrode distance, plate angle, and feed rate to maintain separation efficiency and achieve the desired product purity and recovery.

These separators are employed across a range of applications. In heavy mineral sand processing, they separate conductive minerals like ilmenite and rutile from non-conductive zircon and quartz. In rare earth processing, they are used to concentrate monazite and xenotime from quartz-rich gangue. Within industrial mineral operations, they remove conductive contaminants such as graphite or iron oxides from feldspar, quartz, and phosphate minerals. They are also used in recycling industries for recovering conductive metals from electronic waste and separating metallic from plastic components. Often, plate and screen separators are paired with High-Tension Roll Separators, with the HTR performing initial separation and the plate or screen unit providing final cleaning or product refinement.

The main advantages of Plate and Screen Electrostatic Separators include their ability to operate in a completely dry environment, eliminating the need for water or chemical reagents. They are particularly suitable for coarse particles that do not adhere well to rotating rolls and have a simple, robust design requiring minimal maintenance. The ability to adjust key parameters such as voltage, plate inclination, and feed rate provides operational flexibility, and when combined with other electrostatic or magnetic systems, these separators can produce high-purity concentrates suitable for downstream processing.

However, there are several limitations. The feed must be completely dry, as even small amounts of moisture can cause particles to clump, reducing charging efficiency and separation quality. The method is less effective for ultrafine particles (smaller than 75 μm), which may become airborne or fail to charge consistently. Separation performance also declines for minerals with similar electrical conductivities, where differential charging is minimal. Additionally, environmental factors such as humidity, dust buildup, and uneven feed distribution can affect efficiency and may require frequent adjustments or cleaning.

Integration and Process Sequencing

In most processing plants, multiple separation techniques are combined sequentially to achieve maximum selectivity and recovery, particularly in rare earth element (REE) and critical mineral beneficiation. The sequence and configuration depend on the physical and chemical properties of the target minerals, as well as the nature of the gangue. For example:

- **Magnetic → Electrostatic:** This sequence is widely used in rare earth element and heavy mineral sand processing circuits. Magnetic minerals such as ilmenite ($FeTiO_3$), magnetite (Fe_3O_4), and ferro-manganese oxides are first removed using low- or high-intensity magnetic separators. The remaining non-magnetic fraction, which contains weakly magnetic and non-conductive minerals, is then treated by electrostatic separation. This second stage separates conductive monazite (a light rare earth phosphate mineral rich in Ce, La, Nd, and Pr) from non-conductive zircon ($ZrSiO_4$) and xenotime (a heavy rare earth phosphate containing Y, Dy, Er, and Tb). In REE placer deposits from Australia and Madagascar, this sequence is crucial for upgrading mixed monazite–xenotime concentrates after removal of ilmenite and rutile.

- **Gravity → Magnetic → Electrostatic:** This multi-stage approach is common in niobium–tantalum (Nb–Ta), tin (Sn), and tungsten (W) mineral circuits. Initially, gravity separation (using spirals, shaking tables, or centrifugal concentrators) concentrates dense minerals such as columbite–tantalite (coltan), cassiterite (SnO_2), and wolframite ($(Fe,Mn)WO_4$). The gravity concentrate is then subjected to magnetic separation to remove magnetic impurities like ilmenite and iron oxides, followed by electrostatic separation to refine conductive coltan from non-conductive cassiterite or quartz gangue. This process is routinely used in tantalum and tungsten operations in central Africa, Brazil, and Western Australia to achieve high-grade concentrates for downstream hydrometallurgical recovery.

- **Gravity → Flotation:** This combination is typically used in lithium and graphite beneficiation circuits, where density-based separation first removes gangue minerals before surface-chemistry separation. In spodumene ($LiAlSi_2O_6$) processing, dense media separation (DMS) concentrates the lithium-bearing mineral from low-density waste such as quartz and feldspar. The pre-concentrated material is then subjected to flotation, which exploits differences in surface hydrophobicity to separate spodumene from mica and feldspar. Similarly, in graphite operations (e.g., natural flake graphite deposits in Canada and Mozambique), gravity concentration removes coarse gangue before froth flotation upgrades the product to >95% carbon purity by floating graphite flakes away from silicate minerals.

This sequencing of gravity, magnetic, and electrostatic methods—and sometimes flotation—provides a flexible and efficient strategy for refining rare earth and critical mineral concentrates, ensuring both high recovery and purity levels essential for downstream refining and advanced material production.

Flotation Fundamentals and Reagent Use

Flotation is one of the most important and versatile mineral separation processes in modern mineral processing. It works by exploiting differences in the surface chemistry of mineral particles, allowing valuable minerals to be selectively separated from gangue material. Through this method, particles that have been made hydrophobic attach to air bubbles and float, while hydrophilic particles remain submerged and are discarded. Flotation is extensively used in the beneficiation of ores containing base metals (such as copper, lead, zinc, and nickel), precious metals (such as gold and platinum), industrial minerals (including phosphate, fluorite, and potash), and critical minerals such as graphite, lithium, and rare earth elements (REEs).

The principle of flotation is based on surface hydrophobicity, which refers to a mineral's tendency to repel water. Finely ground ore is mixed with water and specific reagents to form a slurry, or pulp, and air is introduced into the system in the form of small bubbles. Hydrophobic mineral particles attach to these air bubbles and rise to the surface, forming a froth layer, while hydrophilic minerals remain in the water and sink to the bottom. The froth, which contains the valuable minerals, is skimmed off as the concentrate, while the remaining slurry is removed as tailings. The success of flotation depends on manipulating the mineral surface properties through careful control of reagent types and dosages.

Flotation proceeds through several fundamental steps. During conditioning, the ground ore is mixed with reagents that modify its surface chemistry. Aeration introduces air bubbles into the pulp, and attachment occurs when hydrophobic particles adhere to these bubbles. Froth formation follows, as the mineral-laden bubbles rise to the surface, and finally, collection and cleaning remove the froth, which is reprocessed through several stages to improve grade and recovery.

Flotation takes place inside specially designed flotation cells, which are fitted with mechanical agitators or air injectors. The process occurs in three main zones: the pulp zone, where the slurry is conditioned and bubbles are dispersed; the froth zone, where hydrophobic particles accumulate; and the tailings zone, where hydrophilic particles settle and are discharged. Modern circuits often employ multiple flotation stages—rougher, cleaner, and scavenger cells—to optimize both recovery and concentrate purity. Throughout the process, critical variables such as pH, pulp density, air flow rate, and reagent dosage are carefully controlled to ensure stable froth formation and efficient separation.

Flotation relies heavily on chemical reagents to achieve selective separation. There are four primary classes of reagents: collectors, frothers, modifiers (regulators), and pH modifiers. Collectors are organic chemicals that selectively make target mineral surfaces hydrophobic, allowing them to attach to air bubbles. For sulphide minerals like chalcopyrite, galena, and sphalerite, collectors such as xanthates (e.g., potassium amyl xanthate – PAX), dithiophosphates, and thionocarbamates are commonly used. For oxide and silicate minerals—including apatite, cassiterite, and rare earth minerals—collectors such as fatty acids, hydroxamates, and amines are employed. Naturally hydrophobic minerals like graphite and coal often need little or no collector addition.

Collectors are the most critical class of reagents used in flotation because they determine which minerals become hydrophobic and attach to air bubbles for recovery. Their main function is to selectively modify the surface properties of target minerals, rendering them hydrophobic (water-repelling) so that they can attach to rising air bubbles, while unwanted gangue minerals remain hydrophilic (water-attracting) and sink. In essence, collectors are the "surface conditioners" that make flotation selective.

The effectiveness of a collector depends on how well it adsorbs (attaches) to the mineral surface. Most mineral surfaces are naturally hydrophilic due to hydroxyl (–OH) groups and water molecules bonded to their surfaces. Collectors replace or displace these water molecules by forming a chemical or physical bond with surface atoms, creating a hydrophobic layer.

Once this layer forms, the mineral surface has a strong affinity for air rather than water. During aeration, these hydrophobic particles readily attach to air bubbles and are carried upward into the froth phase, where they can be collected as concentrate.

There are two main mechanisms by which collectors interact with mineral surfaces:

Chemisorption (Chemical Adsorption): In this mechanism, the collector forms a chemical bond with metal ions on the mineral surface. This occurs mainly in sulphide minerals such as chalcopyrite ($CuFeS_2$), galena (PbS), or sphalerite (ZnS). The collector's polar (reactive) group reacts directly with metal cations, forming a stable compound that imparts hydrophobicity.

- Example: Xanthates, commonly used sulfide collectors, chemically react with metal ions to form metal-xanthate complexes, which are hydrophobic.

- Reaction:

$$2R\text{–}O\text{–}CS_2^- + M^{2+} \rightarrow M(R\text{–}O\text{–}CS_2)_2 + 2e^-$$

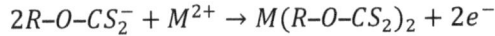

Here, M^{2+} is a metal ion (e.g., Cu^{2+}, Pb^{2+}, Zn^{2+}), and $R\text{–}O\text{–}CS_2^-$ is the xanthate ion.

Physisorption (Physical Adsorption): This occurs mainly on oxide, silicate, and carbonate minerals where collectors attach via weaker van der Waals forces or electrostatic interactions, rather than forming chemical bonds. Collectors such as fatty acids, amines, and hydroxamates physically adsorb onto surfaces modified by pH or activators.

- Example: Fatty acid collectors adsorb onto calcium-bearing minerals like apatite or scheelite in slightly alkaline pH conditions.

A collector molecule consists of two key functional components that work together to alter the mineral's surface properties during flotation. The polar head group is the chemically active end of the molecule that attaches to the mineral surface. Its chemical composition determines the type of mineral it can bond with—such as xanthate groups for sulfide minerals or carboxyl and hydroxamate groups for oxide minerals. The nonpolar hydrocarbon tail, usually a long hydrocarbon chain, extends outward from the mineral surface. This part of the molecule is hydrophobic, meaning it repels water, which helps render the mineral surface water-repellent and promotes attachment to air bubbles during flotation.

The overall effectiveness and selectivity of a collector depend on the balance between these two components. Longer hydrocarbon chains generally increase the hydrophobic character of the mineral surface, improving bubble attachment and froth stability. However, longer chains can also reduce selectivity and solubility in water, requiring careful reagent design and dosage control.

Different types of collectors are used depending on the mineral group being processed. Sulphide collectors (chemical collectors) are commonly used for metallic sulphide minerals such as chalcopyrite, galena, sphalerite, and pentlandite. Among these, xanthates ($ROCS_2^-$) are the most widely used for copper, lead, and zinc sulphides, while dithiophosphates and thionocarbamates offer stronger and more selective bonding, particularly for complex or partially oxidized sulphides. A common example is potassium amyl xanthate (PAX), widely applied in chalcopyrite flotation.

For oxide and silicate collectors (physical collectors), which are used on non-sulphide minerals such as cassiterite (SnO_2), apatite ($Ca_5(PO_4)_3(F,Cl,OH)$), scheelite ($CaWO_4$), and rare earth phosphates like monazite and xenotime, the chemistry is different. Fatty acids such as oleic acid are effective for calcium-bearing minerals, while hydroxamates are preferred for rare earth, tungsten, and tin oxides due to their ability to form stable surface complexes at low pH. Amines are used in reverse flotation circuits—for example, removing quartz from iron ore.

Some minerals, such as graphite, coal, and molybdenite (MoS_2), are naturally hydrophobic and do not require chemical modification. In these cases, small amounts of hydrocarbon oils such as kerosene or diesel can act as simple collectors to improve bubble attachment and froth stability.

The performance of collectors in flotation depends on several operating parameters. The pH of the pulp is one of the most important factors, as it affects both the surface charge of the mineral and the degree of collector ionization. For instance, xanthates perform best under moderately alkaline conditions (pH 8–10), while fatty acids show optimal results under slightly alkaline conditions. Mineral surface composition also plays a key role, as oxidized or coated surfaces may require activators—such as copper sulphate ($CuSO_4$)—to promote collector adsorption.

Collector concentration must be carefully optimized: insufficient dosage can lead to poor recovery, whereas excessive amounts may reduce selectivity by floating unwanted gangue minerals. Other influencing factors include pulp potential (Eh), which affects the oxidation state of both collectors and mineral surfaces in sulphide systems, and conditioning time and temperature, which govern adsorption kinetics and the strength of the collector film formed on the mineral surface.

Collectors play a central role in processing critical and rare earth minerals. For rare earth minerals such as monazite and bastnäsite, hydroxamate collectors form surface complexes with rare earth ions under mildly acidic conditions (pH 5–6), making the particles hydrophobic and allowing them to float away from quartz and carbonate gangue. In lithium ore flotation, fatty acid or amine collectors are used on spodumene ($LiAlSi_2O_6$) after thermal activation at around 1000°C, which converts the α-phase to β-spodumene, improving collector adsorption. Graphite, naturally hydrophobic, requires only small doses of diesel oil or kerosene to enhance bubble attachment and froth stability. For tungsten ores such as scheelite and wolframite, hydroxamate collectors form chelate complexes with surface calcium ions (Ca^{2+}), allowing for selective flotation from silicate gangue.

Frothers control the bubble size and froth stability, ensuring effective bubble-particle attachment and a stable froth capable of carrying the minerals to the surface. Common frothers include methyl isobutyl carbinol (MIBC), pine oil, and polypropylene glycol ethers (Dowfroth series). The dosage must be balanced—too much frother can lead to excessive entrainment of unwanted fine particles.

Frothers are essential flotation reagents used to control the formation, stability, and characteristics of froth within a flotation cell. While collectors determine which minerals attach to air bubbles, frothers govern how those bubbles behave—how large they are, how stable the froth becomes, and how effectively mineral-laden bubbles rise through the pulp to the surface for recovery. In simple terms, frothers create the right environment for hydrophobic particles to travel efficiently from the pulp zone to the froth layer without collapsing prematurely.

When air is introduced into a flotation cell, it forms bubbles of varying sizes. Without frothers, these bubbles tend to coalesce, forming large, unstable bubbles that burst easily and reduce recovery. Frothers work by reducing surface tension at the air–water interface, encouraging the formation of numerous small, uniform, and stable bubbles. Smaller bubbles offer a greater total surface area, increasing the likelihood of collision and attachment between hydrophobic particles and air bubbles.

In addition to bubble formation, frothers stabilize the froth at the top of the flotation cell. This stable froth maintains its structure long enough for valuable, mineral-bearing bubbles to rise and be skimmed off effectively. However, frother dosage must be carefully controlled. Insufficient frother results in weak froth that collapses quickly, while excessive frother can over-stabilize the froth, leading to entrainment of unwanted gangue and reduced concentrate quality.

Frothers perform several important functions in the flotation process. They assist in bubble formation by generating fine, uniform bubbles during aeration, thereby increasing the number of potential attachment sites for hydrophobic particles. They promote froth stability by reducing bubble coalescence and drainage, creating a persistent froth layer that allows consistent, selective recovery of minerals. Frothers also influence selectivity and transport by controlling bubble size and froth texture, ensuring mineral-laden bubbles rise smoothly while minimizing gangue entrainment. Finally, they enhance concentrate recovery by maintaining froth stability long enough for the bubbles to reach the collection launder without collapsing.

Frothers are typically organic compounds containing both polar and nonpolar groups, allowing them to interact effectively at the air–water interface. They are generally categorized into three main types.

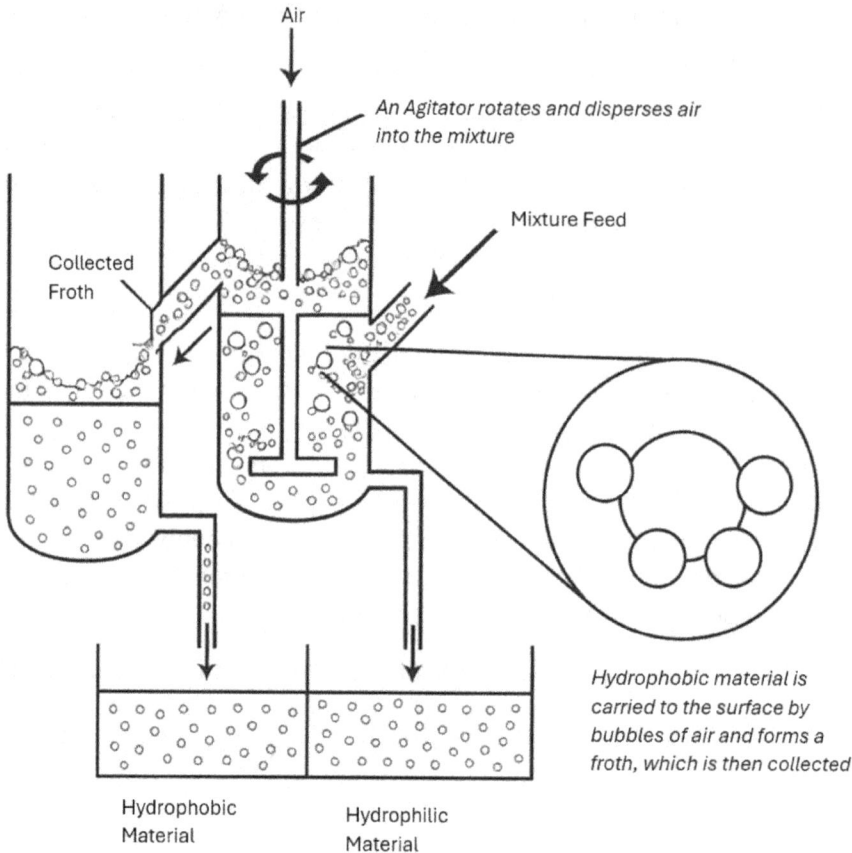

Figure 50: Flotation Froth.

Alcohol-based frothers, such as methyl isobutyl carbinol (MIBC) and hexanol, are the most common. They produce small to medium-sized bubbles and provide moderate froth stability, making them ideal for base and precious metal flotation. Polyglycol and polypropylene glycol ether frothers, such as those in the Dowfroth series (Dowfroth 250, 400), generate finer and more stable bubbles. These are used in complex sulphide, rare earth, and oxide ore flotation where fine particle recovery is important. Pine oil and other natural frothers are derived from plant sources and create coarser bubbles with moderate froth stability. These are often preferred in coal and precious metal flotation, where selective froths are advantageous. Each frother type offers a different balance of bubble size, stability, and drainage. The selection depends on ore characteristics, particle size distribution, and downstream processing requirements.

Several factors influence frother performance. Frother concentration must be carefully optimized—too little creates unstable froth, while too much causes over-stabilization and entrainment of gangue. Pulp density affects bubble behaviour, with denser pulps often requiring stronger or higher frother dosages. Air flow rate also influences froth structure; as aeration increases, frother dosage must be adjusted to maintain consistent bubble size and texture. pH and temperature alter pulp surface tension and frother solubility, affecting bubble formation and stability. Lastly, mineral type and particle size play a major role—fine or highly hydrophobic particles require smaller, more stable bubbles, while coarser particles need larger, less persistent froth for effective recovery.

In graphite flotation, frothers such as MIBC or pine oil are used to create stable, low-density froth that carries naturally hydrophobic graphite flakes to the surface. For rare earth elements like monazite, bastnäsite, and xenotime, Dowfroth frothers are often used alongside hydroxamate collectors to generate fine, persistent bubbles ideal for ultrafine REE recovery. In lithium flotation, particularly for spodumene ($LiAlSi_2O_6$), MIBC is preferred for producing small, uniform bubbles that improve attachment efficiency and recovery following collector activation. For base metal sulphides such as copper, lead, and zinc, frothers like MIBC or polyglycol ethers are selected to balance froth stability and selectivity, ensuring clean separation of complex sulphide minerals.

Modifiers, also known as regulators, control the pulp chemistry, ensuring that only desired minerals float. pH regulators such as lime (CaO) or sodium carbonate (Na_2CO_3) maintain the optimal chemical environment. Activators, like copper sulphate ($CuSO_4$), can enhance collector adsorption—for example, activating sphalerite (ZnS) to make it float with xanthates. Depressants—including sodium cyanide (NaCN) for pyrite, starch or dextrin for silicates, and sodium silicate for quartz—suppress unwanted minerals. Dispersants and flocculants are sometimes added to control fine particle behaviour and improve slurry dispersion.

Modifiers, also known as regulators, are a vital group of flotation reagents that control the chemical environment of the pulp and regulate the surface properties of minerals. While collectors make specific minerals hydrophobic and frothers stabilize the froth, modifiers fine-tune the conditions that determine how collectors interact with mineral surfaces. Their primary role is to ensure that only the desired minerals float while unwanted gangue remains depressed. In this sense, modifiers act as the "chemical managers" of the flotation system, maintaining both selectivity and efficiency in the separation process.

The principle behind the action of modifiers lies in their ability to alter the surface chemistry of minerals and the composition of the liquid phase. They can change the electrical charge, ion concentration, or surface potential of mineral particles, influencing how collectors adsorb onto these surfaces. By adjusting pH, dissolving surface films, activating certain minerals, or preventing collector adsorption on gangue minerals, modifiers control which particles become hydrophobic and which remain hydrophilic. They also help stabilize the pulp environment by preventing undesirable chemical reactions—such as oxidation or precipitation—that could interfere with reagent performance. Through this selective control, modifiers enable flotation circuits to target specific minerals even within complex ore mixtures.

Modifiers are typically grouped into four main categories based on their function: pH regulators, activators, depressants, and dispersants/flocculants. Each plays a distinct role in conditioning the pulp for selective flotation. pH regulators control the acidity or alkalinity of the pulp, optimizing collector and reagent performance. pH influences mineral surface charge, collector ionization, and reagent solubility. Common alkaline regulators include lime (CaO) and sodium carbonate (Na_2CO_3), which raise the pH in sulphide and oxide flotation, promoting selective adsorption of xanthates on sulphide minerals while depressing iron-bearing gangue such as pyrite. In contrast, acids such as sulfuric acid (H_2SO_4) and hydrochloric acid (HCl) are used to lower pH in oxide, phosphate, and rare earth flotation systems, where mildly acidic conditions improve the performance of collectors like hydroxamates and fatty acids. Maintaining the correct pH range is essential to ensure collectors remain in the correct ionic form for effective mineral attachment.

Activators enhance the ability of collectors to adsorb onto mineral surfaces, particularly when a mineral would otherwise be unresponsive. They achieve this by chemically modifying the mineral surface—often by introducing or exchanging metal ions that create active bonding sites. A classic example is copper sulphate ($CuSO_4$), which activates sphalerite (ZnS) by depositing copper ions (Cu^{2+}) on the surface, allowing xanthate collectors to attach. Lead nitrate ($Pb(NO_3)_2$) is another activator, commonly used to promote flotation of pyrite and oxidized minerals in gold and base metal circuits. In rare earth and oxide flotation, metal salts like ferric chloride ($FeCl_3$) or aluminium sulphate ($Al_2(SO_4)_3$) can activate surfaces for effective adsorption of hydroxamate collectors. Without such activators, many valuable minerals would remain unresponsive, resulting in reduced recovery.

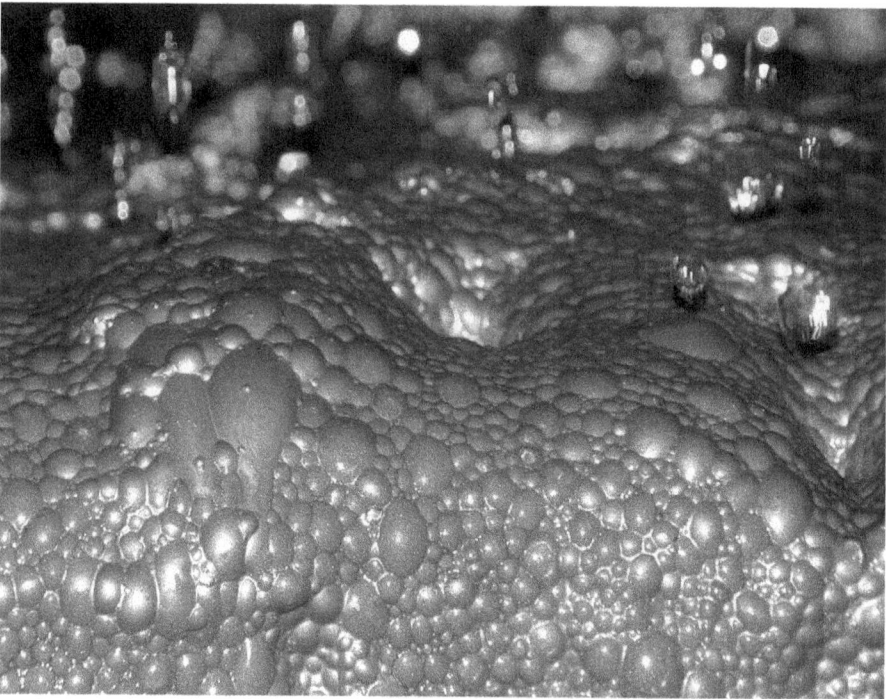

Figure 51: Copper-sulphide foam on a Jameson Cell, Prominent Hill, South Australia. Geomartin, CC BY-SA 4.0, via Wikimedia Commons.

Rare Earth and Critical Mineral Operations and Processing

Depressants act in the opposite way—they prevent unwanted minerals from floating by inhibiting collector adsorption or rendering the mineral surface hydrophilic. This is particularly important when multiple minerals respond to the same collector. For example, sodium cyanide (NaCN) and zinc sulfate ($ZnSO_4$) are used to depress pyrite and sphalerite in copper flotation circuits, allowing chalcopyrite to float selectively. Organic depressants such as starch, dextrin, and guar gum coat the surfaces of silicate and carbonate gangue, making them water-attractive and less likely to float. These are widely used in phosphate, lithium, and rare earth flotation. Sodium silicate (Na_2SiO_3) serves both as a depressant and dispersant, preventing slime coating and depressing quartz in non-metallic and oxide ore systems. Depressants are often used alongside pH regulators and activators to achieve precise mineral separation in complex ores.

Dispersants and flocculants influence pulp rheology and particle behavior during flotation. Dispersants such as sodium silicate, tetrasodium pyrophosphate, and polyphosphates prevent fine particles and clays from agglomerating or coating valuable mineral surfaces. This improves pulp dispersion, enhances selectivity, and reduces slime-related problems. In contrast, flocculants are sometimes used before or after flotation to encourage fine particles to aggregate, simplifying their removal or settling. Maintaining proper dispersion and pulp stability is particularly important in fine-grained rare earth and lithium ore systems, where slimes can easily interfere with separation efficiency.

The performance of modifiers depends on several operational and chemical factors. The pH and redox potential (Eh) influence the chemical species present in solution and the charge on mineral surfaces. Mineral composition and oxidation state determine whether specific activators or depressants are needed. The concentration and addition order of modifiers are also critical—adding reagents in the wrong sequence can cause interference or reduced selectivity. Temperature and ionic strength affect reaction rates and the solubility of reagent complexes, influencing overall flotation efficiency. Balancing these factors ensures that modifiers work synergistically with collectors and frothers to produce effective and selective separation.

In the processing of critical and rare earth minerals, modifiers are indispensable. For example, in monazite, bastnäsite, and xenotime flotation, sodium silicate and starch are used as depressants to suppress quartz and carbonate gangue, while mildly acidic conditions are maintained for hydroxamate collectors. In lithium processing, particularly spodumene flotation, sodium carbonate and lime control pulp alkalinity, and sodium silicate depresses mica and feldspar to enhance selectivity. For tungsten ores like scheelite, sodium carbonate regulates pH, sodium silicate suppresses silicate gangue, and lead nitrate acts as an activator. In graphite flotation, organic depressants such as starch or carboxymethyl cellulose (CMC) are used during cleaning stages to minimize fine gangue entrainment while maintaining graphite's natural hydrophobicity.

Overall, modifiers are the chemical regulators of flotation, ensuring that the pulp environment and mineral surfaces are conditioned for optimal separation. Through careful control of pH, activation, and depression, they make selective flotation possible even in complex ore systems. In modern mineral processing—especially for fine-grained, rare earth, or energy-critical minerals—precise use of modifiers, in combination with collectors and frothers, is essential for achieving high recovery, product purity, and process stability.

Finally, pH modifiers play a crucial role in determining reagent activity. Alkaline conditions (pH 9–11) are typically used for sulphide mineral flotation with xanthate collectors, while acidic environments (pH 4–6) are preferred for oxide and rare earth minerals using hydroxamate collectors.

pH modifiers are reagents used in flotation to control the acidity or alkalinity of the pulp. Since the surface chemistry of minerals and the behaviour of flotation reagents depend heavily on pH, these modifiers play a crucial role in maintaining optimal conditions for selective separation. The pH of the pulp influences how collector molecules ionize, how minerals acquire surface charge, and how various ions interact in solution—all of which determine whether a mineral surface becomes hydrophobic or remains hydrophilic.

In flotation systems, different minerals respond best under specific pH conditions. For example, sulphide minerals such as chalcopyrite or galena typically float best under alkaline conditions (pH 8–11), where xanthate collectors form stable, hydrophobic films on the mineral surface. In contrast, oxide and rare earth minerals often require acidic to neutral pH levels (pH 4–7) for collectors such as hydroxamates or fatty acids to perform effectively. Maintaining the correct pH range ensures that collectors are in their most active chemical form and that unwanted gangue minerals are either depressed or rendered unreactive.

Common alkaline pH modifiers include lime (CaO), sodium carbonate (Na_2CO_3), and sodium hydroxide (NaOH). Lime is widely used because it not only raises pH but also precipitates metal ions that might otherwise activate unwanted minerals like pyrite. Sodium carbonate provides smoother pH control in oxide and non-sulphide flotation systems, while sodium hydroxide offers rapid pH adjustment when strong alkalinity is required.

Acidic pH modifiers, such as sulfuric acid (H_2SO_4) and hydrochloric acid (HCl), are used to lower the pulp pH in flotation of minerals like phosphates, fluorite, and rare earth elements. Acidic environments can improve collector adsorption for oxide minerals, enhance selectivity, and dissolve surface films that inhibit flotation.

The use of pH modifiers also affects other reagents. For instance, the stability of frothers and collectors depends on the ionic balance in solution, and changes in pH can enhance or suppress their effectiveness. Additionally, pH interacts with depressants and activators—for example, in copper–zinc systems, lime at high pH depresses sphalerite, while a lower pH allows copper sulphate to activate it.

In critical and rare earth mineral processing, pH control is especially important. For monazite and bastnäsite, a mildly acidic pH (around 5–6) enhances adsorption of hydroxamate collectors while allowing sodium silicate or starch to depress gangue minerals like quartz and calcite. In spodumene flotation, an alkaline pH (9–10) using sodium carbonate or lime enhances the effectiveness of fatty acid collectors after thermal activation of the ore. For scheelite (tungsten ore), maintaining a pH of around 9 ensures optimal adsorption of collectors while allowing sodium silicate to suppress silicate gangue.

Flotation is particularly important in critical mineral processing. In rare earth element (REE) beneficiation, minerals such as monazite, bastnäsite, and xenotime are floated using hydroxamate collectors at mildly acidic pH, often with starch depressants to suppress quartz and carbonate gangue. In lithium processing, spodumene ($LiAlSi_2O_6$) is floated using amine or fatty acid collectors after thermal activation and desliming, with sodium hydroxide or lime used to adjust pH. Graphite flotation takes advantage of its natural hydrophobicity, enhanced by small additions of kerosene or diesel oil as collectors and MIBC as a frother. For tungsten and tin ores, oxide minerals such as wolframite and cassiterite are floated using hydroxamate or phosphonic acid collectors, often after gravity concentration to improve efficiency.

Factors Affecting Recovery and Selectivity

The efficiency of flotation processes in mineral processing depends critically on the key performance metrics of recovery and selectivity. Recovery refers to the percentage of valuable minerals successfully collected in the concentrate, while selectivity reflects the effectiveness of separating valuable minerals from gangue. Attaining a balance between high recovery and selectivity is vital for generating a marketable concentrate, as maximizing one metric can often compromise the other [186, 187].

Mineralogical Characteristics: The intrinsic characteristics of ores significantly influence flotation outcomes. Key attributes include mineral type, particle size, and liberation degree. For instance, very fine particles (<20 μm) exhibit poor collision efficiency with bubbles due to low momentum, leading to reduced recovery. Conversely, coarse particles (>150 μm) are prone to detachment from bubbles because of their greater mass, hence optimal flotation typically occurs within the 20-150 μm range [188, 189]. In addition, incomplete liberation, where valuable and gangue minerals co-exist within a single particle, can adversely affect selectivity since collectors may adsorb onto mixed surfaces instead of the target valuable minerals [190]. Furthermore, surface oxidation can impede collector adsorption by decreasing the hydrophobicity of mineral surfaces [191, 192].

Pulp Chemistry: The chemical properties of the flotation pulp are paramount in dictating recovery and selectivity. The pH level plays a crucial role in determining the surface charge of minerals and therefore influences collector ionization and the solubility of various metal ions, which directly affects collector adsorption efficiency [193]. Redox potential (Eh) also affects flotation performance; sulphide minerals require particular Eh conditions to maintain collector-active sites. Inadequate management of these chemical parameters, particularly in recycled process waters, can lead to inconsistent flotation outcomes and reduced selectivity [194, 195].

Reagent Type and Dosage: The selection and application of flotation reagents are critical for optimizing flotation outcomes. Collectors must be matched to the mineral's surface chemistry, and both insufficient and excessive dosages can have detrimental effects. Insufficient dosage can lead to low hydrophobicity, while excessive usage might result in gangue flotation, undermining selectivity [196]. Frothers also play a significant role by controlling bubble size and froth stability, which influences both recovery and selectivity. Overly stable froths may entrap gangue particles, while insufficient froth stability can cause valuable minerals to collapse back into the pulp [197, 198].

Air Flow Rate and Bubble Characteristics: The rate of air inflow and the characteristics of bubbles directly impact the likelihood of particle-bubble collisions, which is essential for effective flotation. Smaller bubbles improve particle attachment due to their larger surface area; however, excessively fine bubbles can hinder froth drainage, thereby trapping unwanted gangue within the froth phase [197, 199]. Proper optimization of air flow rates is essential; low rates can limit bubble availability while high rates can generate turbulence, destabilizing the froth or detaching particles prematurely [200].

Pulp Density and Hydrodynamics: Pulp density and the dynamics of mixing are significant factors influencing flotation efficiency. Increased pulp density enhances collision frequency, thereby improving attachment but may hinder bubble rise and stability [191]. Conversely, lower density can result in better particle-bubble interactions but may dilute reagent concentrations, impacting recovery. Therefore, achieving optimal mixing and residence times is critical to maintain effective flotation processes [196, 201].

Froth Stability and Structure: The properties of the froth itself are crucial in determining the overall effectiveness of the flotation process. A stable froth layer can promote high recovery by retaining valuable minerals, but over-stability can lead to the inclusion of fine gangue particles, compromising selectivity [202]. Optimal froth texture, bubble size, and drainage rates must be continuously monitored and adjusted to achieve a clean and high-yield concentrate [197, 199].

7. Circuit Design and Operating Conditions

The overall design of the flotation circuit greatly influences performance. Employing a multi-stage arrangement with rougher, cleaner, and scavenger steps facilitates higher recovery in initial stages while enabling better selectivity in later stages [203, 204]. The strategic arrangement and management of recycle streams and reagent distribution play a vital role in maintaining consistent conditions throughout the flotation process [196, 200].

8. Temperature and Water Quality

Lastly, temperature and the quality of water used in flotation processes cannot be neglected. Temperature variations can affect pulp viscosity, reagent solubility, and reaction kinetics, thus impacting recovery and selectivity. High temperatures tend to increase reaction rates but may destabilize froth or lead to reagent degradation [192, 193]. The quality of water, including salinity and hardness, needs to be managed meticulously, especially in systems using recycled or seawater to ensure reliable flotation chemistry [196, 201].

In summary, achieving optimal flotation involves navigating a complex interplay of mineral properties, pulp chemistry, reagent management, and operational parameters. Given that high recovery often conflicts with selectivity, it is essential to finely tune each contributing factor to maximize both recovery and the purity of the concentrate, particularly in processing complex ores such as rare earth elements and lithium [205, 206].

Example Flowsheets for REE and Lithium Ores

Flowsheets for rare earth element (REE) and lithium ore processing are designed to match the mineralogical characteristics of each deposit, with circuits tailored for optimal liberation, selectivity, and recovery. These flowsheets typically integrate comminution, classification, and beneficiation processes, with specific choices of equipment and grind size based on ore hardness, texture, and target minerals.

The processing of REE ores begins with primary crushing using a jaw or gyratory crusher, followed by secondary and tertiary crushing through cone crushers to achieve a 10–12 mm feed size. This crushed material is then ground in a ball mill to a target size of P80 around 75–150 µm. After grinding, the slurry is deslimed using hydrocyclones or classified through air separators to remove fine clays and slimes that reduce flotation performance and recovery efficiency.

For bastnäsite and monazite carbonatite ores, the primary beneficiation step is froth flotation, often combined with stirred-mill regrinding to achieve fine liberation where gangue minerals are complex or interlocked. In contrast, monazite (LREE phosphate) ores may follow similar comminution but proceed to magnetic or electrostatic separation or even direct chemical cracking instead of flotation, depending on mineral associations.

Xenotime (HREE phosphate) requires much finer liberation, typically achieved by fine or ultrafine grinding below 50 µm using stirred mills. The concentrated pulp is then processed by high-intensity magnetic or electrostatic separation to isolate heavy rare earth minerals. Ion-adsorption REE clays, common in southern China, are treated differently—they require little or no grinding and rely instead on desliming and direct leaching, since mechanical processing can destroy adsorption sites essential for recover.

This progression—from crushing to grinding, classification, and finally flotation, magnetic, or leaching separation—illustrates the flexibility and mineral-specific nature of REE flowsheets.

Processing of hard-rock lithium ores, particularly spodumene pegmatites, follows a similar structure but emphasizes thermal and chemical treatment. The ore is first crushed by jaw or gyratory crushers, followed by cone or impact crushers for size reduction. Grinding circuits may involve SAG, ball, and stirred mills, producing a tightly controlled particle size distribution managed by hydrocyclones, spirals, or air classifiers.

A critical step unique to lithium is roasting at around 1000°C to convert α-spodumene into β-spodumene, which has higher reactivity. After roasting, the material may be sent through flotation circuits using fatty acid or amine collectors, or directly to acid or carbonate leaching for lithium extraction. In all cases, the goal is precise particle size control and selective reagent use to maximize lithium recovery while minimizing contamination from gangue minerals.

Both REE and lithium flowsheets emphasize mineral-specific processing: comminution and classification are finely tuned to ore texture and hardness, while beneficiation pathways—flotation, magnetic or electrostatic separation, or leaching—are selected based on mineral chemistry. REE circuits focus on complex multi-stage physical separation, while lithium circuits combine mechanical and thermal activation with selective flotation or leaching to achieve high-purity recovery.

Below are compact flowsheet "patterns" you can adapt. Each shows the main unit operations (left → right), typical setpoints, and the critical control notes that make or break performance.

Rare Earth Elements (REE)

1) Bastnäsite-Dominant Carbonatite (e.g., Mountain Pass–style)

Crushing & Grind:

Jaw/Gyratory → Cone → Ball mill (P80 75–150 μm) → Hydrocyclone (deslime <10–20 μm)

Upstream cleaning:

LIMS (remove magnetite/Fe-oxides) → Deslime thickener (clay control)

Flotation train (fatty acid / hydroxamate blend):

Rougher (pH 8–9 or mildly acidic for hydroxamates) → Regrind (stirred mill; P80 25–40 μm) → Cleaner 1–3 → Scavenger

Solids handling & concentrate upgrade:

Dewater (thickener + pressure filter) → Calcine (carbonate/organic removal, optional) → Magnetic/E-static trim (if needed)

Key notes: Tight desliming before flotation; reagent regime balances calcium carbonates vs bastnäsite; regrind sharpens liberation and reduces entrained barite/calcite.

2) Monazite/Xenotime in Heavy-Mineral Sands (dry plant emphasis)

Mining & primary prep:

Scrub → Screen (2–3 mm) → **Spirals** (gravity rougher)

Magnetic/electrostatic split:

LIMS (ilmenite) → HIMS/HGMS (weakly magnetic fractions) → High-tension roll (HTR) / Plate-screen e-stat (conductors vs non-conductors)

Finishing:

Cleaner e-stat passes → Hand magnet / rare-earth magnets (polish) → Product blends (monazite/xenotime/zircon/rutile)

Key notes: Dry circuits demand strictly controlled moisture; splitters and recycle loops tune monazite vs xenotime purity; radioactivity management dictates product handling.

3) Ionic-Adsorption Clay REE (South China–style)

Minimal comminution:

Rip/loam → Slurry → **Deslime** (cyclone / rake thickener)

Leach & capture:

Salt leach ($NH_{42}SO_4$ or Na_2SO_4; pH ~4.5–5.5) in counter-current → CCD/clarification → IX/solvent extraction (group separation) → Precipitation ($RE(OH)_3$ or carbonate)

Residue handling:

Washed tails → Paste/thickened disposal

Key notes: PSD control protects permeability; ammonium management and effluent polishing are core; essentially no grinding—recovery hinges on ion exchange kinetics.

Lithium

4) Spodumene Pegmatite (Hard-Rock, Flotation Route)

Front end:

Jaw/Gyratory → Cone → SAG/ball (option) → DMS (2.6–2.8 SG cut) to reject quartz/feldspar

Roast & conversion:

Kiln: $\alpha \rightarrow \beta$ at ~1000–1100 °C → Crush/trim (HSI/VSI) → Ball/stirred mill (P80 75–106 μm)

Flotation (fatty acid or amine system):

Condition/deslime → Rougher → Regrind (P80 25–40 μm) → Cleaners (2–4) → Concentrate (5.5–6.0% Li_2O)

Refining options:

Sulfation/bicarbonation leach (chemical grade) **or** saleable conc → converter/chemical plant

Key notes: DMS improves mill throughput; roast mineralogy drives collector choice; strict slimes control prevents mica carryover; Na_2CO_3/lime for pH 9–10.

5) Spodumene Pegmatite (Dense-Media + Direct Leach Variant)

Comminution & DMS:

As above → High-grade DMS concentrate

Thermal activation:

Roast α→β

Direct leach:

Acid bake (H_2SO_4 200–250 °C) → Water leach → Impurity removal (Fe/Al/Mg) → Li_2CO_3/LiOH precipitation → Crystallization

Key notes: Fewer flotation reagents; energy shifts from milling to thermal/chemical; gypsum/sulphate handling and silica gel control are the pain points.

6) Lepidolite/Mica-Rich Lithium (Mica Flowsheet)

Front end:

Crush → Ball mill (P80 75–106 μm) → Deslime

Mica concentration:

Froth flotation (cationic/amine or specialty collectors) → Mica-Li concentrate

Conversion & leach:

Roast with lime/Na_2SO_4 (lithium metasilicate route **or** sulfate route) → Water leach → IX/SX polish → Li_2CO_3/LiOH crystallization

Key notes: Reagent suite suppresses quartz/feldspar; K/Rb/Cs credits possible; viscosity and slimes management are critical for stable flotation.

7) Lithium Brine (High-Level)

Brine conditioning:

Pretreat (Fe/Mn removal, filtration) → Evaporation ponds or DLE (sorbent/IX/solvent membranes)

Concentration & conversion:

Li-rich eluate → Polishing (boron/magnesium removal) → Li_2CO_3 precipitation (soda ash) **or** LiOH via electrolysis/caustic routes → Drying/granulation

Rare Earth and Critical Mineral Operations and Processing

Key notes: Choose ponds vs DLE by climate and Mg/Li ratio; DLE cuts residence time but adds reagent/energy complexity; water balance and impurity bleed are decisive.

Key Terms and Concepts	

Physical Separation: A mineral-processing approach that uses differences in physical properties—such as density, magnetism, or electrical conductivity—to separate valuable minerals from gangue.

Gravity Separation: A process that relies on differences in specific gravity to concentrate heavier minerals using devices such as jigs, spirals, and shaking tables.

Magnetic Separation: A technique that exploits differences in magnetic susceptibility between minerals, ranging from low-intensity separators for ferromagnetic minerals to high-gradient systems for weakly magnetic ones.

Electrostatic Separation: A dry-process method that separates minerals based on electrical conductivity and surface charge using high-voltage fields.

Flotation: A surface-chemistry-based process that separates minerals by attaching hydrophobic particles to air bubbles, producing a froth enriched in valuable minerals.

Collector: A chemical reagent that selectively renders the surface of target minerals hydrophobic to facilitate bubble attachment during flotation.

Frother: An additive used in flotation to create and stabilise air bubbles in the froth layer, improving mineral recovery.

Depressant: A reagent that prevents specific minerals from floating, enhancing the selectivity of the flotation process.

Activator: A reagent that modifies mineral surfaces to increase their floatability or restore flotation response after depression.

Selectivity: The ability of a separation process to distinguish between valuable and non-valuable minerals under specific operating conditions.

Key Terms and Concepts

Recovery: The percentage of the valuable mineral or metal recovered from the ore during a beneficiation process.

Flowsheet: A diagram showing the sequence of physical separation steps—such as gravity, magnetic, and flotation circuits—used to process an ore.

Concentrate: The enriched product containing a high proportion of valuable minerals following separation.

Tailings: The waste material remaining after valuable minerals have been removed during processing.

Reagent Regime: The combination and dosage of flotation chemicals optimised to achieve target recovery and grade.

Chapter 5 Review Questions

Principles of Physical Separation

1. Which of the following best describes the principle behind physical separation methods?

(a) Exploiting differences in chemical composition

(b) Exploiting differences in physical properties such as density, magnetism, and conductivity

(c) Using acids and bases to dissolve gangue minerals

(d) Melting minerals for selective recovery

2. Explain the fundamental purpose of physical separation processes in mineral beneficiation.

3. Describe how physical properties such as density, magnetic susceptibility, and electrical conductivity are used to design separation processes in rare earth and critical mineral circuits.

Gravity Separation

4. Gravity separation works primarily on which property?

(a) Surface charge

(b) Magnetic susceptibility

(c) Specific-gravity (density) difference

(d) Chemical reactivity

5. Name three types of gravity-separation equipment and briefly describe their mode of operation.

6. What particle-size ranges are generally suited to jigs, spirals, and shaking tables?

7. Compare the design and operating principles of jigs, spiral concentrators, and shaking tables, explaining which ore types each is best suited for.

Magnetic Separation

8. Which separator operates below 0.2 Tesla and is used for strongly magnetic minerals such as magnetite?

(a) High-Intensity Magnetic Separator (HIMS)

(b) High-Gradient Magnetic Separator (HGMS)

(c) Low-Intensity Magnetic Separator (LIMS)

(d) Drum-type Electrostatic Separator

9. Differentiate between LIMS, HIMS, and HGMS in terms of magnetic-field strength and applications.

10. Why are high-gradient separators preferred for fine rare-earth minerals such as xenotime?

11. Explain how magnetic separation is integrated into a rare-earth processing flowsheet and why desliming is critical before magnetic stages.

Electrostatic Separation

12. Electrostatic separation depends on differences in:

(a) Density

(b) Electrical conductivity

(c) Magnetic susceptibility

(d) Particle shape

13. Describe the principle of operation of a High-Tension Roll (HTR) separator.

14. Why must feed for electrostatic separation be completely dry?

15. Discuss how electrostatic separation complements magnetic separation in heavy-mineral and rare-earth processing circuits.

Froth Flotation

16. Which flotation reagent renders mineral surfaces hydrophobic?

(a) Frother

(b) Collector

(c) Depressant

(d) Activator

17. List the four main classes of flotation reagents and describe their primary functions.

18. What is the effect of excessive frother dosage?

19. Explain how collectors, frothers, and modifiers interact to control flotation selectivity in rare-earth or lithium-ore processing.

Recovery, Selectivity, and Process Efficiency

20. Which of the following factors most strongly affects selectivity in gravity separation?

(a) Collector dosage

(b) Density contrast between minerals

(c) Pulp viscosity

(d) Froth depth

21. List three physical or chemical parameters that influence recovery and selectivity.

22. Discuss how particle size, density, and surface chemistry determine the efficiency of separation in gravity, magnetic, and flotation methods.

Process Flowsheets and Integration

23. Outline the major stages in a typical rare-earth element (REE) processing flowsheet.

24. Describe how comminution and classification prepare lithium ores for flotation or leaching.

25. Explain the sequence "Gravity → Flotation" in lithium processing and the function of each stage.

Safety and Environmental Considerations

26. Which of the following is a major safety consideration in operating electrostatic separators?

(a) Excessive vibration

(b) High-voltage discharge and grounding

(c) Oxygen deficiency

(d) Slurry spillage

27. List two environmental-management practices required when operating flotation circuits.

28. Discuss safety and environmental measures that should be in place when operating magnetic, electrostatic, and flotation equipment.

Richard Skiba

Chapter 6

Hydrometallurgical Techniques

The previous chapter examined how physical separation methods—such as gravity, magnetic, electrostatic, and flotation techniques—concentrate valuable minerals by exploiting differences in their physical properties. These beneficiation steps yield mineral concentrates enriched in rare earths, lithium, and other critical metals but still containing chemical impurities that limit their direct use in manufacturing. To achieve the high purity demanded by modern technologies, these concentrates must undergo further refinement through chemical processing. Hydrometallurgical techniques form the next critical stage in this transformation. By using aqueous chemical reactions, hydrometallurgy dissolves target metals from their mineral matrices, selectively separates them through solvent extraction and ion exchange, and finally recovers them as purified solids or oxides. This chapter introduces the scientific and industrial foundations of these methods, illustrating how leaching, purification, and recovery processes together convert beneficiated ores into high-value products. It bridges the gap between physical beneficiation and final refining, emphasizing hydrometallurgy's role in achieving high recovery, purity, and sustainability in modern critical-mineral production.

Learning Outcomes	

This chapter aims to give you the ability to: 1. Explain the fundamental principles of hydrometallurgy and its role in extracting and refining rare earth and critical minerals from ore concentrates and secondary sources. 2. Differentiate between various leaching methods—including acid, alkaline, pressure, and bioleaching—and evaluate their applicability to different mineral systems and desired products. 3. Describe the chemical reactions and parameters that control leaching efficiency, such as temperature, pH, reagent concentration, redox potential, and particle size.	

Learning Outcomes	

4. Identify and explain the principles of solvent extraction, including phase equilibria, selectivity, and the use of extractants, diluents, and modifiers for the separation of rare earth elements.
5. Describe the use of ion exchange processes for purification and recovery, and compare them with solvent extraction in terms of selectivity, cost, and scalability.
6. Explain the precipitation techniques used for product recovery, purification, and conversion of dissolved species into oxides, hydroxides, or salts.
7. Interpret process flow diagrams that integrate leaching, extraction, and precipitation stages in hydrometallurgical circuits for REE, lithium, and cobalt production.
8. Assess the environmental and safety considerations associated with reagent use, effluent management, and waste treatment in hydrometallurgical operations.
9. Relate bench-scale and pilot-scale hydrometallurgical testing to process design, optimisation, and commercial-scale implementation.
10. Demonstrate an understanding of how hydrometallurgical techniques contribute to achieving higher recovery, purity, and sustainability compared to traditional pyrometallurgical methods.

Leaching Principles: Acid, Alkaline, and Pressure Leaching

Leaching is a hydrometallurgical process that involves dissolving valuable minerals from an ore using a suitable solvent, typically an acid, alkali, or other chemical reagent. It is particularly important in the extraction and refining of rare earth elements (REEs), lithium, nickel, cobalt, uranium, and other critical minerals. The main goal is to convert the target element into a soluble compound that can then be recovered from solution by precipitation, solvent extraction, or ion exchange.

A hydrometallurgical process is a method used to extract and purify metals from ores, concentrates, or recycled materials through aqueous (water-based) chemical reactions. It is one of the three main metallurgical routes—alongside pyrometallurgy (high-temperature smelting) and electrometallurgy (electrochemical methods). Hydrometallurgy operates at relatively low temperatures and relies on chemical dissolution, making it suitable for treating both oxide and low-grade ores, as well as complex or refractory materials.

The process typically involves three main stages:

1. **Leaching** – The valuable metals are dissolved from the solid ore using acidic, alkaline, or neutral aqueous solutions (e.g., sulfuric acid, hydrochloric acid, or sodium hydroxide). This forms a pregnant leach solution (PLS) containing dissolved metal ions.

2. **Solution Concentration and Purification** – The leach solution is treated to separate impurities and concentrate the target metals using techniques such as solvent extraction, ion exchange, or precipitation.

3. **Metal Recovery** – The purified metal ions are recovered from solution through electrowinning, precipitation, or crystallization, yielding the final metallic product or compound.

Hydrometallurgical processing is widely used for nickel, cobalt, copper, uranium, zinc, rare earth elements (REEs), and lithium. It offers several advantages, including high metal recovery, selectivity, and lower energy consumption compared to pyrometallurgical methods. However, it also requires careful management of reagents and waste streams to minimize environmental impacts.

In essence, hydrometallurgy enables efficient and selective metal extraction through liquid-phase chemistry, making it a cornerstone of modern critical mineral and rare earth element beneficiation.

Acid Leaching

Acid leaching uses mineral acids such as sulphuric acid (H_2SO_4), hydrochloric acid (HCl), or nitric acid (HNO_3) to dissolve metal oxides, carbonates, or sulphides. This method is commonly used for lateritic nickel ores, rare earth phosphates (monazite, bastnäsite), and uranium ores.

The acid reacts with metal-bearing minerals to form soluble metal salts and insoluble residues. For example:

$$MCO_3 + 2HCl \rightarrow MCl_2 + H_2O + CO_2$$

Acid leaching is widely applied in the extraction of several critical minerals, including rare earth elements (REEs), lithium, and uranium. In rare earth processing, acids such as hydrochloric acid (HCl) or sulfuric acid (H_2SO_4) are used to dissolve REE phosphates and oxides. These acids effectively break down minerals like monazite and bastnäsite, converting the contained rare earths into soluble chlorides or sulphates that can then be separated and purified.

In lithium processing, acid leaching follows a crucial roasting stage that transforms α-spodumene into β-spodumene at around 1000°C. This phase change enhances the mineral's reactivity, allowing lithium to be efficiently extracted using sulphuric or hydrochloric acid. The resulting lithium sulphate or lithium chloride solutions form the basis for downstream production of lithium carbonate or lithium hydroxide.

For uranium ores, sulphuric acid leaching converts uranium oxide (U_3O_8) into soluble uranyl ions (UO_2^{2+}), which are then recovered through solvent extraction or ion exchange. This process is especially common in sandstone-hosted and calcrete uranium deposits, where acid strength and oxidant control are critical for maximizing yield while minimizing reagent consumption.

Acid leaching offers high recovery rates and is particularly effective for oxide ores, making it one of the most versatile hydrometallurgical methods. However, its limitations include the generation of acid waste streams and the need for corrosion-resistant equipment to handle aggressive solutions and elevated temperatures during processing

Process steps:

1. **Comminution and Preparation:** The ore is crushed and ground to increase surface area, ensuring efficient contact between the acid and mineral particles.

2. **Acid Contacting:** The prepared ore is mixed with the acid in leach tanks, heaps, or autoclaves. The choice of acid and concentration depends on ore type and mineralogy.

3. **Leaching Reaction:** The acid dissolves the target metals while leaving most gangue minerals intact. Temperature, acid concentration, and agitation are controlled to optimize dissolution rates.

4. **Solid-Liquid Separation:** After leaching, the slurry is filtered or settled to separate the pregnant leach solution (PLS), which contains dissolved metals, from the solid residue or tailings.

5. **Metal Recovery:** The dissolved metals are recovered from the PLS using methods such as solvent extraction, ion exchange, precipitation, or electrowinning.

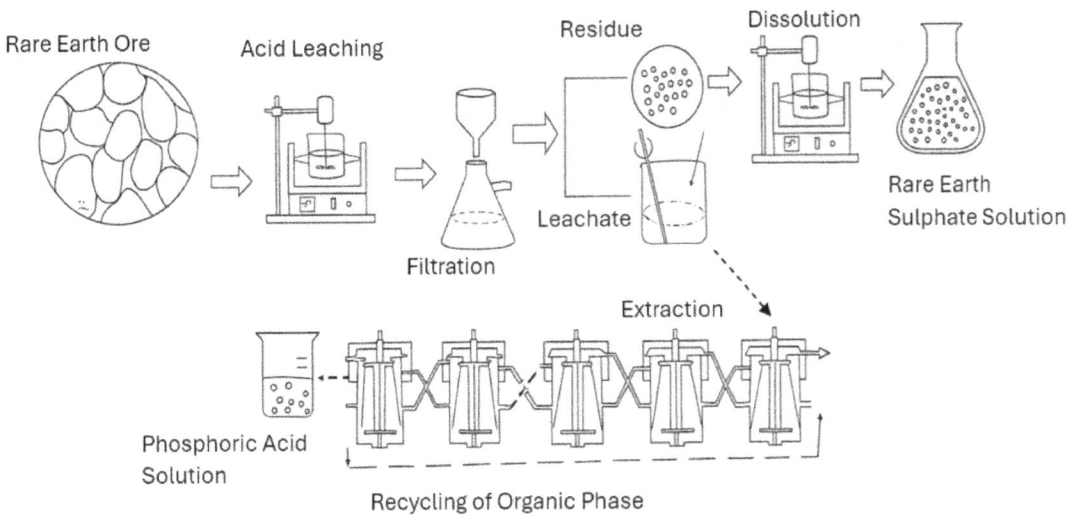

Figure 52: Effect of phosphoric acid on leaching of monazite during low-temperature sulfuric acid cyclic leaching process,

In the processing of rare earth elements (REEs), the selection of acid and the method of acid contacting depend on the mineral type and desired product stream. Minerals such as monazite and bastnäsite, which are phosphate and carbonate minerals respectively, typically undergo leaching with sulfuric acid (H_2SO_4) at elevated temperatures ranging from 200°C to 250°C. This leaching process effectively converts rare earth phosphates into soluble sulphates, allowing for the recovery of metals. The Bayan Obo deposit in China employs a method where bastnäsite ore is initially calcined to decompose its carbonate matrix, followed by H_2SO_4 leaching [207, 208]. Additionally, hydrochloric acid (HCl) is sometimes chosen when chloride-based feedstocks are preferred for downstream solvent extraction [209].

In the case of ion-adsorption REE clays from southern China, milder treatments with ammonium sulphate ((NH_4)$_2SO_4$) or ammonium carbonate ((NH_4)$_2CO_3$) at ambient temperatures release REE^{3+} ions through ion exchange without significant damage to the clay structure [208].

Similarly, in lithium ore processing, the choice of acid varies according to the specific mineral. For spodumene ($LiAlSi_2O_6$), which is roasted at approximately 1000°C to convert the α-phase into the more reactive β-phase, leaching is performed with sulfuric acid (H_2SO_4) at temperatures between 90°C and 250°C. This process yields lithium sulphate (Li_2SO_4), which can be processed into lithium carbonate or lithium hydroxide [210, 211]. In contrast, clay-based lithium ores like hectorite and montmorillonite may be leached with either HCl or H_2SO_4, depending on mineralogical characteristics, particularly magnesium content [212]. An illustrative example of such processing can be found in the Thacker Pass project in the United States, where efficient lithium recovery is achieved through H_2SO_4 leaching at around 200°C [213, 214].

In nickel and cobalt laterite operations, high-pressure acid leaching (HPAL) with sulfuric acid is conducted under extreme conditions—approximately 250°C and 40–50 atmospheres. This process efficiently dissolves nickel and cobalt oxides into soluble sulphates while precipitating iron as hematite [215, 216]. A related sector is uranium processing, where sulfuric acid is frequently used alongside oxidizing agents such as ferric iron (Fe^{3+}) or hydrogen peroxide (H_2O_2) to facilitate the conversion of U^{4+} to U^{6+}, resulting in soluble uranyl ions (UO_2^{2+}) [217, 218]. Acid heap leaching is particularly common in extracting uranium from low-grade deposits in regions such as Namibia and Kazakhstan [219].

For titanium and zirconium minerals like ilmenite ($FeTiO_3$), concentrated sulfuric acid is utilized in a digestion phase to manufacture titanium oxysulphate, serving as a precursor for titanium dioxide (TiO_2) pigment [220, 221]. Alternatively, the chloride process, which employs hydrochloric acid (HCl), provides a route to achieve high-purity TiO_2 for industrial applications [222]. While tungsten ores like scheelite are primarily processed under alkaline conditions, acidic systems utilizing HCl or H_2SO_4 may be applied post-pre-treatment under controlled conditions [209, 223]. Moreover, in the case of vanadium and graphite ores, acid leaching is instrumental in removing impurities or recovering valuable metals. The leaching of vanadium-bearing shales occurs using sulfuric acid or nitric acid (HNO_3), while graphite can be purified using HCl to eliminate silicate and iron oxide impurities [224].

Acid leaching offers several significant advantages in mineral processing, particularly for oxide and phosphate ores. It achieves high metal recovery rates by efficiently dissolving target metals into solution, often surpassing the performance of other leaching methods. This process is also highly versatile—it can be adapted for a wide variety of minerals, including rare earth elements (REEs), lithium, and uranium. The conditions of acid leaching—such as acid type, concentration, temperature, and oxidation potential—can be precisely adjusted to enhance selectivity and optimize recovery for specific ores. This flexibility makes it an essential technique in modern hydrometallurgy, especially for processing complex or refractory mineral systems.

Despite its effectiveness, acid leaching presents several challenges. The process generates acidic waste streams that require careful neutralization and treatment before disposal to prevent environmental contamination. The corrosive nature of the leach solution also necessitates the use of specialized, corrosion-resistant materials for tanks, piping, and processing equipment, which increases capital and maintenance costs. Moreover, acid leaching is less effective for sulphide ores, which typically need pre-oxidation or roasting to break down the sulphide matrix before metals can be dissolved. As a result, while acid leaching remains a powerful and adaptable extraction method, it demands careful management of waste, equipment integrity, and process chemistry to ensure safe and sustainable operation.

Alkaline Leaching

Alkaline leaching uses reagents such as sodium hydroxide (NaOH), sodium carbonate (Na_2CO_3), or ammonia (NH_3) to dissolve amphoteric metals or silicates that are not readily soluble in acids.

Metals such as aluminium, tungsten, and certain rare earths form soluble complexes under alkaline conditions:

$$Al_2O_3 + 2NaOH + 3H_2O \rightarrow 2Na[Al(OH)_4]$$

Alkaline leaching has several important applications across different mineral systems, particularly where acid-based processes would be ineffective or overly corrosive. One of the most notable uses is in the extraction of tungsten from scheelite ($CaWO_4$). In this process, sodium carbonate (Na_2CO_3) solution is used to dissolve the tungsten-bearing mineral, forming soluble sodium tungstate, which can then be further processed to produce tungsten compounds or metal.

Another major application is in the production of alumina from bauxite using the Bayer process. In this method, sodium hydroxide (NaOH) selectively dissolves alumina (Al_2O_3) from the bauxite ore under elevated temperature and pressure, leaving behind an insoluble residue known as "red mud." The dissolved alumina is later precipitated and calcined to produce pure aluminium oxide, which serves as the feedstock for aluminium smelting.

Alkaline leaching is also critical in the processing of ion-adsorption rare earth element (REE) clays, which are abundant in southern China and other tropical regions. These ores contain REEs weakly bound to clay minerals, and mild leaching with ammonium sulphate or ammonium carbonate solutions can effectively release the rare earth ions into solution without damaging the clay structure. This method allows for environmentally gentler recovery compared to traditional acid systems.

The main advantages of alkaline leaching include its high selectivity for specific minerals and the production of less corrosive waste streams, making it suitable for environmentally sensitive operations. However, its reaction kinetics are generally slower than acid leaching, and it is less effective for sulphide or carbonate ores, which do not readily dissolve in alkaline media.

The basic principle of alkaline leaching is chemical dissolution through hydrolysis or complexation. Under strongly basic conditions, metal ions form soluble complexes with hydroxide (OH^-) or carbonate (CO_3^{2-}) ions. This reaction converts insoluble metal oxides or salts into soluble forms that can be separated from undissolved gangue.

For example:

- In the Bayer process for alumina extraction, sodium hydroxide dissolves aluminium oxide (Al_2O_3) from bauxite, forming sodium aluminate ($NaAl(OH)_4$), while impurities such as iron oxides remain solid.

- In tungsten extraction, sodium carbonate reacts with scheelite ($CaWO_4$) to form soluble sodium tungstate (Na_2WO_4), which can then be recovered by precipitation.

- In rare earth clay deposits, ammonium carbonate or bicarbonate leaching releases adsorbed rare earth ions (REE^{3+}) into solution without breaking down the clay lattice.

Process steps:

1. **Ore Preparation:** The ore is crushed, ground, and sometimes pretreated (e.g., calcined) to enhance reactivity and expose reactive mineral surfaces.

2. **Leaching:** The prepared material is mixed with an alkaline solution in tanks, heaps, or autoclaves. Temperature and pressure are controlled depending on the mineral system—ranging from ambient conditions for clays to over 150°C for refractory ores.

3. **Chemical Reaction:** The alkaline reagent dissolves metal-bearing minerals by forming soluble metal-hydroxide or metal-carbonate complexes.

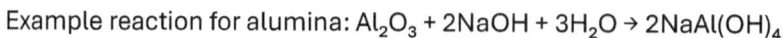

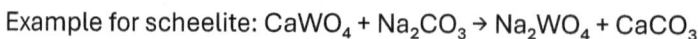

    Example reaction for alumina: $Al_2O_3 + 2NaOH + 3H_2O \rightarrow 2NaAl(OH)_4$

    Example for scheelite: $CaWO_4 + Na_2CO_3 \rightarrow Na_2WO_4 + CaCO_3$

4. **Solid–Liquid Separation:** The resulting slurry is filtered or settled to separate the pregnant leach solution (PLS), containing the dissolved metals, from the residue, which mainly consists of inert gangue minerals.

5. **Metal Recovery:** The dissolved metals are recovered from the PLS by precipitation, crystallization, ion exchange, or solvent extraction. In some cases, the alkaline solution is recycled to minimize reagent consumption.

Alkaline leaching offers several significant benefits compared to acid-based methods. It enables selective dissolution, targeting specific minerals while leaving unwanted gangue materials unaffected, thereby improving product purity. The process also results in reduced corrosion, as alkaline solutions are far less aggressive on processing equipment than strong acids, leading to longer equipment life and lower maintenance costs. Additionally, alkaline leaching produces cleaner waste streams, generating residues that are less toxic and easier to handle and dispose of compared to acidic tailings.

Despite its advantages, alkaline leaching also has notable limitations. It generally exhibits slower reaction kinetics, meaning leaching reactions occur at a slower rate and often require elevated temperatures and extended residence times to achieve satisfactory recoveries. The process also has limited applicability, as many sulphide and carbonate minerals do not readily dissolve under alkaline conditions, restricting its use to specific ore types. Furthermore, reagent costs can be significant, as large volumes of sodium hydroxide (NaOH) or sodium carbonate (Na_2CO_3) are often required, increasing operational expenses.

Overall, while alkaline leaching provides a selective and environmentally cleaner alternative to acid leaching, its economic and kinetic constraints necessitate careful process optimization and ore selection.

Pressure (Autoclave) Leaching

Pressure leaching (also called autoclave leaching) involves leaching under elevated temperature and pressure to accelerate dissolution kinetics and improve recovery. Both acid and alkaline systems can be used.

By increasing pressure and temperature, the solubility of gases and reaction rates increase, allowing difficult minerals to dissolve faster. For example, refractory sulphides or complex oxides can be oxidized and leached simultaneously.

Pressure leaching, also known as autoclave leaching, is applied in several key mineral processing operations where conventional atmospheric leaching is insufficient to achieve high recoveries. One of its most significant applications is in the treatment of nickel-cobalt laterites through the high-pressure acid leach (HPAL) process. In this method, sulphuric acid (H_2SO_4) is used at elevated temperatures of around 250°C and pressures of approximately 40 atmospheres to dissolve nickel and cobalt from their oxide and silicate minerals. The HPAL process enables rapid and efficient metal extraction compared with traditional methods, making it essential for processing low-grade lateritic ores.

In gold and copper operations, pressure oxidation is used to treat refractory sulphide ores that are resistant to direct leaching. Under high pressure and temperature, sulphide minerals such as pyrite and arsenopyrite are oxidized to form soluble oxides or sulphates. This oxidation step breaks down the sulphide matrix, liberating encapsulated gold or copper for subsequent recovery through cyanide or acid leaching.

Pressure leaching is also employed in the processing of refractory rare earth element (REE) ores, where the minerals exhibit strong lattice bonding or association with resistant gangue materials. By using elevated temperature and pressure, these systems enhance reagent penetration and dissolution rates, improving overall rare earth recovery where conventional acid or alkaline leaching would be ineffective.

The key advantages of pressure leaching include high metal recovery rates and significantly reduced leaching times compared with atmospheric processes. However, these benefits come at the cost of high capital investment, complex equipment requirements, and stringent safety standards due to the need to operate under extreme pressure and corrosive conditions.

The process relies on increasing temperature and pressure to alter the thermodynamics and kinetics of leaching reactions. Under pressurized conditions—typically 150–270°C and 1–5 MPa—the solubility of gases (like oxygen or air) and the reactivity of acids are enhanced. This allows rapid oxidation and dissolution of target metals.

Pressure leaching can be conducted in three main environments:

1. Acidic (H_2SO_4 or HCl-based) for sulphides, laterites, and rare earth minerals.

2. Alkaline (NaOH or Na_2CO_3-based) for bauxite or scheelite.

3. Ammoniacal systems for selective leaching of copper, cobalt, and nickel.

The process may also involve oxidizing agents such as oxygen, air, or ferric ions to convert metals into soluble forms (e.g., $Fe^{2+} \rightarrow Fe^{3+}$ or $U^{4+} \rightarrow U^{6+}$).

Process steps:

1. **Feed Preparation:** The ore or concentrate is finely ground and often subjected to pre-treatment (e.g., roasting or desliming) to improve permeability and reaction efficiency.

2. **Slurry Conditioning:** The feed is mixed with a leaching solution—usually an acid or alkaline reagent—forming a pulp that is pumped into the autoclave.

3. **Autoclave Leaching:** Inside the autoclave, the slurry is heated under pressure while oxygen or air is introduced. Elevated temperature and pressure accelerate dissolution reactions, liberating metals into solution.

4. **Discharge and Cooling:** After the leach period, the slurry is depressurized, cooled, and discharged for solid–liquid separation. The resulting pregnant leach solution (PLS) contains the dissolved metal values.

5. **Metal Recovery:** Metals are recovered from the PLS via precipitation, solvent extraction, or ion exchange, depending on the product desired.

Pressure (autoclave) leaching is applied across several critical mineral and metal systems where conventional leaching methods are ineffective or too slow. In nickel–cobalt laterite processing, the High-Pressure Acid Leach (HPAL) process uses sulfuric acid (H_2SO_4) at approximately 250°C and 40–50 atmospheres to dissolve nickel and cobalt oxides into solution as sulphates. During this reaction, iron is simultaneously precipitated as hematite, simplifying downstream purification and improving overall recovery.

In gold and copper sulphide ores, pressure oxidation leaching is used to convert refractory sulphide minerals into oxides or sulphates in the presence of oxygen. This oxidation process breaks down the sulphide matrix, liberating gold or copper for subsequent recovery using cyanide or acid leaching methods. The technique is especially valuable for ores that are resistant to standard oxidation or roasting processes.

For refractory rare earth minerals, autoclave leaching helps to open dense or chemically stable mineral structures that hinder conventional acid leaching. By subjecting the ore to high pressure and temperature, the process enhances the dissolution of rare earth elements, leading to higher recovery rates and improved efficiency in downstream separation.

The primary advantage of pressure leaching lies in its significantly faster reaction kinetics compared with atmospheric leaching. Elevated pressure and temperature accelerate the dissolution of metals, enabling shorter processing times and higher throughput. It also provides high recovery rates, even from refractory or complex ores that would otherwise yield low extraction efficiency. Additionally, due to improved reaction control and mass transfer, reagent consumption is often reduced, resulting in lower overall chemical costs and more sustainable operation.

However, pressure leaching also presents several limitations. The process demands high capital investment for autoclaves, pressure vessels, and corrosion-resistant materials capable of withstanding extreme conditions. Operationally, it requires precise control of temperature, pressure, and oxidation parameters, as deviations can affect product quality or equipment integrity. Finally, due to the combination of high temperatures, elevated pressures, and corrosive reagents, safety management is complex, necessitating specialized training, strict procedural controls, and advanced monitoring systems to ensure safe and stable operation.

Applications

Hydrometallurgical processes play a crucial role in the extraction and purification of rare earth elements (REEs), lithium, nickel, and other critical minerals, particularly after the initial physical beneficiation stages such as flotation, magnetic separation, or gravity separation. Hydrometallurgy is employed primarily for its ability to effectively and selectively dissolve various metals into an aqueous solution, allowing for further purification and recovery, thereby enabling high recovery rates and enhanced selectivity of targeted elements [225, 226].

For rare earth elements, the hydrometallurgical process begins with leaching techniques following preliminary concentration processes. Acidic leaching using sulfuric or hydrochloric acid is commonly applied to ores such as bastnäsite and monazite, typically at elevated temperatures, to efficiently extract REEs into solution. The leaching of ion-adsorption clays, on the other hand, can be conducted at ambient temperatures using ammonium sulphate or carbonate solutions, capitalizing on the unique solubility characteristics of these elements [227, 228].

In the case of lithium extraction from spodumene, a two-step process is utilized. Initially, spodumene concentrates are roasted to convert them to β-spodumene, which significantly enhances their solubility in acids. Subsequently, sulfuric or hydrochloric acid leaching is employed to dissolve lithium, producing soluble lithium sulphate or chloride that can be precipitated as lithium carbonate or lithium hydroxide. This method has been well-documented and is recognized for achieving substantial recovery yields [229, 230].

Similar approaches are observed in the processing of nickel-cobalt laterites, where high-pressure acid leaching (HPAL) employs sulfuric acid under pressure and elevated temperatures to maximize the dissolution of nickel and cobalt oxides from lateritic ores [231-233]. HPAL is often preferred due to its proven effectiveness, despite the associated higher operational costs.

When discussing the specific equipment integral to hydrometallurgical circuits, various leach tanks and reactors are utilized for atmospheric leaching, while specialized autoclaves are employed for pressure leaching processes. Solid-liquid separation post-leaching is facilitated through filters and thickeners, allowing for effective recovery of metals through ion exchange columns or solvent extraction units. Crystallizers and precipitation reactors are also utilized to refine and recover final products, such as rare earth oxalates and lithium carbonate, underscoring the technological sophistication involved in these processes [231, 234].

Hydrometallurgical processes vary greatly in scale and are adaptable to laboratory, pilot, and industrial operations. Laboratory-scale processes are essential for initial metallurgical testing and optimization; pilot-scale operations often simulate industrial conditions to validate the efficacy of processes under near-full-scale conditions. Industrial-scale hydrometallurgy is characterized by the integration of continuous leaching circuits, equipped to process large volumes of ore, facilitating economies of scale and improved resource utilization across the mineral extraction industry [235].

Case Study Scenario: Acid Leaching of Bastnäsite Ore for Rare Earth Element Recovery

At the *Bayan Obo* rare earth deposit in Inner Mongolia, China, one of the world's largest and most complex REE operations, hydrometallurgical processing plays a central role in extracting and refining rare earth elements after initial physical beneficiation. The ore primarily consists of bastnäsite ($REECO_3F$), a carbonate-fluoride mineral rich in light rare earth elements (LREEs) such as cerium, lanthanum, and neodymium, alongside associated gangue minerals like barite, fluorite, and iron oxides.

Figure 53: Bayan Obo. Squishyhippie, CC BY-SA 4.0, via Wikimedia Commons.

Stage 1: Comminution and Preparation

The mined ore undergoes crushing and grinding to a fine particle size (typically P80 <150 μm), increasing surface area for subsequent leaching. After classification, the fine material is calcined at approximately 500–600°C to decompose the carbonate matrix, liberating the bastnäsite mineral and making it more reactive to acid dissolution.

Stage 2: Acid Leaching

The calcined ore is mixed with concentrated sulfuric acid (H_2SO_4) in large, agitated leach tanks at elevated temperatures of 200–250°C. This step dissolves the rare earth elements by converting the REE carbonates and fluorides into rare earth sulphates, forming a pregnant leach solution (PLS). The reaction can be represented as:

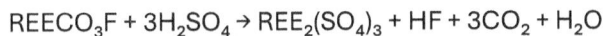

$$REECO_3F + 3H_2SO_4 \rightarrow REE_2(SO_4)_3 + HF + 3CO_2 + H_2O$$

Temperature, acid concentration, and residence time are carefully controlled to ensure high dissolution efficiency while minimizing acid consumption and silica gel formation.

Figure 54: Bastnaesite. Christian Rewitzer, CC BY-SA 3.0, via Wikimedia Commons.

Stage 3: Solid–Liquid Separation

After leaching, the slurry is cooled and transferred to counter-current decantation (CCD) thickeners or pressure filters to separate the solid residue (mainly iron and barite) from the liquid PLS. The clarified solution, rich in dissolved rare earth sulphates, moves to the purification stage, while the residue undergoes neutralization before safe disposal.

Stage 4: Solution Concentration and Purification

The PLS is purified using solvent extraction (SX) systems, where an organic extractant—typically an organophosphorus compound such as P204 (di-2-ethylhexyl phosphoric acid)—selectively transfers REE ions from the aqueous phase into an organic solvent. Through staged extraction, scrubbing, and stripping, individual REEs are separated based on their chemical affinities and ionic radii. For example, lighter REEs like cerium and lanthanum are extracted first, followed by neodymium and praseodymium.

Stage 5: Metal Recovery

After separation, the REE-bearing strip solutions are treated through precipitation to recover the elements as oxalates or carbonates, which are then calcined at 800–1000°C to produce high-purity rare earth oxides (REOs)—the final commercial product used in magnets, catalysts, and electronic components.

Equipment and Scale of Operation

At Bayan Obo, the hydrometallurgical operation is conducted on a large industrial scale, processing thousands of tonnes of concentrate daily. The main equipment includes:

- Acid-resistant leach tanks made of stainless steel or lined with PTFE for high-temperature leaching.

- CCD thickeners and pressure filters for slurry clarification.

- Solvent extraction columns and mixer-settlers for REE purification.

- Calcination kilns for final oxide production.

Pilot-scale plants are also maintained for testing reagent formulations and optimizing process variables before industrial implementation.

Outcome and Significance

This hydrometallurgical circuit achieves over 90% REE recovery, demonstrating the effectiveness of acid leaching combined with solvent extraction in processing complex ores. The process highlights hydrometallurgy's advantages—high selectivity, scalability, and adaptability—in rare earth extraction. However, it also underscores key challenges, including acid waste management, corrosion control, and reagent recycling to ensure environmental sustainability.

Summary

This case study illustrates how acid leaching, as part of a hydrometallurgical process, transforms rare earth minerals into soluble compounds for recovery and refining. It combines chemical precision with industrial engineering to achieve efficient, selective, and environmentally managed metal extraction—making it a cornerstone of critical mineral processing worldwide.

Solvent Extraction and Ion Exchange

In hydrometallurgical processing, both solvent extraction (SX) and ion exchange (IX) play pivotal roles as effective techniques for the purification and separation of valuable metals from leach solutions after leaching. These processes are essential for enhancing the efficiency of metal recovery and are widely utilized across various applications in the mining and recycling of metals.

The process of leaching is fundamental in hydrometallurgy, where valuable metals are dissolved from solid materials, typically using acidic solutions such as sulfuric or hydrochloric acid, among others [236, 237]. Following leaching, the resultant solution contains a mixture of metal ions that must be selectively separated to recover specific metals efficiently. This is where SX and IX come into play. SX is recognized for its effectiveness in selectively isolating metal ions based on their solubility and chemical affinity through the use of organic solvents

[238, 239]. The adaptability of SX allows for the separation of metals from complex mixtures, making it an essential method in various hydrometallurgical operations, including lithium-ion battery recycling [240].

On the other hand, ion exchange operates by utilizing resins that selectively adsorb certain metal ions from the leach solution. Metals such as cobalt and nickel, which may coexist in the solution, can be effectively extracted using IX methods, which are characterized by high selectivity and efficiency [239, 241]. IX is particularly advantageous in scenarios where metal ions exhibit similar properties and require careful separation to ensure high purity of the recovered metals [241]. The combination of SX and IX in hydrometallurgical processing often results in a more robust purification system, as both methods complement each other's capabilities, allowing refined metal recovery strategies to be developed [242].

Moreover, advancements in both techniques have enhanced their operational efficiency and environmental sustainability. Recent research shows that integrating various hydrometallurgical methods, such as employing new types of solvent systems or optimizing operational conditions, can significantly amplify separation efficiencies and reduce environmental impacts [243, 244]. This trend is not only crucial for the recovery of metals from primary resources but also plays a substantial role in the recycling sector, particularly in recovering metals from spent lithium-ion batteries, which is increasingly vital given the global push for sustainable resource utilization [245, 246].

The integration of solvent extraction and ion exchange in hydrometallurgical processes forms a comprehensive approach to the purification and separation of metals. The synergy between these methods facilitates the efficient recovery of valuable materials from leach solutions, addressing both economic and ecological goals in the modern metallurgical landscape.

Solvent Extraction (SX)

Solvent extraction is a liquid–liquid separation process that transfers metal ions from an aqueous phase (the leach solution) into an immiscible organic phase containing a selective extractant. The metal ions form chemical complexes with the extractant and move into the organic solvent. Once extracted, they are stripped (re-extracted) back into a new aqueous phase for recovery or further processing.

Stages:

1. **Extraction:** The pregnant leach solution (PLS) is mixed with an organic solvent containing extractant molecules (e.g., organophosphorus acids or amines). Metals selectively transfer to the organic phase.

2. **Scrubbing:** The loaded organic phase is washed to remove co-extracted impurities.

3. **Stripping:** The purified metals are transferred back into an aqueous solution using a strong acid or base.

4. **Regeneration:** The organic phase is recycled for further use.

Solvent extraction is widely used for separating and purifying rare earth elements (REEs), uranium, copper, cobalt, and nickel. In REE processing, SX enables group separation—for example, light REEs (La–Nd) from heavy REEs (Dy–Y)—due to subtle chemical differences between their ionic species. In lithium refining, solvent extraction can polish eluates after leaching or ion exchange to remove impurities such as iron and magnesium

Industrial SX circuits employ mixer–settlers, pulsed columns, or centrifugal contactors. These systems operate in continuous counter-current flow to achieve efficient phase contact. SX operations range from pilot-scale units (tens of litres per hour) to full-scale plants processing thousands of cubic metres per day, such as in REE and copper refineries.

Figure 55: Solvent extraction plant. Palagiri, CC BY-SA 3.0, via Wikimedia Commons.

Solvent extraction (SX) is a fundamental technique in hydrometallurgy, widely used to selectively separate and purify metal ions from aqueous solutions. The process hinges on the transfer of metal ions from an aqueous phase—typically a leach solution—into an organic phase that contains a selective extractant. This phase interaction enables precise control over which metals are captured and how they are isolated.

The aqueous phase, known as the pregnant leach solution (PLS), is generated by leaching ore with acid or other reagents. It contains dissolved metal ions such as REE^{3+}, Ni^{2+}, and Co^{2+}, along with various impurities. These ions are hydrated and exist in equilibrium with other species, with their behaviour influenced by factors like pH and ionic strength.

The organic phase consists of a water-immiscible solvent—typically kerosene or aliphatic hydrocarbons—in which an extractant is dissolved. A common extractant is D2EHPA (di-2-ethylhexyl phosphoric acid), denoted as HA. This acidic compound can donate protons and bind metal ions to form stable complexes.

At the interface between the two phases, a ligand exchange reaction occurs. For example, the reaction:

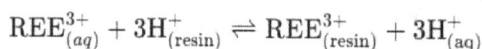

$$REE^{3+}_{(aq)} + 3H^+_{(resin)} \rightleftharpoons REE^{3+}_{(resin)} + 3H^+_{(aq)}$$

illustrates how a rare earth ion (REE^{3+}) in the aqueous phase reacts with three molecules of HA in the organic phase to form a neutral complex $REE(A)_3$. This complex is soluble in the organic phase, while the released protons (H^+) return to the aqueous phase, lowering its pH.

The resulting metal-extractant complex is hydrophobic, which facilitates its partitioning into the organic phase. This phase transfer is governed by several factors: pH control (where lower pH favours stripping and higher pH favours loading), extractant concentration (which affects capacity and selectivity), and temperature and ionic strength (which influence equilibrium and kinetics).

This mechanism is crucial for achieving selective separation of metals based on their affinity for the extractant. It is particularly effective for rare earth elements (REEs), which form stable complexes with acidic extractants. Moreover, the process is reversible, allowing the metal to be stripped from the organic phase into a clean aqueous solution for further refinement or recovery.

Scrubbing is a critical intermediate step in the solvent extraction (SX) process, designed to enhance the purity of the final product by removing unwanted co-extracted impurities from the organic phase. After the target metal ions have been successfully transferred from the aqueous phase into the organic phase, the resulting "loaded organic" solution may still contain trace amounts of other metals that were unintentionally extracted. These impurities—commonly iron (Fe), manganese (Mn), or uranium (U)—can interfere with downstream separation or contaminate the final product if not properly removed.

To address this, the loaded organic phase undergoes a scrubbing operation. This involves contacting the organic solution with a weak acid solution, typically a dilute mineral acid such as sulfuric or hydrochloric acid. The acid acts as a selective stripping agent, preferentially removing the loosely bound or less stable impurity complexes from the organic phase while leaving the desired metal-extractant complexes intact. The effectiveness of scrubbing depends on the chemical affinity of the extractant for different metal ions, as well as the pH and composition of the scrubbing solution.

During scrubbing, the impurities are transferred back into the aqueous phase, which is then discarded or treated separately. The scrubbed organic phase—now enriched in the target metal and free of major contaminants—is ready for the final stripping stage, where the metal is recovered in a purified aqueous form. This step is especially important in refining rare earth elements (REEs), where high purity is essential for downstream applications in electronics, magnets, and catalysts. By removing co-extracted metals early in the process, scrubbing improves selectivity, reduces reagent consumption, and enhances overall process efficiency.

Stripping, also known as re-extraction, is the final and essential step in the solvent extraction (SX) cycle, where the target metal is recovered from the organic phase into a clean aqueous solution. After the organic phase has been loaded with the desired metal ions and scrubbed to remove impurities, it contains a purified metal-extractant complex. However, to isolate the metal in a usable form—whether for precipitation, crystallization, or further refining—it must be transferred back into an aqueous phase. This is achieved through the stripping process.

In stripping, the loaded organic solution is contacted with a strong aqueous reagent, typically a concentrated acid or base, depending on the chemistry of the metal-extractant complex. This reagent disrupts the equilibrium of the original extraction reaction, effectively reversing it. For example, if the metal was extracted using an acidic extractant like D2EHPA, then a strong acid such as hydrochloric or sulfuric acid may be used to protonate the extractant and release the metal ion. Conversely, for metals extracted with basic or chelating extractants, a strong base might be used to break the complex and recover the metal.

The reaction drives the metal ions out of the organic phase and into the aqueous phase, where they are now free of extractant and ready for downstream processing. The efficiency of stripping depends on factors such as reagent concentration, temperature, phase contact time, and the specific binding strength of the metal-extractant complex. Proper control of these variables ensures high recovery rates and minimizes reagent consumption.

Stripping is not only crucial for metal recovery but also for regenerating the organic phase. Once the metal is removed, the organic solution can be recycled back into the extraction circuit, reducing waste and operational costs. This cyclical use of the organic phase makes SX a highly sustainable and scalable method for refining metals such as rare earth elements (REEs), cobalt, nickel, and uranium—especially from complex or low-grade ores.

Regeneration and recycling of the organic solvent are essential components of a well-designed solvent extraction (SX) circuit, contributing significantly to both operational efficiency and environmental sustainability. After the target metal has been stripped from the organic phase, the solvent—typically a mixture of a water-immiscible carrier like kerosene and a dissolved extractant such as D2EHPA—still retains chemical functionality and can be reused. However, before re-entering the extraction cycle, the solvent must be cleaned and regenerated to maintain its performance and selectivity.

During SX operations, the organic phase can accumulate degradation products, residual impurities, and traces of aqueous contaminants. These may include entrained water, dissolved salts, or oxidized extractant species that reduce extraction efficiency or interfere with metal selectivity. Regeneration involves removing these unwanted components through washing, phase separation, and sometimes chemical treatment. For example, the organic solvent may be washed with deionized water to remove residual acid or base, or treated with activated carbon to adsorb organic degradation products.

Once cleaned, the regenerated organic phase is reconditioned—often by adjusting the extractant concentration or pH—and recycled back into the extraction stage. This closed-loop approach minimizes solvent losses, reduces the need for fresh chemical inputs, and lowers the overall environmental footprint of the process. It also ensures consistent extraction performance across multiple cycles, which is critical for maintaining product quality and process economics.

In large-scale hydrometallurgical plants, solvent regeneration and recycling are integrated into continuous flow systems, allowing for uninterrupted operation and real-time monitoring of solvent health. This makes SX not only a powerful tool for metal separation but also a highly sustainable one, especially when refining rare earth elements, lithium, cobalt, and other strategic metals from complex or low-grade ores. By maximizing solvent reuse and minimizing waste, regeneration reinforces the long-term viability of solvent extraction in modern metallurgical practice.

Applications in Rare Earth Element (REE) Processing

Rare earth elements (REEs) have nearly identical ionic radii and chemical behaviour, which makes their separation extremely difficult using conventional methods. Solvent extraction provides the fine selectivity required to separate them into high-purity individual oxides.

1. Bastnäsite and Monazite Leach Liquors

After acid leaching of bastnäsite ($REECO_3F$) or monazite ($REEPO_4$) ores using sulfuric or hydrochloric acid, the resulting leach solution contains a mixture of light rare earth elements (LREEs) such as cerium (Ce), lanthanum (La), neodymium (Nd), and praseodymium (Pr).

- The solution is treated in multi-stage mixer–settler SX circuits using organophosphorus extractants, such as:

    o **D2EHPA (di-2-ethylhexyl phosphoric acid)** – for group separation of LREEs.

    o **PC88A (2-ethylhexyl phosphonic acid mono-2-ethylhexyl ester)** – for better selectivity and kinetics.

    o **Cyanex 272 (bis(2,4,4-trimethylpentyl)phosphinic acid)** – used for specific separations or purification.

Rare Earth and Critical Mineral Operations and Processing

Typical process flow:

- Extraction separates Ce–La from Nd–Pr groups.

- Subsequent stages further separate Nd and Pr, leading to individual rare earth oxides after precipitation and calcination.

This approach is used at large-scale REE refineries in China, Malaysia, and Australia (e.g., Lynas Advanced Materials Plant).

2. Ion-Adsorption Clays

For ion-adsorption rare earth clays from southern China, the leach solution obtained using ammonium sulfate contains medium and heavy rare earths (e.g., Y, Dy, Tb, Er, Yb). Solvent extraction refines this mixed solution using:

- HEHEHP (2-ethylhexyl phosphonic acid mono-2-ethylhexyl ester) for Y/Dy separation.

- Multi-step extraction–scrubbing–stripping cascades that isolate individual heavy REEs.

This system can involve hundreds of SX stages due to the subtle chemical differences among the heavy rare earths.

Applications in Critical Mineral Processing

1. Lithium

After sulphuric acid or hydrochloric acid leaching of spodumene or clay-based lithium ores, the leach liquor contains lithium sulphate or chloride mixed with impurities such as magnesium, calcium, and iron.

Solvent extraction is used as a polishing stage to purify lithium solution before crystallization:

- Organophosphorus extractants (Cyanex 923, Cyanex 272) or crown ethers selectively remove Mg^{2+}, Ca^{2+}, or Fe^{3+} while leaving Li^+ in the aqueous phase.

- This step produces high-purity lithium sulphate suitable for conversion into battery-grade lithium carbonate or hydroxide.

2. Nickel and Cobalt

In laterite processing (HPAL circuits), the pregnant leach solution contains both Ni^{2+} and Co^{2+} ions. Solvent extraction is used to separate cobalt from nickel using extractants such as:

- Cyanex 272 or Versatic 10 acid, which preferentially extract Co^{2+} over Ni^{2+}.

- After separation, cobalt is stripped and recovered as cobalt hydroxide, and nickel is recovered as nickel sulphate or metal powder.

3. Uranium

In uranium hydrometallurgy, tri-n-butyl phosphate (TBP) dissolved in kerosene is a standard extractant for uranyl ions (UO_2^{2+}):

- The process transfers uranium from a sulfuric acid leach solution into the organic phase.

- Stripping with ammonium carbonate or nitric acid yields purified uranium, later precipitated as yellowcake (U_3O_8).

Ion Exchange (IX)

Ion exchange is a solid–liquid separation process where metal ions in solution are exchanged with ions bound to an insoluble resin or adsorbent material. The resins contain functional groups that attract specific cations or anions based on charge and size, allowing highly selective recovery of metals.

Stages:

1. **Loading:** The leach solution passes through a resin column where target metal ions are adsorbed.

2. **Washing:** The column is rinsed to remove impurities.

3. **Elution:** The metal ions are stripped using a suitable eluent (e.g., ammonium sulphate, acid, or base).

4. **Regeneration:** The resin is regenerated and reused in the next cycle.

IX is particularly important in processing ion-adsorption clays rich in rare earth elements (southern China style deposits). In these cases, REEs are leached using mild ammonium sulphate or carbonate, and the solution is passed through IX columns to capture REE^{3+} ions, followed by elution to produce a purified concentrate. It is also used in direct lithium extraction (DLE) from brines, where lithium ions are selectively adsorbed on sorbent materials or ion-exchange membranes before purification.

IX systems use fixed-bed or fluidized-bed columns packed with synthetic polymer resins or natural zeolites. They are widely used at pilot and industrial scales, often integrated into continuous counter-current configurations to optimize metal recovery and resin utilization.

The loading step in ion exchange (IX) is the initial and foundational stage where metal recovery begins. In this phase, the metal-bearing leach solution—typically acidic and derived from ore processing—is passed through a column packed with ion exchange resin beads. These resins are engineered with functional groups that can selectively bind metal ions based on their charge, size, and chemical affinity. For rare earth elements (REEs) and critical minerals, cation exchange resins are commonly used, as they are designed to attract and hold positively charged ions like La^{3+}, Nd^{3+}, or Co^{2+}.

Rare Earth and Critical Mineral Operations and Processing

As the leach solution flows through the resin bed, target metal ions come into contact with the active sites on the resin surface. Through electrostatic attraction or coordination chemistry, these ions are adsorbed onto the resin, displacing hydrogen or sodium ions that were previously occupying those sites. This exchange is reversible and governed by equilibrium dynamics, meaning the resin will preferentially bind metals with higher affinity under the given conditions—typically influenced by pH, ionic strength, and temperature.

The efficiency of the loading step depends on several factors: the concentration of target metals in the leach solution, the flow rate through the column, the resin's selectivity profile, and the presence of competing ions. For example, in a solution containing both neodymium and calcium, a well-chosen resin can selectively adsorb neodymium while allowing calcium to pass through. Once the resin is saturated with the desired metal ions, the column is considered "loaded" and ready for the next stage—washing and elution—where the metals are stripped off for recovery.

This step is crucial not only for capturing valuable metals but also for setting the stage for high-purity separation. By concentrating the target ions onto a solid phase, ion exchange enables precise control over downstream purification, making it especially effective for refining REEs and critical minerals from complex or low-grade sources.

The washing step in ion exchange (IX) is a crucial intermediate phase that follows the loading of metal ions onto the resin. Once the leach solution has passed through the resin column and the target metal ions have been adsorbed, the resin bed may still retain unwanted impurities—such as loosely bound ions, entrained solution, or fine particulates—that were co-transported with the feed. Washing is performed to remove these residual contaminants before the elution (stripping) stage, ensuring that the final recovered product is as pure as possible.

To carry out washing, a rinse solution—typically deionized water or a dilute acid—is passed through the resin column. This solution flushes away unbound or weakly adsorbed species that did not form strong interactions with the resin's functional groups. For example, in rare earth element (REE) recovery, washing can help remove alkali or alkaline earth metals like sodium, calcium, or magnesium, which may have been present in the leach solution but are not strongly retained by the resin.

The effectiveness of the washing step depends on the flow rate, volume, and composition of the rinse solution. It must be sufficient to displace impurities without prematurely eluting the target metal ions. In some cases, a pH-adjusted rinse is used to stabilize the resin-metal complexes while selectively removing contaminants. This careful control ensures that the resin remains fully loaded with the desired metals and that the subsequent elution step yields a clean, concentrated product.

By incorporating a well-designed washing step, the ion exchange process improves selectivity, reduces cross-contamination, and enhances the overall efficiency of metal recovery. It is especially important in multi-stage purification flowsheets where high-purity outputs are required, such as in the separation of individual REEs or the refining of critical battery metals.

Elution is the third and final core step in the ion exchange (IX) process, where the target metal ions—previously adsorbed onto the resin—are stripped off and recovered into a clean aqueous solution. This step is essential for isolating the desired metals in a usable form and regenerating the resin for reuse. The process involves passing a carefully selected eluent through the loaded resin column to reverse the ion exchange reaction and displace the bound metal ions.

The choice of eluent depends on the type of resin and the specific metal being recovered. Common eluents include ammonium sulphate, mineral acids (such as hydrochloric or sulfuric acid), and alkaline solutions like sodium hydroxide. These reagents alter the chemical environment around the resin's functional groups, either by changing the pH or introducing competing ions that have a stronger affinity for the resin. As a result, the metal ions are released from the resin and enter the eluent solution in a purified form.

For example, in the recovery of rare earth elements (REEs), dilute hydrochloric acid may be used to protonate the resin and displace trivalent REE ions such as Nd^{3+} or Dy^{3+}. In other cases, ammonium sulphate can be used to elute transition metals like cobalt or nickel by forming soluble complexes that are weakly retained by the resin. The elution process is typically optimized to maximize recovery while minimizing reagent consumption and avoiding damage to the resin structure.

Once elution is complete, the resulting solution contains the target metal in a concentrated and purified state, ready for downstream processing such as precipitation, crystallization, or solvent extraction. Meanwhile, the resin—now free of metal ions—can be rinsed and reused in subsequent loading cycles. This cyclical use of resin and eluent makes ion exchange a highly efficient and sustainable method for recovering critical minerals and rare earths from complex feedstocks.

Regeneration is the final step in the ion exchange (IX) cycle, where the resin—having completed its role in metal loading, washing, and elution—is restored to its original functional state for reuse. This step is essential for maintaining the long-term efficiency and cost-effectiveness of the IX process, especially in continuous or multi-batch operations. Without proper regeneration, the resin's capacity to bind metal ions would diminish over time due to fouling, chemical degradation, or residual contaminants.

During regeneration, the resin is treated with a chemical solution that reactivates its exchange sites. For cation exchange resins, this typically involves flushing the column with a strong acid (such as hydrochloric or sulfuric acid) to replace residual metal ions with hydrogen ions. For anion exchange resins, a strong base like sodium hydroxide may be used to restore the resin's functional groups to their original form. The choice of regenerant depends on the resin type and the metals previously processed.

This chemical treatment not only displaces any remaining metal ions but also helps remove adsorbed impurities and restore the resin's ionic balance. In some cases, additional rinsing with deionized water is performed to remove excess regenerant and stabilize the resin before the next loading cycle. The effectiveness of regeneration is monitored by checking parameters such as resin capacity, breakthrough curves, and metal recovery rates.

Rare Earth and Critical Mineral Operations and Processing

By enabling repeated use of the same resin, regeneration significantly reduces operating costs and minimizes waste generation. It also supports sustainable processing by extending the lifespan of resin materials and reducing the need for frequent replacement. In rare earth element (REE) and critical mineral recovery, where high selectivity and purity are essential, consistent resin performance through proper regeneration is key to achieving reliable and scalable separation outcomes.

Ion exchange (IX) offers several key advantages in the recovery of rare earth elements (REEs) and critical minerals, making it a preferred technique in modern hydrometallurgical flowsheets. One of its most valuable attributes is high selectivity. IX resins can distinguish between closely related elements, enabling precise separation of species with similar chemical properties—for example, neodymium (Nd) from praseodymium (Pr), or dysprosium (Dy) from terbium (Tb). This level of resolution is essential for producing high-purity materials used in advanced technologies.

Another strength of ion exchange is its scalability. The process is suitable for both continuous and batch operations, allowing it to be adapted to a wide range of plant sizes and throughput requirements. Whether used in pilot-scale testing or full-scale industrial production, IX systems can be configured to meet specific recovery goals efficiently.

From an environmental standpoint, IX is compatible with sustainable processing. It operates at ambient temperatures and avoids the use of organic solvents, reducing energy consumption and minimizing hazardous waste. This makes it an attractive option for green metallurgy initiatives and for processing ores in regions with strict environmental regulations.

Finally, ion exchange integrates seamlessly with other unit operations. It works well with upstream leaching, where metals are solubilized from ore, and with downstream solvent extraction (SX), which can further purify or fractionate the recovered metals. This compatibility enables multi-stage purification strategies that maximize recovery, selectivity, and product quality across a wide

Rare Earth Element Examples

- **Lanthanum (La^{3+}), Neodymium (Nd^{3+}), and Dysprosium (Dy^{3+})**: These REEs are commonly recovered from acid leach solutions using cation exchange resins. The resin selectively binds REE ions based on ionic radius and charge density. Elution is often done with dilute hydrochloric acid or complexing agents like EDTA.

- **Yttrium (Y^{3+}) and Heavy REEs**: These show stronger binding to resins due to their smaller ionic radii and higher charge density. Fractionation columns can be used to separate light and heavy REEs based on their elution profiles, achieving high purity—up to 98.4% for heavy REEs like Y and Dy under optimized condition.

Critical Mineral Examples

- **Lithium (Li⁺):** Recovered from brines or leachates using specialized resins that selectively bind lithium over sodium and potassium. Elution is typically done with dilute acid or water, depending on the resin chemistry.

- **Nickel (Ni²⁺) and Cobalt (Co²⁺):** These transition metals are recovered from laterite or sulphide ores. Ion exchange resins can separate Ni and Co from impurities like magnesium and calcium. Selective elution with ammonia or sulfuric acid enables downstream refining.

- **Uranium (UO₂²⁺):** Often present as anionic complexes in acidic leach solutions. Anion exchange resins are used to capture uranium selectively, followed by elution with sodium chloride or carbonate solutions.

Integration in Hydrometallurgical Circuits

In modern hydrometallurgical plants, ion exchange (IX) and solvent extraction (SX) are often employed in a sequential combination to optimize metal recovery and purification. The process typically begins with ion exchange, which is used for the bulk capture and concentration of metals from dilute leach solutions. This step efficiently removes large volumes of target metals from low-concentration feeds, setting the stage for more refined processing.

Following ion exchange, solvent extraction is applied to achieve fine purification and separation. This technique isolates individual elements or produces product-grade solutions, ensuring that the final outputs meet stringent purity requirements. The synergy between IX and SX delivers high selectivity, purity, and operational efficiency—qualities that are especially critical when refining rare earth elements (REEs), lithium, nickel, cobalt, and uranium from complex or low-grade ores.

Case Study Scenario: Integrated Ion Exchange and Solvent Extraction in Rare Earth Element Refining

Background

A rare earth processing facility in Southeast Asia was established to recover and refine mixed rare earth elements (REEs) from ion-adsorption clays and bastnäsite concentrate. The operation's main objective was to achieve high-purity separated rare earth oxides (REOs) — such as neodymium, praseodymium, and dysprosium — for use in permanent magnet manufacturing. Because the ore feed contained a complex mixture of REEs with variable concentrations, the plant adopted an integrated hydrometallurgical circuit combining Ion Exchange (IX) and Solvent Extraction (SX) processes to maximize recovery and purity.

Stage 1: Leaching and Pregnant Solution Preparation

The raw ore underwent acid leaching with dilute sulfuric acid (H_2SO_4) at controlled pH and temperature to dissolve the rare earth ions (REE^{3+}) into solution, forming a pregnant leach solution (PLS). The solution also contained impurities such as Fe^{3+}, Al^{3+}, and Ca^{2+}. Solid–liquid separation was carried out using filtration, producing a clarified PLS ready for hydrometallurgical treatment.

Stage 2: Ion Exchange (IX) – Bulk Recovery and Concentration

The first hydrometallurgical step employed strong-acid cation exchange resins (e.g., sulfonic-type resin) packed in large columns.

As the PLS was passed through the columns, REE^{3+} ions were selectively adsorbed onto the resin, displacing H^+ ions:

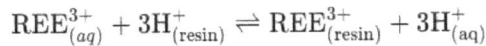

$$REE^{3+}_{(aq)} + 3H^+_{(resin)} \rightleftharpoons REE^{3+}_{(resin)} + 3H^+_{(aq)}$$

This step allowed for bulk extraction of rare earths from a large volume of dilute solution, concentrating them onto the resin.

Once the resin was saturated, it was eluted using hydrochloric acid (HCl), producing a concentrated rare earth chloride solution (typically 20–30 g/L total REE) with greatly reduced impurities.

This intermediate product was then sent to the solvent extraction stage for further separation.

Stage 3: Solvent Extraction (SX) – Selective Separation and Purification

The concentrated REE chloride solution served as the feed for a multi-stage solvent extraction circuit using D2EHPA (di-2-ethylhexyl phosphoric acid) as the extractant, dissolved in kerosene.

The extraction process involved a series of mixer-settlers operating in counter-current flow:

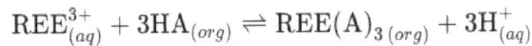

$$REE^{3+}_{(aq)} + 3HA_{(org)} \rightleftharpoons REE(A)_{3\,(org)} + 3H^+_{(aq)}$$

Through careful control of pH gradients and extractant concentration, individual REEs were separated based on their ionic radius and complex stability.

For example:

- Light REEs (La, Ce, Nd, Pr) were extracted first at higher pH.

- Heavy REEs (Tb, Dy, Y) were extracted at lower pH due to stronger complex formation.

Stripping stages using clean acid regenerated the extractant and produced purified aqueous REE solutions, which were subsequently precipitated as rare earth oxalates and calcined into REO powders exceeding 99.9% purity.

Stage 4: Process Integration and Optimization

The integration of IX and SX created a synergistic flow:

- IX provided efficient bulk capture and concentration, reducing the volume of solution entering SX and lowering reagent costs.
- SX delivered fine separation and purification, enabling production of high-purity individual REEs suitable for magnet and battery-grade applications.

Automation of pH control, flow rates, and resin regeneration cycles further improved recovery yields (≥98%) and operational reliability.

Waste acid and eluates were neutralized and partially recycled, minimizing environmental impact and chemical consumption.

Outcomes and Benefits

- **High Recovery and Purity:** Combined IX–SX processing achieved over 98% total REE recovery with product purity exceeding 99.9%.
- **Reduced Chemical Use:** Pre-concentration via IX reduced SX solvent requirements by approximately 30%.
- **Operational Efficiency:** Integration shortened residence times and decreased the number of SX stages required for separation.
- **Environmental Performance:** The closed-loop design enabled partial acid recycling and reduced effluent discharge.

Summary

This case illustrates how integrating Ion Exchange (IX) and Solvent Extraction (SX) within a hydrometallurgical circuit enables efficient, scalable, and environmentally sustainable recovery of rare earth elements. IX acts as a front-end concentrator for dilute feeds, while SX provides the fine-tuned selectivity necessary for high-purity separation. Together, they form the backbone of modern critical mineral refining, applicable not only to REEs but also to lithium, nickel, cobalt, uranium, and vanadium processing operations worldwide.

Precipitation and Purification

Rare Earth and Critical Mineral Operations and Processing

Precipitation and Purification are the final stages of many hydrometallurgical processes, where dissolved metals in a solution—often recovered through leaching, solvent extraction (SX), or ion exchange (IX)—are converted into solid compounds and refined into high-purity products. These steps play a crucial role in producing saleable materials such as metal oxides, hydroxides, carbonates, or sulphates from complex leach solutions containing multiple dissolved species.

After metals have been selectively leached and concentrated, the goal is to:

- Recover the dissolved metal ions from solution as solid products.

- Remove or minimize impurities (such as Fe, Al, Mg, Ca).

- Produce a pure, stable, and easily filterable solid suitable for further processing or direct sale.

Precipitation is a chemical reaction in which dissolved metal ions are converted into insoluble compounds (solids) by adding suitable reagents that alter pH, oxidation state, or ionic composition.

Precipitation occurs when the solubility limit of a metal compound in solution is exceeded. This can be achieved by:

- Changing the pH (adding acids or bases).

- Adding a precipitating agent (such as carbonate, hydroxide, sulphate, or oxalate).

- Controlling redox conditions (using oxidizing or reducing agents).

- Temperature adjustments, which can influence solubility and crystal growth.

For example, if a rare earth ion (REE^{3+}) in solution reacts with oxalate ions ($C_2O_4^{2-}$), an insoluble rare earth oxalate precipitate forms:

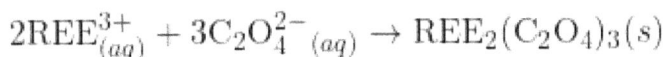

$$2REE^{3+}_{(aq)} + 3C_2O_4^{2-}{}_{(aq)} \rightarrow REE_2(C_2O_4)_3(s)$$

Common types of precipitation reactions:

a. Hydroxide Precipitation

Metal hydroxides are formed by adjusting the solution pH with an alkali (NaOH, NH_4OH, $Ca(OH)_2$).

Example:

$$Ni^{2+}_{(aq)} + 2OH^-_{(aq)} \rightarrow Ni(OH)_2(s)$$

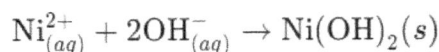

Used for metals like nickel, cobalt, and aluminium.

b. Carbonate Precipitation

Adding carbonate ions (from Na_2CO_3 or $(NH_4)_2CO_3$) forms metal carbonates:

$$Li^+_{(aq)} + CO_3^{2-}{}_{(aq)} \rightarrow Li_2CO_3(s)$$

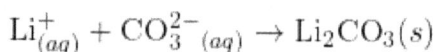

Common in lithium production, where lithium carbonate is the primary product from brines or leach solutions.

c. Oxalate Precipitation

Oxalate reagents ($H_2C_2O_4$ or $(NH_4)_2C_2O_4$) selectively precipitate rare earth elements:

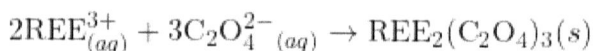

$$2REE^{3+}_{(aq)} + 3C_2O_4^{2-}{}_{(aq)} \rightarrow REE_2(C_2O_4)_3(s)$$

This is widely used in rare earth processing to obtain REE oxalates, which are later calcined into oxides (REO).

d. Sulphide Precipitation

Sulphide ions (H_2S, Na_2S) react with metal ions to form insoluble metal sulphides:

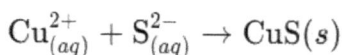

$$Cu^{2+}_{(aq)} + S^{2-}_{(aq)} \rightarrow CuS(s)$$

This is used for selectively removing impurities such as Cu, Zn, and Pb.

Purification involves removing unwanted elements from the product or solution before final recovery. This is achieved through:

- **Sequential pH adjustment:** Precipitating impurities at specific pH values before the target metal.

    - Example: Iron and aluminium precipitate around pH 3–4; nickel and cobalt at pH 7–8.

- **Selective reagent addition:** Using specific chemicals that react only with target ions.

- **Re-precipitation:** Dissolving and re-precipitating solids to enhance purity.

- **Washing and filtration:** Removing soluble impurities trapped in precipitates.

Sequential pH adjustment is a strategic technique used in hydrometallurgical precipitation to selectively remove impurities before recovering the target metal. The principle relies on the fact that different metal ions precipitate at different pH levels due to variations in their solubility and hydrolysis behaviour. By carefully controlling the pH of the solution in stages, operators can induce the precipitation of unwanted metals while keeping the desired metal ions in solution until a later step.

For example, iron and aluminium—common impurities in leach solutions—begin to precipitate as hydroxides at relatively low pH values, typically around pH 3 to 4. By adjusting the solution to this range using alkaline reagents such as sodium hydroxide or lime, these metals can be selectively removed as insoluble $Fe(OH)_3$ and $Al(OH)_3$. Once these impurities are filtered out, the solution can be further adjusted to a higher pH to target the precipitation of valuable metals.

Nickel and cobalt, for instance, remain soluble at low pH but begin to precipitate around pH 7 to 8. After removing iron and aluminium, the pH is raised again to induce the formation of $Ni(OH)_2$ and $Co(OH)_2$, which can then be recovered as solid products. This stepwise approach enhances selectivity, reduces contamination, and improves the purity of the final metal precipitate.

Sequential pH adjustment is especially useful in complex feedstocks where multiple metals coexist, such as laterite leachates or recycled battery materials. It allows for staged purification without the need for expensive reagents or elaborate separation equipment, making it a practical and scalable method in industrial metal recovery.

Selective reagent addition is a targeted approach in hydrometallurgical precipitation that enhances the purity and efficiency of metal recovery. This method involves introducing specific chemical reagents that are designed to react exclusively—or preferentially—with the desired metal ions in solution, forming insoluble compounds that can be easily separated. Unlike broad-spectrum precipitation techniques that affect multiple species, selective reagents minimize co-precipitation and reduce contamination from unwanted elements.

The effectiveness of this strategy depends on the chemical affinity between the reagent and the target ion. For example, oxalic acid ($H_2C_2O_4$) is commonly used to selectively precipitate rare earth elements (REEs) as oxalates, while leaving other metals like calcium or magnesium in solution. Similarly, sodium sulphide (Na_2S) can be added to selectively precipitate copper or lead as metal sulphides, without affecting more stable ions like sodium or potassium. The selectivity arises from differences in solubility products (Ksp), coordination chemistry, and reaction kinetics.

To implement selective reagent addition successfully, operators must carefully control the reagent dosage, solution pH, temperature, and mixing conditions. These parameters influence the reaction pathway and determine whether the target metal forms a stable precipitate. In some cases, chelating agents or complexing ligands are used to fine-tune selectivity, allowing for the separation of closely related elements such as cobalt and nickel or individual REEs.

This technique is especially valuable in flowsheets where high-purity products are required or where feed solutions contain a mix of valuable and interfering ions. By isolating specific metals with minimal reagent use and waste generation, selective reagent addition supports both economic and environmental goals in modern metallurgical processing.

Re-precipitation is a purification technique used in hydrometallurgical processing to improve the chemical purity of a precipitated solid. The method involves two main steps: first, dissolving the initially precipitated material back into solution; second, re-precipitating it under more controlled or selective conditions. This approach is particularly useful when the first precipitation step captures both the target metal and unwanted impurities, resulting in a product that does not meet purity specifications.

The dissolution step typically uses an acid, base, or complexing agent to solubilize the crude precipitate. For example, a mixed hydroxide containing nickel and trace iron might be redissolved in dilute acid. Once in solution, conditions such as pH, temperature, and reagent concentration are carefully adjusted to selectively re-precipitate the desired metal while minimizing co-precipitation of contaminants. This second precipitation often yields a purer product because the process can be fine-tuned to favour the formation of the target compound while leaving impurities in solution.

Re-precipitation is especially valuable in the production of high-purity rare earth oxalates, carbonates, or hydroxides, where even trace levels of other metals can compromise downstream applications. For instance, in rare earth separation, an initial oxalate precipitate may contain small amounts of calcium or iron. By redissolving and re-precipitating the REE oxalate under optimized conditions, these impurities can be effectively excluded, resulting in a cleaner product suitable for calcination into rare earth oxides.

This method is relatively simple to implement and integrates well with existing precipitation circuits. However, it does require additional reagents and process time, and the handling of intermediate solutions must be carefully managed to avoid losses. When applied judiciously, re-precipitation serves as a powerful tool for refining product quality and meeting stringent purity standards in critical mineral and rare earth processing.

Washing and filtration are essential post-precipitation steps in hydrometallurgical processing, designed to remove soluble impurities that may be physically trapped within or adsorbed onto the surface of precipitated solids. When metal ions are precipitated—whether as hydroxides, carbonates, oxalates, or sulphides—the resulting solids often co-precipitate with unwanted ions or retain traces of the original solution. These residual impurities can compromise product purity and interfere with downstream refining steps.

The washing stage involves rinsing the precipitate with a clean liquid, typically deionized water or a dilute acid or base, depending on the chemistry of the system. This rinse solution helps dissolve and flush away soluble contaminants such as sodium, calcium, chloride, or sulphate ions that may be loosely bound to the precipitate. In some cases, multiple washing cycles are used to ensure thorough removal, especially when high-purity products are required—such as rare earth oxalates destined for calcination into oxides.

Filtration follows washing and serves two purposes: separating the cleaned solid from the liquid phase and concentrating the product for drying or further processing. The choice of filtration method—vacuum filtration, pressure filtration, or membrane-based separation—depends on the particle size, slurry viscosity, and required throughput. Efficient filtration ensures that fine particles are retained while the wash solution and dissolved impurities are removed.

Together, washing and filtration enhance the chemical quality of the precipitate, reduce contamination, and improve the consistency of the final product. These steps are especially critical in the production of battery-grade materials, high-purity rare earth compounds, and specialty chemicals, where even trace impurities can affect performance or regulatory compliance. Proper design and control of these operations are key to achieving reliable and scalable purification in modern hydrometallurgical flowsheets.

Examples in rare earth and critical mineral processing:

a. Rare Earth Elements (REEs): After solvent extraction, REEs are precipitated as oxalates using ammonium oxalate or oxalic acid. The REE oxalate precipitates are then filtered, washed, and calcined at 800–1000°C to yield high-purity REE oxides (REO).

b. Lithium: In lithium refining, lithium carbonate (Li_2CO_3) is precipitated by adding sodium carbonate to a purified lithium sulphate or chloride solution. The product is filtered, washed, and dried to battery-grade purity.

c. Nickel and Cobalt: In HPAL (High Pressure Acid Leach) circuits, after impurity removal, nickel and cobalt are recovered as hydroxides or sulphides through controlled precipitation. For battery materials, these intermediate hydroxides are further processed to produce nickel sulphate or cobalt sulphate.

d. Uranium: After solvent extraction, uranium is precipitated as ammonium diuranate (ADU) using ammonia or hydrogen peroxide. This is later calcined to produce U_3O_8, the standard uranium oxide concentrate.

Precipitation and purification techniques offer several advantages in hydrometallurgical processing, particularly for the recovery of rare earth elements and critical minerals. One of the key benefits is the production of solid, storable products that are suitable for further refining or direct sale. These precipitates—such as hydroxides, carbonates, or oxalates—can be easily handled, transported, and processed into high-value materials like oxides or metals.

Another major advantage is the ability to selectively remove impurities while recovering high-purity target metals. By carefully choosing reagents and controlling reaction conditions, operators can isolate specific elements and minimize contamination. Additionally, precipitation reactions typically operate at relatively low temperatures and pressures, which reduces energy consumption and simplifies equipment requirements. These methods also integrate well with upstream hydrometallurgical processes such as ion exchange (IX) and solvent extraction (SX), forming part of a cohesive and efficient separation strategy.

Despite these strengths, precipitation and purification methods also present several challenges. Achieving selectivity requires precise control over pH and reagent concentrations; otherwise, co-precipitation of unwanted species can occur, reducing product purity. The process also generates solid waste—such as tailings or filter residues—that must be properly managed to avoid environmental impact. Operational issues like scaling and fouling in reactors and filtration systems can impair efficiency and increase maintenance costs. Furthermore, the quality of reagents and the effectiveness of washing steps directly influence the purity and consistency of the final product, making process optimization essential for reliable performance.

Case Studies in Rare Earth Separation and Purification

Innovative Pathways in Rare Earth Element Separation: Integrating Solvent Extraction, Hydrometallurgy, and Polymer Chemistry

The extraction and purification of rare earth elements (REEs) is a complex process due to the similarities in their chemical and physical properties. This case study explores various methodologies and innovations relevant to the separation and purification of REEs, emphasizing both traditional and novel techniques utilized in recent research.

One promising method for the separation of rare earth elements involves solvent extraction techniques. Traditional solvent extraction has been refined through the use of various chemical extractants, such as organophosphorus compounds. Studies have indicated that P204 and Cyanex272 can effectively separate light rare earth elements from chloride media, allowing for the selective isolation of rare earth ions from complex mixtures in aqueous solutions under optimized conditions [247]. Furthermore, the use of multi-stage extraction columns in chromatography mode has shown to enhance separation efficiency by utilizing commercial extractants that target heavy rare earth elements specifically [248].

Hydrometallurgical techniques also play a crucial role in the recovery of rare earths from waste materials. A notable example is the recovery of REEs from Nd–Fe–B scrap, where iron impurities were efficiently separated to yield high-purity hematite nanoparticles through a one-step hydrothermal method using fructose as an auxiliary reagent [170]. This process illustrates innovative waste management practices and aligns with the increasing push towards a circular economy in rare earth element usage [249].

Additionally, complexation with water-soluble polymers offers an effective path for the separation of rare earth ions. Research on phosphorylated chitosan has demonstrated its potential to form stable complexes with REEs, which can then be separated via ultrafiltration techniques [250]. This method addresses the growing need for eco-friendly extraction methods, leveraging biodegradable materials that minimize environmental impact compared to conventional extractants.

Another cutting-edge approach involves the application of glycopolymers tailored for selective separation of middle rare earth elements. Adjustments in the glycopolymer's properties, such as charge density and degree of polymerization, allow for a distinct U-shaped selectivity profile, markedly enhancing the separation efficiency between REEs like samarium and europium compared to others [251]. This innovative methodology signifies an essential advancement in the selective extraction of REEs, particularly given the industrial demand for these elements in high-technology applications.

Overall, ongoing research in the field of rare earth separation and purification emphasizes the synergy of established techniques with novel approaches. By integrating advancements in polymer chemistry, hydrometallurgy, and solvent extraction, significant improvements in the efficiency and sustainability of rare earth element recovery can be achieved, which are essential for meeting the increasing global demand for these critical materials.

Advanced Materials and Sustainable Methods in Rare Earth Element Separation: Innovations in Extraction Science

This case study examines innovative techniques for the separation and purification of rare earth elements (REEs), focusing on recent advancements in extraction methods and materials science. Leveraging modern methodologies, researchers have developed more efficient and sustainable practices while addressing environmental challenges associated with traditional techniques.

One notable advancement in REE separation involves a hybrid extraction process combining magnetic and hydrometallurgical methods. The VIM-HMS (Vapor-Induced Mass Transfer Horizontal Method) technique is highlighted in the extraction of rare earth elements from permanent magnet scraps. In this approach, alloy powders undergo hydrolysis, leading to the rapid reaction of rare earth carbides with water, forming rare earth hydroxides. These hydroxides can be efficiently isolated from iron residues using magnetic separation. This method effectively reduces the complexity often associated with traditional separation processes while facilitating the recovery of valuable materials [252].

Another promising strategy is the utilization of water-soluble polymers such as phosphorylated chitosan for selective complexation of rare earth ions. The synthesis of this complexant enables the formation of large molecular coordination compounds with REEs, thereby enhancing separation efficacy. Studies indicate that using polymers with differing functional groups can yield greater selectivity and efficiency compared to conventional extractants, addressing challenges related to environmental toxicity and waste management [250, 253].

Research into nanomaterials such as nitrogen-doped nanoporous graphene reveals significant potential for membrane-based separations. This innovative approach exploits the unique structural properties of graphene to achieve selective permeability for various REEs. By designing membranes that target specific ionic sizes and charge characteristics, enhanced separation efficiency is obtained, potentially translating to lower energy costs and reduced environmental impact [254].

Furthermore, the development of functional ionic liquids represents another innovative direction for REE separation. Unlike traditional solvent extraction methods, ionic liquids can enhance the recovery of rare earth elements while reducing ammonia emissions, addressing environmental concerns prevalent in conventional methods. These ionic liquids have shown effectiveness in selectively extracting REEs, providing a more sustainable alternative in the industry [255, 256].

In summary, the evolving landscape of rare earth separation and purification is characterized by an integration of advanced materials science, sustainable practices, and innovative extraction methodologies. The combination of magnetic techniques, polymer chemistry, membrane technology, and ionic liquids showcases the commitment to improving separation efficiencies while mitigating environmental impacts. As technology continues to advance, these methods highlight the potential for achieving high-purity rare earth elements necessary for various industrial applications.

Key Terms and Concepts	

Hydrometallurgy: A branch of extractive metallurgy that uses aqueous chemistry to recover metals from ores, concentrates, and recycled materials through leaching, solution concentration, and metal recovery.

Leaching: The process of dissolving valuable metals from solid minerals into a liquid solution using chemical reagents such as acids, bases, or oxidising agents.

Acid Leaching: A leaching process employing acids—typically sulphuric, hydrochloric, or nitric acid—to dissolve target metals from oxides or carbonates.

Alkaline Leaching: A process that uses basic solutions, such as sodium hydroxide or ammonium carbonate, to selectively extract metals like aluminium or uranium.

Pressure Leaching (Autoclave Leaching): A high-temperature, high-pressure leaching technique that accelerates dissolution reactions and improves metal recovery.

Bioleaching: The use of microorganisms to promote metal dissolution from ores, offering an environmentally friendly alternative to chemical leaching.

Solvent Extraction (SX): A separation process that transfers metal ions from an aqueous solution into an organic solvent using specific extractants for selective recovery.

Ion Exchange (IX): A purification and separation method that exchanges target ions between a solution and a solid resin, often used to recover rare earths and other metals.

Precipitation: The process of converting dissolved metal ions into solid compounds (e.g., hydroxides, oxalates, or carbonates) for further processing or sale.

Purification: The removal of impurities or unwanted elements from solution to produce high-purity metal salts or oxides.

Key Terms and Concepts	

Redox Reaction: A chemical process involving reduction and oxidation, fundamental to many leaching and precipitation reactions in hydrometallurgy.

Solution Chemistry: The study of ionic equilibria, pH, and complexation reactions that determine the behaviour of metals in leaching and extraction systems.

Residue Management: The safe handling, neutralisation, and disposal of solid or liquid waste products generated during hydrometallurgical operations.

Process Flow Diagram (PFD): A schematic representation showing the sequence of hydrometallurgical operations, including leaching, extraction, and precipitation stages.

Selective Extraction: The ability of a reagent or process to target specific metals while minimising co-extraction of impurities, critical for rare earth separation.

Richard Skiba

Chapter 6 Review Questions

Principles and Role of Hydrometallurgy

1. Explain the fundamental principle of hydrometallurgy.

2. Describe the main stages in a typical hydrometallurgical process.

3. Discuss how hydrometallurgy contributes to refining rare earth and critical minerals compared with pyrometallurgy.

Leaching Methods and Applicability

4. Define leaching and name four main types used in hydrometallurgy.

5. Which leaching method is most effective for refractory nickel–cobalt laterites?

6. Differentiate between acid and alkaline leaching in terms of reagents and ore suitability.

7. Why is roasting necessary before acid leaching spodumene for lithium extraction?

8. Identify two advantages and two limitations of acid leaching.

Parameters Controlling Leaching Efficiency

9. List five parameters that influence leaching efficiency.

10. Explain how temperature and redox potential affect uranium leaching performance.

11. How does particle size impact leaching kinetics?

12. Describe how reagent concentration and pH can be adjusted to optimise recovery.

Solvent Extraction (SX) Principles

13. What is the main function of the extractant in solvent extraction?

14. Write the chemical equation showing extraction of a rare-earth ion (REE^{3+}) using an acidic extractant (HA).

15. Name the four principal stages in a solvent-extraction circuit.

16. List two factors that influence phase transfer and selectivity in SX.

17. Explain why solvent extraction is especially suited to separating rare earth elements.

Ion Exchange (IX)

18. Describe how ion-exchange resins function in metal purification.

19. Compare solvent extraction and ion exchange in terms of selectivity, cost, and scalability.

20. Give one example of an ion-exchange resin used in rare-earth processing and its role.

Rare Earth and Critical Mineral Operations and Processing

Precipitation and Purification

21. Define precipitation in hydrometallurgy.

22. Write the balanced chemical equation for rare-earth oxalate precipitation.

23. Why is sequential pH adjustment used in multi-metal purification?

24. List three common precipitating agents used in hydrometallurgical circuits.

25. Explain how temperature can influence precipitation efficiency and product quality.

Process Flow Interpretation

26. Outline the three key stages visible in a hydrometallurgical flow diagram for REE recovery.

27. Describe how leaching, extraction, and precipitation are integrated in a lithium-processing circuit.

28. Identify one piece of equipment used in each of the three stages of a hydrometallurgical circuit.

Environmental and Safety Considerations

29. Identify two major environmental challenges in hydrometallurgical plants.

30. Explain two measures used to mitigate environmental impacts from acid or alkali effluents.

31. Describe one occupational safety risk specific to pressure leaching and how it is controlled.

Bench-Scale and Pilot-Scale Testing

32. What is the purpose of bench-scale hydrometallurgical testing?

33. What is the role of pilot-scale testing before full commercial deployment?

34. Name one piece of pilot-scale equipment used for high-pressure acid leaching trials.

Sustainability and Comparison with Pyrometallurgy

35. List three sustainability advantages of hydrometallurgy over pyrometallurgy.

36. Explain why hydrometallurgy is better suited to low-grade or complex ores.

37. Discuss how hydrometallurgy supports the circular economy and metal recycling.

Chapter 7

Pyrometallurgy and Thermal Processing

Pyrometallurgy represents one of the oldest and most powerful techniques in extractive metallurgy, yet it remains at the forefront of modern critical mineral and rare earth processing. As global demand for strategic materials such as lithium, nickel, cobalt, and rare earth elements continues to rise, thermally driven processes have become essential to transform raw minerals into high-purity products used in batteries, magnets, and advanced alloys. This chapter introduces the scientific and practical foundations of pyrometallurgy, focusing on how controlled heat, atmosphere, and reaction kinetics are used to drive key transformations such as roasting, calcination, reduction, and thermal decomposition. It also explores how these processes integrate with hydrometallurgical circuits to create efficient, low-waste hybrid systems. By examining major equipment types—including rotary kilns, multiple-hearth furnaces, and fluidized-bed roasters—learners will develop a clear understanding of how thermal processing underpins the production of many critical technologies central to a low-carbon future.

Learning Outcomes	

This chapter aims to give you the ability to: 1. Explain the fundamental principles of pyrometallurgy and its role in the extraction, upgrading, and refining of rare earth and critical minerals. 2. Identify and describe key pyrometallurgical processes—including roasting, calcination, smelting, and reduction—and differentiate their applications based on mineral type and desired product. 3. Describe the chemical and thermodynamic reactions involved in common thermal treatments and interpret phase changes using Ellingham diagrams or related thermochemical tools.	

Learning Outcomes	

4. Recognise the functions and operating parameters of major pyrometallurgical equipment, including rotary kilns, fluidised bed reactors, electric arc furnaces, and induction furnaces.
5. Discuss the importance of temperature control, atmosphere composition, and residence time in achieving desired reaction outcomes and product purity.
6. Compare and contrast pyrometallurgical and hydrometallurgical routes in terms of energy efficiency, scalability, environmental impact, and product quality.
7. Interpret basic flowsheets showing the integration of roasting, leaching, and reduction stages in critical mineral processing circuits.
8. Assess the environmental, safety, and occupational health considerations associated with high-temperature operations, including dust, gas emissions, and waste handling.
9. Evaluate emerging low-carbon and hybrid pyrometallurgical technologies, such as plasma processing, microwave-assisted heating, and hydrogen-based reduction.
10. Apply principles of process optimisation to improve energy efficiency, yield, and product quality in thermal processing operations.

Roasting, Calcination, and Reduction

Pyrometallurgy is one of the three main metallurgical routes (alongside hydrometallurgy and electrometallurgy) used to extract and refine metals. It involves the use of high-temperature processes to chemically transform metal-bearing minerals into pure metals or intermediate compounds. These reactions typically occur in furnaces, kilns, or reactors operating at temperatures ranging from several hundred to over 2000 °C. Common pyrometallurgical operations include roasting, calcination, smelting, and reduction.

Pyrometallurgical and other thermal processing methods are used to:

- Decompose minerals or drive off volatile components.

- Oxidize or reduce compounds to change their chemical state.

- Produce intermediates suitable for leaching in hydrometallurgical stages.

In contrast to hydrometallurgy, which relies on aqueous chemistry at low temperatures, pyrometallurgy depends on thermal energy and gas–solid reactions to achieve separation and purification. It is widely used for iron, copper, nickel, cobalt, zinc, and rare earth ores, especially when high-grade concentrates or specific mineral transformations are required before leaching.

Roasting

Roasting is a thermal treatment process conducted in the presence of oxygen or air. Its purpose is to oxidize, sulphidize, chlorinate, or volatilize components in the ore, thereby converting them into forms suitable for further processing.

Examples include:

- Oxidizing sulphide minerals (e.g., ZnS → ZnO + SO_2) before leaching or smelting.

- Decomposing carbonates (e.g., $CaCO_3$ → CaO + CO_2).

- Activating lithium ores: α-spodumene is roasted at about 1000 °C to form β-spodumene, which is more reactive to acid leaching.

Roasting often serves as a pre-treatment step before hydrometallurgical leaching, improving mineral reactivity or releasing volatile impurities.

Roasting is a thermal treatment process used in extractive metallurgy to alter the chemical composition of an ore or concentrate by heating it in the presence of oxygen (air) or other gases. The goal is to convert metal compounds into more stable or soluble forms, remove volatile components such as sulphur, carbon dioxide, or water, and prepare the material for subsequent processing steps like smelting or leaching.

Roasting serves several key functions in mineral processing and metallurgy:

- **Oxidation of sulphide ores:** Converts metal sulphides (e.g., ZnS, PbS, $CuFeS_2$) into oxides or sulphates that are easier to leach or reduce.

  Example:

$$2ZnS + 3O_2 \rightarrow 2ZnO + 2SO_2$$

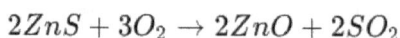

- **Decomposition of carbonates and hydroxides:** Drives off volatile components such as CO_2 or H_2O to form oxides.
  Example:

$$CaCO_3 \rightarrow CaO + CO_2$$

- **Removal of impurities:** Volatile impurities such as arsenic, antimony, or sulphur can be eliminated as oxides or gases.

- **Phase transformation or activation:** Alters the crystal structure to increase mineral reactivity—this is common in lithium processing, where α-spodumene ($LiAlSi_2O_6$) is converted to β-spodumene at about 1000°C, making it more amenable to acid leaching.

Roasting relies on gas–solid reactions that occur at elevated temperatures, typically between 400°C and 1200°C, depending on the ore and desired outcome. The process takes place in an oxidizing, chloridizing, or sulphidizing atmosphere, and the reactions are generally exothermic once initiated.

Main chemical pathways include:

- **Oxidation:** Conversion of sulphides to oxides or sulphates.

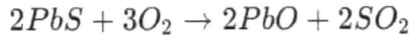

$$2PbS + 3O_2 \rightarrow 2PbO + 2SO_2$$

- **Sulphation:** Formation of metal sulphates at moderate temperatures (used for zinc, copper, and nickel).

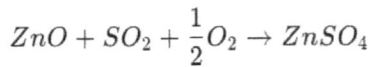

$$ZnO + SO_2 + \frac{1}{2}O_2 \rightarrow ZnSO_4$$

- **Chlorination:** Conversion of oxides to chlorides using chlorine or HCl gas.

$$TiO_2 + 2Cl_2 + C \rightarrow TiCl_4 + CO_2$$

- **Reduction (limited oxygen):** Partial oxidation of sulphides producing elemental sulphur or mixed products.

Roasting can be tailored based on its purpose and the ore characteristics:

Table 3: Roasting types.

Type of Roasting	Atmosphere	Purpose / Example
Oxidizing Roasting	Air/O_2	Converts sulphides to oxides (e.g., ZnS → ZnO).
Sulphating Roasting	O_2 + SO_2	Forms soluble sulphates for leaching (e.g., CuO → $CuSO_4$).
Chloridizing Roasting	Cl_2/HCl + O_2	Converts metals to chlorides (e.g., TiO_2 → $TiCl_4$).
Reducing Roasting	CO or H_2	Removes oxygen to form partial metals or reduced oxides.
Volatilizing Roasting	Air or Cl_2	Removes volatile impurities like As, Sb, or S.

Different types of furnaces are used for roasting operations, selected based on factors such as ore type, particle size, and the level of reaction control required. A multiple-hearth furnace is commonly employed for fine sulphide concentrates such as zinc and copper ores. It consists of several circular hearths stacked vertically, allowing the material to pass slowly through different temperature zones for uniform roasting.

The fluidized-bed roaster is designed for efficient heat transfer and excellent gas–solid contact. In this system, fine particles are suspended in an upward flow of air or gas, creating a turbulent "fluidized" state that promotes rapid and uniform reaction—making it ideal for sulphide oxidation and sulphate formation. A rotary kiln is typically used for coarser or mixed materials, including spodumene and nickel laterites. The kiln's continuous rotation ensures steady mixing and exposure of the ore to the heated atmosphere, allowing controlled thermal decomposition or phase transformation.

Finally, the reverberatory furnace is used for large-scale oxidation and smelting pre-treatment. It operates with radiant heat reflected from the furnace roof onto the ore bed, enabling bulk processing of concentrates where uniform oxidation or partial fusion is required before smelting.

Roasting is applied across a range of industrial processes to prepare ores and concentrates for further extraction and refining.

In zinc roasting, zinc sulphide (ZnS) concentrates are oxidized to zinc oxide (ZnO) in the presence of air. This transformation facilitates subsequent leaching in sulfuric acid, where the ZnO dissolves to form a zinc sulphate solution that serves as the feed for electrolytic zinc production.

In copper roasting, ores containing chalcopyrite and other copper–iron sulphides are partially oxidized to remove sulphur as sulphur dioxide gas. This step not only reduces the sulphur content but also produces an oxide–sulphide mixture that melts more easily during smelting, improving metal recovery and furnace efficiency.

In the case of lithium roasting, α-spodumene ($LiAlSi_2O_6$) is thermally treated at approximately 1000°C to convert it into β-spodumene, a more reactive crystal form. This phase transformation dramatically enhances the mineral's solubility, allowing efficient extraction of lithium through subsequent sulfuric acid leaching.

For rare earth element (REE) production, bastnäsite concentrates are roasted to drive off carbon dioxide and fluorine from carbonate and fluoride compounds within the mineral. This pre-treatment step makes the REE oxides more accessible to acid leaching, thereby improving recovery rates and reducing reagent consumption during downstream hydrometallurgical processing.

In the processing of rare earth elements (REEs), roasting is a critical step that employs various furnace types tailored to the specific mineral composition, particle size distribution, and desired reaction control mechanisms. The primary minerals of interest in REE extraction include bastnäsite, monazite, and xenotime, each necessitating distinct roasting techniques for optimal preparation before further processing, such as leaching or separation.

Rare Earth and Critical Mineral Operations and Processing

Bastnäsite Processing: This REE fluorocarbonate presents unique processing challenges that are often addressed using multiple-hearth furnaces. These furnaces allow for precise temperature control and staged heating, which is vital for the effective removal of carbon dioxide (CO_2) and fluorine (F) from the mineral structure. Such controlled oxidation processes ensure that the roasted bastnäsite transforms into a chemically reactive product that is conducive to subsequent acid leaching, significantly improving the yield of valuable rare earth oxides [257].

Monazite Processing: In the case of monazite, which is a REE phosphate, fluidized-bed roasters are favoured due to their superior gas-solid contact and uniform heat distribution, providing efficient abatement of organic compounds and associated carbonates [258]. This design enhances temperature control for selective oxidation, making it suitable for processing fine-particle feeds such as flotation concentrates. Such efficiencies lead to higher roasting performance and consistent product quality, which are essential for downstream chemical extraction processes [259].

Xenotime Processing: For xenotime, which comprises yttrium and heavy rare earth phosphates, rotary kilns are employed due to their capability to achieve the high and sustained temperatures necessary for facilitating phase transformation or partial decomposition of the ore. This is particularly relevant for preparing xenotime and other refractory ores prior to leaching [259]. Rotary kilns can handle coarser or mixed-particle feeds, thus making them versatile for a variety of REE-bearing intermediates, including hydroxides and carbonates.

Miscellaneous REE Concentrates: In instances where mixed REE concentrates or by-products are derived from other processes—such as ion-adsorption clays or industrial residues—other furnace types, like reverberatory or rotary-hearth furnaces, may be employed to achieve pre-oxidation or removal of volatile impurities [121]. These systems support continuous operation and can accommodate fluctuations in feed composition, an essential feature for integrated REE processing operations that must adapt to diverse raw material feeds [260].

Roasting offers several advantages in metallurgical processing, but it also presents some limitations that must be carefully managed. One of its main advantages is that it effectively removes sulphur and other volatile components from ores, producing cleaner and more reactive materials for subsequent extraction stages. This pre-treatment enhances the efficiency of leaching or smelting, as oxidized products such as metal oxides are more readily processed than their sulphide or carbonate counterparts. Additionally, roasting helps produce stable and reactive oxide forms, which not only improve metal recovery but also make downstream chemical reactions more predictable and controllable.

However, roasting also has limitations. It is an energy-intensive process that requires precise temperature regulation to prevent over-oxidation or sintering of materials. The high operating temperatures result in substantial fuel consumption, increasing operational costs. Moreover, roasting generates gaseous pollutants, particularly sulphur dioxide (SO_2), which must be captured and treated to comply with environmental regulations and reduce atmospheric emissions. Lastly, roasting is unsuitable for highly refractory or heat-sensitive ores, as excessive heating can cause phase changes or decomposition that reduce metal recoverability or alter mineral structure in undesirable ways. Overall, while roasting remains a critical process in the preparation of many ores, its application must balance metallurgical benefits with energy efficiency, cost, and environmental impact.

Roasting is a key thermal pre-treatment that bridges the gap between mineral beneficiation and metallurgical extraction. By oxidizing, decomposing, or activating ores, roasting transforms raw minerals into chemically favourable forms for hydrometallurgical or pyrometallurgical recovery. Its control of temperature, atmosphere, and reaction kinetics determines both metal yield and environmental performance in modern metallurgical plants.

Calcination

Calcination involves heating a mineral or compound in the absence or limited supply of air to cause thermal decomposition, phase transformation, or removal of volatile substances such as CO_2, H_2O, or organic matter.

Typical reactions include:

- Decomposition of carbonates (e.g., $CaCO_3 \rightarrow CaO + CO_2$).

- Dehydration of hydroxides (e.g., $Al(OH)_3 \rightarrow Al_2O_3 + 3H_2O$).

In rare earth processing, calcination at 500–600 °C is used to decompose carbonate minerals (e.g., bastnäsite), making them more reactive during subsequent acid leaching

Calcination differs from roasting mainly by the lack of oxidizing gas—it is used to prepare oxides, remove volatiles, or achieve structural changes without chemical oxidation.

Calcination is a thermal treatment process used in metallurgy and mineral processing to drive off volatile substances, decompose carbonates or hydrates, and produce chemically or physically altered solid products—usually oxides. Unlike roasting, calcination is carried out in the absence or limited supply of air or oxygen, meaning no significant oxidation occurs. The primary purpose of calcination is decomposition, phase transformation, or removal of bound moisture and gases from the ore or concentrate to make it more suitable for subsequent processing steps such as leaching, reduction, or smelting.

Calcination operates on the principle of thermal decomposition. When a mineral or compound is heated to a temperature below its melting point but high enough to cause chemical or physical changes, volatile components such as CO_2, H_2O, or organic matter are released. The remaining material—typically an oxide—is more thermally stable and reactive for downstream processes.

A common reaction example is the decomposition of calcium carbonate:

$$CaCO_3(s) \rightarrow CaO(s) + CO_2(g)$$

Rare Earth and Critical Mineral Operations and Processing

Here, limestone ($CaCO_3$) is converted to lime (CaO), releasing carbon dioxide gas.

Calcination is generally performed at temperatures between 400°C and 1200°C, depending on the chemical composition and thermal stability of the material. The process is typically conducted in rotary kilns, shaft furnaces, or fluidized-bed calciners, where heat is applied indirectly to prevent oxidation.

Temperature control is critical—too low, and decomposition remains incomplete; too high, and sintering may occur, reducing surface area and reactivity.

Calcination plays a crucial role across a range of industrial processes, preparing raw minerals for further refinement and use in manufacturing.

One of the most common examples is the conversion of limestone ($CaCO_3$) to lime (CaO). This reaction, which drives off carbon dioxide, is fundamental in the production of cement and serves as a flux in steelmaking, where lime helps remove impurities during smelting.

In the aluminium industry, bauxite ($Al(OH)_3$) is calcined as part of the Bayer process to remove chemically bound water. This produces anhydrous alumina (Al_2O_3), a critical intermediate that is later reduced electrolytically to produce pure aluminium metal.

For lithium extraction, spodumene ($LiAlSi_2O_6$) undergoes calcination at approximately 1000°C. This high-temperature treatment converts the mineral from its naturally occurring α-phase to β-spodumene—a more reactive crystal structure that significantly improves lithium recovery during subsequent acid leaching.

Another important example is the calcination of gypsum ($CaSO_4 \cdot 2H_2O$) to produce plaster of Paris ($CaSO_4 \cdot \frac{1}{2}H_2O$). Through controlled partial dehydration, this process yields a versatile material used in construction, ceramics, and medical casting applications.

Finally, rare earth carbonates and hydroxides are often calcined to remove residual carbon dioxide, water, and organic matter. This treatment produces stable rare earth oxides, which serve as the purified feedstocks for further metallurgical refining, reduction, or solvent extraction in the production of rare earth elements.

Together, these examples illustrate how calcination enhances reactivity, stability, and purity—making it an indispensable step in the processing of a wide variety of industrial and critical minerals.

Calcination offers several important advantages in mineral and metallurgical processing. One of its key benefits is the removal of moisture, carbon dioxide, and other volatile impurities, which purifies the material and prepares it for subsequent processing stages. The process also produces chemically stable and reactive oxides, making the calcined material more amenable to later treatments such as leaching, reduction, or smelting. Furthermore, calcination significantly enhances leaching and reduction efficiency by increasing the surface reactivity of the treated minerals, thereby improving overall metal recovery rates.

Richard Skiba

Despite these advantages, calcination also has several limitations. It is an energy-intensive process, as it requires sustained high temperatures to achieve the desired chemical transformations. This contributes to higher operational costs and environmental impacts if energy sources are not well-managed. The process also demands precise temperature control, since excessive heating can lead to sintering or unwanted phase transformations that reduce reactivity and recovery efficiency. Additionally, calcination is unsuitable for ores that emit toxic or hazardous gases—such as fluorine or sulphur compounds—unless advanced emission control and gas treatment systems are in place.

In essence, calcination is a controlled heating process that prepares ores and minerals for downstream metallurgical treatment. By altering their physical and chemical structure—typically through the removal of volatile components—it increases reactivity and purity. The process is fundamental in producing oxides such as alumina, lime, and rare earth oxides, which serve as essential intermediates in modern industrial and metallurgical operations.

Reduction

Reduction is a pyrometallurgical process that converts metal oxides into metallic form by removing oxygen. This is achieved through reaction with a reducing agent such as carbon (C), carbon monoxide (CO), hydrogen (H_2), or other metals.

Examples:

- Ironmaking: $Fe_2O_3 + 3CO \rightarrow 2Fe + 3CO_2$.

- Tungsten recovery: $WO_3 + 3H_2 \rightarrow W + 3H_2O$.

- Rare earth oxides can also be reduced using calcium or aluminium in metallothermic reduction.

Reduction is central to smelting operations and final metal recovery in high-temperature circuits, providing the metallic products that hydrometallurgical routes later refine to high purity.

Reduction is a fundamental metallurgical process used to convert metal oxides or other compounds into their metallic form by removing oxygen (or another electronegative element such as sulphur). It is the counterpart to oxidation and lies at the heart of pyrometallurgy—the high-temperature branch of metal extraction. Reduction reactions are thermochemical processes that rely on a reducing agent (such as carbon, carbon monoxide, hydrogen, or even metals like aluminium) to supply electrons that break metal–oxygen bonds, thereby freeing the pure metal.

In chemical terms, reduction involves the gain of electrons by a substance. When applied to metal oxides, the goal is to reduce the oxidation state of the metal to zero, producing the metallic form. A general reduction reaction can be expressed as:

Rare Earth and Critical Mineral Operations and Processing

Metal oxide + Reducing agent → Metal + Oxidized product

For example, in the reduction of iron(III) oxide with carbon monoxide:

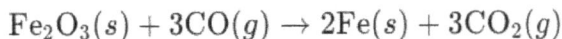

$$Fe_2O_3(s) + 3CO(g) \rightarrow 2Fe(s) + 3CO_2(g)$$

Here, carbon monoxide acts as the reducing agent, donating electrons and converting Fe^{3+} ions to metallic Fe^0, while itself being oxidized to carbon dioxide.

Different reduction methods are used depending on the nature of the ore, the metal's reactivity, and the desired purity.

Carbon Reduction (Carbothermic Reduction): The most common method for base metals such as iron, zinc, and lead. Carbon or carbon monoxide reacts with metal oxides at high temperatures in a furnace.

Example: Ironmaking in a blast furnace uses coke to reduce Fe_2O_3 and Fe_3O_4 to molten iron.

Hydrogen Reduction: Hydrogen gas serves as a cleaner reducing agent, producing only water as a byproduct.

Example: The reduction of tungsten oxide (WO_3) or molybdenum oxide (MoO_3) to metal powder.

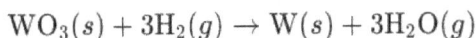

$$WO_3(s) + 3H_2(g) \rightarrow W(s) + 3H_2O(g)$$

Metallothermic Reduction: A reactive metal, such as aluminium, magnesium, or calcium, is used to reduce another metal oxide.

Example: In the aluminothermic (thermite) process, aluminium reduces iron oxide:

$$Fe_2O_3 + 2Al \rightarrow 2Fe + Al_2O_3$$

This highly exothermic reaction is used for producing high-purity metals and in field welding.

Electrolytic Reduction: For very reactive metals like aluminium and sodium, electrical energy is used to reduce metal ions in a molten or aqueous electrolyte rather than using chemical agents.

Example: The Hall–Héroult process reduces alumina (Al_2O_3) dissolved in molten cryolite to aluminium metal.

Reduction is widely applied across various metallurgical industries to extract and purify metals from their oxides or sulphide precursors.

In ironmaking, the process occurs in blast furnaces, where carbon monoxide (CO) generated from coke combustion serves as the primary reducing agent. The CO reacts with iron oxides in the ore, reducing them to molten metallic iron, which is then refined further to produce steel. This carbothermic reduction process remains one of the most important and large-scale industrial applications of reduction chemistry.

For nickel and cobalt, reduction is typically used after hydrometallurgical stages such as leaching and precipitation. In these cases, intermediate oxides or sulphides are reduced to yield high-purity metallic products. This process is crucial in the production of battery-grade nickel and cobalt, which are essential for modern energy storage and superalloy applications.

The reduction of tungsten and molybdenum is carried out using hydrogen gas, which converts their oxides into fine metallic powders. These powders are then compacted or sintered to form components used in high-temperature alloys, cutting tools, and electronic devices. Hydrogen reduction is favoured for these metals due to its clean by-products and ability to produce high-purity results.

In the case of titanium and zirconium, reduction is achieved through the Kroll process, which uses magnesium or sodium as the reducing agent under vacuum or inert atmosphere conditions. This environment prevents re-oxidation and ensures the production of reactive, high-purity metal. These metals are valued for their strength, corrosion resistance, and low density, making them essential in aerospace, defence, and nuclear applications.

Together, these examples demonstrate how reduction processes are tailored to the specific chemical and physical properties of different metals, enabling efficient production across a broad range of critical industrial sectors.

Reduction offers several key advantages in metallurgical processing. It enables the direct production of high-purity metals from their oxide forms, making it a fundamental step in the extraction of both base and critical metals. The process allows precise control over temperature and atmospheric conditions, which is essential for achieving selective reduction and preventing unwanted side reactions or re-oxidation. Because of its versatility and scalability, reduction is well-suited to large-scale industrial operations, such as those used in the production of iron, nickel, and other structural or high-value metals.

However, the process also presents significant limitations. Reduction generally requires very high operating temperatures, which leads to substantial energy consumption and increased operational costs. In addition, many reduction processes—particularly those relying on carbon as a reducing agent—produce greenhouse gas emissions, primarily carbon dioxide (CO_2), which must be managed or captured to minimize environmental impact. Furthermore, certain metals such as aluminium and titanium demand highly specialized and costly reduction methods, often involving complex equipment, controlled atmospheres, and high-purity reagents.

Reduction is the final and decisive stage in many metallurgical routes, transforming processed oxides or concentrates into metallic form. Whether achieved through chemical agents like carbon and hydrogen or through electrolysis, the essence of reduction lies in reversing oxidation to isolate the pure metal. The choice of reducing agent and technology depends on the thermodynamics of the metal–oxygen system, the desired purity, and environmental considerations—making reduction a cornerstone of both traditional and modern metallurgical production.

Thermal Decomposition and Product Upgrading

Thermal Decomposition and Product Upgrading are important stages in pyrometallurgical processing, used to convert minerals or chemical intermediates into more stable, reactive, or pure forms that are suitable for downstream refining. These processes rely on controlled heating to induce chemical transformations—such as the removal of volatile components, the breakdown of complex compounds, or the conversion of intermediates into oxide or metallic forms—without necessarily melting the material.

Thermal Decomposition

Thermal decomposition involves heating a compound to a temperature where it breaks down into simpler substances, usually releasing gases such as carbon dioxide (CO_2), sulphur dioxide (SO_2), water vapour (H_2O), or hydrogen fluoride (HF). This process is typically conducted in furnaces, kilns, or retorts, and is vital for preparing ores and concentrates for further processing.

For example:

- Bastnäsite (REE fluorocarbonate) is thermally decomposed to remove CO_2 and F, forming rare earth oxides that are more amenable to acid leaching.

- Monazite (REE phosphate) can undergo controlled decomposition to release volatile phosphates and prepare the material for sulfuric acid digestion.

- Spodumene ($LiAlSi_2O_6$) undergoes phase transformation from α- to β-spodumene at about 1000°C, increasing its reactivity for lithium extraction.

- Nickel and cobalt oxalates or hydroxides produced during hydrometallurgical purification are decomposed thermally to produce oxides prior to reduction.

The objective of thermal decomposition is to liberate volatile constituents, increase reactivity, and produce stable intermediates with predictable chemical and physical properties.

Step-by-step: Running a Thermal Decomposition (Roast/Calcine) Safely and Effectively

1. Define the goal and chemistry

- Identify the species you want to remove (CO_2, H_2O, SO_2, HF, NH_3, organics) and the desired solid product (e.g., REE oxide, β-spodumene, metal oxide).

- Write the target decomposition reaction(s) and expected gases. Do a quick mass balance to estimate off-gas volumes.

2. Characterise the feed

- Measure moisture, LOI, PSD (particle size distribution), bulk density.

- Use **TGA/DSC** to find onset/peak temperatures and whether the reaction is endo/exothermic.

- Check minerals/phases (XRD) and hazardous constituents (e.g., F in bastnäsite/monazite; S in sulphides).

3. Choose the thermal unit

- **Multiple-hearth/rotary kiln**: versatile, good residence-time control for mixed feeds.

- **Fluidised-bed roaster**: fine feeds, excellent heat/mass transfer.

- **Muffle/box furnace/retort**: batches, R&D to pilot.

- Size loading based on required residence time and throughput.

4. Pick the atmosphere

- **Oxidising** (air/O_2): burn off S/organics; form oxides.

- **Inert** (N_2/Ar): avoid oxidation (e.g., some oxalates).

- **Controlled humidity/HCl/HF scrubbing** as needed downstream—never inject corrosives into the hot zone without engineering controls.

5. Engineer off-gas handling before you heat

- For **CO_2/H_2O**: cyclone + condenser/mist eliminator is often enough.

- For **SO_2**: alkali scrubber (NaOH/lime) or convert to H_2SO_4 in purpose plants.

- For **HF/HCl**: packed-bed scrubber with caustic; HF requires HF-rated alloys/linings and spent-liquor management.

- Add HEPA or baghouse where dust is expected.

6. Prepare the feed

- Dry to a consistent moisture; de-agglomerate/screen to a target PSD (typically 75 μm–10 mm depending on unit).

- If sticky or fine, consider pelletising or mixing with inert recycle to improve bed permeability.

7. Set the temperature profile

- Use TGA/DSC data to define:

  o **Ramp**: 1–10 °C/min to avoid decrepitation/sintering.

  o **Soak**: hold at the peak decomposition window (examples below).

  o **Cool**: controlled to prevent rehydration/carbonation.

- Typical soak ranges:

  o Bastnäsite (remove CO_2/F): 500–800 °C (often staged).

  o Monazite pre-digest: 400–700 °C (process-specific).

  o Spodumene α→β: ~1,000–1,100 °C, 20–60 min.

  o Metal oxalates/hydroxides → oxides: 350–700 °C.

    o   Sulphide oxidation (roasting): 500–900 °C (gas control critical).

8. Load and start up

- Purge with the chosen gas to remove air (for inert runs) or confirm uniform air flow (oxidising).

- Start burners/heaters at the defined ramp. Monitor bed/solids, not just gas temperature.

9. Control residence time and material movement

- Rotary kiln: set rpm and slope for the target minutes/hours.

- Fluidised bed: set superficial velocity for stable fluidisation without elutriation.

- Hearth/box: load depth ≤ 2–5 cm for uniform heat penetration (scale as proven).

10. Monitor in real time

- **Off-gas**: CO_2/H_2O/SO_2/HF with IR/FTIR or electrochemical cells; O_2 for oxidising runs.

- **Mass loss**: continuous feed/weight signals or periodic grab-weigh samples.

- **Differential pressure** across bed (fluidised units) to detect agglomeration.

- Record temperature at multiple thermocouples (inlet, mid-bed, outlet).

11. Decide end-point

- Continue the soak until: mass stabilises ($\Delta m < 0.1$–0.2 % over 10–15 min), off-gas target species plateau at baseline, and interim XRD/LOI checks confirm phase conversion.

12. Cool and discharge safely

- Cool under the same protective atmosphere to <150–200 °C before exposure to air (prevents flash hydration/carbonation).

- Use enclosed handling if HF/HCl/SO_2 could desorb during cooldown.

13. Post-process QA/QC

- **XRD**: confirm phase (e.g., β-spodumene, REO).

- **LOI** at 1,000 °C: confirms volatile removal.

- **Chemistry** (ICP-OES/MS): F, S, C residuals; target metal grade.

- **Reactivity tests**: leachability (e.g., β-spodumene acid leach rate), surface area (BET), PSD.

14. Close the mass and emissions balance

- Compare calculated vs measured gas loads; adjust scrubber caustic strength/flow or residence time as needed.

- Manage spent scrubbing liquors (gypsum/fluorides/sulphites) per your site EHS plan.

15. Troubleshooting guide (quick hits)

- **Sintering/lumps** → lower soak T, slower ramp, finer PSD/pelletise, add bed diluent.

- **Incomplete decomposition** → extend soak, increase T slightly, improve mixing/gas contact.

- **High HF/SO$_2$ in stack** → verify seals, raise scrubber reagent flow/alkalinity, check leaks.

- **Fluidised bed defluidises** → reduce fines, increase gas velocity slightly, lower bed temperature.

- **Re-carbonation on discharge** → cool under inert, use sealed transfer to next stage.

Examples

Each of the following examples represents a thermal decomposition or calcination process tailored to a specific mineral system, illustrating how controlled heating transforms ores or intermediates into more reactive or purified products suitable for subsequent leaching or refining.

Bastnäsite Concentrate → Rare Earth Oxide–Rich Calcine: In bastnäsite (REE fluorocarbonate) processing, the concentrate is heated in an air atmosphere with a gradual temperature ramp of approximately 5 °C per minute up to around 650 °C, where it is held for about two hours. This heating drives off carbon dioxide and volatile fluorine compounds, decomposing the carbonate matrix and converting the bastnäsite into an oxide-rich product. The gases released—primarily CO$_2$ and HF—must be captured and neutralized using a dual-stage caustic scrubber system to prevent environmental release. The resulting calcine contains reactive rare earth oxides that are highly amenable to subsequent acid leaching or solvent extraction refining.

Monazite → Pre-Digest Calcine: Monazite, a phosphate mineral rich in light rare earths, undergoes controlled heating in an oxidizing or limited-oxygen environment at about 550 °C for one to two hours. The goal is to decompose minor organic or sulphate impurities and partially open the phosphate structure without causing sintering or excessive phase change. During this process, phosphorus-bearing fumes may evolve, which must be captured in a fume-scrubbing system. The partially decomposed calcine becomes significantly more reactive toward sulfuric acid, which is later used in the digestion stage to convert the rare earth phosphates into soluble sulphates.

Spodumene α → β Transformation: Spodumene (LiAlSi$_2$O$_6$), the main lithium-bearing mineral in hard-rock pegmatites, must be thermally activated to convert its dense, inert α-form into the more open, acid-soluble β-form. This transformation is achieved by heating the concentrate in a rotary kiln under an air atmosphere at approximately 1,050 °C for 30–45 minutes, typically at a kiln rotation speed of 2–4 rpm. Rapid but controlled cooling in air to below 200 °C is then performed to stabilize the β-phase and prevent reversion. The resulting product has increased lattice openness, enabling efficient acid leaching with sulfuric acid to extract lithium as soluble lithium sulphate.

Cobalt Oxalate → Cobalt(II) Oxide (CoO): In cobalt refining, cobalt oxalate—a precursor often obtained from solution purification—is decomposed thermally to cobalt oxide. This process is carried out in a muffle furnace under an inert nitrogen atmosphere at 400–500 °C for approximately one hour. The oxalate breaks down, releasing carbon dioxide and carbon monoxide while forming a fine, stable CoO powder. The off-gas is passed through a condenser or mist eliminator to capture water and residual vapours. If a partially reduced or non-oxidized product is desired, the inert atmosphere prevents further oxidation to Co_3O_4. The resulting CoO serves as a high-purity feedstock for battery materials or metallic reduction.

Overall, these examples demonstrate how controlled thermal decomposition conditions—temperature, atmosphere, and residence time—are precisely tuned to mineral chemistry and downstream process requirements. Each treatment step enhances the reactivity, purity, and stability of the intermediate product, enabling efficient hydrometallurgical or pyrometallurgical recovery of critical and rare earth elements.

Product Upgrading

Product upgrading refers to the thermal improvement or purification of a mineral product's composition or phase before final recovery or reduction. It can include roasting, calcination, or partial reduction to remove impurities and enhance downstream efficiency.

In rare earth processing, upgrading through controlled roasting or calcination can eliminate volatile impurities (such as fluorides or carbonates), improve oxide purity, and prepare the feedstock for solvent extraction or reduction. For critical minerals such as lithium, tungsten, and cobalt, upgrading ensures that the intermediate oxides or salts meet the required chemical specifications for battery or alloy production.

Thermal upgrading may also enhance surface area, particle morphology, and sinter resistance, improving performance in subsequent leaching or refining operations.

Step-by-Step Process for Thermal Product Upgrading

1. Define the upgrade goal

- Remove volatiles (CO_2, HF, H_2O, organics), drive phase change (e.g., α→β-spodumene), oxidize/decompose sulphides, or partially reduce oxides.

- Set measurable targets: LOI loss (wt%), impurity thresholds (F, P, S, Na, Cl), phase purity (% oxide), surface area (BET), PSD, and downstream extraction performance (e.g., SX loading, leach kinetics).

2. Characterise the feed

- Assay: major/trace elements (ICP-OES/MS), F/S/P/Cl, carbonates, LOI.
- Mineralogy/phase: XRD (Rietveld), SEM-EDS.
- Thermal behaviour: TGA/DTG/DSC with evolved-gas analysis (FTIR/MS) to identify decomposition onsets and gases (CO_2, HF, SO_2).
- PSD, moisture, bulk density, flowability.

3. Select the thermal route

- Roasting (oxidative) to remove S/F/C and open lattices (e.g., bastnäsite, chalcopyrite pre-oxidation).
- Calcination (dehydration/ decarbonation/ dehydroxylation) for REE carbonates/hydroxides, bauxite, gypsum, cobalt oxalate.
- Phase transformation (e.g., spodumene α→β at ~1,000–1,100 °C).
- Partial reduction (e.g., Fe-bearing intermediates) when required for downstream metallurgy.

4. Choose atmosphere and gas management

- Air/O_2-enriched for oxidation; N_2/Ar for inert calcination; controlled H_2/CO for reduction; limited-O_2 for sensitive materials.
- Specify off-gas capture: dual caustic scrubbers for HF/CO_2 (REE fluorocarbonates), wet scrubbers for SO_2, baghouse for particulates, condenser/mist eliminator for organics.

5. Select equipment and scale

- Bench (muffle tube furnace), pilot (rotary kiln, fluidised-bed, multiple-hearth), plant (rotary kiln, fluidised-bed, multiple-hearth, shaft or circulating roaster).
- Match to PSD/throughput/heat & mass transfer:
  - Fine sulphides → fluidised-bed or multiple-hearth.
  - Coarse/mixed feeds & phase change → rotary kiln.
  - Uniform powders for precise holds → muffle/tube furnace.

6. Design the thermal profile

- Ramp rate: typically 2–10 °C/min (slower if exothermic or to avoid sintering).
- Hold/soak temperature & time: determined from TGA/DSC/XRD endpoints (e.g., bastnäsite ~600–700 °C; monazite pre-digest ~500–600 °C; α→β spodumene ~1,050 °C for 30–45 min).
- Cooling regime: controlled/rapid air quench for metastable phases (β-spodumene); slow cool for dense oxides to avoid cracking.

7. Prepare the feed for heating

- Dewater/dry to <1–2% H_2O if needed.
- Optional agglomeration (micro-pellets) for fines to improve bed permeability.
- Homogenise lots; remove tramp metal.

8. Validate safety & environmental controls

- Confirm gas seals, purge protocols, interlocks, LEL/O_2 monitoring for H_2/CO systems.
- Verify scrubber caustic strength and blowdown, SO_2 conversion/neutralisation capacity, dust control, and heat-recovery loops.

9. Execute thermal cycle

- Purge (if inert/reducing), then ramp per profile.
- Hold at target temperature; maintain gas flow and residence time.
- Track off-gas (flow, temperature, composition) to confirm expected devolatilisation/oxidation.

10. In-process monitoring (QA/QC)

- Pull timed samples (pilot/plant) for quick LOI and XRD spot checks.
- Watch bed temperature uniformity (multi-point thermocouples) to avoid hot/cold zones.
- Adjust gas rate, O_2 %, and rotation/bed depth to keep reactions complete without sintering.

11. Controlled cooling & discharge

- Apply required cool-down profile (e.g., rapid to <200 °C for β-spodumene).
- Prevent air in-leakage on reduced products; use inert cooldown if oxidation must be avoided.

12. Post-process conditioning

- De-agglomerate and mill if needed to restore PSD.
- Screen and recycle oversize; collect fines with appropriate dust control.
- Blend for lot uniformity ahead of hydrometallurgy.

13. Verify product upgrade (release tests)

- Chemical: major/trace, F/S/P/Cl, Na/K/Cl carryover.
- Phase: XRD phase purity (e.g., REO content, β-spodumene fraction).
- Physical: BET surface area, PSD, tap/bulk density, flow index.
- Processability: standard leach test (rate/extent), SX loading curve, IX kinetics as applicable.

14. Close the loop (optimisation)

- If targets missed: adjust soak time/temperature, ramp, atmosphere, or bed depth; add staged heating for multi-step decomposition.
- Balance "reactivity vs. sintering": maximise devolatilisation/phase change while preserving surface area and accessible porosity.

15. Document recipe & control limits

- Capture the full "thermal recipe": feed specs; equipment; atmosphere; ramp/soak/cool; gas treatment; acceptance criteria; sampling plan; deviations & corrective actions.
- Define SPC limits for plant control (temperatures, residence time, O_2 %, scrubber pH, pressure drop, emissions).

Examples

Quick recipe templates (illustrative):

- Bastnäsite → REO-rich calcine: Air; ramp 5 °C/min to 650 °C; 2 h hold; off-gas to dual caustic scrubbers (HF/CO_2). Target: ≥90% REO phase, F <0.3 wt%, improved H_2SO_4 leach rate.
- Monazite pre-digest: Air or limited O_2; 550 °C, 1–2 h; capture P-bearing fumes. Target: improved acid digestibility; no sintering.
- Spodumene α→β: Rotary kiln, air; 1,050 °C, 30–45 min at 2–4 rpm; rapid cool to <200 °C. Target: ≥95% β-phase; leach extraction ↑.
- Cobalt oxalate → CoO: N_2 purge; 400–500 °C, 1 h; condenser/mist eliminator on vent. Target: CoO purity ≥99%; minimal Co_3O_4.

The efficient recovery of industrially critical metals through upgraded processes demonstrates significant economic benefits and advances in environmental sustainability. This synthesis reviews pertinent examples of process improvements related to various elements such as rare earths, lithium, tungsten, cobalt, and nickel.

In the case of rare earths, advancements in leaching techniques, especially using cleaner, oxide-rich calcines, have been shown to enhance the kinetics of sulfuric and hydrochloric acid leaching. This improvement not only accelerates the extraction process but also markedly reduces the consumption of reagents such as sulfuric acid and hydrochloric acid. Moreover, by minimizing the introduction of fluorides and carbonates, the subsequent solvent extraction (SX) and ion exchange (IX) processes become more efficient, yielding purer solutions and significantly lowering the environmental footprint of processing operations [261].

Focusing on lithium extraction, recent studies indicate that utilizing β-spodumene facilitates higher extraction efficiencies of lithium through acid leaching while simultaneously shortening residence times in reactors [262]. By optimizing the leaching process, the overall operational costs can be significantly reduced, leading to improved profitability indicators for lithium processing facilities [263].

Tungsten processing presents another compelling case where innovative methods have resulted in substantial payoffs. Recent research indicates that calcined intermediates, particularly the conversion of wolframite to scheelite, enhances the dissolution rates in alkaline leach solutions [263]. This selective dissolution leads to a higher yield of cleaner ammonium paratungstate (APT), a precursor for tungsten metal, thereby underscoring the benefits of developing refined leaching methodologies [264]. Moreover, techniques such as sodium carbonate roasting followed by aqueous leaching have also shown promise in maximizing tungsten recovery from both low-grade and mixed concentrates [265].

For cobalt and nickel, the recovery of these metals has benefited from advancements in the characterization and processing of calcined oxalates and hydroxides. Enhanced calcination methods allow for the controlled formation of oxide phases. This enables precise precipitation and reduction during battery-grade preparations, streamlining the production chain and ensuring higher purity levels of final products [266].

In essence, thermal decomposition and product upgrading bridge the gap between raw mineral concentrates and refined products. By controlling temperature, atmosphere, and residence time, operators can achieve phase transformation, impurity removal, and compositional refinement, leading to higher recovery rates, greater purity, and improved process efficiency in both rare earth and critical mineral production circuits.

Equipment and Temperature Control

Pyrometallurgical Equipment

A range of furnace designs and configurations are used depending on the ore type, particle size, and reaction goals. Common types include:

- **Rotary kilns** — Long, slightly inclined cylindrical furnaces that rotate slowly, ensuring uniform heating and mixing. Widely used for roasting lithium (spodumene) and calcining rare earth carbonates and hydroxides before leaching.

- **Multiple-hearth furnaces** — Suitable for fine sulphide concentrates such as those containing zinc or copper. The multi-layer design allows controlled oxidation and high throughput.

- **Fluidized-bed roasters** — Provide intense gas–solid contact, ideal for sulphide ores and thermal decomposition reactions. They offer superior heat transfer and uniform temperature profiles.

- **Reverberatory and electric furnaces** — Used for large-scale oxidation, reduction, and smelting of copper, nickel, and iron ores. They can handle coarse materials and complex feeds.

- **Muffle or tube furnaces** — Common in laboratory and pilot-scale work for controlled atmosphere heating of small samples, such as cobalt oxalate decomposition or REE precursor calcination.

Each equipment type incorporates refractory linings, burners, and exhaust systems designed to withstand thermal stress and corrosive gases. Off-gas systems such as scrubbers, baghouses, and electrostatic precipitators are essential for environmental control, particularly where SO_2, CO_2, or HF gases are evolved.

Rotary Kilns

A rotary kiln is a large, cylindrical furnace that rotates slowly about its longitudinal axis while being slightly inclined. Feed material is introduced at the higher end, and as the kiln rotates and the cylinder slopes downward slightly, the material gradually moves toward the discharge end. Hot gases flow either co-current (in the same direction) or counter-current (opposite direction) to the material, providing heat and facilitating the chemical/physical transformation required (calcining, roasting, phase change, partial reduction).

Internally, the kiln shell is lined with refractory bricks or castable linings to protect from high temperatures and corrosive materials. Flights or lifters inside the kiln ensure that material is lifted and cascades, promoting mixing and uniform heat transfer.

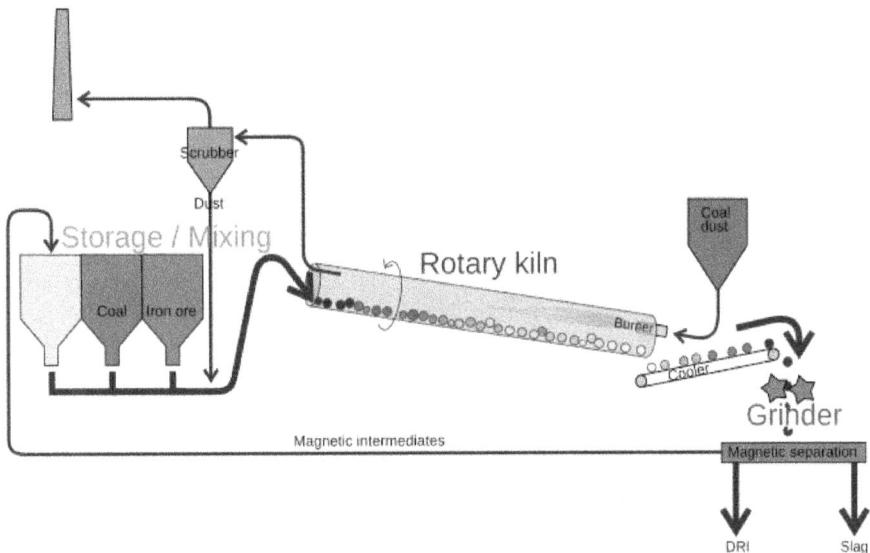

Figure 56: Krupp-Renn Process diagram. Borvan53, CC BY-SA 3.0, via Wikimedia Commons.

The typical process steps inside a rotary kiln:

1. Feeding of dried/pre-treated material at the inlet.

2. Material travels slowly down the channel while being heated.

3. End reactions occur (decomposition of carbonates, phase change, oxidation/reduction).

4. Discharge of the roasted/calcined product at the lower end.

5. Off-gas removal, dust capture, and gas scrubbing/scrubber systems.

Rotary kilns exhibit a wide range of dimensions and throughput capacities, tailored to the specific requirements of their applications. Smaller units typically operate at around 800 °C and can handle between 12 to 100 tons per day. These are often used for pilot-scale or specialized thermal processes.

In contrast, mineral-processing rotary kilns listed by Indian manufacturers span lengths from approximately 10 to 60 meters, with diameters ranging from 1 to 3.6 meters. Their throughput varies significantly, from as little as 150 kilograms per hour to as much as 10 metric tons per hour, depending on the feed material and process conditions.

Figure 57: Overall view rotary kiln plant. U.S. Department of Energy, CC0 1.0, via Rawpixel.

At the industrial scale, a magnetic-reduction roasting rotary kiln designed for iron ore processing might measure around 3.6 meters in diameter and 50 meters in length, with a capacity of roughly 800 tons per day. This highlights the scale achievable for high-throughput metallurgical operations.

Specialized applications, such as rare earth oxalate calcination, typically require more moderate capacities. One example from YingYong Machinery describes a rotary kiln capable of processing 1 to 2 tons per hour, optimized for controlled thermal decomposition and phase transformation.

Overall, rotary kilns can be configured for a broad spectrum of throughput—from a few tons per hour in pilot setups to several hundred or even thousands of tons per hour in full-scale industrial installations. The final design depends heavily on factors such as feed composition, mineralogy, target temperature, and required residence time.

The capital cost of establishing a rotary kiln system is influenced by a multitude of factors including kiln size, temperature rating, materials of construction, and the requisite auxiliary equipment such as feeders and gas-handling systems. Industry analysis consistently points out that kiln size and capacity are the principal determinants of cost; larger kilns necessitate sophisticated refractory linings and advanced control systems to maintain operational efficiency under extreme conditions [267, 268]. Additionally, specific environmental standards significantly impact design and installation costs, particularly when significant emission controls like scrubbers are mandated [269].

From an investment perspective, a rotary kiln unit can be quoted at approximately Rs 70,00,000 (around USD 900,000), with notable capacity variations ranging from 150 kg/hour to 10 metric tons/hour, depending on the kiln dimensions, which can extend between 10 to 60 meters in length and 1 to 3.6 meters in diameter. Maintenance costs are another critical aspect; for instance, studies indicate that the replacement of worn refractory linings in larger kilns can incur expenses between USD 40,000 and 110,000, influenced heavily by both design specifics and the choice of materials used in the linings [268].

When assessing operational costs, fuel or energy consumption emerges as the most substantial factor due to the high operational temperatures of rotary kilns, typically within the range of 800 °C to 1,100 °C [270]. Maintenance plays a similar role, significantly affecting the budget through the wear and tear of linings, bearings, rollers, and critical drive components, which necessitate ongoing attention and resources. Furthermore, the upkeep of emissions control systems, including scrubbers and dust collectors, is both a financial and operational consideration, ensuring compliance with environmental regulations [271]. Finally, labour costs and the extent of automation also contribute significantly, impacting not just efficiency, but also the overall economies of scale through improved process monitoring and adjustments during kiln operation [272].

In summary, the capital and operational costs associated with rotary kilns are substantial and multifaceted. In light of these considerations, effective site engineering and process design are crucial to optimizing performance and prolonging the equipment's lifespan. Specifically, attention must be given to utility infrastructure, the durability of refractory materials, and comprehensive environmental compliance systems, all of which are vital to ensure the safe, efficient, and sustainable operation of rotary kiln installations [269, 273].

The staffing requirements for rotary kilns vary depending on the size, level of automation, and complexity of the operation. In small pilot or research-scale installations, only one or two operators are typically required to handle feeding, temperature monitoring, and product discharge. These smaller kilns often have simpler control systems and manual oversight, making them ideal for testing or low-throughput operations.

In contrast, large industrial rotary kilns—such as those used in lithium or rare earth processing plants—are usually integrated with automated feeding, temperature control, and gas handling systems. Such installations typically operate on a shift basis, with a small team per shift (for example, one to two operators) responsible for monitoring operations, supported by a maintenance crew and control-room personnel who oversee automated systems.

Rare Earth and Critical Mineral Operations and Processing

Additional staff are required for specialized tasks. Maintenance technicians handle routine inspections, lubrication, and replacement of critical components such as refractory linings, bearings, gears, and rollers. Process engineers monitor and optimize temperature profiles, reaction atmospheres, and feed rates to maintain product quality and efficiency. Instrumentation and control specialists ensure accurate data collection and process automation, while quality control analysts conduct sampling and testing of intermediate and final products.

While there is no standardized or published "operators per tonne" ratio for rotary kilns, the staffing levels are generally comparable to other continuous processing units of similar scale. In large-scale operations, this often equates to three to five personnel per shift, including operators, technicians, and process controllers, with additional support available for maintenance and safety functions. Effective workforce planning ensures continuous, safe, and efficient kiln operation while minimizing downtime and maximizing throughput.

In rare earth processing, rotary kilns are extensively used for calcination of REE precursors. For instance, in Myanmar, rare earth oxalates are converted into rare earth oxides at temperatures between 800°C and 1,100°C, producing high-purity feedstock for solvent extraction or reduction. Rotary kilns are also employed in the roasting of bastnäsite and monazite concentrates, which are carbonate, fluoride, or phosphate minerals. The roasting process removes volatile components such as CO_2 and fluorine, resulting in an oxide-rich product that is more reactive and suitable for acid leaching. A patented design describes a specialized rotary kiln system incorporating drying and calcination units specifically tailored for rare earth mineral roasting, ensuring optimal temperature control and off-gas handling.

In the case of critical minerals, rotary kilns are indispensable for several high-temperature transformations. One of the most well-known applications is the α- to β-spodumene conversion in lithium processing. This transformation occurs at around 1,000°C and is a crucial step in preparing spodumene concentrate for acid leaching and lithium recovery. Similarly, tungsten-bearing ores and other heavy minerals may undergo roasting or reductive roasting in rotary kilns to alter mineral phases and improve subsequent leaching performance.

Rotary kilns are also used in the pre-treatment of nickel and cobalt laterites, where ores are pre-roasted before high-pressure acid leaching (HPAL) or smelting. In certain flowsheets, direct reduction roasting is applied to nickel ores within kilns to produce intermediate products with enhanced metal recovery potential.

Overall, rotary kilns are a cornerstone of pyrometallurgical and hybrid processing routes in both the rare earth and critical minerals industries. Their versatility, ability to handle a wide range of feed types, and controlled temperature environment make them indispensable for achieving high-purity intermediate products essential to the manufacture of magnets, batteries, and high-performance alloys.

Multiple-hearth Furnaces

A multiple-hearth furnace (MHF) is a vertical, cylindrical thermal processing unit consisting of several stacked refractory-lined hearths with a central rotating shaft. Attached to this shaft are rabble arms equipped with blades that slowly rotate, continuously stirring and moving the feed material from the centre of each hearth to the periphery—or vice versa—depending on the hearth configuration. This staged movement ensures excellent gas–solid contact, uniform heat distribution, and controlled residence time for the material being processed.

Feed material (such as a concentrate, precipitate, or calcined intermediate) is introduced at the top hearth of the furnace. As the central shaft rotates, rabble arms push the material toward drop holes that lead it to the next hearth below. Depending on the process design, the direction alternates between inward and outward on successive hearths, ensuring thorough exposure to heat and gas flow.

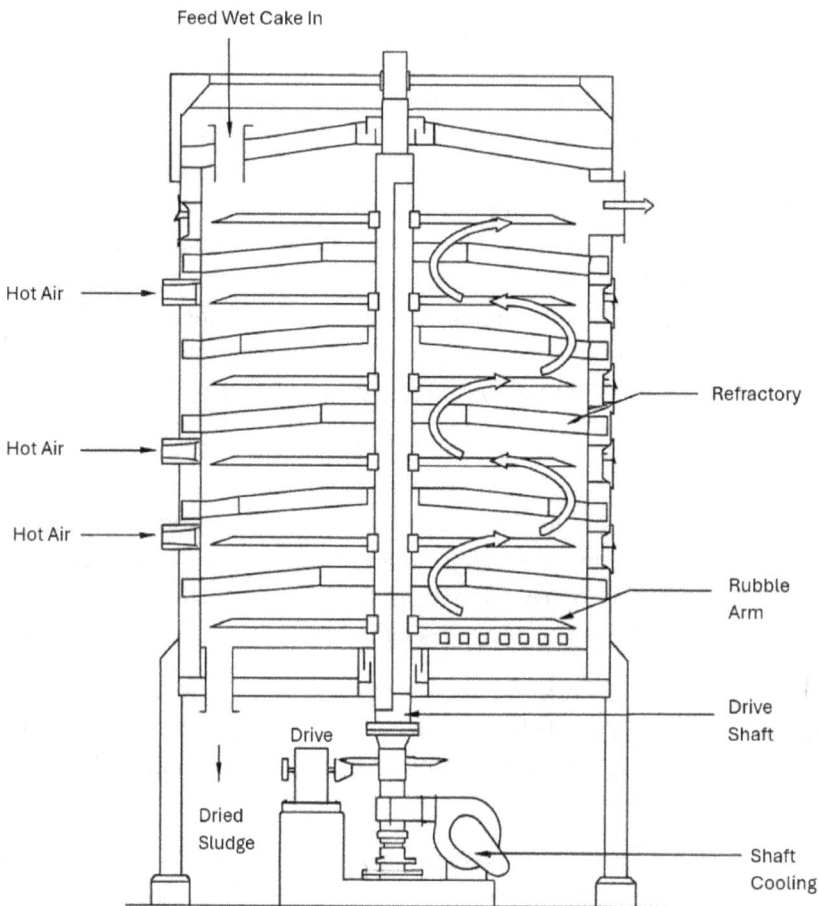

Figure 58: Multiple-hearth Furnace.

Hot gases—produced either by internal burners or introduced from an external source—flow upward countercurrent to the descending solids. This counterflow arrangement provides efficient thermal transfer and allows for sequential temperature zones, from drying at the top to oxidation, reduction, or calcination near the bottom. Typical operating temperatures range from 500°C to 1,200°C, depending on the material and process requirements.

Industrial multiple-hearth furnaces can have 4 to 12 hearths and diameters ranging from 3 to 9 metres, with overall heights up to 25 metres. Throughput capacity varies widely—from a few hundred kilograms per hour in pilot units to hundreds of tonnes per day in full-scale installations. The number of hearths, rotation speed (typically 0.5–2 rpm), and gas velocity are all adjusted to optimize the desired reaction (e.g., oxidation, dehydration, or reduction).

The capital cost of establishing a multiple-hearth furnace depends on its size, refractory materials, burner system, and gas-handling infrastructure. A medium-sized industrial furnace suitable for mineral roasting or calcination may cost between USD 1–3 million for a turnkey setup, with larger, high-capacity units exceeding USD 5 million when including off-gas cleaning systems (scrubbers, cyclones, or baghouses).

Operating costs are dominated by energy consumption, maintenance (particularly of rabble arms, gear drives, and refractory linings), and emission control systems. The energy demand typically ranges between 2–5 GJ per tonne of feed, depending on the process temperature and material type.

The number of operators required for a multiple-hearth furnace depends primarily on the scale of the unit, the degree of automation, and how well it is integrated into the plant's overall control system.

In small pilot-scale operations, the process is relatively straightforward, often requiring only one or two operators to oversee feeding, temperature adjustments, gas flow control, and product discharge. These units tend to have semi-automatic or manual control systems, allowing for flexibility during testing or small-batch production.

In contrast, large industrial multiple-hearth furnaces (MHFs) are designed for continuous 24-hour operation and typically feature extensive automation, including programmable logic controllers (PLCs) and real-time monitoring systems. These installations are usually managed by a control room operator who supervises process parameters and one field technician per shift who handles physical inspections, sampling, and adjustments at the furnace itself. The operation is further supported by a maintenance team responsible for mechanical servicing, refractory repair, and periodic shutdown maintenance.

Overall, a large-scale MHF facility running around the clock may employ three to five staff per shift, a staffing level comparable to other continuous thermal processing lines such as rotary kilns or fluidized-bed roasters. This structure ensures adequate coverage for safe operation, equipment reliability, and process optimization while maintaining high throughput and product quality.

Multiple-hearth furnaces are particularly valuable in rare earth and critical mineral processing due to their ability to handle fine feed materials and control multiple temperature zones within a single unit.

- **Rare Earth Oxalate to Oxide Conversion**: MHFs are used to thermally decompose rare earth oxalates ($REE_2(C_2O_4)_3 \cdot nH_2O$) into rare earth oxides (RE_2O_3) while releasing CO_2 and CO gases. The staged heating minimizes sintering and allows efficient off-gas scrubbing for carbon oxides and fluorides.

- **Monazite Roasting**: In processing **monazite concentrates**, MHFs can be used for **decomposition and oxidation** to prepare the material for subsequent acid digestion, removing organic residues and volatile phosphates.

- **Nickel and Cobalt Intermediates**: Used for oxidizing or reducing lateritic ores prior to leaching or smelting, especially when uniform temperature control and long residence times are required.

- **Cobalt Oxalate or Hydroxide to Oxide Conversion**: MHFs offer excellent control for producing high-purity CoO or Co_3O_4, vital in battery precursor materials.

Multiple-hearth furnaces provide exceptional flexibility, temperature control, and throughput, making them ideal for continuous thermal processing in rare earth, lithium, nickel, and cobalt refining. Although more complex and costly than simpler furnaces, their energy efficiency, modular temperature zoning, and high product uniformity make them essential for processes that demand precise control of volatile removal, oxidation, or phase transformation.

Fluidized-bed Roasters

A fluidized-bed roaster (FBR) is a highly efficient thermal processing system designed to roast, oxidize, or calcine finely divided mineral concentrates or chemical intermediates. It is widely used in metallurgy, chemical processing, and increasingly in the treatment of rare earth and critical minerals, due to its excellent heat transfer, reaction control, and scalability.

A fluidized-bed roaster operates by suspending fine solid particles in an upward stream of hot gas, typically air or oxygen-enriched air. When the gas velocity is high enough, it causes the solid bed to behave like a boiling liquid, creating intense mixing and uniform temperature distribution.

The fluidized-bed roasting process operates through a sequence of well-controlled steps designed to ensure efficient thermal treatment and chemical conversion of fine mineral feeds.

The process begins with feed introduction, where finely ground concentrate or precipitate—such as sulphides, carbonates, or oxalates—is fed continuously into the reactor through a feed chute or screw feeder. Maintaining a consistent feed rate is essential for steady-state operation and uniform bed conditions.

Next is fluidization, the defining feature of the process. Hot air, oxygen, or another process gas enters the reactor through a distributor plate located at the bottom of the vessel. As the gas flows upward, it lifts and suspends the solid particles, creating a vigorously agitated, fluid-like bed. This motion provides excellent gas–solid contact, ensuring that heat and reactants are evenly distributed throughout the material.

During the roasting reaction stage, the suspended particles are exposed uniformly to the hot gas, leading to rapid oxidation, decomposition, or chlorination, depending on the desired reaction. Because of the constant mixing and even heat distribution, reactions occur quickly and uniformly. A typical example is the oxidation of zinc sulphide in air:

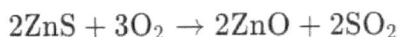

$$2ZnS + 3O_2 \rightarrow 2ZnO + 2SO_2$$

In this reaction, zinc sulphide (ZnS) is converted into zinc oxide (ZnO), while sulfur dioxide (SO_2) gas is released as a by-product.

After the roasting stage, gas–solid separation occurs. The exhaust gases, which contain reaction products such as SO_2, CO_2, and water vapour, are directed through cyclones, bag filters, or wet scrubbers to capture any entrained particles and prevent emissions from escaping into the atmosphere.

Finally, the discharge phase removes the roasted or calcined product continuously from the bottom or side of the reactor. The material is then cooled, collected, and sent to downstream processing—such as leaching or reduction—depending on the specific mineral system being treated.

This carefully controlled sequence enables high reaction efficiency, uniform product quality, and minimal environmental impact, making fluidized-bed roasting one of the most advanced thermal treatment methods used in modern mineral processing.

Fluidized-bed roasters are vertically oriented vessels lined with refractory material and typically constructed from carbon steel or stainless steel. These units are engineered for continuous operation and offer precise process control, making them suitable for a wide range of industrial applications.

Operating temperatures within fluidized-bed roasters generally range from 400°C to 1,000°C, depending on the specific ore or chemical system being processed. To maintain effective fluidization, gas velocities are controlled between 0.3 and 1.5 meters per second. The depth of the fluidized bed itself typically falls between 0.5 and 2.5 meters.

In terms of physical dimensions, reactor heights can vary from 6 to 20 meters, while diameters range from 0.5 to 5 meters, depending on the desired throughput. These systems are scalable, with throughput capacities spanning from a few hundred kilograms per hour in pilot-scale setups to several hundred tonnes per day in full-scale industrial operations.

To accommodate higher processing demands, fluidized-bed roasters are often designed as modular systems. Multiple units can be operated in parallel, allowing for flexible expansion and increased total capacity without compromising process control or efficiency.

The capital and operating costs of fluidized-bed roasters can vary significantly depending on the plant's capacity, the materials used in construction, and the complexity of the gas-handling and emission control systems. These factors determine not only the upfront investment but also the long-term maintenance and operating expenses.

For small pilot-scale units processing around 0.5 to 1 tonne per day, the cost typically ranges between USD $500,000 and $1.5 million. Such systems generally include the basic reactor, heating source, feed mechanism, and dust collection setup, and are used primarily for research, testing, and small-batch production.

Medium industrial plants, capable of handling between 50 and 200 tonnes per day, usually cost between USD $3 million and $10 million. These systems incorporate advanced features such as automated feed control, temperature regulation, and off-gas treatment facilities to comply with environmental standards.

At the upper end of the scale, large commercial installations, such as those used for zinc or nickel roasting exceeding 500 tonnes per day, can cost over USD $25–50 million. These facilities often require comprehensive environmental compliance systems, including sulphur dioxide (SO_2) scrubbers and acid recovery plants, which add substantially to both capital and operational expenditure.

The operating costs of fluidized-bed roasters are driven by several key factors. Energy consumption—whether from fuel or electricity—is a major contributor, as maintaining the high temperatures required for roasting demands continuous heat input. Air compressors and blowers used for fluidization also consume significant power. Maintenance costs include the regular inspection and replacement of refractory linings, distributor plates, and gas filters, all of which experience wear under high-temperature and abrasive conditions. Additionally, emission control systems must be operated and maintained to manage particulates and gases such as SO_2. Finally, labour and process supervision contribute to ongoing costs, though modern automation systems can reduce staffing requirements.

While fluidized-bed roasters require a substantial investment, their high efficiency, continuous operation capability, and environmental performance often justify the cost, especially in large-scale rare earth and critical mineral processing plants.

Staffing requirements for fluidized-bed roasters vary significantly depending on the level of automation and the scale of the plant. Smaller pilot or laboratory-scale roasters typically operate with 1 to 2 personnel per shift. These operators are responsible for manually managing feed input, conducting sampling, and controlling temperature settings.

In medium-sized industrial systems, staffing generally includes a control room operator and a field technician per shift. The control room operator oversees continuous operation and monitors system parameters, while the field technician handles tasks such as gas flow regulation and product discharge.

Large, fully integrated roaster installations require a more comprehensive team, typically comprising 3 to 5 personnel per shift. This team includes a control operator who monitors digital control systems, a field operator responsible for inspections and routine maintenance, and a dedicated maintenance crew tasked with servicing blowers, filters, and refractory linings. Additionally, a process engineer or shift supervisor is usually present to oversee quality assurance and safety compliance.

Advanced automation systems, such as programmable logic controllers (PLC) and distributed control systems (DCS), play a critical role in minimizing manual intervention. These technologies enhance operational stability and safety while reducing the need for extensive hands-on management.

Fluidized-bed roasters play a crucial role in the processing of rare earth elements (REEs) and other critical minerals, where precise temperature control, efficient gas–solid contact, and uniform heat distribution are essential for achieving high reaction efficiency and product quality. Their ability to handle fine materials and maintain consistent reaction environments makes them particularly valuable in oxidation, decomposition, and phase transformation processes.

In the case of rare earth elements (REEs), fluidized-bed roasters are commonly used for the roasting of bastnäsite and monazite. These minerals, which are carbonate and phosphate-based respectively, are treated to remove carbon dioxide (CO_2), fluorine (F), and organic impurities before acid leaching. The highly turbulent environment within the fluidized bed ensures complete decomposition and uniform oxidation throughout the feed material. A representative reaction for bastnäsite decomposition is:

$$CeCO_3F \rightarrow CeO_2 + CO_2 + F_2$$

Additionally, fluidized beds are used for REE oxalate decomposition, where hydrated oxalates ($REE_2(C_2O_4)_3 \cdot nH_2O$) are thermally decomposed between 600°C and 800°C to yield fine-grained rare earth oxides (REE_2O_3). The process releases CO_2, CO, and water vapour while producing a high-purity oxide that is highly suitable for downstream solvent extraction and refining stages.

In lithium processing, fluidized-bed roasters are being developed for the α- to β-phase transformation of spodumene ($LiAlSi_2O_6$). Although rotary kilns are traditionally used for this conversion, fluidized beds offer superior heat transfer efficiency and shorter residence times— typically around 30 minutes. The resulting β-spodumene phase is significantly more reactive and readily leached with sulfuric acid, enhancing lithium extraction efficiency.

For nickel, cobalt, and zinc, fluidized-bed roasting is used to convert metal sulphides into oxides through oxidative roasting. Compounds such as NiS, CoS, and ZnS are oxidized in an air or oxygen-enriched atmosphere, releasing sulphur dioxide (SO_2) gas that can be captured and converted into sulphuric acid. A typical example of such a reaction is:

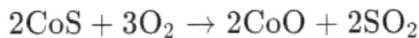

$$2CoS + 3O_2 \rightarrow 2CoO + 2SO_2$$

The even gas distribution and stable temperature profile of a fluidized bed prevent localized overheating and allow for very high conversion efficiencies, often exceeding 98%.

Finally, in the uranium and vanadium industries, fluidized-bed roasters are employed for the oxidation of uraninite and vanadinite. These processes convert uranium from the U^{4+} state to the soluble U^{6+} (uranyl) form, and vanadium from V^{3+} to V^{5+}, both of which improve recovery in subsequent acid leaching steps. The high oxygen availability and efficient mixing in the fluidized bed ensure complete and uniform oxidation, making this approach particularly effective for large-scale hydrometallurgical operations.

Overall, fluidized-bed roasting offers a versatile and efficient solution for the preparation and upgrading of rare earth and critical mineral concentrates, providing high throughput, precise control, and compatibility with modern emission management systems.

Reverberatory and Electric Furnaces

Reverberatory and electric furnaces are high-temperature metallurgical units used for roasting, smelting, and reduction of ores and concentrates. Although both serve similar purposes—to provide controlled thermal energy for chemical reactions—they differ significantly in how heat is generated and transferred. Reverberatory furnaces rely on radiant heat from a fuel flame, while electric furnaces use resistive or arc-generated heat for direct and highly controllable temperature control. Both have niche but important roles in rare earth and critical mineral processing, especially where large-scale oxidation, reduction, or smelting is required.

A reverberatory furnace is a large, refractory-lined chamber where the fuel—commonly coal, oil, or natural gas—is burned above the material charge. Heat is reflected ("reverberated") off the roof and walls onto the material surface below. The process relies primarily on radiant and convective heat transfer, allowing the ore or concentrate to heat uniformly without direct contact with the flame.

The furnace typically has a shallow hearth where the feed material spreads out to ensure maximum surface exposure. The combustion gases flow across the chamber and exit through a flue at the opposite end, sometimes passing through waste-heat boilers or gas scrubbers to recover energy and control emissions.

In mineral processing, reverberatory furnaces are commonly used for oxidation roasting, smelting, and calcination—particularly for sulphide, carbonate, and oxide materials.

An electric furnace generates heat through electrical energy rather than fuel combustion. There are two main types:

- Resistance furnaces, where electric current passes through heating elements (made of graphite or silicon carbide) to heat the charge indirectly.

- Arc furnaces, where an electric arc between electrodes and the charge produces extremely high temperatures—often exceeding 2,000°C.

Electric furnaces allow precise temperature control and can operate in oxidizing, reducing, or inert atmospheres, making them ideal for processing reactive or high-purity materials like rare earth oxides, tungsten, and titanium. Because there is no combustion gas flow, contamination from fuel impurities is minimal—critical for producing high-purity REE oxides or alloys.

Figure 59: Electric Arc Furnace. Arnoldius, CC BY-SA 3.0, via Wikimedia Commons.

Both furnace types are available in a wide range of scales depending on their intended use:

- **Laboratory and pilot-scale units**: 10–100 kg/hour throughput, often bench-mounted electric resistance types for research or process development.

- **Medium industrial furnaces**: 10–100 tonnes/day (e.g., rare earth oxide calcination, lithium roasting).

- **Large reverberatory or arc furnaces**: 500–1,000 tonnes/day, especially in smelting applications for base or critical metals like nickel, cobalt, and rare earth alloys.

Typical dimensions for large reverberatory furnaces are 20–40 metres long, 6–10 metres wide, and 3–6 metres high, depending on capacity and design. Electric furnaces tend to be more compact but heavily insulated and mechanically reinforced due to the high operating temperatures and electrical loads.

The capital cost of reverberatory and electric furnaces varies significantly depending on factors such as production capacity, construction materials, level of automation, and environmental compliance requirements. Small pilot-scale electric furnaces capable of processing up to one tonne per day typically cost between USD $200,000 and $700,000. These units are commonly used for research, pilot testing, and specialty metal production where precision temperature control is essential.

Medium-sized reverberatory furnaces, designed to handle throughputs between 50 and 200 tonnes per day, generally range from USD $3 million to $10 million, including auxiliary systems such as gas-handling, refractory linings, and emission control infrastructure. These systems are common in industrial-scale operations where fuel-based heating is preferred due to its lower upfront cost.

At the large industrial end, electric arc or smelting furnaces with capacities exceeding 500 tonnes per day can cost between USD $25 million and $60 million. These large-scale furnaces require advanced components such as off-gas cleaning systems, electrode assemblies, and high-capacity power supply infrastructure. While electric furnaces demand a higher capital investment, they offer lower operational emissions and improved control over temperature and chemical environment. In contrast, reverberatory furnaces are less expensive to establish but require extensive gas-handling systems to meet modern environmental regulations, increasing their long-term operating costs.

The operating costs of these furnaces depend primarily on the energy source—fuel for reverberatory furnaces or electricity for electric furnaces—as well as the frequency of maintenance and the sophistication of emission control systems. Energy consumption is typically the largest cost driver, with electric furnaces consuming between 300–600 kWh per tonne of processed material, while fuel-fired reverberatory furnaces require approximately 8–12 GJ per tonne.

Maintenance is another major expense, especially the replacement of refractory linings, which generally occurs every 12 to 24 months depending on furnace conditions. For arc furnaces, the consumption of electrodes adds to operating costs, while both furnace types require air and gas treatment systems such as scrubbers, filters, and acid plants to control emissions and comply with environmental standards.

Staffing levels vary with furnace size and the degree of automation. Pilot-scale or batch electric furnaces typically require one to two operators per shift for feeding, monitoring, and product discharge. Medium-sized furnaces may need three to five operators per shift, including a furnace operator, control room technician, and maintenance support staff. Large, continuous smelting operations often employ eight to ten personnel per shift, encompassing process controllers, mechanical and electrical technicians, and supervision roles.

Rare Earth and Critical Mineral Operations and Processing

Modern facilities increasingly rely on automation and digital control systems—including thermocouples, gas analysers, and SCADA (Supervisory Control and Data Acquisition) systems—which have greatly reduced manual labour requirements. These technologies not only improve safety and operational efficiency but also ensure consistent product quality and tighter process control, making them essential for modern rare earth and critical mineral processing plants.

Applications in rare earth and critical minerals:

Rare Earth Elements (REEs): Reverberatory and electric furnaces play a key role in the thermal processing of rare earth minerals, especially during roasting and calcination stages that prepare the ore or concentrate for leaching and refining. In the processing of bastnäsite and monazite, reverberatory or electric furnaces are employed to decompose carbonates, fluorides, and phosphates prior to hydrometallurgical recovery. This controlled heating removes volatile components such as carbon dioxide (CO_2) and hydrogen fluoride (HF), producing an oxide-rich concentrate that is more reactive during subsequent leaching processes.

Another important application is rare earth oxalate calcination, where electric resistance or induction furnaces are used to convert rare earth oxalates ($REE_2(C_2O_4)_3$) into oxides (REE_2O_3) while releasing carbon monoxide (CO) and carbon dioxide (CO_2) gases. Electric furnaces are preferred in this stage because they provide precise temperature control, uniform heating, and contamination-free processing, ensuring high product purity and consistent particle morphology—qualities critical for downstream solvent extraction and separation.

In addition, certain high-purity rare earth metals such as neodymium and dysprosium are produced through metallothermic reduction in small-scale electric or induction furnaces operating under vacuum or inert gas conditions. These systems enable the safe reduction of rare earth halides or oxides using reactive metals (such as calcium or magnesium), producing refined rare earth metals for use in magnets and advanced alloys.

Lithium: In lithium processing, furnaces are central to the phase conversion of spodumene ($LiAlSi_2O_6$)—the main hard rock source of lithium. The mineral is roasted in electric or gas-fired furnaces at around 1,050°C, transforming the naturally occurring α-spodumene into β-spodumene, a metastable form that is significantly more reactive during acid leaching. This step is essential for improving lithium extraction efficiency. Electric furnaces are often preferred for fine-grained or recycled feedstock because they offer precise temperature control, uniform heat distribution, and reduced contamination risk compared to fuel-fired systems.

Nickel and Cobalt: Reverberatory and electric furnaces are also integral to the processing of nickel and cobalt, particularly during reduction smelting. In this process, electric arc furnaces reduce nickel and cobalt oxides into metallic alloys or matte products, which are later refined through hydrometallurgical techniques such as pressure leaching or solvent extraction. These furnaces can sustain the high temperatures required for oxide reduction while maintaining a controlled atmosphere that minimizes impurity formation.

For nickel laterite ores, reverberatory furnaces are sometimes used for pre-treatment, either by oxidizing or partially reducing the ore before high-pressure acid leaching (HPAL). This step improves leaching efficiency by altering mineral phases and removing moisture and volatile components, making the material more amenable to acid dissolution.

Tungsten, Molybdenum, and Titanium: In the production of tungsten and molybdenum, electric furnaces are employed for the reduction of metal oxides under hydrogen or inert gas atmospheres. These reactions yield fine metal powders that serve as feed materials for alloys, cutting tools, and electronic components. The controlled environment of the electric furnace ensures precise reduction and prevents oxidation of the product.

For titanium and zirconium, electric furnaces are essential to the Kroll and Hunter reduction processes, in which metal halides ($TiCl_4$ or $ZrCl_4$) are reduced using magnesium or sodium under vacuum or inert gas conditions. These electrically heated vacuum furnaces maintain the stringent temperature and atmosphere control required to produce high-purity titanium and zirconium metals used in aerospace, medical, and high-performance alloy applications.

Muffle or Tube Furnaces

Muffle furnaces and tube furnaces are precision thermal processing systems designed for small- to medium-scale applications where temperature uniformity, atmosphere control, and product purity are critical. They are widely used in research laboratories, pilot plants, and specialized industrial operations—particularly in rare earth and critical mineral processing, where clean, controlled heating is essential for calcination, decomposition, reduction, or sintering of high-value materials.

While they operate on similar principles, muffle furnaces are box-type enclosures for bulk or batch heating, whereas tube furnaces feature a cylindrical reaction chamber that allows for continuous or semi-continuous operation under controlled atmospheres.

Muffle Furnaces: A muffle furnace consists of an insulated chamber—usually made from refractory ceramics or alumina—with electrical heating elements embedded around it. The "muffle" (a barrier or liner) isolates the sample chamber from the heating elements, preventing direct contact between the charge and combustion gases or resistive elements.

This design ensures contamination-free heating, ideal for analytical work or high-purity material production. Heat transfer occurs primarily through radiation and convection, allowing precise temperature control typically up to 1,200–1,700°C, depending on the furnace type.

Figure 60: Muffle Furnace. Cjp24, CC BY-SA 3.0, via Wikimedia Commons.

The atmosphere inside can be air, inert gas (e.g., nitrogen or argon), or reducing (e.g., hydrogen)—controlled via sealed ports or gas inlet systems. Modern muffle furnaces often include programmable controllers for temperature ramping, dwell times, and cooling cycles.

Tube Furnaces: A tube furnace uses a long, refractory or quartz tube as the reaction chamber, heated externally by coils or elements. Gases can flow through the tube, allowing precise control of the atmosphere composition—a major advantage over open-air furnaces.

Feed material (powder, pellet, or small crucible) is placed inside the tube, either stationary or conveyed slowly through the heated zone using a push-rod or belt system. This configuration allows continuous processing, ideal for calcination, reduction, or decomposition under controlled gas flow.

Tube furnaces operate within the same temperature range as muffle furnaces (typically 800–1,600°C) but provide superior atmosphere control and heating uniformity, making them the preferred choice for research, small-batch production, or high-purity oxide preparation.

Muffle and tube furnaces are notably compact when compared to larger thermal processing systems such as rotary kilns or multiple-hearth furnaces. Their typical batch capacities range from just a few grams to several kilograms, making them suitable for small-scale applications. However, industrial-scale tube furnaces are capable of continuous processing, handling feed rates of approximately 10 to 50 kilograms per hour.

Laboratory units are designed for benchtop use, with internal volumes typically between 1 and 5 litres. These compact systems are ideal for research and small-batch testing. Pilot-scale units, on the other hand, offer increased capacity—ranging from 20 to 200 litres—and are generally floor-mounted. They often include programmable controllers and integrated gas inlet systems to simulate industrial conditions and support process development.

At the industrial level, continuous tube furnaces can reach lengths of 3 to 6 metres with internal diameters between 100 and 200 millimetres. Their throughput depends on factors such as feed density and residence time, but they commonly process 10 to 50 kilograms per hour. Despite their modest physical footprint, these furnaces are optimized for precision rather than volume. They provide meticulous control over temperature and gas composition, which is critical for producing consistent, high-purity materials in both research and production environments.

The cost of establishing a muffle or tube furnace depends on its size, temperature rating, and atmosphere control capability.

- Laboratory-scale furnaces (≤5 L capacity, up to 1,200°C): USD $5,000–$15,000.

- High-temperature research-grade models (up to 1,700°C with gas atmosphere): USD $20,000–$50,000.

- Pilot or small industrial tube furnaces (continuous operation, 10–50 kg/h): USD $80,000–$250,000, including power supply, gas-handling systems, and automation.

Operating costs are relatively low compared to large kilns or roasters, dominated by electricity consumption, heating element wear, and gas supply (argon, nitrogen, hydrogen) for atmosphere control. Maintenance mainly involves replacing heating elements and periodic refractory relining.

Because of their high level of automation, muffle and tube furnaces require very little direct staffing. Most modern systems are equipped with programmable controllers and safety interlocks, allowing them to operate unattended for extended periods once temperature and atmosphere parameters have been set.

In laboratory environments, these furnaces are typically managed by a single technician, who may oversee multiple units simultaneously. The tasks usually involve preparing samples, loading crucibles, starting heating cycles, and monitoring data through digital interfaces. The high degree of automation means that operator involvement is largely limited to setup and shutdown procedures.

At the pilot or small industrial scale, one or two operators per shift are generally sufficient to handle the loading of feed material, monitoring gas flow and temperature stability, and ensuring that safety systems such as gas purging and venting function correctly. These operators may also assist with product discharge and sampling, depending on the process requirements.

Rare Earth and Critical Mineral Operations and Processing

Routine maintenance—such as heating element replacement, calibration, and chamber cleaning—may require assistance from a small maintenance team or technical support personnel. Even in a modest pilot plant running several furnaces simultaneously, the total staffing requirement remains low, with approximately three to five personnel per shift capable of managing operations effectively. This combination of automation, efficiency, and low manpower demand makes muffle and tube furnaces particularly suitable for high-precision, small-scale mineral and materials processing environments.

Muffle and tube furnaces are widely used in rare earth processing, particularly for calcination, decomposition, and phase stabilization steps where precision and cleanliness are essential. One of their most common uses is in the calcination of rare earth oxalates ($REE_2(C_2O_4)_3 \cdot nH_2O$) to produce rare earth oxides (REE_2O_3). The controlled heating profile in these furnaces minimizes contamination and allows precise control of particle size, morphology, and purity—factors critical for manufacturing high-quality rare earth materials used in magnets, phosphors, and catalysts.

They are also used for fluoride and carbonate decomposition, where REE carbonates and fluorides are heated in air or controlled oxygen environments to release carbon dioxide (CO_2) and hydrogen fluoride (HF). This process yields highly reactive REE oxides, which can then undergo acid leaching or solvent extraction to recover individual rare earth elements. In addition, muffle and tube furnaces are used for phase stabilization of specific oxides, such as ceria (CeO_2) and yttria-stabilized zirconia (YSZ), ensuring that the desired crystalline structure is achieved for use in fuel cells, catalysts, and optical materials.

Tube furnaces are integral to the preparation of advanced lithium compounds and battery cathode materials. They are used for the thermal decomposition of lithium hydroxide and lithium carbonate, as well as for sintering mixed-metal precursors like $LiNiMnCoO_2$ (NMC) and $LiFePO_4$ (LFP) under controlled atmospheres. These conditions ensure uniform phase formation and crystal growth, which are crucial for achieving consistent electrochemical performance in lithium-ion batteries.

In spodumene processing, tube furnaces are employed during α→β phase conversion trials at the research and pilot stages before large-scale production in rotary kilns. The β-spodumene phase produced at approximately 1,050°C is significantly more reactive, improving lithium recovery rates during subsequent acid leaching.

For cobalt, muffle furnaces operated under nitrogen atmospheres are used to convert cobalt oxalate into cobalt oxide (CoO) through controlled calcination. The resulting CoO is later reduced to metallic cobalt, which is essential in the production of batteries and superalloys. Similarly, nickel and tungsten oxides are processed in tube furnaces under hydrogen or inert gases, where reduction reactions yield metallic powders used in catalysts, high-performance alloys, and hard materials. The precision of atmosphere control and uniform heating ensures minimal contamination and high metal purity.

Muffle and tube furnaces are also indispensable in the research and production of advanced rare earth-based materials. For instance, tube furnaces operating under argon or vacuum atmospheres are used for sintering and pre-treatment of rare earth alloys used in high-strength permanent magnets such as NdFeB and SmCo. The same systems are utilized for post-calcination treatments that refine microstructure and enhance magnetic or mechanical performance.

In the nanomaterials and optical materials fields, these furnaces are used for precise particle synthesis and doping. The fine temperature control and small reaction volumes enable the production of rare earth-doped oxides and phosphors with consistent particle sizes and luminescent properties, which are essential for display technologies, lighting, and lasers.

Temperature Control and Monitoring

Precise temperature management is critical in pyrometallurgical and thermal processing operations because reaction outcomes—such as oxidation state, phase formation, and particle morphology—depend on narrow thermal windows.

Key aspects of temperature control include:

- **Ramp and soak programming:** Controlled heating and holding periods (e.g., 5 °C/min to 650 °C, 2 h soak) prevent thermal shock and allow complete decomposition or phase transition.

- **Real-time thermocouple arrays:** Multiple thermocouples along the kiln or furnace length ensure even heat distribution and detect hot or cold zones.

- **Atmosphere regulation:** Gas composition (air, O_2, N_2, H_2, CO, or mixtures) is maintained through flow controllers and oxygen sensors to promote oxidation, inert calcination, or reduction as required.

- **Feedback automation:** PLC (Programmable Logic Controller) systems with PID loops continuously adjust burner input, air-fuel ratios, and rotation speeds to maintain stable thermal profiles.

- **Cooling control:** Controlled quenching or gradual cooling preserves desired crystal phases—rapid cooling for β-spodumene, slow cooling for dense oxide stability.

Effective temperature control not only ensures consistent product quality but also minimizes energy consumption and equipment wear, while reducing the risk of sintering, over-oxidation, or unwanted phase reversion. In industrial practice, operators integrate thermal sensors, flow meters, and digital process models to achieve fine control over reaction kinetics and heat transfer—essential for optimizing throughput and ensuring safe, sustainable pyrometallurgical performance.

Ramp and soak programming is a precise temperature control method used in furnaces, kilns, and thermal reactors—such as muffle and tube furnaces—to manage how heat is applied to a material over time. It involves two key stages: the ramp (controlled heating or cooling rate) and the soak (a period of holding at a constant temperature). This method is essential for preventing thermal damage, ensuring uniform reactions, and achieving desired phase or structural transformations during processes such as calcination, decomposition, and sintering.

During the ramp phase, the furnace temperature increases gradually—commonly at rates like 5 °C per minute—to reach the target temperature. This controlled rate ensures that the entire sample heats evenly, reducing the risk of thermal shock or cracking that can occur if materials expand too quickly. The ramping process also allows gases (such as CO_2, H_2O, or HF) released during decomposition to escape gradually, minimizing pressure build-up and avoiding structural damage to the material or crucible.

Once the desired temperature is reached, the system enters the soak phase, where the temperature is held constant (for example, 650 °C for 2 hours). This "soaking" period allows sufficient time for chemical reactions or phase transitions to reach completion. In rare earth and critical mineral processing, this ensures full decomposition of compounds like oxalates, carbonates, or fluorides, and promotes uniform particle morphology or desired oxide formation.

After the soak, the furnace may either cool naturally or follow a programmed controlled cooling ramp, which further prevents stress or unwanted phase changes. Modern programmable logic controllers (PLCs) or PID systems automate these ramp-and-soak cycles, allowing operators to tailor heating profiles for specific materials and reactions.

Real-time thermocouple arrays are an essential tool for monitoring and controlling temperature profiles inside furnaces, kilns, and reactors used in mineral processing. These systems consist of multiple thermocouples—temperature-sensing probes—strategically placed along the length and cross-section of the heating chamber. Their primary purpose is to ensure that heat is distributed evenly, that each process zone maintains the correct temperature, and that any deviations or irregularities are detected immediately for correction.

Each thermocouple continuously measures the local temperature and sends real-time data to a temperature control system—often a programmable logic controller (PLC) or distributed control system (DCS). The system compares these readings against the programmed temperature profile (for example, ramp and soak stages) and automatically adjusts the heating elements, burners, or gas flow to maintain uniform thermal conditions. This feedback control prevents overheating or underheating, both of which can lead to inconsistent product quality, incomplete reactions, or material damage.

In rotary kilns, muffle furnaces, and tube furnaces, thermocouple arrays are typically installed at multiple points: near the feed inlet, mid-section, and discharge end. This allows operators to detect hot or cold zones—areas where heat transfer is uneven due to factors such as gas flow variation, refractory wear, or feed buildup. Identifying these zones helps optimize energy efficiency, throughput, and product consistency.

For rare earth and critical mineral processing, precise temperature control is particularly important. For example, in rare earth oxalate calcination, uniform heating ensures complete decomposition without sintering or contamination. In spodumene α→β conversion, even temperature distribution guarantees consistent phase transformation across the batch.

Atmosphere regulation in furnaces and kilns is the process of controlling the type and composition of gases inside the heating chamber to create the desired chemical environment—oxidizing, reducing, or inert—depending on the material and the target reaction. This is critical in pyrometallurgical and thermal processing of rare earths and critical minerals, where specific gas atmospheres determine whether a metal oxide forms, decomposes, or remains stable.

In practice, gas composition—such as air, oxygen (O_2), nitrogen (N_2), hydrogen (H_2), carbon monoxide (CO), or mixed gases—is controlled using mass flow controllers, pressure regulators, and automated valves. These devices deliver precise flow rates of each gas into the furnace. A gas distribution manifold ensures even delivery throughout the heating zone, while oxygen sensors or gas analyzers continuously monitor the internal atmosphere. The data is fed into a programmable logic controller (PLC) or PID temperature controller, which adjusts gas flows in real time to maintain the desired conditions.

Different gas environments are used depending on the process objective:

- Oxidizing atmosphere (air or O_2) is used for roasting and calcination, promoting reactions like the oxidation of sulfides or carbonates. For example, converting $CeCO_3F$ → $CeO_2 + CO_2 + F_2$ in rare earth roasting requires an oxygen-rich atmosphere.

- Inert atmosphere (N_2 or argon) prevents unwanted oxidation, commonly used in the calcination of oxalates or thermal decomposition of materials sensitive to air.

- Reducing atmosphere (H_2, CO, or mixtures) enables metal reduction from oxides—for example, producing tungsten (W) or cobalt (Co) from their respective oxides.

Safety systems, such as backflow preventers, gas detectors, and exhaust scrubbers, are also essential to handle reactive gases safely.

In rare earth and critical mineral operations, precise atmosphere control ensures that reactions proceed under optimal chemical conditions—achieving higher purity, phase stability, and recovery rates. For instance, the transition of spodumene from α to β phase, the calcination of RE oxalates to oxides, or the hydrogen reduction of tungsten oxide all rely on tightly regulated atmospheres to achieve uniform and reproducible results.

Feedback automation refers to the use of programmable logic controllers (PLCs) and proportional–integral–derivative (PID) control loops to continuously monitor and adjust key operating parameters within furnaces, kilns, and roasters. This automated feedback system ensures that temperature, gas flow, and mechanical motion remain stable and optimized throughout the thermal process—critical for maintaining consistent product quality and energy efficiency in mineral processing operations.

A PLC serves as the central control unit, receiving real-time data from temperature sensors, thermocouples, pressure transmitters, and flow meters. It then compares the actual readings to the programmed setpoints. If any deviation is detected—such as a drop in temperature or a change in air-fuel ratio—the PID control loop automatically calculates the necessary correction and adjusts system outputs to bring the process back to equilibrium.

For example, in a rotary kiln, feedback automation continuously regulates burner fuel input, air or oxygen supply, and kiln rotation speed. If one zone becomes cooler than desired, the PID loop increases burner intensity or airflow to restore balance. Similarly, if the kiln overheats, the system reduces energy input or increases exhaust gas flow. This real-time feedback prevents thermal drift, improves product uniformity, and minimizes fuel waste.

In rare earth and critical mineral processing, such as bastnäsite roasting, spodumene conversion, or REE oxalate calcination, precise thermal control is essential to achieving consistent chemical conversion and phase formation. Automated feedback systems also reduce operator workload, allowing a small team to supervise complex, multi-zone thermal equipment safely and efficiently.

Modern PLCs integrate with SCADA (Supervisory Control and Data Acquisition) systems, providing visual dashboards, alarms, and data logging. This enables operators and engineers to track long-term performance trends, diagnose issues quickly, and continuously optimize the process—ensuring stable operation, reduced downtime, and higher product yield.

Feedback automation refers to the use of programmable logic controllers (PLCs) and proportional–integral–derivative (PID) control loops to continuously monitor and adjust key operating parameters within furnaces, kilns, and roasters. This automated feedback system ensures that temperature, gas flow, and mechanical motion remain stable and optimized throughout the thermal process—critical for maintaining consistent product quality and energy efficiency in mineral processing operations.

A PLC serves as the central control unit, receiving real-time data from temperature sensors, thermocouples, pressure transmitters, and flow meters. It then compares the actual readings to the programmed setpoints. If any deviation is detected—such as a drop in temperature or a change in air-fuel ratio—the PID control loop automatically calculates the necessary correction and adjusts system outputs to bring the process back to equilibrium.

For example, in a rotary kiln, feedback automation continuously regulates burner fuel input, air or oxygen supply, and kiln rotation speed. If one zone becomes cooler than desired, the PID loop increases burner intensity or airflow to restore balance. Similarly, if the kiln overheats, the system reduces energy input or increases exhaust gas flow. This real-time feedback prevents thermal drift, improves product uniformity, and minimizes fuel waste.

In rare earth and critical mineral processing, such as bastnäsite roasting, spodumene conversion, or REE oxalate calcination, precise thermal control is essential to achieving consistent chemical conversion and phase formation. Automated feedback systems also reduce operator workload, allowing a small team to supervise complex, multi-zone thermal equipment safely and efficiently.

Modern PLCs integrate with SCADA (Supervisory Control and Data Acquisition) systems, providing visual dashboards, alarms, and data logging. This enables operators and engineers to track long-term performance trends, diagnose issues quickly, and continuously optimize the process—ensuring stable operation, reduced downtime, and higher product yield.

In the thermal processing of rare earth elements (REEs) such as lanthanides (La–Lu), yttrium (Y), and scandium (Sc), precise control over temperature profiles is essential to obtain the desired product characteristics, including purity and uniformity. The ramp-and-soak programming technique, which involves carefully controlled heating rates and soak times, plays an integral role during critical stages such as calcination, reduction, and phase transformation. Research indicates that controlled ramp rates, typically ranging from 3 to 5 °C/min, are vital in preventing adverse thermal effects like sintering or cracking during the decomposition of oxalates and carbonates of these elements. For instance, cerium undergoes a notable transformation from cerium hydroxide ($Ce(OH)_4$) to cerium dioxide (CeO_2) within a temperature range of 500 to 700 °C. This gradual heating is necessary to manage byproducts such as CO_2, which can complicate the thermal decomposition process [274].

Additionally, elements like samarium and europium require extended soak times, often 2 to 3 hours at temperatures between 700 and 800 °C, to ensure complete conversion to their respective oxides without causing residual carbon or moisture. This careful programming ensures optimal particle morphology and specific surface area, which are imperative for applications in optics and ceramics involving scandium and yttrium oxides [275].

A sophisticated array of thermocouples is employed throughout the thermal processing setup to monitor temperature distribution, which is essential when working with mixed REE feedstocks, such as bastnäsite or monazite, which contain varying thermal behaviours among the elements. For example, neodymium and praseodymium oxides necessitate more exact temperature control, often within a narrow range of ±5 °C, to maintain the required oxidation states (Nd_2O_3, Pr_6O_{11}) and avoid detrimental effects on phase purity [276]. By using thermocouple arrays, manufacturers can ensure that all REE components within the processed material experience uniform thermal conditions, thus preserving phase uniformity and minimizing the need for reprocessing.

The atmosphere in which thermal processing occurs significantly affects the stability and oxidation states of the various rare earth compounds. Oxidizing atmospheres (air or O_2) are utilized for elements like cerium, praseodymium, and terbium to stabilize higher oxidation states, crucial for catalytic applications. Conversely, inert atmospheres (such as nitrogen or argon) are preferred during the calcination of oxalates from samarium and europium compounds to prevent excessive oxidation or volatilization, while reducing atmospheres (H_2/CO) facilitate the conversion of REE oxides to metals, which is necessary for fabricating NdFeB magnets [277]. The regulation of the atmosphere is critical, as excessive O_2 can result in non-stoichiometric defects in CeO_2, whereas insufficient oxygen can hinder complete oxidation in praseodymium and neodymium compounds [278].

Lastly, advanced automated feedback systems are employed to maintain precise operational parameters, including temperature and gas flow rates, tailored specifically to meet the thermal characteristics of each REE. For instance, a PID-controlled furnace can dynamically adjust burner output to sustain temperatures within the critical 650–700 °C range for processes involving bastnäsite concentrates. Off-gas monitoring, such as CO_2 levels, serves as an additional feedback mechanism to determine the completion of reactions, ensuring that soak durations and ramp rates are optimally controlled for reproducibility in crystal phase formation, essential for high-value ceramic and electronic materials [279].

The combination of meticulous temperature control through ramp-and-soak programming, the use of thermocouple arrays for temperature distribution, atmospheric regulation, and automated feedback mechanisms are essential elements in the thermal processing of rare earth elements. These processes collectively achieve high phase purity, uniform particle distribution, and improved reproducibility, which are crucial for meeting the demanding specifications of modern applications involving REEs.

Comparison of Pyro- vs. Hydro-Processing Routes

The extraction and refining of rare earth elements (REEs) and critical minerals can follow two principal pathways—pyrometallurgy (thermal processing) and hydrometallurgy (aqueous processing). While both aim to liberate, concentrate, and purify valuable metals from ores or intermediates, they differ fundamentally in mechanisms, equipment, operating conditions, environmental footprint, and product characteristics.

Pyrometallurgy (Thermal Route): Pyrometallurgy involves high-temperature chemical reactions—such as roasting, calcination, reduction, and smelting—that modify the physical or chemical composition of ores and concentrates. These reactions typically occur in the solid or molten phase, using equipment such as furnaces, kilns, or roasters. The process is driven by heat, often supplied by fuel combustion or electrical resistance, and is designed to remove volatile impurities, oxidize or reduce compounds, and produce more reactive or pure products for further processing.

For example, during the treatment of bastnäsite—a rare earth fluorocarbonate mineral—the following thermal decomposition reaction occurs:

$$2CeCO_3F \rightarrow 2CeO_2 + 2CO_2 + F_2$$

In this reaction, bastnäsite is decomposed into cerium oxide (CeO_2), releasing carbon dioxide (CO_2) and fluorine gas (F_2). The resulting oxide is a more stable and reactive form, suitable for subsequent acid leaching and purification in hydrometallurgical circuits.

Hydrometallurgy (Aqueous Route): Hydrometallurgy, by contrast, relies on chemical dissolution in aqueous solutions—most commonly acids, bases, or salts—to extract metals from ores, concentrates, or intermediate products. The process operates at much lower temperatures than pyrometallurgy and typically involves three main steps: leaching, solution purification, and metal recovery.

For example, rare earth oxalates can be dissolved in sulfuric acid to produce soluble rare earth ions through the following reaction:

$$REE_2(C_2O_4)_3 \cdot nH_2O + H_2SO_4 \rightarrow 2REE^{3+}_{(aq)} + 3C_2O_4^{2-}{}_{(aq)} + SO_4^{2-}{}_{(aq)} + H_2O$$

In this case, the rare earth oxalates ($REE_2(C_2O_4)_3 \cdot nH_2O$) are converted into rare earth ions (REE^{3+}) in solution. These dissolved ions can then be selectively separated and recovered using techniques such as precipitation, solvent extraction, or ion exchange, yielding high-purity rare earth compounds suitable for refining into final oxides or metals.

Together, these two routes—thermal and aqueous—represent complementary strategies in modern mineral processing: pyrometallurgy prepares or upgrades the feed through heat, while hydrometallurgy refines and purifies it through controlled chemical reactions in solution.

Operating Conditions

The operating conditions for pyrometallurgical and hydrometallurgical processes differ significantly due to their distinct physical and chemical principles.

In pyrometallurgy, reactions occur at high temperatures, typically ranging from 500°C to 1600°C, depending on the ore type and process objective. These operations can take place at either atmospheric or elevated pressures, especially when high-temperature oxidation or reduction is required. The energy source is predominantly thermal, supplied through fuel combustion or electrical heating in furnaces, kilns, or roasters. Reactions occur in solid, molten, or gas–solid phases, often involving phase transitions that help drive the separation of metals from impurities.

In contrast, hydrometallurgy operates at much lower temperatures, generally below 300°C, as it relies on aqueous chemistry rather than melting or smelting. Pressure conditions range from atmospheric to about 5 MPa, the latter typically seen in High-Pressure Acid Leaching (HPAL) used for lateritic nickel and cobalt ores. The primary energy input is chemical, driven by acidic or alkaline reagents such as sulfuric acid, hydrochloric acid, or sodium hydroxide. The physical environment is liquid–solid, where the target metals are dissolved into solution from the solid ore matrix, allowing for selective extraction and purification through subsequent chemical processing.

Overall, pyrometallurgy is characterized by intensive heat and rapid kinetics, while hydrometallurgy emphasizes chemical selectivity and controlled dissolution under relatively mild conditions.

Table 4: Operating conditions summary.

Parameter	Pyrometallurgy	Hydrometallurgy
Temperature	500–1600 °C (high)	<300 °C (low to moderate)
Pressure	Atmospheric or high (in autoclaves)	Atmospheric to 5 MPa (for HPAL)
Energy Source	Thermal (combustion or electrical)	Chemical (acid/base reactions)
Physical Phase	Solid, molten, or gas–solid	Liquid–solid (aqueous)

Process Stages

In pyrometallurgical processing, the sequence of operations is primarily driven by high-temperature reactions that alter the physical and chemical composition of the ore. The process typically begins with drying or dehydration, which removes moisture and prepares the feed material for efficient heat transfer during subsequent steps. This is followed by roasting or calcination, where ores are thermally treated in the presence or absence of oxygen to decompose volatile compounds, oxidize sulphides, or convert carbonates into oxides. The next stage involves reduction or smelting, in which metal oxides are reduced to their metallic form using reducing agents such as carbon, carbon monoxide, or hydrogen. Finally, refining or slag separation removes impurities, producing a purified metal or alloy. The slag—an oxide-rich by-product—is separated based on density and chemical composition, leaving behind the molten metal for casting or further treatment.

In hydrometallurgical processing, the operations follow a liquid–solid chemistry pathway, beginning with leaching, where metals are dissolved from the ore using acidic, alkaline, or pressurized solutions depending on mineral type and reactivity. The resulting pregnant leach solution (PLS) contains dissolved metal ions along with impurities. These are removed during solution purification, which commonly employs solvent extraction or ion exchange to selectively isolate the target metal. Once purified, the metal is recovered from solution through processes such as precipitation, electrowinning, or crystallization, producing solid metal compounds or pure metals ready for refining. The final stage involves effluent treatment and recycling, where waste solutions are neutralized, and valuable reagents or water are recovered for reuse, minimizing environmental impact and operational costs.

Together, these process stages illustrate the contrast between thermal and aqueous approaches—pyrometallurgy focuses on high-temperature transformations, while hydrometallurgy emphasizes selective dissolution, purification, and recovery in controlled chemical systems.

Advantages and Disadvantages

Pyrometallurgy offers several operational and technical advantages, particularly in the processing of high-grade or refractory ores. One of its major strengths is the rapid reaction kinetics, allowing for much shorter processing times compared to chemical leaching routes. Pyrometallurgical systems are also highly effective for treating materials that are resistant to dissolution in acids or alkalis, as the high temperatures can break down stable mineral lattices. Additionally, these processes can produce metallic or oxide products directly, often eliminating the need for further refining steps.

However, the disadvantages of pyrometallurgy are primarily related to its energy intensity and environmental impact. The high temperatures required (often between 800°C and 1600°C) result in significant energy consumption, leading to higher operational costs and carbon emissions. Furthermore, thermal reactions frequently generate gaseous pollutants such as CO_2, SO_2, and HF, necessitating complex off-gas treatment systems to comply with environmental regulations. The method also tends to be less selective, as co-occurring metals may co-reduce or co-melt, making separation more challenging and requiring additional downstream purification.

In contrast, hydrometallurgy operates at lower temperatures and requires significantly less energy, making it a more environmentally and economically efficient alternative for many applications. Its high selectivity allows for the precise separation of metals based on their chemical properties, making it particularly well-suited to low-grade, complex, or mixed ores where multiple metals coexist. The liquid-based nature of the process also simplifies the management of solid residues, which can often be neutralized and disposed of with less environmental risk than gaseous emissions.

Nonetheless, hydrometallurgy also presents certain limitations. Reaction rates are generally slower than in thermal methods, often requiring extended leaching times to achieve complete recovery. The process produces liquid waste streams that must be carefully treated to remove residual reagents and dissolved metals before discharge. Additionally, reagent costs—including acids, bases, and organic extractants—can be substantial, especially in large-scale or continuous operations. Finally, hydrometallurgical systems are highly sensitive to feed variability and solution chemistry, requiring tight process control to maintain consistency and avoid losses in recovery or product purity.

Applications to Rare Earths and Critical Minerals

Pyrometallurgical processing plays a crucial role in preparing and upgrading rare earth and critical mineral feedstocks for subsequent recovery and refining. One major application is the roasting of bastnäsite and monazite, which removes volatile components such as carbon dioxide (CO_2), fluorine (F), and organic matter, producing oxide-rich materials more amenable to leaching. Another key process is the α-to-β phase conversion of spodumene—a critical step in lithium extraction—conducted at approximately 1050°C in rotary kilns or furnaces. This transformation significantly increases the mineral's reactivity during subsequent acid leaching. Similarly, rare earth oxalate calcination is widely employed to convert hydrated oxalates into high-purity rare earth oxides, which serve as feedstock for solvent extraction or direct metal reduction processes. In the case of base metals, nickel and cobalt reduction smelting is used to produce intermediate alloy or matte products, which are later refined through hydrometallurgical or electrolytic routes.

In contrast, hydrometallurgical processing focuses on chemical dissolution and selective recovery of metals through aqueous reactions. It includes acid or alkaline leaching of rare earth phosphates, carbonates, and oxides, effectively dissolving the desired elements while leaving impurities in the residue. Once dissolved, the resulting pregnant leach solution is treated through solvent extraction or ion exchange, allowing the precise separation and purification of individual rare earth elements (REEs) based on their ionic radii and chemical affinities. In lithium processing, β-spodumene and lithium brines are leached using sulfuric acid to produce lithium sulphate or chloride solutions that can be refined into lithium carbonate or hydroxide. Additionally, High-Pressure Acid Leaching (HPAL) is extensively used in the nickel-cobalt laterite industry, where elevated temperature and pressure in sulfuric acid systems enable efficient dissolution of nickel and cobalt oxides, achieving high recoveries from low-grade ores.

Together, these pyro- and hydro-based approaches form complementary pathways in modern mineral processing—pyrometallurgy provides the initial phase transformation and impurity removal, while hydrometallurgy delivers high-purity metal separation and recovery.

Integration and Hybrid Processing

Modern rare earth element (REE) and critical mineral refineries increasingly adopt integrated or hybrid processing approaches that combine the strengths of both pyrometallurgical and hydrometallurgical methods. In these systems, pyrometallurgy serves primarily as a pre-treatment stage, where roasting or calcination is used to modify the mineral structure and remove volatile components such as carbonates, fluorides, or organic matter. This step converts the minerals into more chemically reactive oxide forms, enhancing the efficiency of subsequent leaching and extraction processes.

Following this, hydrometallurgy is employed to selectively dissolve and separate the target metals. The processed oxide feed from the thermal stage is leached in controlled acidic or alkaline solutions, forming a pregnant leach solution (PLS) that contains the dissolved metal ions. These are then purified and fractionated using solvent extraction, ion exchange, or precipitation techniques, enabling recovery of high-purity metal products or compounds.

A typical example of this hybrid approach is found in bastnäsite processing. The ore is first roasted to produce a cerium oxide (CeO_2)-rich calcine, effectively driving off CO_2 and F and converting the rare earth minerals into an oxide phase. The calcined material is then subjected to acid leaching, dissolving the rare earth elements into solution as REE^{3+} ions. Finally, solvent extraction is applied to separate and purify individual rare earth oxides such as neodymium, praseodymium, and dysprosium. This integrated method exemplifies how combining thermal and aqueous processes maximizes recovery, selectivity, and overall process efficiency in modern REE and critical mineral refining.

Environmental and Sustainability Aspects

From an environmental and sustainability perspective, pyrometallurgy and hydrometallurgy present contrasting advantages and challenges. Pyrometallurgical processes typically produce high gaseous emissions, including carbon dioxide (CO_2), sulphur dioxide (SO_2), and hydrogen fluoride (HF), which require advanced off-gas cleaning systems to mitigate their environmental impact. The solid waste generated from these processes is generally in the form of slag, which is chemically stable but often produced in large volumes, posing challenges for storage and disposal. Furthermore, pyrometallurgy is highly energy-intensive, relying on thermal or electrical energy to sustain temperatures often exceeding 1000°C. Although it uses relatively small quantities of reagents, its overall carbon footprint remains high due to fuel combustion and power consumption.

In contrast, hydrometallurgy operates at much lower temperatures, resulting in significantly reduced greenhouse gas emissions compared with thermal routes. However, this method generates liquid waste streams—including leach residues and effluents—that must be treated before disposal to prevent contamination of water systems. Hydrometallurgical operations tend to consume large volumes of reagents such as acids, bases, and organic extractants, as well as substantial quantities of water, which can increase both environmental impact and operational costs if not properly managed.

While pyrometallurgy's major sustainability concern lies in its energy consumption and emissions, hydrometallurgy's challenges are centred on effluent management and resource consumption. Nevertheless, hydrometallurgical systems can achieve a lower overall carbon footprint when reagent and water recycling strategies are employed effectively. As global demand for critical and rare earth minerals grows, both approaches are evolving toward cleaner, more energy-efficient, and circular processing models designed to minimize environmental impact while maximizing resource recovery.

Table 5: Environmental and sustainability aspects summary.

Factor	Pyrometallurgy	Hydrometallurgy
Emissions	High (CO_2, SO_2, HF)	Lower gaseous, higher liquid waste
Waste Form	Slag (stable but voluminous)	Leach residue and effluents
Energy Intensity	Very high	Moderate
Reagent Use	Low	High

Rare Earth and Critical Mineral Operations and Processing

Factor	Pyrometallurgy	Hydrometallurgy
Water Use	Minimal	High
Carbon Footprint	High	Lower (if reagents are recycled)

In rare earth and critical mineral processing, pyrometallurgy provides the high-temperature pre-treatment needed to activate or decompose minerals, while hydrometallurgy enables selective extraction and purification under controlled aqueous conditions.

Modern integrated operations—such as those used for bastnäsite, monazite, spodumene, and laterites—combine both methods to maximize recovery, purity, and sustainability, balancing energy use, reagent efficiency, and environmental compliance.

Key Terms and Concepts

Pyrometallurgy: A branch of extractive metallurgy that uses high temperatures to extract, refine, or upgrade metals from ores, concentrates, or recycled materials.

Roasting: A thermal treatment process that converts sulphide or carbonate ores into oxides through controlled heating in the presence of oxygen.

Calcination: The heating of ores or minerals in the absence or limited supply of air to drive off volatile components such as carbon dioxide or water.

Smelting: A pyrometallurgical process that involves melting ores to separate metal from impurities using heat and chemical reducing agents.

Reduction: A chemical reaction in which oxygen is removed or electrons are gained, commonly used in metal extraction to convert oxides into pure metals.

Oxidation: The addition of oxygen or removal of electrons from a material, often used to remove sulphur or carbon impurities during metal refining.

Ellingham Diagram: A thermodynamic chart that shows the temperature dependence of the stability of oxides, used to predict the feasibility of reduction reactions.

Key Terms and Concepts

Furnace: A high-temperature reactor used for thermal processing operations such as roasting, smelting, or calcination.

Rotary Kiln: A long, cylindrical furnace that rotates to ensure uniform heating of ores or concentrates during roasting or reduction.

Electric Arc Furnace (EAF): A furnace that melts materials using high-intensity electric arcs between electrodes, suitable for high-temperature smelting.

Induction Furnace: A furnace that heats materials through electromagnetic induction, offering precise temperature control and energy efficiency.

Thermal Decomposition: The breakdown of chemical compounds into simpler substances by heat, often used to produce oxides or remove volatiles.

Slag: The molten waste material formed from gangue and fluxes during smelting, which floats on top of the molten metal and is removed as a by-product.

Flux: A substance added during smelting to combine with impurities and form a fluid slag, facilitating their removal.

Pyrohydrometallurgy: An integrated process that combines high-temperature pre-treatment with subsequent hydrometallurgical leaching to maximise metal recovery.

Emission Control: The use of scrubbers, filters, and gas-cleaning systems to capture particulates and gases (e.g., SO_2, CO_2) from high-temperature operations.

Chapter 7 Review Questions

Principles of Pyrometallurgy

1. Define pyrometallurgy and explain its role in the extraction and refining of rare earth and critical minerals.

2. What is the difference between endothermic and exothermic reactions in pyrometallurgical processes?

3. Why are high-temperature reactions essential for the decomposition and upgrading of minerals?

Major Pyrometallurgical Processes

4. Describe the purpose of roasting and give one example relevant to rare earth processing.

5. How does calcination differ from roasting in terms of atmosphere and reaction type?

6. Explain the reduction process and identify a reducing agent commonly used in metallurgical reactions.

7. What is smelting, and how is it applied to nickel or cobalt ores?

Chemical and Thermodynamic Reactions

8. What information can be interpreted from an Ellingham diagram?

9. Why does carbon reduce iron oxide to iron but not magnesium oxide?

10. Write a balanced reaction showing the oxidation of ZnS during roasting.

Equipment and Operating Parameters

11. Compare the function of rotary kilns and multiple-hearth furnaces.

12. What is the main advantage of a fluidised-bed roaster for fine-grained materials?

13. Describe how electric and induction furnaces provide precise temperature control.

14. What parameters affect residence time and temperature uniformity in rotary kilns?

Process Control

15. Explain what is meant by ramp-and-soak programming in furnace control.

16. How do thermocouple arrays contribute to process stability and product uniformity?

17. Why is atmosphere regulation critical in pyrometallurgical operations?

18. Describe how feedback automation using PLCs and PID loops maintains temperature profiles.

Pyro- vs. Hydro-Processing

19. Identify two major differences between pyrometallurgical and hydrometallurgical routes.

20. Give one example of a mineral that is first roasted (pyro) and then leached (hydro).

21. What are two key advantages and two disadvantages of hydrometallurgical processing compared with pyroprocessing?

Integrated Processing

22. Outline a simple flowsheet showing how roasting, leaching, and solvent extraction may be combined in a critical mineral refinery.

23. How does pre-roasting improve subsequent hydrometallurgical recovery?

Environmental and Safety Aspects

24. List three main environmental pollutants generated by pyrometallurgical operations.

25. What occupational hazards are associated with furnace operation?

26. Suggest two engineering controls that reduce heat and gas exposure risks for operators.

Emerging Technologies

27. Explain how hydrogen-based reduction differs from carbon-based reduction.

28. Describe one advantage of microwave-assisted heating in mineral processing.

29. What is plasma smelting, and how can it contribute to lower emissions?

Process Optimisation

30. Identify two ways to improve energy efficiency in rotary kiln operations.

31. Why is feed particle size control important for thermal efficiency?

32. How can data logging and automation improve process yield and reproducibility?

Chapter 8

Plant Design, Process Flows, and Equipment

Modern mineral processing facilities represent the culmination of decades of technological advancement, engineering design, and process optimization. Whether the goal is to extract rare earth elements, lithium, cobalt, or other critical minerals, the efficiency and sustainability of production depend on how effectively the plant is designed, integrated, and controlled. Plant design is not just about placing equipment—it is about orchestrating a sequence of physical and chemical operations that transform raw ore into high-value products with precision, consistency, and minimal environmental impact.

Chapter 8 introduces the fundamental principles of mineral processing plant design, exploring how equipment, systems, and process stages are arranged to achieve optimal performance. The chapter bridges theory and practice, showing how material flow, space utilization, and process interconnectivity influence productivity, safety, and operational reliability. From ore reception to final product dispatch, every design decision—such as layout, throughput capacity, automation level, and waste management—affects the overall success of a processing operation.

In addition, the chapter details the major equipment and instrumentation used in mineral processing circuits, explaining their roles in achieving precise control over parameters such as temperature, pressure, flow, and pH. Advanced automation and control systems—including Distributed Control Systems (DCS) and Supervisory Control and Data Acquisition (SCADA)—are also examined, highlighting their importance in ensuring consistent product quality, safety, and data-driven process optimization.

Finally, the chapter emphasizes the integration of quality control, sampling, and product specification within the broader design framework. It demonstrates how plant layout, process control, and analytical testing combine to maintain product standards, meet regulatory obligations, and enhance traceability across the value chain. By understanding these interrelated aspects, learners gain a holistic view of how well-designed processing plants underpin the reliable supply of critical minerals essential for clean energy, electronics, and advanced manufacturing industries.

Learning Outcomes	

This chapter aims to give you the ability to:
1. Explain the fundamental principles of mineral processing plant design, including layout, workflow, and integration of unit operations.
2. Interpret and construct basic process flow diagrams (PFDs) and piping and instrumentation diagrams (P&IDs) used in rare earth and critical mineral processing plants.
3. Identify and describe the major equipment and systems used in comminution, separation, and refining processes, and explain their functions within a complete processing circuit.
4. Discuss the design considerations that influence plant capacity, throughput, and efficiency, including ore characteristics, product requirements, and operational constraints.
5. Analyse the relationships between upstream and downstream processes to ensure consistent material balance and process optimisation.
6. Recognise the role of automation, instrumentation, and process control systems in maintaining product quality and operational stability.
7. Apply basic principles of plant layout design, including space utilisation, material flow, safety zones, and accessibility for maintenance and inspection.
8. Describe the importance of quality control, sampling, and product specification, and demonstrate how representative sampling supports process verification and compliance.
9. Identify key safety and environmental factors influencing plant design, such as dust control, waste management, ventilation, and emergency response systems.
10. Evaluate examples of rare earth and lithium processing plant designs and discuss how process configuration is adapted to local resources, market demands, and sustainability requirements.

Key Plant Equipment and Instrumentation

A mineral processing plant integrates a sophisticated array of mechanical and electronic systems to extract, upgrade, and refine valuable elements effectively and sustainably. Each operational stage—including comminution, separation, leaching, purification, and recovery—depends on specialized equipment and precise instrumentation. These systems are designed to ensure energy efficiency, product quality, and process stability across the mineral processing workflow.

Comminution Equipment plays a vital role in reducing ore to a suitable particle size for efficient mineral liberation. Primary crushing equipment, such as jaw and gyratory crushers, initially breaks down run-of-mine material into manageable sizes. Secondary and tertiary crushers further refine this size, typically utilizing cone or impact types. Subsequently, grinding mills—such as ball, rod, or semi-autogenous mills—pulverize the material into a fine slurry necessary for downstream processes. Additionally, screening and hydrocyclones are employed to separate finer particles while allowing oversized materials to be recycled back for further grinding. Instrumentation at this stage includes load cells, belt scales for feed rate monitoring, and sensors for detecting mill conditions through vibration analysis, optimizing operational efficiency and energy consumption [280, 281].

In the Separation and Concentration Units, valuable minerals are isolated from gangue through various methods based on their physical and chemical properties. Gravity separation units exploit density differences, while magnetic separators target minerals such as magnetite and zircon. Flotation cells utilize chemical reagents and air bubbles to enhance mineral separation. The effectiveness of these processes is supported by sophisticated instrumentation, including pulp density meters and flow controllers that maintain optimal conditions within flotation tanks [282, 283].

The Leaching and Chemical Processing Units are crucial for dissolving metals from ores using controlled chemical reactions. Leach tanks provide an environment for acid or alkaline leaching to solubilize metals; for example, sulfuric acid is often employed for rare earth elements. High-pressure leaching in autoclaves is commonly applied in nickel and cobalt circuits to enhance extraction efficiency. Precise control of process variables is achieved using pH sensors, pressure transmitters, and flow meters to track reagent and slurry movements, ensuring operational stability [284, 285].

During Purification and Recovery, the processed solutions undergo treatments to yield high-purity materials. Solvent extraction and ion exchange processes are employed to selectively remove impurities, while precipitation reactors facilitate the recovery of metals as various compounds. Electrowinning is then used to produce metal in its pure form, supported by real-time monitoring through pH sensors and temperature probes to safeguard operational integrity [286, 287].

Thermal and Pyrometallurgical Processes utilize high-temperature operations to decompose, oxidize, or reduce minerals. Equipment such as rotary kilns and fluidized bed roasters optimize reactions critical for metal recovery. Induction and electric furnaces facilitate the smelting of metals, where managing temperature and chemical atmosphere is essential for process efficiency and product quality. Control systems include thermocouple arrays for precise temperature mapping and gas analysers for atmosphere regulation [288, 289].

Finally, Dewatering and Drying Systems are essential for removing excess water from recovered materials to prepare them for shipment or further processing. Thickening, filtration, and drying technologies maintain product specifications by adjusting moisture content as necessary. Instrumentation such as level sensors and pressure gauges enhances the reliability of these systems, ensuring efficient operation throughout the dewatering process [290].

Process Control and Automation Systems are integral to modern mineral processing operations, utilizing Distributed Control Systems (DCS) or Supervisory Control and Data Acquisition (SCADA) platforms. These systems centralize control and monitoring, employing smart sensors and advanced algorithms to optimize reagent use and energy efficiency, thereby facilitating predictive maintenance and operational reliability [291].

Lastly, the emphasis on Environmental and Safety Systems reflects the industry's commitment to sustainable practices. Systems like gas scrubbers and effluent treatment facilities mitigate environmental impact by controlling emissions and wastewater quality. Continuous monitoring instruments ensure that operations remain within compliance standards, safeguarding both human health and ecological integrity [292].

Layout of a Typical Mineral Processing Plant

A typical mineral processing plant layout is designed to efficiently transform raw ore from a mine into a marketable product through a series of mechanical, physical, and chemical processes. The layout is carefully planned to optimise material flow, minimise handling, ensure safety, and facilitate maintenance and environmental control. While plant designs vary depending on the mineral type and production scale, most include several key sections that function in a logical sequence — from ore reception to product dispatch.

1. Ore Reception and Storage

The process begins with the delivery of ore from the mine to the processing plant. The ore is typically transported by truck, conveyor, or rail, depending on the mine's location and production scale. Upon arrival, it is discharged into Run-of-Mine (ROM) bins or stockpiles, which act as a buffer to provide surge capacity and maintain a consistent feed rate to the plant. From there, the ore passes through feed hoppers and grizzlies, which screen out oversized rocks or remove tramp metal to protect downstream equipment from damage. This initial handling stage is designed for safe, continuous operation, incorporating dust suppression systems to control emissions and strategically placed sampling points to monitor feed quality and ensure consistent processing performance.

Figure 61: Feed hopper and feed conveyor. Peter Craven, CC BY 2.0, via Wikimedia Commons.

2. Crushing and Grinding (Comminution Circuit)

Comminution is the process of reducing the size of ore particles to liberate valuable minerals from the surrounding gangue material. It is a critical step in mineral processing because the degree of liberation directly affects the efficiency of downstream separation and recovery processes. The comminution circuit typically includes several stages of crushing and grinding, each designed to progressively reduce particle size.

Figure 62: Wheel loader dumps rock into crusher plant, Germany. Mozzihh, CC BY-SA 4.0, via Wikimedia Commons.

The process begins with primary crushing, where large run-of-mine rocks are broken down using jaw or gyratory crushers into smaller, more manageable fragments. These fragments then move to secondary and tertiary crushers, such as cone or impact crushers, which provide finer reduction and ensure a uniform particle size suitable for grinding. Once the material is sufficiently reduced, it is fed into grinding mills—commonly rod mills, ball mills, or semi-autogenous (SAG) mills—that pulverize the ore into a fine slurry. This fine material is essential for maximizing mineral liberation before concentration or separation.

To maintain consistent product size, screens and hydrocyclones are used for efficient classification, separating oversized particles for regrinding and allowing correctly sized material to proceed to the next stage. Because comminution equipment handles large volumes of material and operates continuously under heavy load, it is typically the largest consumer of power within the processing plant. As such, the comminution area is strategically positioned for accessibility, maintenance, and noise control, ensuring efficient and safe operation.

3. Classification and Separation

After grinding, the finely milled material undergoes a separation process to concentrate the valuable minerals and remove unwanted gangue. This stage exploits differences in the physical or chemical properties of minerals—such as density, magnetic susceptibility, surface chemistry, or electrical conductivity—to achieve selective separation.

Various types of equipment are used depending on the characteristics of the ore. Gravity separation units, such as spirals, jigs, and shaking tables, separate particles based on density differences and are particularly effective for heavy mineral sands, gold, and some rare earth concentrates. Magnetic separators are employed to recover ferromagnetic or paramagnetic minerals like magnetite, ilmenite, or rare earth oxides, while flotation cells are used to exploit differences in surface chemistry. In flotation, selected minerals attach to air bubbles and rise to the surface, allowing efficient recovery of base metals, lithium-bearing minerals, and rare earth elements. Electrostatic separators are also used in dry processing to separate conductive and non-conductive particles, a common method in titanium, zircon, and rare earth beneficiation.

This separation stage is often designed as a modular system, enabling multiple circuits to operate simultaneously or in parallel. This flexibility allows the plant to process varying ore types and grades efficiently, ensuring consistent concentrate quality and maximizing recovery.

4. Thickening and Filtration

The dewatering stage is designed to remove excess water from mineral slurries after separation, preparing the concentrate for drying, further processing, or transport. This step is essential for improving product handling, reducing storage and transport costs, and minimizing environmental impacts associated with water use.

In most plants, thickeners are used first to concentrate the solids through gravity settling, allowing the clarified water to overflow and be recycled. The thickened slurry then passes to filters, such as disc, belt, or pressure filters, which further reduce the moisture content to the desired level. The recovered water from both stages is typically recycled back into the process, minimizing overall water consumption and reducing the volume of liquid waste that must be treated or discharged.

Dewatering systems are generally positioned at a lower elevation within the plant layout, allowing gravity flow of slurry and reducing the need for pumping. This strategic placement not only improves energy efficiency but also supports safe, continuous operation in an environmentally responsible manner.

5. Hydrometallurgical or Pyrometallurgical Section

The hydrometallurgical or pyrometallurgical section of a mineral processing plant is where the concentrated feed is transformed into purified metals or compounds through either chemical or thermal means, depending on the selected processing route.

In a hydrometallurgical plant, operations typically include leaching tanks, where valuable metals are dissolved into solution using acids, bases, or other reagents. The resulting metal-rich solutions are then processed through solvent extraction and ion exchange units to selectively recover and purify target elements such as rare earths, lithium, nickel, or cobalt. These processes require precise control of parameters like pH, temperature, and reagent concentration to ensure high recovery and purity.

In contrast, pyrometallurgical plants employ furnaces, kilns, and roasters for thermal decomposition, reduction, or smelting of mineral concentrates. These systems operate at elevated temperatures—often exceeding 1000°C—to drive reactions that produce metallic or oxide products. The design of such facilities emphasizes strict temperature and atmosphere control, ensuring consistent product quality and efficient reaction kinetics.

Both hydrometallurgical and pyrometallurgical areas are carefully segregated within the plant layout to meet safety and environmental compliance standards. This includes acid containment systems, refractory linings, gas scrubbing units, and ventilation to manage potentially hazardous fumes or emissions. The integration of these systems enables safe, efficient, and sustainable metal recovery from complex ores and concentrates.

6. Product Recovery and Refining

The product recovery and refining stage represents the final step in the mineral processing and extraction chain, where purified metals, oxides, or salts are produced in their market-ready form. This section is dedicated to achieving high product quality, purity, and consistency through a combination of controlled chemical and physical processes.

Depending on the mineral and processing route, several recovery techniques are used. Precipitation or crystallisation is commonly applied to recover purified salts or oxides from solution, such as rare earth oxalates, lithium carbonate, or cobalt hydroxide. In operations where metallic products are the goal, electrowinning or electrorefining is used to deposit pure metals onto cathodes through controlled electrochemical reactions. After recovery, the material typically proceeds to drying and packaging stations, where it is prepared for shipment, storage, or further processing.

This part of the plant is designed with a strong focus on contamination control and quality assurance. Automated systems handle product weighing, sampling, and packaging to maintain consistency and traceability. Environmental and safety controls, including dust collection and air filtration, ensure that high-value products are processed under clean and controlled conditions, meeting both customer and regulatory standards.

7. Tailings Management and Waste Treatment

The tailings management and waste treatment section is a critical component of any mineral processing plant, responsible for the safe handling, disposal, and recycling of waste materials generated throughout the operation. Its primary objective is to minimise environmental impact while maximising resource recovery and water reuse.

The tailings circuit manages the flow of process residues through thickeners and pumps, which concentrate the solids and transport them to tailings storage facilities (TSFs). These engineered structures are designed to safely contain fine mineral residues, often incorporating liners, seepage control systems, and monitoring instrumentation to prevent contamination of surrounding soil and water resources.

Simultaneously, water treatment plants play a vital role in recovering and recycling process water. They remove suspended solids, neutralise acidity, and eliminate contaminants such as heavy metals or reagents before the water is reused or safely discharged. In pyrometallurgical and chemical processing areas, gas scrubbers and dust collectors are employed to capture airborne emissions, including particulates, sulphur oxides, and fluorides, ensuring compliance with air quality regulations.

Comprehensive environmental protection measures—including effluent monitoring, waste characterisation, and emission controls—are integrated into the plant's design. These systems ensure that tailings and waste management align with sustainability objectives, regulatory requirements, and community expectations for responsible mineral processing operations.

8. Utilities, Infrastructure, and Control Systems

Supporting systems form the essential backbone of a mineral processing plant, providing the infrastructure required for continuous, efficient, and safe operation. These auxiliary services ensure that the core processing circuits—such as comminution, separation, and refining—operate smoothly and within design parameters.

Key supporting systems include power supply and distribution networks, which provide reliable electricity for motors, pumps, and heating systems, along with compressed air systems used for instrumentation, flotation, and general plant operations. Process water networks are also critical, supplying and recycling water for grinding, classification, and dust suppression throughout the facility.

In addition, laboratories are strategically located on-site to conduct process control and quality assurance testing. These labs analyse samples from various plant stages to monitor product grade, recovery rates, and reagent consumption, providing feedback for process optimisation. Central control rooms house Distributed Control Systems (DCS) or Programmable Logic Controllers (PLCs) that allow operators to monitor, adjust, and automate plant functions in real time, improving efficiency and safety.

Complementing these are workshops and stores that support maintenance activities, equipment repair, and spare parts management. These facilities are carefully positioned for accessibility and operational safety, ensuring that maintenance teams can respond quickly to breakdowns or scheduled servicing needs. Together, these supporting systems underpin the reliability, stability, and long-term performance of the entire processing operation.

9. Administrative and Safety Areas

Administrative buildings, training rooms, change houses, and emergency facilities (e.g., fire stations, medical rooms) are located away from the main process area to protect personnel. Traffic and pedestrian movement are managed through designated corridors and exclusion zones.

In essence, a mineral processing plant layout follows a logical, gravity-assisted flow:

Ore → Comminution → Separation → Concentration → Refining → Product → Waste Management

Each section is interconnected by conveyors, pumps, or pipelines and monitored by automation systems to optimise efficiency, product quality, and environmental compliance.

Process Flow Diagrams for REE, Lithium, and Cobalt

Process Flow Diagrams (PFDs) for rare earth elements (REEs), lithium, and cobalt illustrate the sequence of key unit operations—such as comminution, concentration, leaching, extraction, and refining—used to transform raw ores or concentrates into marketable products. Each diagram reflects the mineralogy of the ore, the chosen processing route (pyrometallurgical or hydrometallurgical), and the desired final product (oxide, carbonate, hydroxide, or metal). These diagrams serve as essential tools for plant design, process optimisation, and environmental management.

Rare Earth Elements (REEs)

REE processing flow diagrams are often complex due to the diverse mineral forms (bastnäsite, monazite, xenotime, and ion-adsorption clays) and the close chemical similarity of the rare earths. A typical PFD for REE processing includes the following stages:

1. **Comminution and Classification:** The mined ore is crushed and ground to liberate REE-bearing minerals, followed by desliming to remove fine clays that interfere with separation.

2. **Physical Beneficiation:** Gravity, magnetic, and electrostatic separation concentrate REE minerals from gangue. For example, bastnäsite and monazite are separated from quartz or zircon.

3. **Roasting or Calcination:** The concentrate is thermally treated (e.g., bastnäsite roasting at 500–700 °C) to remove CO_2 and F and to convert REEs into acid-soluble oxides.

4. **Leaching:** Acid (H_2SO_4, HCl) or alkaline (NaOH) leaching dissolves REEs into solution.

5. **Purification and Separation:** Solvent extraction or ion exchange separates individual REEs into groups (light, medium, and heavy REEs).

6. **Precipitation and Calcination:** REEs are precipitated as oxalates or carbonates and then calcined to form high-purity oxides (e.g., Nd_2O_3, Dy_2O_3).

Example Output: REE oxide powders for use in magnets, catalysts, and phosphors.

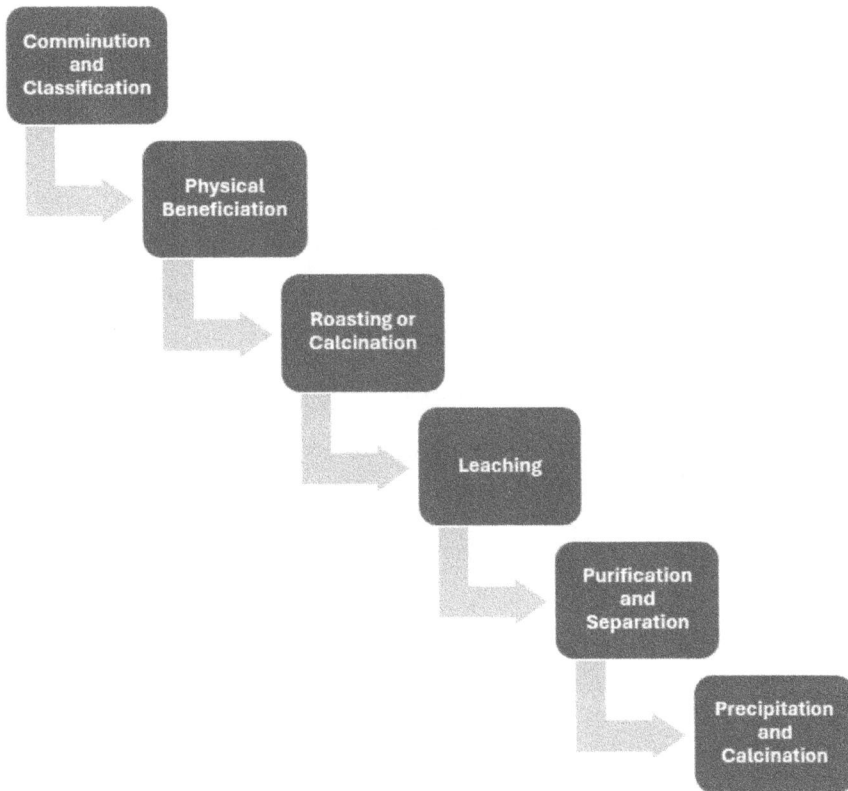

Figure 63: Process flow diagrams for REE.

Lithium

Lithium flow diagrams differ depending on the feedstock—hard-rock spodumene versus brine deposits—but both aim to produce high-purity lithium carbonate or hydroxide for battery production.

a) Hard-Rock (Spodumene) Route:

1. **Comminution:** Crushing and grinding of spodumene-bearing pegmatite.

2. **Beneficiation:** Dense media separation or flotation concentrates spodumene ($LiAlSi_2O_6$).

3. **Thermal Conversion:** Rotary kilns convert α-spodumene to the more reactive β-phase at ~1,050 °C.

4. **Acid Leaching:** β-spodumene is leached with H_2SO_4 at ~200 °C to produce lithium sulphate.

5. **Purification and Precipitation:** Impurities are removed, and lithium is precipitated as lithium carbonate (Li_2CO_3) or hydroxide ($LiOH \cdot H_2O$).

6. **Drying and Packaging:** The final battery-grade product is filtered, dried, and packaged under controlled conditions.

b) Brine Route:

1. **Evaporation:** Concentrated brine from salt flats undergoes solar evaporation to increase lithium concentration.

2. **Precipitation:** Impurities like Mg^{2+} and Ca^{2+} are removed as hydroxides or carbonates.

3. **Chemical Conversion:** The concentrated lithium chloride solution is converted into lithium carbonate or hydroxide via chemical precipitation.

Example Output: Battery-grade lithium carbonate or hydroxide for electric vehicle and energy storage applications.

Cobalt

Cobalt flow diagrams depend on whether the feed is a sulphide ore (e.g., copper-cobalt) or an oxide/laterite ore. Modern processing routes often combine pyrometallurgical pre-treatment with hydrometallurgical refining.

a) Sulphide Ore Route:

1. **Comminution and Flotation:** Concentrates cobalt and copper sulphides.

2. **Roasting or Pressure Oxidation:** Converts sulphides to oxides or sulphates, releasing SO_2 for sulfuric acid production.

3. **Leaching:** Sulfuric acid dissolves cobalt and copper into solution.

4. **Solvent Extraction:** Separates cobalt from copper and other impurities.

5. **Precipitation or Electrowinning:** Produces cobalt hydroxide or metallic cobalt.

b) Laterite Ore (HPAL) Route:

1. **Ore Preparation:** Drying and grinding of nickel-cobalt laterite ores.

2. **High-Pressure Acid Leaching (HPAL):** H_2SO_4 leaching at ~250 °C and 40 atm dissolves Ni and Co.

3. **Neutralisation and Purification:** Iron and aluminium are removed by pH control.

4. **Solvent Extraction or Ion Exchange:** Separates nickel and cobalt selectively.

5. **Recovery:** Cobalt is recovered as hydroxide or carbonate, then refined to metal by reduction or electrolysis.

Example Output: Cobalt hydroxide or metal powders for batteries, superalloys, and catalysts.

Quality Control, Sampling, and Product Specification

Quality control (QC) is a cornerstone of mineral and critical mineral processing, ensuring that the products leaving the plant consistently meet technical, commercial, and regulatory standards. It integrates systematic sampling, analytical testing, and product specification management across every stage of the production chain—from ore feed to final concentrate or refined metal. Effective QC not only guarantees product quality but also supports process optimization, traceability, and compliance with customer and environmental requirements.

The purpose of quality control in a mineral processing plant is to ensure that operations are consistently maintained within the required performance and product specifications. It serves as a systematic check to confirm that each stage of the process— from ore feed to final product—meets established standards of quality, efficiency, and consistency. Quality control ensures that the ore feed entering the plant matches the expected grade and mineral composition, preventing feed variability that could disrupt downstream processes. It also verifies that intermediate and final products conform to precise chemical and physical specifications required by customers and industry standards.

Beyond composition, quality control monitors process efficiency and recovery rates to confirm they remain within target ranges. It also ensures that impurities, moisture levels, and particle size distributions are properly controlled, preventing off-specification shipments that could compromise performance or customer satisfaction. In the context of critical mineral production—such as rare earth elements (REEs), lithium, cobalt, and nickel—quality control is even more essential. The purity and consistency of these materials directly influence their suitability for high-precision applications, including magnets, advanced batteries, and high-performance alloys. Even minor deviations in chemical purity or particle morphology can render a batch unusable for sectors like electronics or renewable energy, where exact specifications define both functionality and value.

Sampling forms the foundation of quality control in mineral and critical mineral processing. Its primary purpose is to obtain representative material for laboratory analysis and decision-making, ensuring that test results accurately reflect the composition and properties of the bulk material. The effectiveness of any quality assurance system depends heavily on the reliability of sampling procedures, as poor or biased sampling can lead to misleading results, process inefficiencies, or even costly product non-compliance. Therefore, great care is taken to design sampling systems that minimize bias, maintain sample integrity, and capture the inherent variability of the material being processed.

There are several key stages of sampling across a processing plant. Run-of-Mine (ROM) sampling occurs immediately after ore delivery from the mine and is used to determine the feed grade and mineral composition. This information supports blending strategies and ensures the feedstock entering the plant meets processing requirements. Process sampling takes place at critical control points such as crushers, mills, flotation cells, leach circuits, and thickener overflows. These samples help operators monitor plant efficiency, metal recovery, and reagent performance. Product sampling focuses on final concentrates, precipitates, or refined materials to confirm compliance with product specifications and customer requirements. Finally, tailings sampling is performed to monitor environmental compliance, assess metal losses, and evaluate opportunities for reprocessing or waste reduction.

A range of sampling methods is employed depending on the material type and process conditions. Cross-stream samplers and rotary cutters are widely used for slurry and pulp streams, providing continuous and representative cross-sectional cuts. For dry solids, grab or scoop sampling may be used, although it is less accurate and generally reserved for preliminary or spot checks. Automatic composite samplers are increasingly common in modern plants, collecting timed increments over production periods to create averaged, statistically valid samples. For ore body characterization or mine feed assessment, drill-core and auger sampling provide detailed insights into ore composition and variability before processing.

To ensure consistency and credibility, representative sampling protocols adhere to internationally recognized standards such as ISO 3082 for iron ores and ASTM D75 for coal and mineral materials. These standards define procedures for equipment design, sample increment size, frequency, and compositing techniques to minimize sampling error. In practice, plants adapt these frameworks to suit the physical properties of specific commodities—such as rare earth concentrates, lithium salts, or cobalt hydroxides—ensuring that each sample collected accurately represents the broader process stream. Through rigorous, standardized sampling, mineral processors gain the reliable data needed to optimize operations, verify quality, and maintain compliance across the production chain.

Once samples have been collected from various stages of the mineral processing circuit, they undergo analytical testing and laboratory analysis to determine their chemical, mineralogical, and physical characteristics. This stage transforms raw sampling data into quantifiable information that guides process optimization, quality control, and compliance verification.

The first analytical focus is chemical composition, which establishes the concentrations of target metals and impurities in the sample. Techniques such as X-ray fluorescence (XRF), inductively coupled plasma optical emission spectroscopy (ICP-OES), and inductively coupled plasma mass spectrometry (ICP-MS) are commonly employed to measure elements across a wide range of concentrations. These analyses are critical for determining the efficiency of beneficiation, leaching, or refining stages, as well as ensuring that final products meet purity requirements.

Next, mineralogical composition is assessed using X-ray diffraction (XRD) and scanning electron microscopy (SEM), often coupled with energy-dispersive spectroscopy (EDS). These techniques reveal the mineral phases present, particle associations, and liberation characteristics—vital for understanding how minerals respond to grinding, flotation, or leaching. In rare earth and critical mineral processing, this data helps distinguish between mineral species such as bastnäsite, monazite, or xenotime, each requiring different treatment conditions.

Particle size distribution is another critical parameter, typically measured by laser diffraction analysers or wet/dry sieving methods. Particle size affects downstream processes such as flotation efficiency, filtration rate, and leach kinetics. Maintaining a consistent grind size is essential for achieving predictable recoveries and product uniformity.

In addition, moisture content and loss on ignition (LOI) tests are performed to determine the presence of water, organics, or carbonates. These measurements are particularly important in thermal processes like calcination, where moisture or volatile components can influence energy demand and reaction behaviour. Physical properties such as bulk density, surface area, and flowability are also assessed to support material handling, blending, and packaging operations.

Analytical results are then compared with in-process targets or customer specifications. Any deviation from these benchmarks can trigger adjustments in grinding, reagent dosing, or furnace temperature to restore control. Increasingly, plants are adopting real-time or online process analysers—for instance, on-stream XRF units for slurry analysis or laser-based particle size sensors for milling circuits. These instruments provide continuous feedback to plant control systems, enabling dynamic process optimization and minimizing the time lag between sampling, analysis, and operational response.

Overall, analytical testing bridges the gap between process performance and product quality. It ensures that mineral processing plants not only operate efficiently but also produce materials that meet the stringent chemical and physical standards demanded by industries such as energy storage, electronics, and advanced manufacturing.

Product specification and quality assurance are critical elements in mineral processing and refining, ensuring that final products consistently meet the stringent chemical and physical standards required by end users, regulatory agencies, and global markets. Each mineral or metal product—whether a concentrate, oxide, carbonate, or refined metal—is produced to a defined specification based on customer requirements and industry benchmarks. These specifications determine not only the material's commercial value but also its suitability for downstream applications such as battery cathodes, permanent magnets, catalysts, or high-performance alloys.

Product specifications vary by commodity and application. For rare earth oxides (REOs), for example, extremely high purity levels are required—often ≥99.9% total REO content—along with controlled ratios of specific elements. A neodymium-praseodymium (NdPr) oxide product may be required to contain Nd_2O_3 ≥ 32% and Pr_6O_{11} ≤ 5%, ensuring predictable magnetic and optical properties. For lithium carbonate, particularly battery-grade Li_2CO_3, the specification typically demands ≥99.5% purity, with trace impurities like iron (Fe <0.005%) and sodium (Na <0.01%) strictly limited to prevent performance degradation in lithium-ion cells. Similarly, cobalt hydroxide and cobalt sulphate products must exhibit minimal contamination from other metals such as nickel, copper, or iron—usually below 50 parts per million (ppm)—to maintain consistent electrochemical behaviour. In nickel sulphate used for battery precursor manufacturing, the purity standard exceeds 99.8%, with strict control of particle size and morphology to ensure uniform precipitation and cathode formation.

The quality assurance (QA) process ensures that every batch leaving the processing plant meets these specifications through systematic verification, documentation, and certification. QA protocols typically include laboratory verification of the product's chemical and physical properties, followed by the issuance of a Certificate of Analysis (CoA). The CoA provides a detailed record of the analytical results, listing concentrations of target elements, impurities, moisture content, and physical parameters such as particle size or bulk density. Each certificate is traceable to specific sampling points, laboratory test numbers, and production lots, ensuring complete transparency and accountability across the supply chain.

Quality assurance also extends to traceability and validation. All sampling, testing, and reporting steps are documented under controlled procedures aligned with international standards such as ISO 17025 (testing and calibration laboratories) and ISO 9001 (quality management systems). In high-value or export-oriented operations, product quality may be independently verified by third-party laboratories or inspection agencies, which provide impartial validation of results prior to shipment. This independent testing is particularly important for critical minerals entering high-technology or energy storage supply chains, where consistency and purity are non-negotiable.

In essence, product specification and quality assurance form the final link in the value chain from ore to refined product. They provide assurance to downstream manufacturers that materials will perform reliably in advanced applications, while also protecting producers' reputations and market competitiveness. Through rigorous specification control, standardized testing, and certified documentation, the mineral processing industry ensures that its products meet the highest global standards for quality, reliability, and traceability.

Rare Earth and Critical Mineral Operations and Processing

Process control integration is a defining feature of modern mineral and metallurgical plants, transforming quality control (QC) from a reactive, end-of-line activity into a dynamic, continuous, and data-driven process. By linking laboratory and online analytical data directly into centralized control systems such as Distributed Control Systems (DCS) or Supervisory Control and Data Acquisition (SCADA) networks, operators can monitor and optimize plant performance in real time. This integration bridges the gap between product quality assurance and process automation, ensuring that every stage of production—from crushing and grinding to refining and drying—remains aligned with quality targets and operational efficiency goals.

At the core of this integration are online analysers and process sensors, which continuously measure key parameters such as metal concentration, particle size, pH, pulp density, gas composition, and temperature. These instruments feed live data directly into the control loop, where it is processed by programmable logic controllers (PLCs) or advanced process control (APC) algorithms. For example, on-stream X-ray fluorescence (XRF) analysers provide continuous readings of elemental composition in flotation or leaching circuits, while laser-based particle size sensors ensure consistent grinding performance. This real-time feedback enables the system to make immediate adjustments—such as modifying reagent dosing, air injection rates, or feed blending ratios—to maintain product specifications and maximize recovery.

Beyond real-time adjustments, integrated control systems use predictive and trend-based analytics to improve process stability and reliability. Historical QC data is stored within the DCS or SCADA database, allowing engineers to identify long-term trends, correlations, and deviations from standard operating conditions. When analysed using statistical or machine learning tools, these data sets can predict potential issues—such as reagent depletion, equipment fouling, or changing ore characteristics—before they impact production. This predictive capability supports proactive maintenance, optimized scheduling, and reduced downtime, all of which contribute to improved throughput and energy efficiency.

Integration also enhances cross-departmental coordination. QC laboratories can automatically upload analytical results to the plant control system, reducing manual data entry errors and ensuring that operators always have access to the most recent quality data. In fully automated systems, deviations in grade or impurity levels can trigger alerts or even initiate automatic corrective actions, such as changing process setpoints or diverting off-spec material for reprocessing. In this way, quality management becomes a continuous feedback cycle rather than a discrete verification step at the end of the process.

Ultimately, process control integration ensures that mineral processing plants maintain consistent output quality even under variable feed conditions—a key challenge in mining and beneficiation. By uniting automation, analytics, and quality control, operators can maintain tighter control over recovery, grade, and purity, while also achieving better resource efficiency and environmental compliance. This fusion of digital and operational intelligence represents a crucial step toward the "smart plant" concept, where data-driven decision-making underpins every aspect of mineral and critical materials production.

Traceability and Reporting are essential components of modern mineral and metallurgical operations, ensuring accountability, product authenticity, and compliance with regulatory frameworks across the entire value chain—from ore extraction to final product export. Comprehensive documentation and digital tracking systems provide transparency, enabling producers and customers to verify the origin, composition, and handling of every batch of material.

At the core of this system is batch tracking, which links each shipment or batch of product to its source ore, processing route, and quality test results. Each batch receives a unique identification code that ties together production data, sampling results, laboratory analyses, and process parameters. This allows any product in the supply chain to be traced back to its original ore body, blending location, and processing conditions—critical for maintaining credibility in the production of high-value or strategic materials like rare earth elements (REEs), lithium, cobalt, and nickel.

A central component of reporting is the Certificate of Analysis (CoA). For each shipment, the CoA provides a detailed breakdown of chemical composition, physical characteristics, and impurity levels. It serves as formal verification to customers that the material meets contractual and technical specifications. These certificates are often required for cross-border trade and are stored in both physical and digital formats for long-term traceability. In some cases, independent laboratories verify the CoA to ensure accuracy and impartiality, especially for export-grade products used in critical technologies and renewable energy systems.

In addition, regulatory record-keeping ensures compliance with national and international laws governing critical minerals. For example, under the EU Critical Raw Materials Act, U.S. Inflation Reduction Act, and similar frameworks in Australia and Canada, producers must document not only product quality but also environmental, ethical, and supply-chain transparency metrics. This includes maintaining logs of emissions, tailings management, worker safety compliance, and evidence of responsible sourcing practices.

To support this growing complexity, mineral producers increasingly rely on digital traceability technologies. Laboratory Information Management Systems (LIMS) automate the recording and retrieval of analytical results, linking test data directly to production batches and Certificates of Analysis. Meanwhile, blockchain-based traceability platforms are being adopted to create tamper-proof digital records of each transaction or transformation along the supply chain. These systems allow downstream manufacturers—such as battery or magnet producers—to verify that the materials they receive originate from compliant, ethical, and sustainable sources.

Quality control (QC) in mineral and metallurgical processing is a dynamic and continuous system rather than a fixed, one-time procedure. It evolves through systematic feedback, performance reviews, and data-driven analysis aimed at maintaining and improving product consistency, process efficiency, and overall plant performance. The goal is not only to detect deviations but to understand their underlying causes and prevent them from recurring, ensuring that each stage of the production chain, from raw ore to final product, operates within its defined quality and efficiency parameters.

Rare Earth and Critical Mineral Operations and Processing

One key component of this evolving system is the use of round-robin tests or inter-laboratory comparisons, where identical samples are analysed by multiple laboratories to verify accuracy, repeatability, and reliability of analytical results. This benchmarking ensures that the plant's laboratory methods remain aligned with international standards, such as ISO 17025, and that instrumentation calibration and operator performance meet expected precision levels. By comparing results with external laboratories or certified reference materials, plants can detect systematic errors, biases, or deviations in analytical procedures that could otherwise compromise quality assurance.

Another powerful tool is Statistical Process Control (SPC), a quantitative method used to monitor ongoing process data for variations and trends. SPC charts and control limits help operators visualize performance over time, distinguishing between normal (random) variation and abnormal deviations that may indicate process instability. For instance, fluctuations in leach solution pH, ore feed grade, or product moisture can be tracked statistically to anticipate when corrective actions are needed. When integrated with the plant's Distributed Control System (DCS), SPC provides real-time feedback that enables immediate adjustments—preventing off-spec product or process inefficiencies before they escalate.

When quality issues do arise, root cause analysis (RCA) is employed to systematically identify the origin of the deviation. This may involve examining raw material quality, reagent dosing, temperature control, equipment wear, or human factors. RCA techniques such as the "5 Whys" or "Fishbone Diagram" help teams dissect complex problems and implement targeted corrective measures. By addressing the root rather than the symptom, plants ensure long-term process stability and continuous improvement.

Ultimately, continuous improvement is the backbone of modern quality management. Plants routinely use feedback from QC results, SPC data, and RCA findings to refine process parameters, upgrade equipment, or optimize reagent use. The outcomes of this cycle include higher recovery rates, lower energy consumption, reduced waste, and improved product uniformity. This iterative approach embodies the principle of operational excellence, where lessons learned are fed back into both technical and managerial systems to foster a culture of precision, accountability, and innovation in mineral processing operations.

Quality control, sampling, and product specification together form an essential closed-loop system in mineral processing. From initial ore feed to final product certification, these practices ensure:

- Product consistency and reliability.
- Efficient process optimization and waste reduction.
- Compliance with environmental and trade standards.
- Customer confidence and market competitiveness.

In rare earth and critical mineral production, where purity and precision define market value, a robust QC framework is the foundation for operational excellence and sustainable resource development.

Key Terms and Concepts	

Plant Design: The engineering and layout planning of a mineral-processing facility to ensure safe, efficient, and cost-effective operation.

Process Flow Diagram (PFD): A schematic showing the sequence of major unit operations and material flows within a processing plant.

Piping and Instrumentation Diagram (P&ID): A detailed drawing that identifies equipment, pipelines, valves, instruments, and control loops used in plant operation.

Unit Operation: A distinct physical or chemical step in the processing sequence—such as crushing, flotation, or leaching—that performs a specific function.

Material Balance: A calculation that accounts for the input, output, and accumulation of materials in a process to ensure mass conservation and identify inefficiencies.

Throughput: The quantity of material processed per unit of time, commonly expressed in tonnes per hour (t/h).

Process Control System: An automated or manual system that monitors and regulates operational variables such as temperature, pressure, flow rate, and pH to maintain stable performance.

Instrumentation: The sensors, transmitters, and control devices used to measure and adjust process parameters in real time.

Plant Layout: The spatial arrangement of equipment, structures, and utilities within a facility to optimise workflow, safety, and maintenance access.

Sampling System: Equipment and protocols designed to collect representative material samples for quality control, metallurgical accounting, and process optimisation.

Key Terms and Concepts

Quality Assurance (QA): Systematic activities that ensure products and processes meet defined quality standards and regulatory requirements.

Quality Control (QC): Routine testing and monitoring of samples or products to verify conformity with specifications and detect process deviations.

Modular Design: A plant-construction approach that uses pre-fabricated, standardised modules for faster assembly, scalability, and easier relocation.

Automation: The use of computer-based systems, sensors, and actuators to operate and control plant processes with minimal human intervention.

Maintenance Planning: The scheduling and management of equipment servicing, inspection, and replacement to prevent downtime and ensure continuous operation.

Safety Zone: A designated area around equipment or process units that provides protection for personnel from potential hazards such as noise, heat, or moving parts.

Chapter 8 Review Questions

Fundamentals of Plant Design

1. Short answer: What are the main objectives of mineral processing plant design?

2. Multiple choice: Which of the following is *not* a typical design consideration in a mineral processing plant?

(a) Material flow efficiency

(b) Noise and vibration control

(c) Randomised equipment placement

(d) Maintenance accessibility

3. True or False: Plant layout design affects both operational safety and energy efficiency.

4. Extended response: Explain how workflow integration between crushing, separation, and refining improves overall process efficiency.

Process Flow Diagrams (PFDs) and Piping & Instrumentation Diagrams (P&IDs)

5. Short answer: What is the main difference between a PFD and a P&ID?

6. Multiple choice: Which symbol is *most likely* to appear on a P&ID but not on a PFD?

(a) Mixer

(b) Flow line

(c) Pressure transmitter

(d) Crusher

7. Applied question: Why are PFDs important tools for plant engineers and operators when designing or troubleshooting a process?

8. Short answer: Describe one way in which automation and instrumentation data are incorporated into a P&ID.

Equipment and Systems

9. Matching: Match each process stage to its corresponding equipment type:

a) Comminution	1) Solvent extraction columns
b) Separation	2) Ball mills
c) Purification	3) Magnetic separators
d) Recovery	4) Electrowinning cells

10. Short answer: What is the primary purpose of hydrocyclones in grinding circuits?

11. Extended response: Explain how flotation cells and magnetic separators contribute to mineral concentration.

12. Multiple choice: Which type of furnace is most suitable for the reduction of metal oxides?

(a) Induction furnace

(b) Rotary kiln

(c) Fluidized-bed roaster

(d) Electric arc furnace

Design Considerations and Capacity

13. Short answer: List three factors that influence the capacity and throughput of a processing plant.

14. Scenario question: A lithium plant is experiencing a bottleneck in its filtration unit. What design or operational modifications could improve throughput?

15. Multiple choice: Which parameter best represents plant throughput?

(a) Total volume of ore mined

(b) Tonnes of material processed per hour

(c) Weight of final product per shipment

(d) Number of operators on shift

Process Integration and Control

16. True or False: Process integration ensures mass balance between upstream and downstream stages.

17. Short answer: Define *Distributed Control System (DCS)* and its purpose in mineral processing.

18. Applied question: Give an example of how online X-ray fluorescence (XRF) analysers support process control.

19. Extended response: Describe how feedback automation improves product uniformity and reduces energy waste.

Quality Control, Sampling, and Product Specification

20. Short answer: What is the main goal of representative sampling in mineral processing?

21. Multiple choice: Which international standard governs sampling of iron ores?

(a) ISO 3082

(b) ISO 14001

(c) ISO 9001

(d) ASTM E122

22. Applied question: How does moisture content analysis affect downstream drying and calcination operations?

23. Extended response: Discuss how Certificates of Analysis (CoA) and Laboratory Information Management Systems (LIMS) enhance product traceability.

Safety and Environmental Factors

24. Short answer: Identify two systems used to control dust and gas emissions in mineral processing plants.

25. Multiple choice: Tailings thickening and filtration primarily aim to:

(a) Reduce product grade

(b) Increase water recovery and minimize waste

(c) Generate power

(d) Control reagent dosage

26. True or False: Hydrometallurgical and pyrometallurgical units are usually co-located to simplify operations.

27. Extended response: Explain how environmental design considerations such as tailings management and water recycling contribute to sustainable plant operation.

Applied Design Evaluation

28. Short answer: Why might a lithium brine plant design differ significantly from a hard-rock spodumene operation?

29. Applied question: What makes REE process flow diagrams more complex than those for base metals?

30. Extended response: Evaluate how regional resources and environmental regulations influence the design of cobalt and lithium processing plants.

Chapter 9

Environmental Management and Safety

The extraction and processing of rare earth elements (REEs) and other critical minerals are essential to global clean-energy and technological advancement—but they also present complex environmental, health, and safety challenges. As demand for these resources grows, so too does the need for responsible management of their ecological and social impacts. This chapter examines the interconnected systems that underpin safe and sustainable mineral production, including radiation protection, tailings and waste handling, water use, emissions control, and workplace safety. It also explores how environmental management systems, regulatory frameworks, and Environmental, Social, and Governance (ESG) principles shape industry best practice. By integrating these technical, regulatory, and ethical dimensions, learners will gain the knowledge required to support operations that safeguard workers, protect the environment, and maintain community trust throughout the life cycle of rare earth and critical mineral projects.

Learning Outcomes	

This chapter aims to give you the ability to:
1. Explain the key environmental and safety challenges associated with rare earth and critical mineral extraction, processing, and waste management.
2. Identify major pollutants and hazards, including radiation from thorium/uranium-bearing minerals, dust, tailings, and chemical residues, and describe their sources within processing operations.
3. Discuss the principles of environmental management systems (EMS) and apply them to mineral processing facilities in accordance with ISO 14001 and related standards.
4. Describe effective tailings and waste management practices, including containment, dewatering, rehabilitation, and long-term monitoring strategies.
5. Explain the importance of water management, including treatment, recycling, and discharge control, in ensuring compliance with environmental regulations.

Learning Outcomes	

6. Recognise the role of air quality and emissions control systems—such as scrubbers, filters, and gas capture units—in maintaining safe and sustainable plant operations.
7. Apply the principles of occupational health and safety (OHS) to identify, assess, and mitigate workplace hazards in mineral processing environments.
8. Evaluate the influence of legislation, regulations, and industry standards on environmental and safety performance in Australia and globally.
9. Discuss the role of Environmental, Social, and Governance (ESG) frameworks and community engagement in building public trust and maintaining a social licence to operate.
10. Demonstrate an understanding of how sustainability, safety culture, and continuous improvement contribute to responsible rare earth and critical mineral operations.

Radiation Hazards (Thorium/Uranium in REE Ores)

The extraction and processing of rare earth elements (REEs), particularly from ores such as monazite, xenotime, and bastnäsite, are associated with significant radiological hazards due to the presence of naturally occurring radioactive elements like thorium (Th) and uranium (U). These actinides are often embedded within the mineral structures and can be released during mining operations such as crushing or leaching [293-295]. Therefore, careful monitoring of radiation hazards is essential not only during mining but also throughout the refining processes [295, 296].

The decay of thorium-232 and uranium-238 leads to the emission of alpha (α), beta (β), and gamma (γ) radiation [294, 297]. Among these, alpha particles pose significant health risks if inhaled or ingested, as they can cause substantial damage at the cellular level, particularly in lung tissue [298, 299]. Conversely, gamma rays primarily contribute to external exposure risks, especially in processing facilities where high concentrations of these ores are present, resulting in radiation exposure concerns for workers [294, 295]. Additionally, the decay chain of these radioactive elements can produce radon gas (Rn), which can accumulate in confined spaces, posing further airborne hazards to workers [297, 298].

In the beneficiation and refining of REEs, several processing techniques can release dust particles containing thorium and uranium [300, 301]. This airborne particulate matter represents an inhalation risk, making dust suppression and suitable ventilation critical in mining and processing environments [302]. Moreover, during acid leaching and other chemical processes, the mobilization of radioactive elements into process liquors necessitates strict waste management protocols to prevent environmental contamination [296, 303]. Tailings generated post-extraction can retain high levels of radioactivity, highlighting the need for secure and long-term disposal methods accompanied by rigorous environmental monitoring [304, 305].

Prolonged exposure to thorium and uranium is correlated with increased cancer risks, including malignancies of the lung and bone, primarily due to the damaging effects of alpha radiation [304, 306]. Furthermore, the environmental consequences of poorly managed tailings can lead to the bioaccumulation of radionuclides in surrounding ecosystems, which further threatens public health through contaminated soil and water resources [296, 307]. Countries like Malaysia and Australia have implemented stringent radiation control measures consistent with International Atomic Energy Agency (IAEA) guidelines to mitigate these risks in REE processing plants [293, 295].

To address these radiation hazards, contemporary REE processing facilities are employing various control measures. These include elevated dust suppression technologies, high-efficiency particulate air (HEPA) filtration, and ongoing radiation surveys to monitor worker exposure levels [308]. Process water and waste containment strategies are additionally being employed to ensure that leachate does not migrate into the surrounding environment [301]. Personal protective equipment (PPE) and restricted access to areas of high radiation exposure are standard protocols aimed at safeguarding worker health in these industrial settings [293, 309]. Continuous monitoring for radon levels in enclosed areas also forms part of comprehensive health and safety management within such facilities [297, 298].

Regulatory oversight of radiation hazards in rare earth element (REE) processing is guided by a comprehensive framework of international and national standards designed to protect workers, the public, and the environment from the harmful effects of ionising radiation. These regulations ensure that all mining, beneficiation, and refining activities involving thorium- and uranium-bearing minerals are carried out under strict safety and monitoring controls.

At the international level, the International Atomic Energy Agency (IAEA) sets the global benchmark through its *Safety Standards Series*, which provides guidance on radiation protection principles, occupational exposure management, waste handling, and environmental monitoring. The IAEA's recommendations are based on the principle of ALARA— "As Low As Reasonably Achievable"—which requires that all radiation exposures be minimised to the lowest practicable levels, considering social and economic factors. These standards are adopted or adapted by national regulatory bodies to establish local compliance frameworks.

In Australia, regulation of radiation exposure is governed by the Australian Radiation Protection and Nuclear Safety Agency (ARPANSA) under the *Radiation Protection and Nuclear Safety Act 1998* and the *Code for Radiation Protection in Planned Exposure Situations (RPS C-1)*. ARPANSA sets strict limits for occupational exposure, defines monitoring and reporting requirements, and mandates radiation management plans for industries handling naturally occurring radioactive materials (NORM), such as monazite sands and other REE-bearing ores.

Each state and territory enforces these codes through its environmental and mining safety authorities, ensuring consistent oversight across all stages of mineral processing.

In the United States, similar regulatory authority lies with the Nuclear Regulatory Commission (NRC) and the Department of Energy (DOE). These agencies establish standards under the *Code of Federal Regulations* (10 CFR Part 20) and DOE *Radiation Protection Standards* for occupational and environmental exposure. They oversee the licensing, transport, and disposal of radioactive materials, requiring detailed risk assessments, radiation protection programs, and regular audits of operational facilities.

Across all jurisdictions, exposure limits are harmonised with recommendations from the International Commission on Radiological Protection (ICRP). For occupational exposure, the typical limit is 20 millisieverts (mSv) per year, averaged over a five-year period, with no single year exceeding 50 mSv. This limit applies to personnel working directly with radioactive materials in mining, processing, or waste management. For the general public, the allowable exposure is much lower—generally 1 mSv per year—to ensure that radiation levels beyond natural background remain negligible.

Compliance with these limits requires continuous personal dosimetry, radiation surveys, and environmental monitoring programs. Facilities must maintain detailed exposure records, implement shielding and ventilation systems, and ensure that any release of radioactive materials—whether to air, water, or solid waste—is within regulated discharge limits.

Tailings and Waste Handling

Tailings and waste handling form a critical part of rare earth element (REE) and critical mineral processing, as these materials often contain residual chemicals, heavy metals, and naturally occurring radioactive materials (NORM) such as thorium and uranium. Effective management ensures environmental protection, operational safety, and regulatory compliance throughout the lifecycle of a processing plant.

Tailings are the finely ground residues that remain after valuable minerals have been extracted from ore using physical or chemical processing methods. In rare earth element (REE) operations, tailings represent a complex mixture of materials that can vary widely in composition, particle size, and chemical reactivity depending on the type of ore and the processing method used. They are an unavoidable by-product of beneficiation, flotation, leaching, and refining operations, and their management is critical to both environmental protection and plant efficiency.

A major component of REE tailings is residual reagents—the chemicals used during the extraction process that remain in the waste material. For example, sulfuric acid, sodium hydroxide, or ammonium sulphate are commonly used in acid or alkaline leaching circuits, and small quantities may persist in the tailings. These reagents can influence pH, solubility, and the potential for secondary chemical reactions within tailings storage facilities.

Rare Earth and Critical Mineral Operations and Processing

Many REE ores, such as monazite and xenotime, naturally contain radioactive elements, including thorium and uranium, along with their decay products. During mineral processing, these radionuclides may become concentrated in the tailings rather than in the recovered product. This introduces a radiological hazard that must be managed carefully to prevent exposure to workers and the environment through dust, groundwater seepage, or radon emissions.

Unreacted gangue minerals—such as silicates, phosphates, carbonates, and oxides—also make up a significant portion of tailings. These minerals are chemically stable but can affect the physical structure of the tailings deposit, influencing settling rates, permeability, and long-term stability. In some cases, these minerals can react with atmospheric oxygen or moisture, leading to acid generation or secondary mineral formation over time.

Tailings also contain fine particulates, which present dust and sedimentation hazards. These ultra-fine particles can become airborne under dry conditions or contribute to siltation in nearby water bodies if not properly contained. Fine particle management is therefore essential to controlling air quality, water quality, and erosion in tailings storage facilities.

The composition of tailings ultimately depends on the ore type and the chosen extraction route. Acid leach circuits typically produce tailings rich in soluble salts and residual acidity, requiring neutralisation and water treatment. In contrast, alkaline leaching or physical separation processes yield coarser, more inert tailings that pose fewer chemical hazards but still require careful physical containment. Understanding this variability is crucial for designing appropriate tailings storage, water treatment, and rehabilitation strategies in REE and critical mineral processing operations.

Tailings Storage Facilities (TSFs) are purpose-built, engineered structures designed to safely contain and manage the large volumes of waste generated during mineral processing. In rare earth element (REE) and critical mineral operations, TSFs play a crucial role in ensuring that finely ground residues, often containing chemical reagents, metals, and naturally occurring radioactive materials, are securely stored for decades or even centuries. Because these wastes can pose environmental and health risks if improperly handled, the design, construction, and monitoring of TSFs are governed by rigorous engineering standards and environmental regulations.

TSFs can take several forms, such as tailings dams, impoundments, or lined ponds, depending on the specific site conditions. The choice of design is influenced by local geology, topography, hydrology, and climatic factors, as well as the chemical and physical properties of the tailings. For example, facilities located in arid regions may favour evaporation ponds, whereas those in high-rainfall areas typically require more robust drainage and water management systems.

Figure 64: Bituminous geomembrane installation on a mine tailings storage facility. Pathfounder, CC BY-SA 4.0, via Wikimedia Commons.

A key element in modern TSF design is the use of liners, which may be composed of compacted clay, geomembrane materials (such as high-density polyethylene, HDPE), or a combination of both. Liners serve as an impermeable barrier, preventing contaminated leachate from seeping into surrounding soils or groundwater systems. Above the liner, decant systems are installed to collect and recover process water from the tailings slurry. This recovered water is often recycled back into the processing plant, reducing both water consumption and the volume of liquid requiring treatment or disposal.

To ensure environmental protection and regulatory compliance, monitoring wells are strategically placed around the perimeter of the TSF. These wells allow operators to detect any early signs of groundwater contamination or seepage, enabling corrective action before significant environmental damage occurs. Continuous monitoring is supported by automated sensors and data logging systems that provide real-time feedback on water quality, pressure, and flow conditions within the facility.

Rare Earth and Critical Mineral Operations and Processing

The embankments and containment walls of TSFs are engineered to withstand long-term stress, extreme weather conditions, and seismic activity. They are constructed using compacted earth, rock, or tailings themselves (in the case of upstream or downstream construction methods) and are carefully designed to maintain structural integrity even during heavy rainfall or flooding events. Periodic inspections, slope stability analyses, and maintenance programs are essential to ensure their continued safety and performance.

For radioactive rare earth tailings, which often contain thorium and uranium residues, multi-barrier containment systems are employed to provide additional layers of protection. These systems combine engineered barriers, such as liners and encapsulation, with natural geological barriers, such as impermeable bedrock, to achieve long-term stability and radiation isolation. In some cases, encapsulation techniques are used, where tailings are immobilised within chemically inert matrices (such as cementitious binders or clay) to minimise the potential for leaching. Alternatively, tailings may be backfilled into mined-out pits or underground voids, which reduces surface exposure and limits the risk of wind or water erosion.

Overall, the safe storage and containment of tailings are fundamental to sustainable REE and critical mineral operations. Effective TSF design not only mitigates environmental risks but also enhances water conservation, improves community trust, and supports the long-term viability of mining projects through responsible waste stewardship.

An essential component of modern tailings management is water recovery and effluent treatment, which aims to minimise water consumption, reduce waste discharge, and maintain environmental compliance. In mineral processing operations—especially those involving rare earth elements (REEs) and other critical minerals—large volumes of water are required for leaching, separation, and slurry transport. Efficient recovery of this water from tailings not only improves sustainability but also reduces operational costs by lowering the demand for fresh water intake from local sources.

In a well-designed tailings management system, water recovery begins within the Tailings Storage Facility (TSF) itself. As tailings settle in thickeners or impoundments, gravity causes the solid particles to sink while clarified water accumulates at the surface. This supernatant water is then collected through decant systems or reclaim pumps and returned to the processing plant for reuse. Recycling water in this way significantly reduces the overall water footprint of the operation and decreases the need for external water supplies—an especially critical advantage in arid or drought-prone regions where water scarcity can limit production capacity.

To ensure the recovered water is suitable for reuse or safe discharge, effluent treatment plants (ETPs) are installed as part of the tailings circuit. These facilities perform multiple treatment steps to remove contaminants and balance the chemical composition of the water. The first step typically involves neutralisation, where acidic or alkaline residues are adjusted to a near-neutral pH using reagents such as lime ($Ca(OH)_2$) or soda ash (Na_2CO_3). This prevents corrosion of pipelines and process equipment while also preparing the water for further treatment.

Following neutralisation, the effluent undergoes metal removal to eliminate dissolved contaminants such as iron, aluminium, or trace heavy metals that may have leached from the ore or reagents. This is achieved through chemical precipitation, where metals form insoluble hydroxides or carbonates that can be filtered out, or through ion exchange systems, which selectively absorb metal ions onto a resin bed for recovery or disposal. These steps are crucial for meeting environmental discharge standards and preventing downstream contamination of surface or groundwater systems.

The final phase of treatment focuses on solid–liquid separation, where remaining fine particles and suspended solids are removed using equipment such as thickeners, clarifiers, and pressure filters. Thickeners concentrate solids through gravity settling, while clarifiers polish the overflow water, and pressure filters mechanically dewater the remaining sludge. The treated water can then be recycled directly into the plant's process circuits or released into the environment under strict quality controls.

In rare earth element (REE) processing, radiation and chemical safety form a critical part of tailings management, particularly when ores contain thorium (Th), uranium (U), or their radioactive decay products. These naturally occurring radionuclides are often concentrated in the residual tailings after mineral extraction, making them a form of Naturally Occurring Radioactive Material (NORM) waste. As such, these materials must be handled under stringent safety and environmental controls to prevent radiological exposure to workers, surrounding communities, and the environment.

The first and most immediate safety priority in managing radioactive tailings is dust suppression. Because REE tailings are often finely milled, airborne dust can transport radioactive particles and chemical residues beyond containment areas. To control this hazard, tailings are routinely moisture-conditioned to keep the surface damp, reducing dust generation. In some cases, polymer sprays or binders are applied to form a thin crust over the tailings surface, or vegetation cover is established to stabilise the surface over time. These measures effectively reduce both particulate dispersion and radon gas release.

A key component of radiological protection is shielding and containment. Radioactive tailings are typically buried or capped with layers of inert materials such as clay, soil, or gravel. This not only provides a physical barrier that reduces radiation dose rates at the surface but also limits the infiltration of rainwater, which could otherwise leach radioactive isotopes into groundwater systems. In long-term containment designs, multi-layer capping systems may be used—combining low-permeability clay, geomembranes, and vegetated soil layers—to ensure durable radiation shielding and environmental stability over decades.

Continuous radiation monitoring is also essential for maintaining safety and compliance. Tailings facilities are equipped with sensors and sampling systems that track radon gas emissions, gamma dose rates, and radioactive particulates in air or water. These data are regularly reviewed to verify that radiation levels remain below regulatory thresholds and to detect any changes that might signal containment degradation. In some operations, remote sensing and automated logging systems are used to provide real-time data to plant operators and radiation safety officers.

Rare Earth and Critical Mineral Operations and Processing

Worker protection is guided by internationally recognised exposure standards established by the International Atomic Energy Agency (IAEA) and national authorities such as the Australian Radiation Protection and Nuclear Safety Agency (ARPANSA). These frameworks limit occupational exposure to an average of 20 millisieverts (mSv) per year over a five-year period, with lower limits for the general public. To comply with these standards, operators implement time, distance, and shielding principles—minimising time spent near radioactive materials, maximising distance from sources, and using protective barriers where necessary. Personal dosimeters, radiation training, and medical surveillance are also part of comprehensive worker protection programs.

Chemical safety is managed in parallel with radiological safety. REE tailings can contain residual acids, alkalis, or complexing agents from leaching processes, posing additional risks such as chemical burns, corrosion, or toxic exposure. Tailings management systems therefore include pH neutralisation, containment bunding, and personal protective equipment (PPE) requirements for workers handling slurry, reagents, or effluent.

Together, these measures—dust suppression, shielding, monitoring, and exposure control—form an integrated approach to radiation and chemical safety. By combining engineering controls with strict regulatory oversight and continuous monitoring, REE operations can effectively protect workers and the environment while maintaining compliance with international best practices for NORM waste management.

Waste Reprocessing and Rehabilitation represent the final stages in responsible rare earth element (REE) and critical mineral operations, aiming to reduce long-term environmental impact while maximising resource recovery. Modern REE projects increasingly recognise that tailings and process residues still contain valuable materials that can be economically and sustainably recovered using improved extraction technologies. Reprocessing these tailings allows operators to extract additional rare earths or by-products that were either uneconomical or technically challenging to recover during the initial processing phase. This approach not only reduces the overall waste volume but also enhances the efficiency of resource utilisation, contributing to circular economy principles within the mining sector.

Tailings reprocessing typically involves re-mining or hydraulic re-slurrying of existing tailings deposits, followed by re-treatment using advanced separation or leaching technologies. For example, historical tailings from monazite or bastnäsite operations may still contain significant concentrations of light rare earths (such as cerium, lanthanum, and neodymium) that can be recovered using modern hydrometallurgical techniques. In addition to REEs, other valuable metals, such as scandium, zirconium, or titanium, may also be extracted from previously discarded materials. Beyond the economic benefits, reprocessing tailings can improve environmental outcomes by stabilising older waste deposits, removing hazardous elements, and reducing the potential for acid or radioactive drainage.

Once the potential for reprocessing has been fully evaluated and the operation nears its end of life, rehabilitation becomes the primary focus. The rehabilitation of tailings storage facilities (TSFs) and processing areas is designed to restore the site to a safe, stable, and environmentally acceptable condition. This process typically begins with covering the tailings surface using clean soil, clay, or rock layers. These covers serve multiple purposes: they prevent erosion by wind or rain, minimise water infiltration that could lead to leaching of contaminants, and provide a suitable substrate for vegetation growth.

Following the application of cover materials, re-vegetation is carried out to stabilise the surface and encourage ecological recovery. Native plant species are often selected to re-establish local biodiversity, prevent erosion, and reduce dust emissions. Over time, vegetation helps to improve soil quality and creates a self-sustaining ecosystem that integrates with the surrounding environment. In some cases, progressive rehabilitation is implemented, where inactive areas of the TSF are capped and vegetated even while other parts remain in operation, to spread costs and enhance early environmental recovery.

A significant component of post-closure management is long-term monitoring. After rehabilitation, the site continues to be monitored for radiation levels, groundwater quality, and structural stability to ensure that containment systems remain effective and that no contaminants migrate beyond the controlled area. Monitoring wells, surface water sampling points, and remote sensing tools are used to detect any deviations from acceptable environmental standards. If issues are identified, corrective actions, such as additional capping, drainage improvements, or vegetation reinforcement, can be implemented.

Ultimately, effective waste reprocessing and rehabilitation practices ensure that REE and critical mineral projects achieve not only economic viability but also social licence to operate through responsible environmental stewardship. By recovering residual value from waste and restoring disturbed land to a safe and productive condition, these processes help align the industry with global sustainability goals and regulatory frameworks for long-term ecological protection.

Environmental and regulatory oversight plays a central role in ensuring that tailings and waste handling in rare earth element (REE) and critical mineral projects are conducted safely, transparently, and in compliance with both national and international standards. Given the potential risks associated with radioactive residues, chemical reagents, and long-term environmental impacts, regulators impose stringent requirements to protect ecosystems, workers, and surrounding communities. These frameworks not only dictate how waste must be stored and monitored but also establish clear expectations for accountability, rehabilitation, and post-closure stewardship.

At the international level, several key organisations provide the foundation for best-practice management of tailings and radioactive materials. The International Atomic Energy Agency (IAEA) sets the benchmark through its *Safety Standards Series*, which outlines requirements for the safe management of radioactive waste, including Naturally Occurring Radioactive Materials (NORM) commonly found in REE tailings. These standards guide countries in developing legislation that addresses radiation protection, waste storage, and long-term containment integrity. Similarly, the International Council on Mining and Metals (ICMM) provides industry-specific guidelines for tailings management, emphasising risk assessment, stakeholder engagement, and continuous improvement in operational practices.

One of the most significant developments in recent years is the Global Industry Standard on Tailings Management (GISTM), introduced in 2020 following several catastrophic tailings dam failures worldwide. Developed jointly by the ICMM, the United Nations Environment Programme (UNEP), and the Principles for Responsible Investment (PRI), the GISTM establishes a comprehensive framework for preventing tailings-related disasters. It sets out clear expectations for transparency, independent oversight, community engagement, and governance, requiring operators to design, construct, and manage tailings storage facilities with the utmost priority on human and environmental safety. Under the GISTM, companies must demonstrate that tailings facilities are resilient to extreme events—such as earthquakes and floods—and maintain effective monitoring and emergency response systems.

At the national level, governments enforce these international principles through regulatory frameworks that mandate rigorous planning, monitoring, and reporting processes. Before a project can commence, proponents are required to conduct a detailed Environmental Impact Assessment (EIA), which evaluates potential effects on land, water, air, and biodiversity. The EIA must also outline proposed mitigation strategies and demonstrate how the project will comply with applicable environmental and radiation protection laws. For REE projects that generate radioactive waste, a Radiation Management Plan (RMP) is compulsory, detailing how radiation exposure will be minimised for workers and the public, and how radioactive materials will be stored, monitored, and eventually decommissioned.

In addition to operational approvals, regulators also require Closure and Post-Closure Monitoring Programs. These programs ensure that after the mine or processing plant ceases operation, tailings storage facilities remain stable, groundwater remains uncontaminated, and radiation levels stay within safe limits. Long-term monitoring commitments—often extending decades beyond closure—reflect a growing recognition of the need for sustained environmental stewardship in the mining sector.

Collectively, these frameworks—spanning from international safety standards to site-specific monitoring programs—ensure that REE and critical mineral projects operate within strict environmental and safety boundaries. They promote transparency, accountability, and continuous improvement, reinforcing public trust and supporting the industry's alignment with global sustainability and responsible resource development goals.

Water Use, Recycling, and Emissions Control

Water use, recycling, and emissions control are essential components of sustainable mineral processing and environmental management in rare earth element (REE) and critical mineral operations. These activities ensure that plants minimise their ecological footprint, maintain regulatory compliance, and preserve valuable natural resources such as freshwater and clean air. Because REE and lithium processing circuits often involve aqueous leaching, solvent extraction, and chemical precipitation, efficient water and emissions management are vital for both operational performance and long-term environmental protection.

Water is a critical resource throughout the mineral processing chain, from ore grinding and slurry transport to leaching, washing, and product purification. In REE and lithium plants, process water serves multiple purposes: dissolving reagents, suspending particles in slurries, and cooling thermal systems. The amount of water required depends on ore type, process configuration, and the degree of recycling implemented. Without proper management, large volumes of water may be consumed, leading to competition with local communities and ecosystems, particularly in arid mining regions such as Western Australia or Inner Mongolia.

To address these challenges, modern facilities are designed with closed-loop water circuits that reduce the demand for fresh water. Process water from thickeners, filters, and tailings dams is collected, clarified, and recycled back into the plant for reuse in grinding, flotation, or leaching. This not only conserves water but also reduces the need to discharge effluent, limiting potential contamination of nearby rivers or groundwater systems. Advanced water management systems use flow meters, automated valves, and conductivity sensors to monitor usage in real time, optimising recycling efficiency and maintaining chemical balance in process streams.

Even with high levels of recycling, some portion of process water becomes contaminated with residual acids, alkalis, dissolved metals, or fine particulates. Before discharge, this effluent must be treated to neutralise chemical residues and remove suspended solids. Treatment processes typically include lime neutralisation to balance pH, precipitation to remove heavy metals, and filtration or clarification to separate fine solids. In REE operations where leaching involves sulfuric or hydrochloric acid, effluent treatment may also include ion exchange systems or membrane filters to capture valuable dissolved metals or radionuclides before disposal. Treated water is often reused for dust suppression, irrigation of rehabilitated areas, or released under strict environmental discharge permits.

Air emissions are another major consideration in REE and critical mineral processing, especially during roasting, calcination, or smelting stages. These processes can release gases such as CO_2, SO_2, HF, HCl, and dust particulates, depending on the feed material and reagents used. To control these emissions, plants are fitted with gas scrubbers, baghouses, and electrostatic precipitators (ESPs) that capture and neutralise harmful substances before they are vented to the atmosphere. For example, wet scrubbers use alkaline solutions to remove acid gases like HF and SO_2, producing a neutralised slurry that can be treated in the water circuit. Bag filters and cyclones trap fine dust particles from dryers, crushers, and transfer points, preventing airborne contamination and improving workplace air quality.

In addition to particulate and gas control, greenhouse gas (GHG) reduction strategies are increasingly integrated into plant design. These include switching to renewable energy sources, using high-efficiency motors and heat exchangers, and optimising combustion efficiency in furnaces and kilns. Hydrogen-based reduction and electrified roasting are emerging technologies that can dramatically reduce CO_2 emissions in pyrometallurgical processes.

Effective water and emissions management rely on continuous monitoring and data reporting. Modern plants use integrated Supervisory Control and Data Acquisition (SCADA) or Distributed Control Systems (DCS) to monitor flow rates, pH, temperature, dissolved oxygen, and emission levels in real time. These systems provide early warning of any deviation from environmental or regulatory limits and allow for automatic corrective actions.

National and international standards—such as those from the International Finance Corporation (IFC) Environmental, Health, and Safety Guidelines, ISO 14001 Environmental Management Systems, and local environmental protection agencies—define permissible discharge and emission levels. Compliance with these frameworks ensures that REE and lithium producers maintain their social licence to operate and contribute to the global transition toward more sustainable critical mineral supply chains.

Workplace Safety and Regulatory Compliance

Workplace safety and regulatory compliance are foundational elements in mineral processing and mining operations, particularly in sectors dealing with rare earth elements (REE) and critical minerals. These industries often involve complex mechanical processes that operate under high temperatures and utilize hazardous materials, presenting elevated risks that necessitate robust safety management systems. The application of stringent safety management frameworks is essential for preventing injuries and occupational diseases, thereby protecting not just workers but also surrounding communities and the environment [310].

At the international level, organizations like the International Labour Organization (ILO) and the International Organization for Standardization (ISO) establish frameworks for workplace safety. The ILO's Convention 176 on Safety and Health in Mines is a vital document outlining risk assessment, emergency preparedness, and essential training programs [310]. Similarly, the ISO 45001:2018 standard establishes a comprehensive framework for occupational health and safety management systems, focusing on hazard identification and risk management [311]. These international guidelines often inform national legislation, reinforcing compliance through necessary inspections and penalties for non-conformance [312].

In Australia, workplace safety within the mining sector is heavily regulated under the Work Health and Safety (WHS) Act 2011, along with specific regulations relevant to the mining and processing industries. These laws mandate the implementation of Safety Management Systems (SMS), essential for identifying hazards, assessing risks, and clarifying responsibilities for safety performance [313]. For example, Lynas Rare Earths has implemented radiation monitoring, emission controls, and comprehensive training programs in accordance with the guidelines provided by the Australian Radiation Protection and Nuclear Safety Agency (ARPANSA) [314].

China has significantly focused on enhancing workplace safety in the REE sector, instituting stricter regulatory measures following notable past accidents. The revised Safety Production Law (2021) emphasizes real-time monitoring and mandatory safety assessments, which have improved safety standards within REE operations [315]. Such measures have not only enhanced worker safety but also reduced the risk of environmental incidents, a notable shift from previous oversight conditions that garnered considerable criticism [315].

In the United States, regulatory oversight is primarily exercised by the Occupational Safety and Health Administration (OSHA) and the Mine Safety and Health Administration (MSHA). OSHA's regulations encompass various aspects of workplace safety in mineral processing, emphasizing chemical handling and confined spaces, while MSHA focuses on safety at mining sites [310]. Facilities like MP Materials' Mountain Pass site exemplify a comprehensive approach to safety, including mandatory training and dust suppression measures that adhere to established chemical safety standards [316].

The European Union integrates safety protocols into its environmental regulations, demonstrated by the Seveso III Directive. This directive mandates comprehensive safety management policies for facilities handling hazardous substances [317]. REE processing plants must maintain emergency response plans and conduct thorough hazard analyses to minimize risks associated with industrial accidents [318].

In emerging economies like Malawi and Tanzania, initiatives supported by international bodies are underway to develop national safety standards for mining operations. Engaging local communities and maintaining ongoing environmental monitoring form the cornerstone of responsible resource development, even amid often limited regulatory enforcement capacities [319].

Introduction to ESG and Community Engagement

In the modern mining and mineral processing industry—particularly within the rare earth elements (REE) and critical minerals sectors—the concepts of Environmental, Social, and Governance (ESG) responsibility and community engagement have become central to sustainable project development. ESG frameworks define how companies manage environmental performance, social responsibility, and governance transparency. Community engagement, as part of the "Social" pillar, ensures that local stakeholders are informed, consulted, and benefit from mining activities that affect their land, livelihoods, and environment. Together, ESG and community engagement practices not only protect corporate reputation and investor confidence but also form the foundation for ethical and long-term resource development.

Environmental (E) – Stewardship and Sustainability

The environmental pillar of ESG focuses on reducing the ecological impact of mining and processing operations. Producers of rare earth elements (REEs) and critical minerals bear a significant responsibility that extends beyond mere compliance with existing regulations; they are called to embrace proactive stewardship as a core operational philosophy. This includes comprehensive management of water resources, energy consumption, waste production, emissions output, and biodiversity conservation. Effective climate change mitigation strategies must also be integrated to ensure the sustainability of critical material extraction and processing practices [320, 321].

One prevalent approach within the mining sector to enhance environmental performance includes the implementation of advanced water recycling and treatment systems. These systems not only minimize freshwater withdrawals but also help prevent contamination of local water sources, which has been a historical concern for mining operations. For instance, both the Mountain Pass mine in the United States and the Lynas processing facility in Malaysia have faced environmental scrutiny due to past pollution incidents; thus, modern initiatives aim to mitigate these risks through innovative technology [320]. Additionally, efforts to boost energy efficiency and adopt renewable energy sources are becoming standard practice. Facilities are increasingly transitioning to rely on solar and wind energy, significantly lowering greenhouse gas emissions linked to fossil fuel use [322].

Tailings and waste management practices have also evolved, aligning with international standards such as the Global Industry Standard on Tailings Management (GISTM). These standards promote safe engineering practices to prevent structural failures of storage facilities that can lead to catastrophic environmental disasters. Comprehensive rehabilitation and closure plans are now mandated, including strategies to restore disturbed lands, prevent soil erosion, and encourage natural re-vegetation post-mining activities [323]. The goal of these extensive environmental initiatives is to ensure that mining operations can coexist with ecological integrity, thus fostering long-term sustainability.

Real-world exemplars highlight the effectiveness of these practices in operation. Lynas Rare Earths Ltd. has implemented rigorous environmental controls to manage radioactive residues from its processing, ensuring compliance with international health and safety standards [322]. Simultaneously, MP Materials has pioneered the use of closed-loop water systems alongside renewable energy integration in its operations, achieving a near-zero waste processing model [321, 324]. Such initiatives illustrate a growing trend where mining companies are not only adhering to regulatory requirements but are also actively contributing to global sustainability goals, thus reinforcing the environmental foundation of ESG within the critical minerals sector.

Social (S) – Community, Labour, and Human Rights

The social component of ESG (Environmental, Social, and Governance) is critically significant in the realm of critical minerals projects, particularly as these operations often impact Indigenous and local communities. The complex interactions between community relations, workforce welfare, diversity, and human rights necessitate a robust social license to operate (SLO) for the success of these projects. SLO is defined by continuous stakeholder engagement, where community engagement becomes both an ethical obligation and a business imperative [325, 326].

Effective community engagement must encompass the principle of Free, Prior, and Informed Consent (FPIC). This principle ensures that Indigenous communities are actively involved in decision-making processes well before any projects commence, thereby acknowledging their rights and sovereignty [327, 328]. Stakeholder consultations are essential throughout the project lifecycle to address concerns surrounding land access, environmental impacts, and the long-term benefits available to communities [329, 330]. Additionally, local employment initiatives, training programs, and enterprise development are crucial for creating sustainable economic opportunities that extend beyond the operational life of a mine [330].

Moreover, health and safety partnerships, including medical outreach and environmental health monitoring, reinforce the importance of community welfare in project areas. For example, empirical studies highlight that the means and content of community engagement can address local grievances while enhancing overall project acceptability [331]. Successful engagement is not merely a procedural formality but integral to sustaining the social fabric of the communities involved [332, 333].

Case studies illustrate commendable practices of social engagement. The Arafura Nolans Rare Earth Project in Australia exemplifies this by actively involving Indigenous employment and establishing agreements to protect cultural heritage [330]. Similarly, Boliden's operations in Sweden underline the importance of continuous dialogue with local municipalities concerning environmental standards and land utilization [330]. These instances reveal the effectiveness of genuine stakeholder collaboration, which nurtures relationships and reduces resistance.

Thus, the convergence of ethical and operational considerations in critical minerals projects manifests a broader understanding of the social dimension in ESG frameworks. Adopting approaches that elevate community participation and embed FPIC principles into corporate social responsibility (CSR) strategies not only mitigates risks but also fosters trust and support from local populations [334].

Governance (G) – Transparency, Ethics, and Compliance

The governance pillar within corporate frameworks plays a vital role in ensuring that companies conduct their operations with integrity, accountability, and transparency. This pillar encompasses various mechanisms and practices that not only promote ethical behaviour but also establish robust frameworks for decision-making, risk management, and compliance. As companies face increasing pressure from stakeholders, particularly in Europe and North America, to disclose their Environmental, Social, and Governance (ESG) performance, effective governance practices are becoming increasingly indispensable.

One critical aspect of governance is the implementation of corporate ethics policies and anti-corruption measures. Such measures are essential in fostering an environment of integrity and trust within corporations. Enhancing ESG performance requires companies to establish robust management systems, underscoring the significant role of ethical practices in corporate governance [335]. These practices help align corporate behaviour with stakeholder expectations, which is increasingly demanded by investors looking to mitigate risks related to poor governance.

Additionally, transparent reporting of ESG performance is facilitated through various established standards, such as the Global Reporting Initiative (GRI), the Task Force on Climate-related Financial Disclosures (TCFD), and the Sustainability Accounting Standards Board (SASB). These frameworks provide guidelines for companies to benchmark their ESG performance and communicate this information clearly to stakeholders. The OECD has been at the forefront of providing guidance on responsible business conduct, underscoring best practices for sustainable development [336].

Board oversight and diversity have also emerged as crucial governance elements. Diverse boards are generally more effective in decision-making, as they provide a range of perspectives that can enhance accountability [337]. The increasing awareness of diversity in governance structures aligns with global trends aimed at improving corporate decision-making processes. The proposal of the EU's Corporate Sustainability Due Diligence Directive is a notable development in this area, mandating companies to identify and mitigate negative impacts on human rights and the environment, emphasizing the importance of board diversity and competent oversight [338].

Supply chain due diligence, specifically concerning critical minerals used in clean energy technologies, has taken centre stage in governance discussions. Compliance with supply chain regulations, such as those outlined in the EU's Conflict Minerals Regulation, is essential for ensuring that materials are ethically sourced and free from human rights abuses, including child labour [339]. The Responsible Minerals Initiative (RMI) plays an instrumental role in promoting compliance with these standards, thereby pushing companies to enhance their supply chain governance [340].

Investors, particularly in Europe and North America, are increasingly stringent regarding ESG performance. Legislative initiatives like the European Union's Critical Raw Materials Act (2023) and the U.S. Inflation Reduction Act (2022) represent a growing trend to prioritize funding and trade agreements with companies exhibiting strong ESG practices. These policies not only incentivize companies to enhance their governance structures but also reflect a broader expectation from the investment community that ethical considerations are integral to corporate success.

Community Engagement and the Social Licence to Operate

Community engagement in the mining and mineral processing sector goes far beyond meeting regulatory requirements—it is about fostering trust, transparency, and long-term partnership between companies and the communities that host their operations. Effective engagement acknowledges that mining activities can significantly affect local people, land, and livelihoods, and that maintaining positive relationships is essential for sustainable project development. The concept of a Social Licence to Operate (SLO) captures this principle: it is not an official permit granted by governments, but rather earned continuously through open dialogue, mutual respect, and the fair distribution of benefits. Without community trust and participation, even technically successful projects can face opposition, delays, or closure.

Meaningful engagement involves proactive and inclusive strategies throughout the project lifecycle. Community baseline studies are conducted before mining begins to assess existing socioeconomic and environmental conditions, helping to identify potential risks and opportunities. Participatory monitoring programs empower local stakeholders by involving them directly in the tracking of environmental indicators such as water quality, air emissions, and land rehabilitation progress, ensuring transparency and accountability.

Benefit-sharing agreements form another cornerstone of engagement, creating tangible value for local populations. These agreements can include royalty payments, community development funds, or joint infrastructure projects—for example, the construction of schools, healthcare clinics, or roads that benefit both the operation and the broader community. Finally, robust conflict resolution mechanisms are essential to address grievances promptly and fairly, preventing small issues from escalating into larger disputes.

Through continuous and respectful engagement, mining companies can build enduring relationships with communities, securing not only operational stability but also shared prosperity and sustainable development outcomes. A notable example is Greenland's Kvanefjeld Rare Earth Element (REE) project, where developers have prioritized ongoing consultations with the community to mitigate apprehensions surrounding radiation and environmental safety. This endeavour emphasizes the significance of transparency and open communication in fostering public trust and stakeholder participation, which aligns with environmental justice principles. Studies indicate that effective communication can help alleviate fears related to radiation exposure, as local community sentiments are largely influenced by the perceived risks associated with environmental changes brought about by mining activities [341].

In Greenland, the Kvanefjeld project has incorporated stakeholder perspectives into its planning and operational processes. According to research on the governance of uranium mining in Greenland, the principles of openness and public participation have been fundamental in addressing community concerns around environmental impacts [341]. The project illustrates how the acknowledgment of local voices contributes to a more balanced approach to resource development. Transparency regarding health and environmental assessments is essential not only for compliance but also for cultivating long-term relationships with affected communities.

Similarly, the Thacker Pass Project developed by Lithium Americas in the United States showcases a proactive approach to Indigenous engagement. The project has involved early and transparent discussions with Indigenous groups. Transparent environmental reporting practices allow communities to understand the implications of mining activities on their lands and resources, fostering a shared understanding of both risks and benefits [342]. This notion is supported by governance frameworks which assert that early consultation and involvement of local stakeholders are instrumental in enhancing the social acceptance of resource extraction projects.

The Strategic Importance of ESG in Critical Minerals

As the world transitions to renewable energy, electric vehicles, and low-carbon technologies, REEs, lithium, nickel, and cobalt have become essential to global decarbonisation. However, the industries producing these minerals face growing scrutiny over environmental and social impacts. ESG frameworks provide the roadmap to align mineral development with global sustainability goals.

Investors, regulators, and end-users, such as manufacturers of electric vehicles and wind turbines, now demand verifiable ESG performance as part of supply chain due diligence. Companies that demonstrate excellence in ESG not only gain competitive advantage but also attract sustainable finance, access global markets, and strengthen their long-term social licence.

ESG and community engagement represent the modern standard of accountability and sustainability in mineral processing and extraction. By integrating environmental stewardship, social responsibility, and sound governance into every stage of project planning and operation, companies can achieve both economic success and ethical legitimacy. Meaningful community engagement ensures that the benefits of critical mineral development are shared equitably, while environmental and governance commitments safeguard ecosystems and future generations.

As the demand for clean energy materials accelerates, strong ESG performance is no longer optional—it is the foundation of responsible growth in the global rare earth and critical minerals industry.

Key Terms and Concepts	
Environmental Management System (EMS): A structured framework that helps organisations monitor, manage, and improve their environmental performance in accordance with standards such as ISO 14001.	

Key Terms and Concepts

Tailings: The residual materials left after valuable minerals have been extracted from ore, typically stored in engineered containment facilities.

Tailings Dam: A specially designed embankment or impoundment structure used to store tailings and process water safely, preventing environmental contamination.

Waste Rock: Non-ore material removed during mining that must be managed to minimise environmental impacts such as acid mine drainage.

Acid Mine Drainage (AMD): The acidic water formed when sulphide minerals in waste rock or tailings are exposed to air and water, leading to metal leaching.

Effluent Treatment: The process of treating wastewater from mineral processing operations to remove contaminants before discharge or reuse.

Air Emission Control: The use of technologies such as scrubbers, filters, and baghouses to capture particulates and gases released from plant operations.

Radiological Hazard: Exposure risk associated with radioactive elements such as thorium or uranium commonly found in rare earth ores.

Occupational Health and Safety (OHS): The discipline focused on protecting worker health and safety through hazard identification, risk assessment, and control measures.

Risk Assessment: A systematic process of identifying hazards, analysing their likelihood and consequences, and determining appropriate mitigation strategies.

Hazard Identification: The recognition of potential sources of harm—physical, chemical, biological, or ergonomic—within a workplace or process.

Personal Protective Equipment (PPE): Clothing or gear such as gloves, helmets, respirators, and goggles used to protect workers from exposure to workplace hazards.

Environmental, Social, and Governance (ESG): A framework used by organisations to evaluate their sustainability performance across environmental, social, and ethical dimensions.

Key Terms and Concepts	

Community Engagement: The process of consulting, informing, and involving local communities in decision-making to build trust and maintain a social licence to operate.

Rehabilitation: The restoration of disturbed mining or processing areas to a safe and stable condition, often involving revegetation and environmental monitoring.

Sustainability Reporting: The public disclosure of an organisation's environmental, social, and governance performance in line with frameworks such as the Global Reporting Initiative (GRI).

Chapter 9 Review Questions

Environmental and Safety Challenges

1. Short Answer: What are the key environmental and safety challenges associated with rare earth and critical mineral extraction and processing?

2. Multiple Choice: Which of the following minerals often contain naturally occurring radioactive materials (NORM)?

(a) Quartz

(b) Monazite

(c) Bauxite

(d) Galena

3. True/False: The main sources of radiation exposure in REE processing are alpha particles emitted from uranium decay.

4. Extended Response: Explain how poor waste and tailings management can affect both worker safety and the surrounding environment.

Radiation and Chemical Hazards

5. Short Answer: Define "NORM" and explain why it is significant in rare earth processing.

6. Multiple Choice: Which radiation type poses the greatest risk if inhaled or ingested?

(a) Alpha

(b) Beta

(c) Gamma

(d) Neutron

7. Short Answer: What is the principle of ALARA, and how does it guide radiation protection in REE facilities?

8. Extended Response: Describe how radiation hazards are monitored and controlled in REE processing plants, referencing specific technologies or protocols.

Environmental Management Systems (EMS)

9. Multiple Choice: ISO 14001 is a standard for:

(a) Quality Management

(b) Radiation Safety

(c) Environmental Management Systems

(d) Occupational Health and Safety

10. Short Answer: Identify three key components of an effective EMS in mineral processing.

11. Scenario Question: A processing plant wants to reduce chemical discharge and improve water recycling. Which EMS strategies could be applied to achieve this goal?

Tailings and Waste Management

12. Short Answer: What are tailings, and why are they considered an environmental concern in REE processing?

13. Multiple Choice: The main function of a geomembrane liner in a Tailings Storage Facility (TSF) is to:

(a) Improve drainage

(b) Prevent groundwater contamination

(c) Increase storage volume

(d) Reduce evaporation

14. Extended Response: Explain the main steps involved in tailings rehabilitation after mine closure.

Water Use and Emissions Control

15. Short Answer: Why is water recycling important in REE and lithium processing operations?

Rare Earth and Critical Mineral Operations and Processing

16. Multiple Choice: Which technology is most effective for removing HF and SO_2 gases from process emissions?

(a) Electrostatic precipitator

(b) Wet scrubber

(c) Baghouse filter

(d) Cyclone separator

17. True/False: All effluent from mineral processing plants must be discharged to nearby rivers after neutralisation.

18. Extended Response: Discuss how water recovery and emissions control systems contribute to sustainability and regulatory compliance.

Occupational Health and Safety (OHS)

19. Short Answer: Name three key components of an Occupational Health and Safety Management System (OHSMS).

20. Multiple Choice: Which international standard governs occupational health and safety management systems?

(a) ISO 9001

(b) ISO 14001

(c) ISO 45001

(d) ISO 31000

21. Scenario Question: An REE plant worker reports dizziness after exposure to fumes during acid leaching. Describe the immediate safety response steps that should be taken.

Regulatory and Legislative Frameworks

22. Short Answer: Which Australian agency regulates radiation exposure from REE processing?

23. Multiple Choice: The Global Industry Standard on Tailings Management (GISTM) was developed in response to:

(a) Economic downturns in the mining industry

(b) Tailings dam failures worldwide

(c) Community protests

(d) Climate change agreements

24. Extended Response: Compare how radiation safety is regulated in Australia and the United States.

ESG and Community Engagement

25. Short Answer: Define the three pillars of ESG and provide one example of each within the mining context.

26. Multiple Choice: What does "Social Licence to Operate" mean?

a) A government-issued environmental permit

b) Formal community ownership

c) Ongoing acceptance and approval by stakeholders

d) A mining lease agreement

27. Short Answer: What is Free, Prior, and Informed Consent (FPIC), and why is it important?

28. Extended Response: Explain how effective ESG implementation contributes to sustainable rare earth and critical mineral operations.

Chapter 10

Circular Economy, Recycling, and Secondary Sources

The accelerating global transition toward renewable energy, digital infrastructure, and low-carbon technologies has intensified demand for rare earth elements (REEs) and other critical minerals. However, reliance on conventional extractive industries poses significant environmental, economic, and geopolitical challenges. The emerging paradigm of the circular economy offers a transformative framework for addressing these issues by closing material loops, extending product life cycles, and decoupling resource use from environmental degradation. Within this framework, waste is reconceptualised as a resource, and secondary sources—such as end-of-life products, industrial residues, and mine tailings—become essential inputs to mineral supply chains.

This chapter examines the scientific, technological, and policy dimensions of circular mineral systems, with particular emphasis on recycling, reprocessing, and recovery of critical materials from secondary sources. It explores contemporary developments in urban mining, battery and magnet recycling, and tailings revalorisation, as well as the integration of life-cycle assessment (LCA) and sustainability metrics into process design and policy planning. Furthermore, the chapter considers the role of digital traceability, regulatory innovation, and cross-sectoral collaboration in enabling circular supply networks. Collectively, these approaches illustrate how the circular economy redefines mineral production from an extractive activity into a regenerative and knowledge-driven system—supporting long-term resource security, environmental protection, and social responsibility in the global transition to sustainable development.

Learning Outcomes	
This chapter aims to give you the ability to:	

<table>
<tr><td colspan="2">Learning Outcomes</td></tr>
</table>

1. Explain the concept of the circular economy and its relevance to rare earth and critical mineral supply chains.
2. Describe the environmental and economic benefits of recycling and resource recovery from end-of-life products, including electronic waste, magnets, batteries, and catalysts.
3. Identify the main secondary sources of critical minerals and evaluate the technological and logistical challenges associated with their recovery.
4. Discuss the recycling processes applicable to rare earth and battery materials, including mechanical, hydrometallurgical, and pyrometallurgical approaches.
5. Compare the efficiency, cost, and environmental impact of primary production versus secondary recovery routes for key critical minerals.
6. Interpret basic flowsheets illustrating integrated recycling and reprocessing systems for REEs, lithium, and cobalt-bearing products.
7. Explain the role of life-cycle assessment (LCA) in measuring environmental performance and guiding sustainable decision-making in mineral production and recycling.
8. Recognise the policy drivers and international initiatives that promote circular economy practices, such as extended producer responsibility (EPR) and critical material stewardship programs.
9. Analyse case studies demonstrating successful circular economy models in the rare earth and energy storage industries.
10. Evaluate the potential of recycling and secondary sourcing to enhance supply chain resilience, reduce waste, and support net-zero and sustainability targets.

Recovery of Critical Minerals from E-Waste and Batteries

The recovery of critical minerals from electronic waste (e-waste) and spent batteries has become a major focus in modern resource management, offering a sustainable alternative to primary mining. As global demand for rare earth elements (REEs), lithium, cobalt, nickel, and other strategic materials increases, particularly for clean-energy technologies, urban mining of discarded electronics and batteries provides both economic and environmental benefits. This process reduces the pressure on natural resources, mitigates landfill waste, and decreases the carbon footprint associated with mining and ore processing.

E-waste, a rapidly growing global concern, includes a diverse array of discarded electrical and electronic devices such as smartphones, laptops, televisions, and components utilized in renewable energy technologies like wind turbines and electric vehicles. The comprehensive management and recovery of e-waste are complicated by its heterogeneity, which encompasses valuable metals and minerals alongside environmentally hazardous materials.

In terms of valuable minerals, e-waste serves as a significant reservoir for several critical categories of materials. Rare Earth Elements (REEs), for example, are prevalent in components such as permanent magnets, hard drives, and displays. Specifically, neodymium, dysprosium, europium, and terbium are critical for these applications and can be recovered from electronic waste [343]. Furthermore, precious metals including gold, silver, palladium, and platinum are found in abundant quantities in printed circuit boards (PCBs) of electronic devices. Studies suggest that the concentration of these precious metals is notably higher in e-waste compared to their occurrence in natural ores, with significant quantities recoverable from waste mobile phones [344-347].

Base and critical metals such as copper, nickel, tin, cobalt, and lithium are essential components of wiring, capacitors, and batteries within electronic devices. With the rise of electric vehicles and the increasing demand for energy storage solutions, the recovery of battery-grade materials like lithium, cobalt, nickel, and manganese from e-waste, particularly from lithium-ion and nickel-metal hydride batteries, is becoming more urgent [348-350]. These metals not only have monetary value but also play a crucial role in the production of new electronic and energy storage technologies.

Despite the potential benefits of recycling valuable metals, the presence of hazardous materials presents significant challenges. E-waste is laden with plastics, glass, ceramics, and toxic components including lead, mercury, and brominated flame retardants. This complexity complicates the recovery processes and poses risks to environmental and human health [344, 351, 352]. For instance, while e-waste consists of approximately 40% metals, 30% plastics, and 30% refractory oxides, toxic elements necessitate careful handling and advanced recycling technologies [353]. Furthermore, the economic viability of e-waste recycling hinges on overcoming these environmental challenges, which require innovative and sustainable recovery methods [354, 355].

Collection and Pre-Processing

The first and most important stage in recovering critical minerals from electronic waste (e-waste) and spent batteries is collection and pre-processing. This phase determines the overall efficiency and safety of downstream recovery processes by ensuring that the feed material is both representative and safely prepared for mechanical, chemical, or thermal treatment. Because e-waste is highly heterogeneous—containing plastics, metals, ceramics, and potentially hazardous materials—careful segregation and conditioning are essential to protect workers, equipment, and the environment.

The process typically begins with manual dismantling, where workers or robotic systems disassemble devices to remove valuable or hazardous components. Items such as printed circuit boards (PCBs), electric motors, hard disk drives, magnets, and batteries are separated manually due to their high metal content and recyclability. Manual dismantling allows for targeted recovery of high-value parts while minimizing contamination from mixed materials. In some facilities, automated robotic systems with visual recognition are increasingly used to identify and extract components containing critical minerals such as neodymium or cobalt.

Following dismantling, shredding and crushing operations are employed to reduce the size of remaining materials. Mechanical shredders break down casings, wiring, and circuit boards into smaller fragments, liberating metal foils and powders from polymers and ceramics. This liberation is essential for efficient downstream separation, as tightly bound materials can hinder recovery. The crushed material is often screened to classify particle sizes before proceeding to separation steps.

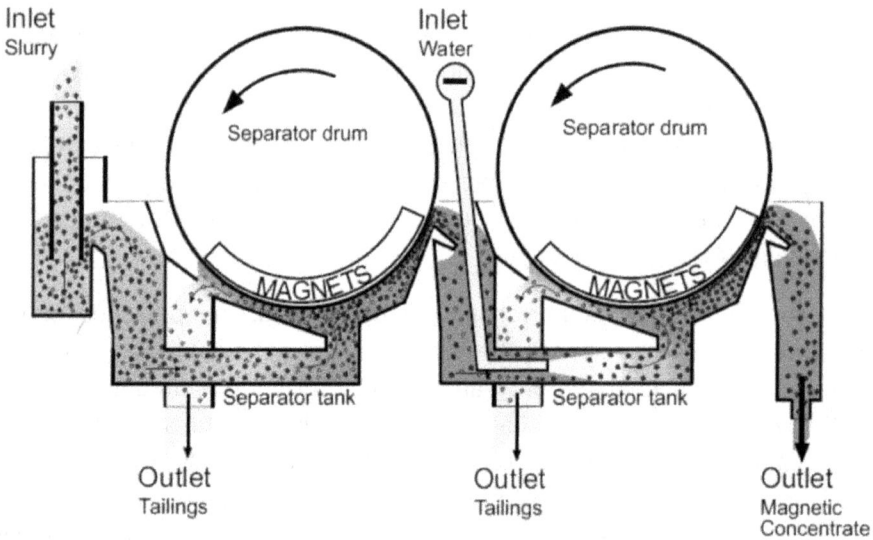

Figure 65: Magnetic Separation. Joellegavazziapril, CC BY-SA 4.0, via Wikimedia Commons.

Magnetic and eddy-current separation are then used to recover metallic fractions. Magnetic separators extract ferrous metals such as iron and steel using permanent magnets or electromagnets, while eddy-current separators induce repulsive currents in conductive non-ferrous metals such as copper, aluminium, and brass, propelling them away from non-conductive waste. These processes significantly enrich the metal content of the recovered stream and are standard in most industrial recycling lines.

Figure 66: Magnetic separator. Nuno Madeira Alves, CC BY-SA 4.0 , via Wikimedia Commons.

Non-metallic materials—such as plastics, ceramics, and glass—are often separated using density and electrostatic separation techniques. Density-based separation, including air classification or water-based flotation, divides materials based on their specific gravity, allowing lighter plastics to float while heavier glass or metal residues sink. Electrostatic separators further refine the output by using high-voltage fields to attract or repel charged particles, effectively isolating conductive from non-conductive materials. These recovered plastics and glasses can then be reprocessed for use in manufacturing or safely disposed of if contaminated.

When processing batteries, additional precautions are necessary due to the presence of reactive chemicals and residual electrical charge. Batteries must first be discharged to eliminate stored energy and prevent short circuits. Chemical neutralization or immersion in brine solutions may be used to render reactive electrolytes safe. Some advanced recycling systems employ cryogenic crushing, where batteries are cooled to extremely low temperatures (using liquid nitrogen) before mechanical processing. This method stabilizes volatile materials and prevents fires or explosions by suppressing exothermic reactions during shredding.

Pyrometallurgical Recovery

Pyrometallurgical routes are essential in the recovery of metals from electronic waste (e-waste), employing high-temperature processes to efficiently separate and concentrate metals. Smelting, a primary method employed in pyrometallurgy, involves melting e-waste in electric or plasma furnaces at temperatures ranging from 1,200 °C to 1,500 °C. This process facilitates the formation of a metallic phase containing copper, nickel, and cobalt, while lighter oxides are removed as slag [352, 356]. The effectiveness of smelting stems from its ability to handle complex and mixed feeds often found within e-waste, making it a robust option for metal recovery [357].

Another crucial pyrometallurgical operation is roasting or thermal decomposition, which primarily targets the removal of plastics and organic materials from e-waste. This operation yields metal-rich residues, which can subsequently be refined further [357, 358]. The coupled operations of smelting and roasting highlight the versatility of pyrometallurgy, not only for metal recovery but also for the efficient transformation of diverse e-waste compositions into recoverable materials [357].

Despite its advantages, the pyrometallurgical route is not without limitations. The process is characterized by high energy consumption and significant CO_2 emissions, raising environmental concerns [359, 360]. Moreover, the temperature conditions required can lead to the loss of volatile elements such as lithium and indium during processing, which are crucial for future technological applications [360]. Therefore, while pyrometallurgical processes demonstrate the potential to recover numerous metals from electronic scrap, the need for improved environmental efficiencies and the recovery of volatile elements presents ongoing challenges [359, 360].

Hydrometallurgical Recovery

Hydrometallurgical recovery is a crucial process for recycling valuable metals from spent lithium-ion batteries (LIBs), involving a series of steps including leaching, selective extraction, and precipitation or electrowinning. This method offers distinct advantages such as lower energy demands and high selectivity in metal recovery, essential attributes for promoting a sustainable circular economy [361-363].

Leaching is often initiated using acids like sulfuric acid (H_2SO_4) or organic acids, which dissolve target metals into the solution. Research suggests that acids are effective leaching agents for recovering lithium (Li), cobalt (Co), and nickel (Ni) from spent batteries, with evolving techniques exploring the use of organic acids for more environmentally friendly outcomes [361-363]. For instance, studies indicate that using gluconic acid can result in recovery efficiencies exceeding 98% for Li, Co, and manganese (Mn) [364]. Additionally, alternatives like citric acid have been analysed, presenting sustainable leaching options while still achieving satisfactory metal recovery rates [365, 366].

Following leaching, selective extraction is employed to separate the dissolved metals. Common methods include solvent extraction (SX) and ion exchange (IX), which distinguish metals based on their chemical properties [367, 368]. For example, solvent extraction techniques utilizing reagents such as Cyanex 272 have been documented for effective cobalt recovery from battery waste [369]. These techniques bolster the recovery process by enabling the separation of metals from impurities, enhancing the purity of recovered products [361, 363, 367].

Precipitation and electrowinning are the final steps in the hydrometallurgical recovery process. Precipitation involves converting dissolved metals back into solid form, often through the addition of precipitating agents, while electrowinning uses electrical current to extract metals from their solutions [348, 370]. This step is crucial for obtaining high-purity metal salts, such as lithium carbonate (Li_2CO_3) and cobalt hydroxide ($Co(OH)_2$), which can subsequently be utilized in new battery production [370, 371]. Despite these benefits, it is essential to manage the liquid effluents generated during the recovery process, particularly through neutralization and recycling efforts, which are necessary to minimize environmental impacts [348, 372].

Hydrometallurgical Recovery

Hydrometallurgy is a low-temperature, solution-based method used to recover valuable metals from solid materials such as ores, concentrates, industrial residues, and electronic or battery waste. It is often applied after mechanical or thermal pre-treatment—such as crushing, roasting, or calcination—to expose metal surfaces and make them more reactive in solution. Compared to pyrometallurgical techniques, hydrometallurgy offers higher selectivity, lower energy consumption, and the ability to process low-grade or complex materials. However, it requires careful management of chemical reagents and liquid effluents to ensure environmental compliance.

The process generally involves three key stages: leaching, selective extraction, and metal recovery. In the leaching stage, metal values are dissolved from the solid matrix into an aqueous solution. Depending on the mineral type and metal of interest, leaching agents may include acids such as sulfuric acid (H_2SO_4) or hydrochloric acid (HCl), bases like sodium hydroxide ($NaOH$), or organic acids such as citric or oxalic acid. For example, lithium, cobalt, and nickel from spent lithium-ion batteries are commonly leached using a combination of sulfuric acid and hydrogen peroxide (H_2O_2). The peroxide acts as a reducing agent, converting metal oxides into soluble sulphates. This step produces a pregnant leach solution (PLS)—a metal-rich liquid that serves as the feed for downstream purification.

Once metals are dissolved, the selective extraction stage isolates individual elements based on their chemical affinities. Two widely used methods are solvent extraction (SX) and ion exchange (IX). In solvent extraction, the metal-bearing solution is contacted with an organic extractant—typically dissolved in a water-immiscible solvent such as kerosene—which selectively binds target metal ions (e.g., Co^{2+}, Ni^{2+}, or REE^{3+}). The metals are then transferred into the organic phase and stripped later into a purified aqueous phase. Ion exchange, by contrast, uses solid resin beads that selectively adsorb metal ions from solution through electrostatic interactions, releasing them upon exposure to a regenerating solution. Both techniques enable precise control over purity and separation, particularly useful for complex mixtures like rare earth elements (REEs).

The final step, metal recovery, converts dissolved metals into solid, saleable forms. This is achieved through precipitation, crystallization, or electrowinning. Precipitation involves adding reagents—such as sodium carbonate or hydroxide—to form insoluble compounds like lithium carbonate (Li_2CO_3), cobalt hydroxide [$Co(OH)_2$], or nickel sulfate ($NiSO_4$). Alternatively, in electrowinning, an electric current is passed through the solution, depositing pure metal onto cathodes while oxygen or hydrogen evolves at the anode. These methods produce high-purity outputs suitable for reuse in batteries, alloys, and advanced technologies.

For example, in the treatment of spent lithium-ion batteries, hydrometallurgical recovery begins with acid leaching using H_2SO_4 and H_2O_2 to dissolve lithium, cobalt, and nickel. The resulting solution undergoes solvent extraction to separate cobalt and nickel, while lithium is later recovered by precipitation as Li_2CO_3. This integrated process achieves high recovery efficiencies (>90%) with far less energy than smelting.

Despite its advantages, hydrometallurgy generates liquid effluents containing residual acids, metals, and salts, which must be neutralized and treated before discharge or reuse. Techniques such as lime neutralization, ion exchange polishing, and water recycling systems are employed to minimize waste. Overall, hydrometallurgy plays a critical role in the sustainable recovery of critical minerals, combining chemical precision with environmental responsibility to support circular economy models in modern resource industries.

Biohydrometallurgy (Emerging Technology)

Biohydrometallurgy is an emerging technology that utilizes microorganisms, such as bacteria and fungi, to solubilize metals, facilitating their recovery from various waste materials. This innovative method demonstrates environmental advantages due to its lower energy requirements compared to traditional extraction techniques and its integration into sustainable waste management systems.

Microbial Mechanisms: Acidithiobacillus ferrooxidans, a prominent acidophilic bacterium, plays a crucial role in bioleaching by oxidizing metal sulfides, thereby releasing metals into solution. The ability of A. ferrooxidans to induce microbial corrosion in metals highlights its significance in biohydrometallurgy, as it enhances the extraction of valuable metals like copper and cobalt from sulphide ores and e-waste [373-375]. Research indicates that A. ferrooxidans not only aids in the solubilization of metals but also has potential for urban mining, where metals from electronic waste could be effectively recovered using this bacterium [373, 375]. Furthermore, advancements in biotechnology utilizing these microorganisms can significantly impact the efficiency of extracting cobalt and nickel from low-grade resources [375, 376].

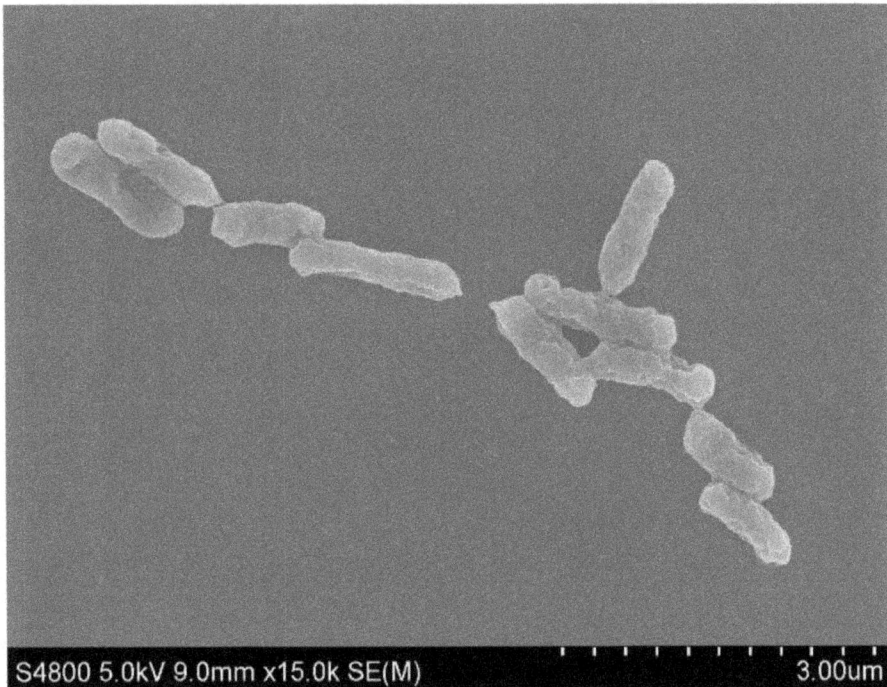

Figure 67: Microscope image of Acidithiobacillus ferrooxidans. Yu Zhang, Shuang Zhang, Dan Zhao, Yongqing Ni, Weidong Wang and Lei Yan, CC BY-SA 4.0, via Wikimedia Commons.

Fungal Contributions: Fungal species, particularly Aspergillus niger, also exhibit promising capabilities in bioleaching. This fungus secretes organic acids that dissolve metals from electronic waste, such as printed circuit boards (PCBs) and other mixed residues, making it a viable complement to bacterial methods [377, 378]. Research has shown that A. niger can recover substantial amounts of copper and nickel from waste electronic materials, with leaching rates of up to 41% for copper and 80% for nickel, highlighting its efficiency in bioleaching applications [377, 379]. The use of fungi alongside bacteria can exploit different biochemical pathways, potentially enhancing metal recovery rates from complex waste matrices.

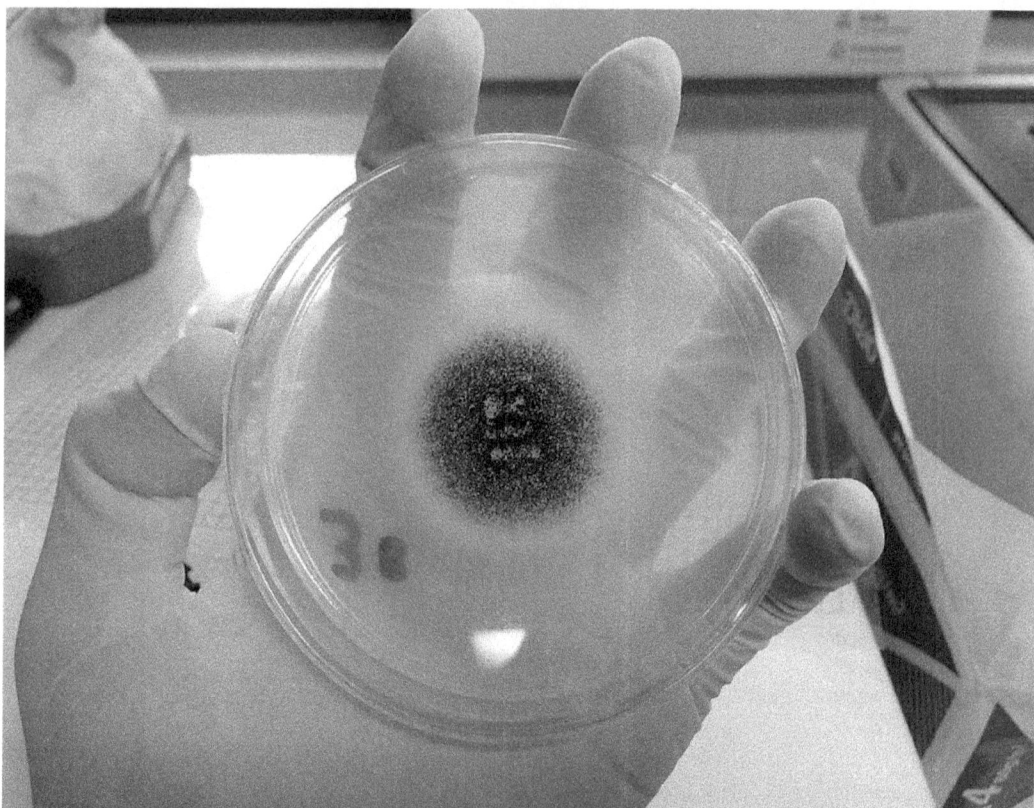

Figure 68: An Aspergillus niger colony growing at 25 degrees Celsius on Potato dextrose agar 3 days after inoculation. Saxon Vinkovic, CC BY-SA 4.0, via Wikimedia Commons.

Environmental and Economic Viability: From an environmental perspective, biohydrometallurgy is considered a greener alternative to traditional hydrometallurgical methods, which often involve harmful chemicals and high-energy processes [374, 380]. Bioleaching can operate effectively at ambient temperatures and pressures, significantly reducing operational costs and environmental footprints. The integration of bioleaching into practices targeting low-grade ores and waste materials aligns well with sustainable development, addressing the growing need for resource recovery while minimizing ecological impact [375, 376, 381].

Challenges and Future Prospects: However, the scalability of bioleaching technologies remains a challenge due to slow reaction kinetics and complexities involved in optimizing microbial performance in large-scale applications [375, 382]. Despite these hurdles, ongoing research focuses on enhancing microbial efficiencies and understanding the interaction between various microbial consortia, which could lead to improved recovery rates and processing times [381, 383]. Continuous advancements in genetic and metabolic engineering of bacteria and fungi may pave the way for more robust biotechnological solutions that fulfill the demands of metal recovery from diverse and complex waste streams [376, 381].

Direct Recycling and Closed-Loop Systems

Direct recycling represents an innovative approach to critical mineral recovery that focuses on preserving the functional integrity of materials rather than breaking them down into base metals. This method is particularly important in the recycling of lithium-ion batteries, where cathode materials such as lithium cobalt oxide ($LiCoO_2$) or nickel–manganese–cobalt oxide (NMC) retain much of their electrochemical structure even after several charge–discharge cycles. Instead of dissolving these materials through hydrometallurgical or pyrometallurgical processes, direct recycling restores and reuses them in their original form, significantly reducing waste and energy consumption.

The key goal of direct recycling is to recover and rejuvenate cathode powders—the most valuable component of a spent battery—without complete chemical decomposition. Techniques such as cathode relithiation and solid-state rejuvenation are employed to replenish lithium lost during battery use. In cathode relithiation, the depleted material is treated with lithium sources (such as lithium carbonate or lithium hydroxide) under controlled temperature conditions to restore its original stoichiometry and structure. Solid-state rejuvenation further enhances this by annealing the material, allowing crystal lattice repair and improving electrochemical performance. These processes can restore up to 90–95% of the original capacity, making the material suitable for new battery production.

One of the major advantages of direct recycling is its efficiency and environmental performance. Since the process avoids complete dissolution or smelting, it eliminates the need for large volumes of chemical reagents and minimizes liquid effluent generation. The energy requirements are also significantly lower than those of conventional hydrometallurgical or pyrometallurgical routes, as it skips high-temperature reduction or extensive purification stages. This makes direct recycling a cleaner, faster, and more cost-effective solution that aligns with circular economy principles.

Closed-loop recycling systems extend this concept further by creating a continuous cycle where materials recovered from end-of-life batteries are directly reintroduced into new battery manufacturing. This "battery-to-battery" model forms the foundation of modern sustainable supply chains for electric vehicles and renewable energy storage systems. Companies such as Tesla, CATL (Contemporary Amperex Technology Co. Limited), and Redwood Materials are leading examples of this approach. Tesla's Nevada Gigafactory integrates recycling within its production ecosystem, while Redwood Materials—founded by a former Tesla executive—focuses on recovering lithium, cobalt, nickel, and copper for reuse in new cells. CATL, one of the world's largest battery manufacturers, has established closed-loop recycling facilities that process used EV batteries and production scrap to recover active materials for next-generation cells.

The benefits of these systems are both economic and environmental. Closed-loop recycling reduces dependence on primary mining—especially for critical minerals sourced from geopolitically sensitive regions—while cutting greenhouse gas emissions associated with material extraction and transport. It also ensures supply chain resilience, enabling manufacturers to secure essential resources locally. From an environmental standpoint, it diverts hazardous waste from landfills and supports long-term sustainability goals in the transition to clean energy technologies.

Environmental and Economic Benefits

Recovering critical minerals from e-waste presents several environmental and economic benefits, particularly by reducing reliance on primary mining, which is often characterized by energy-intensive and environmentally detrimental processes. The recovery of these materials contributes significantly to reducing carbon emissions and minimizing land disturbance, as it substitutes the need for virgin mineral extraction, which is a major contributor to land degradation and ecological disruption [384]. This transition not only fosters a more circular economy but also aligns with sustainable development principles aimed at responsible consumption and production [384].

Moreover, the process of recovering critical minerals from e-waste plays a pivotal role in waste management strategies by minimizing hazardous waste in landfills. The substantial volume of e-waste generated globally, estimated at 53.6 million metric tons in 2019, highlights the urgent need for efficient recycling systems. Only a small fraction, approximately 17.4%, of this waste is adequately recycled, resulting in valuable metals being lost and contributing to environmental degradation [385, 386]. Thus, developing effective recycling systems can enable the recovery of these valuable materials while simultaneously generating secondary raw materials that bolster supply chain resilience and reduce vulnerability to fluctuating market prices [387].

Economically, the recovery of critical minerals from e-waste diminishes the costs associated with raw material acquisition for industries reliant on these metals. Various studies emphasize that efficient metallurgical processes can lead to significant economic advantages by lowering production costs through the use of recovered materials, which are generally more affordable than newly mined resources [387]. In the context of solar energy, for instance, recovering materials from photovoltaic modules can recapture up to 90% of the original material investment cost [388]. Furthermore, consistent recovery practices can lead to the establishment of a stable supply of critical minerals, reducing the overall economic risks tied to raw material shortages and market volatility [386].

Despite these benefits, several challenges hinder the effective recovery and recycling of e-waste, particularly in developing countries. Informal recycling practices, characterized by inadequate safety protocols, can result in significant environmental and health repercussions. These practices raise major concerns regarding the proper handling of hazardous materials found in e-waste streams, as well as the overall efficacy of recycling efforts [389]. Additionally, inconsistent collection systems exacerbate the issues of e-waste management, highlighting the need for integrated approaches that leverage both formal and informal mechanisms to enhance recycling rates [389].

Rare Earth and Critical Mineral Operations and Processing

Global Developments

The principles governing global developments in battery recycling are multifaceted and region-specific. Each prominent market, namely Europe, China, Australia, and the United States, has embarked on structured initiatives aimed at bolstering the recycling of lithium-ion batteries (LIBs) and critical minerals necessary for energy sustainability.

In Europe, the newly established EU Battery Regulation (2023) introduces stringent recycling efficiency requirements as part of an overarching move towards circular economy principles. Specifically, it mandates a minimum recycling efficiency rate of 90% for cobalt and nickel, and 50% for lithium by 2031 [390]. This regulatory framework aims to achieve resource recovery efficiency and promote a manageable lifecycle for batteries, thus aligning with EU goals for sustainability and energy security.

China's focus on battery recycling manifests through its "Urban Mining" initiative, which facilitates the operation of large-scale lithium and rare earth recycling facilities. These facilities specifically target the recovery of critical minerals from urban electronic waste, thereby reducing reliance on virgin material extraction while addressing environmental concerns related to electronic waste. Research suggests that this approach conserves natural resources and mitigates the environmental impacts associated with traditional mining operations [391, 392]. Additionally, the growing market for LIB recycling in China addresses burgeoning demands arising from the electric vehicle sector, as production and e-waste escalate [393].

In Australia, the Commonwealth Scientific and Industrial Research Organisation (CSIRO) has launched the "Critical Minerals Recycling" program. This initiative emphasizes novel hydrometallurgical processes aimed at extracting lithium, cobalt, and neodymium from electronic waste. Research indicates that methods employed within this program can significantly enhance the efficiency of resource recovery from spent batteries, facilitating Australia's position as a critical player in the global battery supply chain [394, 395].

On the other hand, the United States is making strides through the Department of Energy's ReCell Center, which supports research into closed-loop lithium-ion battery recycling. This initiative aims to facilitate the recovery and reuse of battery materials, thereby strengthening domestic supply chains [395, 396]. The centre's focus on advancing recycling techniques aligns with broader national goals surrounding energy independence and environmental sustainability.

The recovery of critical minerals from e-waste and spent batteries is a vital pillar of sustainable resource management and circular economy strategy. By combining mechanical, thermal, and chemical techniques, modern recycling plants can reclaim high-value materials like lithium, cobalt, and rare earths while minimizing environmental impact. As technology advances, integrated hydro-pyro hybrid and bio-assisted processes promise even higher efficiencies, transforming waste streams into renewable sources of strategic materials for the clean-energy future.

Reprocessing of Tailings and By-Products

Reprocessing of tailings and by-products has emerged as a strategic approach to recover valuable minerals from previously discarded materials, particularly in the context of rare earth elements (REEs) and other critical minerals. Historically, tailings—fine-grained residues from mineral processing—were viewed as waste with limited economic potential. However, advances in extraction technologies, coupled with rising global demand for critical minerals, have transformed tailings into secondary resources. Many modern mining operations now treat tailings reprocessing as both a sustainability initiative and a means to enhance overall resource efficiency.

The reprocessing of tailings begins with the reassessment of existing storage facilities to determine the presence and concentration of residual metals. Older tailings, especially from operations conducted before the development of modern beneficiation techniques, often contain recoverable quantities of REEs, lithium, cobalt, or base metals that were not economically viable to extract at the time. Geochemical and mineralogical analyses are conducted to characterize these materials and design suitable reprocessing strategies.

Depending on the nature of the tailings, mechanical, chemical, or combined methods may be employed for recovery. For instance, mechanical reprocessing may involve regrinding and advanced separation techniques such as flotation, magnetic separation, or gravity concentration to recover remaining mineral fractions. In contrast, chemical reprocessing uses leaching techniques—acidic, alkaline, or bioleaching—to dissolve target metals from the fine-grained matrix. Hydrometallurgical methods, including solvent extraction and ion exchange, can then purify and concentrate the recovered metals for reuse. In some operations, pyrometallurgical treatments, such as roasting or smelting, are applied to stabilize hazardous materials and extract specific high-value metals.

A key advantage of tailings reprocessing is its environmental benefit. By removing reactive or toxic components from existing tailings storage facilities (TSFs), reprocessing can reduce long-term contamination risks, including acid mine drainage, metal leaching, and dust emissions. Moreover, recovered process water can be recycled, decreasing the need for fresh water inputs. After valuable materials are extracted, the residual waste can be re-deposited in a safer, more stable form—often as a paste or cemented backfill—to improve containment integrity.

In addition to tailings, the recovery of by-products from mineral and metallurgical processes provides another valuable resource stream. Many critical minerals are produced as secondary outputs—for example, scandium and yttrium from titanium or zirconium processing, or gallium from bauxite refining. By optimizing separation and purification systems, these elements can be extracted efficiently without the need for new mining operations. This approach not only reduces waste generation but also enhances the economic viability of primary production.

Successful examples of tailings reprocessing projects have emerged globally, reflecting a growing trend towards recycling waste from mining operations. In China, there has been a notable focus on recovering rare earth elements (REEs) from tailings produced during iron ore beneficiation. This initiative has represented a crucial development in supplementing the supply of REEs, as these materials are integral to various high-tech applications and are often produced with significant environmental repercussions [397]. The recovery processes typically aim to extract valuable materials from existing waste, thereby reducing the environmental footprint associated with traditional mining for these scarce resources.

In South Africa, the strategic reprocessing of gold tailings has delivered both economic and environmental benefits. The tailings present in the Witwatersrand region still contain notable gold reserves that could potentially be recovered using modern techniques, including re-milling and extraction methods such as carbon-in-leach [398, 399]. While these operations facilitate the recovery of precious metals, there are substantial amounts of gold that remain unrecoverable, which underscores the need for ongoing research and improved techniques [398, 399]. Additionally, tailings reprocessing can aid in the remediation of sites previously burdened with hazardous waste, thus potentially improving public health dynamics in surrounding communities; however, studies have shown mixed results on the direct impact of tailings on health outcomes [400, 401]. The economic and environmental merits of tailings reprocessing highlight its importance as both a viable industrial practice and a public health consideration [402].

Similarly, in Australia, projects focused on re-mining historical monazite-bearing tailings have been pursued, particularly aiming to extract thorium-free rare earth concentrates. These initiatives serve a dual purpose: they mitigate the radiological hazards associated with unprocessed tailings while concurrently extracting valuable REEs [403]. Advanced technologies in processing tailings not only enhance resource recovery but also address longstanding environmental concerns, creating a pathway toward more sustainable mining practices [403, 404]. By emphasizing re-mining efforts, Australia demonstrates the potential for tailings to transition from perceived waste to valuable resources, thus aligning with the principles of a circular economy.

Life-Cycle Assessment and Sustainability Principles

Life-Cycle Assessment (LCA) is a systematic framework used to evaluate the environmental impacts associated with all stages of a product's life, from raw material extraction through production, use, and final disposal. In the context of rare earth and critical mineral industries, LCA provides a holistic understanding of how mining, processing, transport, and manufacturing contribute to overall sustainability performance. Rather than focusing on a single stage—such as extraction or refining—LCA tracks the cumulative effects of energy consumption, emissions, water use, and waste generation across the entire value chain.

The life-cycle approach is typically divided into four key stages:

1. Raw material extraction and beneficiation, which includes mining and initial concentration of ores.

2. Processing and refining, covering pyrometallurgical, hydrometallurgical, and solvent extraction steps.

3. Product manufacturing and use, where materials are incorporated into technologies like magnets, batteries, or electronics.

4. End-of-life management, including recycling, reprocessing, and disposal.

Life cycle assessment (LCA) is a pivotal methodology for evaluating the environmental implications of various production processes, particularly concerning greenhouse gas (GHG) emissions, water pollution, and resource depletion. Key standards, such as ISO 14040 and ISO 14044, provide a framework for these evaluations. The impact of rare earth element (REE) processing exemplifies the significant burdens associated with different stages of production, where refining rare earth oxides emerges as one of the most energy-intensive and emission-heavy phases [405]. Studies indicate that the most considerable emissions often arise during mining, extraction, and initial processing stages, underscoring the need for effective management of these phases to minimize overall environmental impact [405, 406].

Conversely, the recycling and reprocessing of materials, such as mine tailings and electronic waste, present viable strategies for significantly mitigating GHG emissions. Research shows that the recovery of europium (Eu) from recycled fluorescent powders could lead to a reduction in carbon footprint by factors as high as 200 times compared to conventional primary production techniques [407]. LCA studies applied to rare earth recycling highlight that the integration of recycling not only lowers carbon emissions but also promotes resource sustainability by utilizing secondary materials [408]. Furthermore, the concept of closed-loop supply chains, as explored in recent literature, suggests that systemic approaches to recycling can yield economic benefits while contributing to environmental sustainability [408].

The environmental burdens generated during the production of rare earth elements, particularly during the refining processes, point to a pressing need for recycling initiatives and material recovery strategies. By employing LCA methodologies to compare the impacts of different production routes, researchers can quantitatively determine the advantages of recycling over traditional mining practices [405, 409]. The benefits of recycling encompass reductions in GHG emissions and significant decreases in related resource depletion and pollution. Therefore, the strategic expansion of recycling systems within the rare earth elements sector is critical for achieving both economic and environmental sustainability goals.

Sustainability principles build upon the insights gained from LCA by guiding the design and operation of mineral projects toward minimal environmental impact and long-term resource efficiency. These principles emphasize:

- Resource efficiency, ensuring that ores are used fully and waste streams are minimized or revalorized.

- Energy transition, incorporating renewable power sources and waste heat recovery to reduce carbon intensity.

- Water stewardship, including closed-loop water systems and advanced filtration to protect local ecosystems.

- Circular economy integration, where by-products, tailings, and end-of-life materials are reused or reprocessed rather than discarded.

- Social responsibility, ensuring that local communities share in the economic and social benefits of mining and processing activities.

In practice, LCA results often inform environmental management systems (EMS) such as ISO 14001 and corporate Environmental, Social, and Governance (ESG) strategies. These frameworks help identify high-impact areas where process optimization, cleaner technologies, or design changes can deliver the greatest sustainability gains. For instance, implementing solvent extraction with organic reagents that are recyclable, or substituting fossil-fuel-fired kilns with electric or hydrogen-based furnaces, can dramatically reduce emissions and energy demand.

A notable example is the European Union's Raw Materials Initiative, which promotes the use of LCA in evaluating critical raw materials supply chains, ensuring transparency and supporting policy decisions. Similarly, companies like Lynas Rare Earths and MP Materials have integrated life-cycle thinking into their sustainability reporting, using LCA to measure progress toward carbon-neutral operations and to benchmark against global best practices.

Ultimately, Life-Cycle Assessment and sustainability principles provide the quantitative and ethical foundation for responsible mineral production. By evaluating impacts "from mine to magnet" or "from battery to recycling," these tools enable industry leaders to balance economic growth with environmental protection and social responsibility—ensuring that the transition to clean energy and advanced technologies remains genuinely sustainable.

Life-Cycle Assessment (LCA) applied to rare earth elements (REEs)—including lanthanum (La), cerium (Ce), praseodymium (Pr), neodymium (Nd), promethium (Pm), samarium (Sm), europium (Eu), gadolinium (Gd), terbium (Tb), dysprosium (Dy), holmium (Ho), erbium (Er), thulium (Tm), ytterbium (Yb), and lutetium (Lu), as well as yttrium (Y) and scandium (Sc)—provides critical insights into their environmental performance throughout extraction, processing, use, and recycling. Each of these elements plays a distinct role in advanced technologies, and their life-cycle impacts vary depending on the processes used, the ore type, and the end-use applications. The following examples illustrate how LCA principles are applied to evaluate sustainability across the four key life-cycle stages.

Raw Material Extraction and Beneficiation

At the mining stage, LCA focuses on energy inputs, ore grade, overburden removal, and tailings generation.

- **Example – Bayan Obo Mine (China):** The world's largest source of light rare earths (La, Ce, Pr, Nd) produces significant CO_2 and SO_2 emissions during mining and beneficiation. LCAs have identified that the carbon footprint per kilogram of REO (rare earth oxide) produced is high due to low ore grades (typically <6%) and intensive chemical use for bastnäsite and monazite beneficiation.

- **Example – Mount Weld (Australia):** The Lynas operation has demonstrated lower life-cycle energy intensity through efficient open-pit mining and pre-concentration techniques, reducing transport-related impacts by upgrading ore before shipment.

- **Example – Scandium Extraction (Russia and Ukraine):** Scandium is often recovered as a by-product from titanium and uranium processing. LCA highlights that utilizing waste streams (e.g., red mud from alumina refineries) for scandium recovery significantly lowers environmental impacts by avoiding new mining.

Processing and Refining

This stage typically has the highest life-cycle impact due to chemical intensity, high-temperature processes, and waste generation.

- **Example – Cerium (Ce) and Lanthanum (La):** These light REEs are produced in large volumes through acid leaching and solvent extraction. LCA studies show that sulfuric acid roasting and oxalate precipitation contribute heavily to GHG emissions and wastewater toxicity. Substituting with carbonate-based leaching or using ion-exchange resins can reduce acid consumption by 40–60%.

- **Example – Neodymium (Nd) and Dysprosium (Dy):** In magnet manufacturing, Nd and Dy oxides are refined into metal alloys via molten-salt electrolysis or vacuum induction melting. LCA reveals that hydrogen-based reduction and recycling of process gases can lower CO_2 emissions by up to 30%.

- **Example – Yttrium (Y) and Europium (Eu):** These heavy REEs, used in phosphors and LEDs, require multi-stage solvent extraction (hundreds of stages in some cases). LCA-driven process optimization has led to adoption of ionic liquid-based extraction methods that generate less organic waste and improve recovery efficiency.

- **Example – Scandium (Sc):** The LCA of scandium oxide production shows significant improvement when hydrometallurgical recovery from waste residues is used instead of direct ore mining, reducing energy demand by up to 70%.

Product Manufacturing and Use

Rare earths are vital in clean energy and high-performance applications, but manufacturing introduces additional impacts that LCA helps to quantify.

- **Example – Neodymium-Iron-Boron (NdFeB) Magnets:** Used in electric vehicle motors and wind turbines, these magnets are energy-intensive to produce due to Nd and Dy refining. However, LCA indicates that their lifetime use offsets initial impacts by enabling renewable energy generation—one tonne of NdFeB magnets in a 3 MW wind turbine can prevent approximately 3,000 tonnes of CO_2 emissions over its lifespan.

- **Example – Samarium-Cobalt (SmCo) Magnets:** LCA comparisons show that while SmCo magnets require more energy to manufacture than NdFeB magnets, their higher thermal stability extends device lifespan, improving overall life-cycle efficiency.

- **Example – Yttrium (Y) and Europium (Eu) Phosphors:** Used in LED and display technologies, LCA highlights trade-offs—while extraction is chemically intensive, the long operational life and low energy consumption of LEDs result in net environmental benefits over conventional lighting.

- **Example – Scandium-Aluminium Alloys:** Used in aerospace and additive manufacturing, these alloys reduce component weight by up to 20%, leading to major fuel savings and lower CO_2 emissions during aircraft operation.

End-of-Life Management and Recycling

LCA also evaluates how materials can be recovered and reintroduced into the supply chain, reducing total life-cycle impact.

- **Example – NdFeB Magnet Recycling:** Direct recycling through hydrogen decrepitation allows magnet powder to be reused without breaking down to oxides, cutting energy use by 90% compared to virgin production.

- **Example – Phosphor and Lamp Recycling:** REEs like Y, Eu, and Tb are recovered from spent fluorescent lamps through hydrometallurgical processes. LCA shows that this reduces both chemical waste and reliance on primary mining.

- **Example – Lithium-Ion Battery Recycling (for Co, Ni, and REEs in cathodes):** Closed-loop recycling systems recover critical materials such as Li, Co, and Ni, with LCAs indicating up to 60–80% reduction in GHG emissions compared to raw material extraction.

- **Example – Scandium from Industrial Residues:** Reprocessing tailings and red mud to recover scandium contributes to circular economy goals, as LCAs demonstrate large reductions in environmental footprint per kilogram of Sc produced.

Integration with Sustainability Principles

Across all stages, LCAs of REEs emphasize energy transition, water management, and waste minimization.

- **Energy Transition:** The use of renewable energy at Lynas' Kalgoorlie processing facility in Australia demonstrates how LCA findings can guide investment in solar and gas hybrid systems to reduce carbon intensity.

- **Water Stewardship:** Solvent recovery and closed-loop water recycling in REE separation plants have cut freshwater demand by more than 50% in modern operations.

- **Circular Economy Integration:** Projects like Urban Mining Company (USA) and Hitachi Metals (Japan) are reusing NdFeB magnets from discarded electronics, aligning with LCA findings that direct recycling offers the best sustainability performance.

- **Social Responsibility:** The adoption of ISO 14001-based environmental management systems ensures that mining and processing facilities maintain transparent reporting and community engagement throughout the life cycle.

When applied to lanthanides, yttrium, and scandium, Life-Cycle Assessment demonstrates the trade-offs between technological progress and environmental responsibility. It reveals that the refining and separation phases are the most energy- and emission-intensive, while recycling, reprocessing, and substitution with cleaner technologies offer the greatest potential for reducing impacts. Through continual integration of LCA findings into design, regulation, and innovation, REE and critical mineral industries can move toward a more sustainable, circular, and low-carbon future—supporting the global transition to renewable energy, electric mobility, and advanced manufacturing.

Future Trends in Circular Mineral Supply

The ongoing transformation in global mineral resource management, specifically towards a circular economy model, represents a critical pivot from the conventional linear "take–make–dispose" paradigm. This shift emphasizes continuous recovery, reutilization, and reintegration of materials, particularly for rare earth elements (REEs) and critical minerals like lithium, cobalt, and nickel, which are integral for renewable energy systems, electric vehicles, and advanced electronics [100, 410]. The sustainability implications of this transformation are profound, as the increasing demand for these resources necessitates a systemic overhaul in how industries operate, guiding them towards more sustainable industrial growth and climate resilience [411, 412].

The circularity concept is characterized by maximizing resource efficiency, minimizing waste, and prolonging the lifecycle of materials. This systemic approach advocates for recovery of valuable minerals from end-of-life products and process by-products rather than depending solely on primary mining. Hence, urban mining frameworks are emerging as pivotal, allowing extraction of valuable elements from e-waste, batteries, and other discards that have hitherto been regarded merely as trash [413, 414]. Techniques in urban mining not only reduce reliance on virgin resource extraction but also help mitigate environmental damages such as land degradation and contamination [412, 415]. For example, policies focused on enhancing recycling initiatives have become paramount for critical minerals, with advancements in recovery technologies enabling significant portions—up to 95%—of lithium, cobalt, and nickel to be recycled effectively within the same product streams [416, 417].

One of the defining aspects of this circular approach is closed-loop recycling as applied particularly to energy materials like lithium-ion batteries and permanent magnets. Facilities now utilize a combination of mechanical, thermal, and hydrometallurgical processes that ensure that materials can be reclaimed for reuse [411, 418]. These processes not only yield high recovery rates but also result in significant reductions in carbon emissions—up to 90% compared to primary production, signifying a crucial contribution to mitigative climate strategies [419].

Digitalization plays a key role in enhancing traceability in circular mineral supply chains, employing technologies such as blockchain to assure transparent and ethical sourcing of materials, while also leveraging artificial intelligence for optimizing production processes [413, 420]. For instance, initiatives like the EU Battery Passport illustrate how digital platforms can establish clearer accountability in terms of sourcing and supply chains, thereby bolstering consumer trust in sustainably sourced products [411, 421].

Government policy is increasingly aligning with these circular economy principles. Regions like the European Union are adopting recycling targets and secondary sourcing quotas for critical minerals under legislations such as the Critical Raw Materials Act. This establishes a structured framework for advancing toward sustainable and resilient mineral economies [100, 410, 422]. The involvement of national governments and international regulatory bodies is crucial for driving these changes [91].

Technological advancements continue to enhance the viability of circular mineral systems. Innovations such as bioleaching and low-energy separation techniques are providing alternatives to traditional, more environmentally damaging extraction methods [413, 423]. Companies pioneering these technologies are proving that it is possible to recover critical minerals with improved efficiency and a reduced environmental footprint, emphasizing the significance of integrating renewable energy sources within processing systems to enhance sustainability further [413, 420].

Looking ahead, the future landscape of mineral resource management may involve regional circular mineral hubs, integrating mining, refining, recycling, and manufacturing [418, 421]. This will not only boost economic efficiencies but also reduce transportation emissions and alleviate environmental impacts. Collaborations between key stakeholders—governments, companies, and research institutions—will be essential to developing standardized recovery technology and ensuring the sustainable and responsible management of critical mineral resources for future generations [100, 414].

Key Terms and Concepts	

Circular Economy: An economic model that aims to eliminate waste and continually reuse, recycle, and repurpose materials to create a closed-loop production system.

Resource Recovery: The process of extracting valuable materials or energy from waste streams through recycling, reprocessing, or reuse.

Secondary Sources: Non-primary resources such as industrial waste, electronic scrap, and end-of-life products that contain recoverable critical minerals.

Urban Mining: The recovery of valuable metals and minerals from discarded products, buildings, and infrastructure within urban environments.

E-Waste (Electronic Waste): Discarded electrical and electronic equipment, often containing critical minerals such as rare earths, gold, and cobalt.

Battery Recycling: The process of recovering lithium, cobalt, nickel, and other valuable components from spent batteries for reuse in new energy storage systems.

Mechanical Processing: A pre-treatment method that uses physical means—such as shredding, crushing, and sorting—to prepare waste for mineral recovery.

Hydrometallurgical Recycling: The use of aqueous chemical processes, such as leaching and solvent extraction, to recover metals from secondary materials.

Pyrometallurgical Recycling: The recovery of metals from waste materials through high-temperature processes such as smelting or thermal reduction.

Life-Cycle Assessment (LCA): A systematic evaluation of the environmental impacts associated with all stages of a product's life, from raw material extraction to disposal.

Extended Producer Responsibility (EPR): A policy approach that makes manufacturers responsible for the environmental impacts of their products throughout the entire life cycle.

Critical Material Stewardship: The responsible management of critical minerals through design, production, use, and recovery to reduce waste and supply risk.

Key Terms and Concepts

Closed-Loop System: A process where materials are continually recycled and reused within the production cycle, minimising the need for virgin resources.

Sustainable Supply Chain: A system that integrates environmental and social considerations into all stages of material sourcing, production, and distribution.

By-Product Recovery: The extraction of valuable secondary elements or compounds that are produced incidentally during primary processing.

Circular Design: The creation of products intentionally designed for longevity, reparability, and recyclability to support circular economy objectives.

Chapter 10 Review Questions

Concepts of the Circular Economy

1. Short Answer: Define the circular economy and explain how it differs from the traditional linear "take–make–dispose" model.

2. Multiple Choice: Which of the following best describes the main goal of the circular economy?

(a) Maximising profit through mass production

(b) Minimising resource use and closing material loops

(c) Eliminating renewable energy systems

(d) Increasing reliance on virgin resource extraction

3. Extended Response: Discuss the relevance of the circular economy to rare earth and critical mineral supply chains.

Recycling and Resource Recovery

4. Short Answer: List three environmental benefits and two economic benefits of recycling end-of-life products containing rare earth elements (REEs).

5. Multiple Choice: Which type of waste is often referred to as an "urban mine" for critical minerals?

(a) Construction rubble

(b) Electronic waste (e-waste)

(c) Agricultural residues

(d) Radioactive waste

6. Case Study Question: Explain how recovering lithium and cobalt from spent batteries contributes to both carbon reduction and resource security.

Secondary Sources and Recovery Challenges

7. Short Answer: Identify four common secondary sources of critical minerals.

8. True/False: Tailings and by-products are not considered viable sources for critical mineral recovery.

9. Discussion Question: Describe two major technological or logistical challenges associated with recovering critical minerals from secondary sources such as e-waste or tailings.

Recycling Processes

10. Matching Question: Match the recycling method with its key characteristic.

a) Mechanical processing

b) Hydrometallurgical processing

c) Pyrometallurgical processing

d) Direct recycling

- Uses aqueous leaching and solvent extraction.

- Employs high-temperature smelting or roasting.

- Physically liberates materials by crushing or shredding.

- Restores cathode materials without breaking them down chemically.

11. Short Answer: Explain one advantage and one limitation of pyrometallurgical recycling.

12. Applied Question: Describe the steps involved in hydrometallurgical recycling of lithium-ion batteries.

Primary vs. Secondary Production

13. Multiple Choice: Compared with primary mining, secondary recovery generally:

(a) Consumes more energy but produces cleaner outputs

(b) Consumes less energy and generates lower emissions

(c) Requires more reagents and water use

(d) Produces more hazardous waste

14. Short Answer: Explain why secondary recovery routes improve supply chain resilience.

Life-Cycle Assessment (LCA)

15. Short Answer: Define Life-Cycle Assessment and describe its four major stages.

16. Multiple Choice: Which LCA standards are most commonly applied to mineral processing industries?

(a) ISO 9001 and ISO 45001

(b) ISO 14040 and ISO 14044

(c) ASTM C150 and AS/NZS 1170

(d) IEC 61508 and ISO 50001

17. Extended Response: Discuss how LCA findings can inform process design, sustainability reporting, and regulatory decision-making.

Policy and International Frameworks

18. Short Answer: What is Extended Producer Responsibility (EPR), and how does it encourage circular economy practices?

19. Multiple Choice: Which of the following policies mandates recycling targets for critical minerals within the European Union?

(a) REACH Directive

(b) Critical Raw Materials Act (2023)

(c) Basel Convention

(d) Paris Agreement

20. Applied Question: Explain how national initiatives such as Australia's Critical Minerals Strategy or the U.S. ReCell Center support the development of circular mineral supply chains.

Case Studies and Industry Examples

21. Short Answer: Summarise one successful example of a company implementing closed-loop recycling (e.g., Tesla, CATL, or Redwood Materials).

22. Scenario Question: Compare how Europe and China approach battery recycling and discuss the implications for resource sustainability.

Future Trends and Sustainability

23. Short Answer: Identify two technological innovations driving future trends in circular mineral supply.

24. Multiple Choice: Which of the following best describes a "closed-loop system"?

a) Mining and exporting raw minerals for processing overseas

b) Continuous reuse and recycling of materials within production

c) Disposal of waste in sealed tailings facilities

d) Shipping end-of-life products to landfills

25. Essay Question: Evaluate how recycling and secondary sourcing contribute to achieving global net-zero and sustainability targets.

Chapter 11

Industry Operations and Global Market Dynamics

The global rare earth and critical minerals industry stands at the centre of a rapidly changing economic, technological, and geopolitical landscape. As nations accelerate their transitions toward clean energy, digital infrastructure, and advanced manufacturing, demand for these strategically vital materials continues to surge. Rare earths and other critical minerals form the backbone of modern technologies—from electric vehicles and wind turbines to defence systems and electronics—linking resource-rich regions with high-value manufacturing hubs across the world. Yet, behind this growth lies a complex web of interdependencies shaped by production bottlenecks, concentrated refining capacity, and shifting government strategies. This chapter examines how the structure of the global industry, market pricing, and policy frameworks interact to influence supply security, investment, and international competitiveness. Through case studies spanning Australia, China, the United States, and Africa, it explores the full value chain—from exploration and extraction to refining, manufacturing, logistics, and trade—highlighting both the opportunities and challenges of building a resilient, sustainable, and transparent critical minerals ecosystem.

Learning Outcomes	
This chapter aims to give you the ability to: 1. Describe the structure of the global rare earth and critical minerals industry, including key stages from exploration and extraction to refining, manufacturing, and end-use applications. 2. Identify the major producing countries and companies involved in rare earth and critical mineral supply chains, and compare their production capacities and market roles. 3. Explain how market forces, including supply and demand trends, pricing mechanisms, and trade policies, influence global mineral markets.	

Learning Outcomes	

4. Interpret the impact of geopolitical factors—such as trade restrictions, export controls, and strategic alliances—on supply security and international competitiveness.
5. Discuss the role of Australia, China, the United States, Africa, and other regions in global supply and processing, and evaluate their comparative advantages.
6. Explain the influence of government policies and critical minerals strategies on exploration investment, downstream processing, and industrial diversification.
7. Assess the challenges associated with logistics, transport, and export infrastructure in moving critical minerals from mine to market.
8. Analyse the value chain from raw material to advanced manufacturing, including refining, alloying, magnet production, and battery precursor development.
9. Evaluate the market trends driving the transition to clean energy technologies and their implications for future critical mineral demand.
10. Recognise the role of industry standards, certification schemes, and responsible sourcing initiatives in promoting transparency and sustainability in global mineral markets.

Major Mining and Processing Operations (Australia, China, USA, Africa)

Lynas Rare Earths operates the Mount Weld mine, located in Western Australia, which has been identified as one of the most economically viable non-Chinese rare earth deposits globally. The mine boasts a rich concentration of rare earth elements (REEs) and plays a significant role in providing essential materials utilized in various high-technology industries, including clean energy and electronic applications [140, 424]. The ongoing operations at Mount Weld underscore Australia's strategic positioning in the global rare earths market, particularly against the backdrop of Chinese dominance [140, 406].

Australia has established itself as one of the largest producers of lithium, primarily through operations located in Western Australia. In addition to lithium, Australia is a key exporter of various critical minerals necessary for modern technology and energy solutions [425]. Empirical data indicates that Australia contributes significantly to the global lithium supply, reinforcing its role in the pivot towards renewable energy technologies such as electric vehicles [425].

Australia

1.1 Mount Weld Mine + Kalgoorlie Processing (Lynas Rare Earths)

- **Location:** Near Laverton, WA
- **Operator:** Lynas Rare Earths Ltd
- **Deposit/ore type:** World-class carbonatite REE deposit; high-grade, with distinct CLD (Central Lanthanide Deposit).
- **Grade & resource:** CLD ore has been publicly described as very high grade, commonly 5–8% REO, locally higher; Lynas reports one of the highest-grade REE ores globally.
- **Mining/throughput:** Open-pit; ore is concentrated at Mount Weld; concentrate is sent for cracking/leaching (currently in Malaysia; Lynas is adding Australian capacity).
- **Processing status:** Lynas is building a Kalgoorlie cracking & leaching plant so it can do more upstream processing in Australia. Target NdPr output in its "Towards 2030" plan is ≈12,000 t/y NdPr (product), not ROM ore.
- **Status:** Operating, expanding.

1.2 Nolans NdPr Project (Arafura Rare Earths)

- **Location:** 135 km north of Alice Springs, NT
- **Operator:** Arafura Rare Earths
- **Ore/grade:** Proved + probable ore reserve about 29–30 Mt at ~2.9% REO, strongly NdPr-rich.
- **Planned production:** About 4,400 t/y NdPr oxide within >4,300 t/y NdPr equivalent product, plus other REEs, over 38-year mine life.
- **Flowsheet:** Mine → beneficiation → acid bake → hydrometallurgy → SX separation in the NT.
- **Status:** Definitive feasibility level; project is in financing/approvals.

1.3 Eneabba / Rare Earths Refinery (Iluka)

- **Location:** Eneabba, WA
- **Operator:** Iluka Resources
- **Feedstock:** High-grade monazite/xenotime stockpiles from historic mineral-sands operations (not a classic hard-rock mine).
- **Planned capacity:** To produce light and heavy REE separated products from stockpiles; tonnage depends on staged plant build, but Eneabba is seen as a strategic heavy-REE source for Australia.
- **Status:** Refinery under development; feed is secured.

In response to the increasing demand for critical minerals, the Australian government has taken proactive measures, launching the Critical Minerals Strategy 2023–2030. This strategy aims to facilitate value-added processing, enhance downstream manufacturing capabilities, and diversify supply chains to ensure a stable and secure supply of critical minerals [121, 426].

Australia's regulatory environment is favourable, providing assurance to investors and stakeholders alike [426].

Processing of these minerals, however, has historically been focused offshore, particularly in China, due to several logistical and economic factors. Nevertheless, there has been a marked shift in Australia towards greater onshore processing capabilities [427]. The government has provided systematic support to establish domestic facilities for processing rare earth oxides, as well as the downstream manufacturing of essential products such as permanent magnets. This initiative aims to reduce dependence on Chinese processing dominance, highlighting a strategic pivot within Australia's mineral resource management policies [427].

China

China still dominates global mine supply (typically 60–70%+ of mined REO) and >80% of separation capacity. Most REE tonnage is consolidated in a few very large operations and processed through integrated separation plants in Inner Mongolia and southern China.

2.1 Bayan Obo Mine + Baotou Processing Hub

- **Location:** Inner Mongolia
- **Operator:** Baotou Steel Rare Earth (part of the Baotou group / Northern Rare Earth)
- **Deposit/ore:** Giant iron-REE-fluorite orebody.
- **Reserves:** Frequently cited as >1,000 Mt of ore containing ~3–6% REO equivalent; regarded as the largest single REE resource in the world.
- **Production:** Public reporting and Chinese industry summaries often quote ≈100,000–120,000 t/y REO-equivalent concentrate and products from Bayan Obo feed through Baotou separation, though exact split between mine, concentrate and separated oxides varies by year and quota.
- **Processing:** Ore → beneficiation → REE concentrate → acid baking/cracking → multi-stage SX at Baotou to individual oxides (La–Nd, Sm–Eu–Gd, heavy fractions).
- **Status:** Long-life, fully operating anchor asset for China's northern REE quota.

2.2 Southern Ionic Clay REE Operations (Jiangxi, Guangdong, Guangxi)

- **Type:** In-situ leach / heap-style extraction of ion-adsorption clays rich in MREE/HREE (Y, Dy, Tb, Gd).
- **Operators:** Multiple state-controlled and licensed producers; very fragmented historically.
- **Output:** Smaller per-site, but collectively a strategic source of heavy REEs that China then separates in SX parks in the south.
- **Status:** Operating under tighter environmental controls since 2016; volumes controlled by state quota.

2.3 Newer Integrated Projects (e.g. Guangdong, Sichuan)

- Focused on integrating mining + SX + magnet materials to match China's export-control strategy on NdFeB. Figures are modest compared to Bayan Obo but important for product diversity.

Furthermore, the comparative analysis of Australia's rare earth mining capabilities against other global players such as China reveals a crucial divergence in supply chain control. While China dominates both the mining and processing segments, Australia's enhanced focus on local processing aims to mitigate vulnerabilities within the global supply chain [121]. The establishment of domestic processing and value-add activities reflects a strategic move not only to secure resource independence but also to strengthen international partnerships, such as the U.S.–Australia Critical Minerals Agreement, aimed at fostering cooperative supply chains [121].

United States

3.1 Mountain Pass Mine and Processing (MP Materials)

- **Location:** California, USA
- **Operator:** MP Materials
- **Ore/grade:** Carbonatite REE deposit; historical grade ≈7% REO in bastnäsite-rich ore.
- **Current production:** MP reports ~40,000–43,000 t/y REO in concentrate in recent years, making it the largest single REE mine outside China.
- **Processing status:** Produces concentrate on site; MP is restarting U.S. separation (Stage II) so NdPr oxide can be made domestically; DoD is also funding magnet manufacturing ("10X Facility").
- **Target products:** NdPr oxides, separated lights, and eventually magnets for defence/EV supply chains.
- **Status:** Operating mine; downstream build-out in progress.

3.2 U.S. Downstream / Defence-Backed Projects

- **Lynas U.S. Heavy/Light REE plant (Texas):** U.S. DoD-backed, to receive Mt Weld concentrate and produce separated REE products inside the U.S. (important for NdPr and some heavies).
- **NioCorp – Elk Creek (Nebraska):** Niobium-Scandium-REE underground project; received $10m DoD support to develop a domestic scandium chain. Planned REE tonnage is much smaller than Bayan Obo/Mt Weld, but Sc is strategic and high value.
- **Status:** Mostly development / early works, aimed at closing U.S. processing gaps.

The landscape of rare earth element (REE) production and processing is a vital area of focus, particularly in light of the strategic importance of these materials in modern technologies. The global supply of REEs is overwhelmingly dominated by specific regions, particularly China, which significantly influences the design of any global flowsheet for rare earth processing.

China's Bayan Obo deposit and the southern ion-adsorption clay deposits are arguably the most critical sources of REEs globally. The Bayan Obo mine is recognized as one of the largest producers, contributing a substantial portion of the light REEs, while the southern ion-adsorption clays are known to yield most of the heavy REEs [321, 428]. The geographical concentration of these resources necessitates that all flowsheet designs incorporate the Chinese solvent extraction (SX) benchmarks, reflecting the country's established expertise and efficiency in processing these minerals [6, 429].

On the non-China front, Australia has taken a leading role in the REE mining sector, with significant contributions from companies like Lynas and Arafura [279, 430]. Mount Weld stands out as a significant deposit, currently producing REEs at a global scale, and serves as an essential non-Chinese supply source to balance global needs [430]. Furthermore, the U.S. is making strides in this sector with the Mountain Pass mine, which represents a strong orebody and is expanding its capacity, aided by funding from the Department of Defense, aimed at closing the gap from concentrate production to magnet manufacturing [168].

In Africa, the potential for growth in REE production exists, particularly with projects like Ngualla and Kangankunde, which could add between 15 to 25 kilotonnes per year of REO by the 2030s; however, the realization of this potential often faces capital expenditure constraints [23, 406]. Additionally, tailings and secondary projects, such as those at Phalaborwa and Eneabba, are emerging as low-risk, environmentally friendly opportunities to increase REE supply without the challenges associated with traditional mining methods [431, 432].

Overall, the interplay between these mining operations and regional concentrations of resources paints a complex picture of the REE landscape. The reliance on Chinese benchmarks for processing and the potential contributions from new mines and secondary projects create a dynamic environment wherein global supply chains are constantly adjusting to meet the demands of advanced technologies reliant on these critical materials [7, 433].

Africa

Africa is more "project heavy" than "tonnage heavy" right now: several very good deposits, but only limited sustained production (Burundi Gakara was one).

4.1 Ngualla Rare Earth Project (Tanzania)

- **Operator:** Peak Rare Earths
- **Resource/reserve:** Reported ore reserve about 18–20 Mt at ~4.8–4.9% REO—excellent grade for a hard-rock REE carbonatite.

- **Planned production:** DFS-level studies show ~30,000 t/y REO concentrate, translating to ~7–8,000 t/y of separated REO products once the proposed downstream plant is included.
- **Flowsheet:** Mine in Tanzania → concentrate → planned refinery (originally in Teesside/Europe; Peak has looked at alternate locations).
- **Status:** Advanced project; working through financing and offtake.

4.2 Longonjo / Coola (Angola)

- **Operator:** Pensana (UK-listed)
- **Resource:** Measured + indicated resource publicly quoted at >300 Mt at ~1.4% REO for Longonjo, giving a large tonnage, mid-grade asset.
- **Planned output:** Pensana market material shows ~5,000 t/y NdPr oxide equivalent from an integrated mine in Angola + separation plant in the UK/Europe.
- **Status:** Development; staged approvals.

4.3 Kangankunde (Malawi)

- **Operator:** Lindian Resources
- **Resource:** Recently reported as ~261 Mt at 2.19% TREO, with low thorium—commercially very attractive.
- **Planned scale:** Still in project-definition, but at that grade/size it could support 10,000–15,000 t/y REO-equivalent long term.
- **Status:** Exploration/early project; moving toward testwork and scoping.

4.4 Phalaborwa (South Africa) – Tailings REE Recovery

- **Operator:** Rainbow Rare Earths
- **Feed:** Historic phosphogypsum/tailings from the old Phalaborwa phosphate/pyro operation.
- **Resource:** Feeds contain monazite-derived REE in acid-soluble form, with relatively elevated NdPr and limited radioactivity compared with hard-rock monazite.
- **Planned product:** Mixed rare-earth carbonate/oxide from an on-site hydromet circuit.
- **Status:** Development / pilot.

4.5 Gakara (Burundi)

- **Operator:** Rainbow (historically), but export bans and licence issues stopped production in 2021–22.
- **Historical scale:** Small but very high-grade vein REE; shipped mixed concentrate to off-takers. Now effectively on hold.

Recent Agreements (2025)

Richard Skiba

The United States and Australia have established a comprehensive policy framework designed to strengthen cooperation in the mining, processing, and supply of critical minerals and rare earth elements. Signed in Washington, D.C., on 20 October 2025, this agreement reflects both nations' shared commitment to securing reliable access to materials that are essential to advanced technologies, defence systems, renewable energy, and industrial production. The framework builds on existing domestic capabilities while promoting new investments to expand extraction, separation, and processing capacity by 2026.

The agreement recognises that critical minerals and rare earths underpin modern innovation—from electric vehicles and renewable energy systems to advanced defence technologies. Both countries aim to enhance resilience and security across the supply chain, reducing vulnerability to global market disruptions and geopolitical risks. The framework's overarching goal is to create diversified, fair, and transparent markets through coordinated economic policy, investment, and regulatory alignment.

Under the framework, the U.S. and Australia have committed to intensifying cooperation to secure supplies of these vital resources. This includes leveraging tools such as America's industrial demand and stockpiling infrastructure and Australia's Critical Mineral Strategic Reserve. Both governments will mobilise public and private investment, including through guarantees, loans, and equity participation, to develop priority mining and processing projects. Within six months of the agreement, each country will provide at least USD $1 billion in financing for projects that directly contribute to joint supply chain goals.

The framework also prioritises regulatory reform, with both nations seeking to streamline or deregulate permitting processes to accelerate development timelines. New price mechanisms are being explored to counter non-market policies and ensure fair competition, potentially including price floors or trade standards to stabilise markets. In addition, the countries will cooperate on geological mapping, scrap recycling technologies, and the establishment of a U.S.-Australia Critical Minerals Supply Security Response Group, which will monitor supply vulnerabilities and coordinate rapid response measures.

The agreement represents a shift from raw material exports to full value-chain integration, positioning Australia as more than just a supplier of unprocessed resources. With substantial funding commitments—estimated at up to USD $8.5 billion over multiple years—the partnership seeks to build domestic processing infrastructure, improve refining capabilities, and establish downstream manufacturing links. A Mining, Minerals and Metals Investment Ministerial will convene within 180 days to guide investment priorities, while quarterly reviews will track progress and adjust financing strategies as needed.

The U.S.–Australia rare earth framework is part of a broader geostrategic realignment among allied democracies, aimed at reducing dependence on China's near-monopoly over global rare earth processing. It complements similar frameworks between the U.S. and Japan and between the U.S. and Malaysia, collectively forming a network of secure, transparent supply chains that prioritise environmental responsibility and national security.

The U.S.–Japan framework, signed around the same period, shares nearly identical objectives but differs in scope and financial commitments. While both agreements aim to secure supplies through mining cooperation, recycling, and investment in downstream industries, the U.S.–Japan version focuses more heavily on industrial collaboration across the defence, electronics, and energy sectors. Notably, the Japan framework lacks the explicit $1 billion per-country financing commitment found in the U.S.–Australia arrangement. Instead, it promotes a complementary stockpiling strategy, policy coordination, and dialogues to facilitate investment and technology transfer.

In both cases, the frameworks do not create legally binding obligations but instead outline policy intentions and cooperative mechanisms. They symbolise a strategic convergence among like-minded nations seeking to build resilient, sustainable, and market-based critical minerals supply chains essential to future economic and defence capabilities.

Market Pricing, Supply–Demand Trends, and Value Chain Stages

Market Pricing

Prices for rare earth oxides (REO) and alloys are monitored by specialist market sources such as the *Shanghai Metal Market (SMM)*, which provides authoritative Chinese market data. As of July 2025, *Praseodymium-Neodymium (Pr-Nd) oxide* was priced around ¥444,000–445,000 per metric tonne [434]. In contrast, heavy rare earths such as *Dysprosium oxide* remained significantly more expensive, stabilising at approximately ¥1.61–1.62 million per metric tonne [434]. In addition to regional sources, global price assessment services such as *Fastmarkets* provide weekly updates and analytical reports on multiple rare earth elements, offering benchmarks used by industry and investors alike.

Forward-looking forecasts suggest that prices for several rare earths may rise sharply if supply fails to keep pace with growing demand. For example, *Global Mining Review* has projected that Dysprosium oxide could reach USD 1,100 per kilogram (≈ USD 1.1 million per tonne) by 2034, representing an increase of about 340 percent over current levels [435].

One of the dominant factors shaping rare earth pricing is supply concentration and export control. Because the majority of global refining capacity is concentrated in a single country—China—any policy change, trade restriction, or export quota can have a disproportionate effect on world prices. Demand pressures from downstream industries such as electric vehicles, wind turbines, defence systems, and advanced electronics also play a major role, since these sectors rely heavily on high-performance magnets and other components that use rare earth elements like neodymium (Nd), praseodymium (Pr), dysprosium (Dy), and terbium (Tb).

Processing and refinement bottlenecks further add to price volatility. Mining an ore is only the first step; many of the critical, value-adding processes such as separation, alloying, and magnet production are highly specialised and geographically concentrated, creating potential chokepoints. Additionally, regulatory, geopolitical, and environmental risk premiums increase costs, as companies must navigate strict environmental standards, obtain complex permits, and ensure supply chain resilience.

Finally, while recycling and substitution technologies could eventually ease price pressure, their current impact is limited. Recycling infrastructure remains underdeveloped, and viable substitutes for high-performance rare earths are still in the early stages of technological and commercial adoption.

After sharp price surges earlier in the decade driven by export restrictions and geopolitical tension, rare earth prices have recently shown signs of stabilisation. Reports from *SMM* in early 2025 noted that several oxide prices were "stabilising after a decline." Nonetheless, medium- to long-term forecasts remain bullish. Research from *Grand View Research* projects an annual compound growth rate (CAGR) of about 8.6 percent for the rare earth market between 2025 and 2030 [436]. Because global refining capacity remains concentrated in only a few regions, the risk of sudden supply shocks leading to price spikes continues to be a major market concern.

Supply–Demand Trends

Demand for rare earths and other critical minerals is being driven by the rapid global transition to clean energy technologies, electrified transportation, and advanced manufacturing. Permanent magnet applications—especially those used in electric vehicle motors and wind turbines—are among the fastest-growing uses. According to *Mordor Intelligence*, demand for these materials is expected to accelerate as global electrification efforts expand.

Forecasts indicate significant market growth. *Fortune Business Insights* estimates that the global rare earth elements market will increase from around USD 3.39 billion in 2023 to approximately USD 8.14 billion by 2032, representing a CAGR of roughly 10.2 percent. Another projection suggests total production volume could rise from about 196.6 kilotonnes in 2025 to 260.4 kilotonnes by 2030 (CAGR ≈ 5.8 percent) [437]. Geographically, the Asia-Pacific region currently dominates both supply and demand, but growth in North America and Europe is accelerating as these regions invest heavily in domestic mining and processing capacity.

By application, magnets account for 30–40 percent of total rare earth consumption and are projected to grow at more than 8 percent per year [436]. Research by *Adamas Intelligence* forecasts demand for *NdFeB magnets* to expand at 8.7 percent annually through 2040, while the production of magnet-related rare earths is expected to increase by only 5.1 percent annually—suggesting persistent structural shortages [438].

On the supply side, the industry faces numerous challenges. Many rare earth deposits are geographically limited, and few countries possess the full downstream infrastructure for separation and refining. Environmental regulations, technical complexity, and long permitting timelines also restrict output. *Global Mining Review* highlights that bringing a new mine or refinery online can take several years, leaving the market vulnerable to shortages.

Rare Earth and Critical Mineral Operations and Processing

The Asia-Pacific region, particularly China, maintains an overwhelming share of global production and refining—estimated at around 86 percent of output in 2024 according to *Grand View Research* [436]. Analysts project continuing supply deficits, with one forecast predicting a shortage of roughly 2,823 tonnes of Dysprosium by 2034 if new production capacity is not developed [439]. Recycling and secondary recovery remain minimal, meaning primary mining continues to dominate the supply landscape.

As global demand grows faster than new production capacity, most experts expect tightening supply conditions and subsequent upward pressure on prices. Heavy reliance on China for refining introduces significant policy and geopolitical risks—for example, Beijing's 2025 export controls on specific rare earths triggered immediate concern across global markets.

Furthermore, bottlenecks in the mid-stream stages of the value chain—particularly in separation, processing, and alloying—remain a key vulnerability. Additional risks stem from ESG and permitting challenges, infrastructure gaps, and the need for long-term offtake agreements. While nations such as the United States, Australia, Canada, and members of the European Union are investing in new upstream capacity, these projects will take years to reach full operation. As a result, short- to medium-term supply risks remain elevated, and market participants anticipate continued volatility and price sensitivity across the sector.

Value-Chain Stages for Rare Earths & Critical Minerals

Understanding how value is added from raw material to final advanced component helps clarify where risk and opportunity lie.

Stage 1: Exploration & Mining

- Identification of deposits, drilling, resource estimation, permitting and mine construction.
- Ore extracted may contain a mix of multiple rare earth elements and other minerals.
- Physical geography, grade of ore, depth, mining cost, logistics all matter.

Stage 2: Separation & Refinement (Oxide/Carbonate)

- After mining, the ore must be processed: crushed, separated, chemically treated to produce rare earth oxides or carbonates (REO/REC) or specific element concentrates.
- This is technically complex and often environmentally and chemically intensive (use of acids, etc).
- Example: Producing Pr-Nd oxide, Dysprosium oxide etc.
- Because of large fixed capital and technical barriers, many countries outsource this to a few select refiners.

Stage 3: Alloying & Intermediate Products

- From oxides you may produce alloys (e.g., Nd-Fe-B alloy), metals, or other semi-finished products.

- This is a higher value step, closer to final applications.

- Example: Nd-Fe-B magnets require alloy ingots, slicing, magnetising, etc.

Stage 4: Manufacturing & Final Application

- Use of these materials in final products: e.g., permanent magnets for electric vehicle motors or wind turbines; catalysts; electronics; defence hardware; batteries (for some critical minerals).

- The value addition here is highest, but supply of the upstream materials is critical for competitiveness and security.

Stage 5: Recycling / Secondary Supply

- End-of-life products (motors, electronics, magnets) can be processed to recover rare earths or other critical minerals.

- This stage is increasingly important for supply resilience, circular economy, and reducing reliance on primary mining.

- Currently still nascent at global scale for many rare earths.

Stage 6: Logistics, Stockpiling, Strategic Reserves

- Because many of these materials are deemed "critical" for national security, logistics (transport, storage), stockpiling (strategic reserves) and trade flows matter.

- Example: countries may maintain strategic mineral reserves, implement export licensing, or impose pricing/anti-dumping measures.

Key observations across the value chain

The greatest value and technological sophistication within the rare earth and critical minerals industry often lies in the downstream stages—specifically alloying, component manufacturing, and integration into advanced products. However, many regions remain concentrated in the upstream phase, limited to mining and producing oxides. This imbalance leaves resource-rich nations vulnerable to those who control downstream processes such as separation, refining, and manufacturing, where economic and strategic value is highest.

A major bottleneck exists at the separation and refinement stage, as these processes require advanced chemical expertise, substantial capital investment, and stringent environmental controls. Only a handful of countries possess the facilities and technical capacity to refine rare earth elements at scale, which creates significant dependency risks and limits market competition.

Rare Earth and Critical Mineral Operations and Processing

Recent policy frameworks in the United States, Australia, Japan, and the European Union recognise this vulnerability and aim to strengthen the entire value chain—from mine to final product. These strategies promote integrated development across all stages, encouraging not just raw material extraction but also investment in domestic refining, component manufacturing, and recycling infrastructure.

Building supply chain resilience therefore requires more than expanding mining output; it demands diversification across processing, manufacturing, and recycling capacities, ideally distributed among trusted global partners. This geographical diversification reduces the risk of single-source dependence and enhances collective security and market stability.

For education and training, understanding the full value chain provides an essential framework for identifying both risks and opportunities. Educators can use the stages of production to illustrate where vulnerabilities—such as permitting delays, export controls, or environmental constraints—are most likely to occur. Likewise, they can highlight areas for innovation and investment, including new processing technologies, recycling systems, and sustainable production methods that strengthen the overall resilience and efficiency of critical mineral supply chains.

The rare earth and critical minerals market is entering a period of sustained structural growth, driven by soaring demand across key industries such as electric vehicles, renewable energy, defence, and advanced electronics. As nations accelerate their energy transitions and digitalisation strategies, these materials have become indispensable components of technologies ranging from high-efficiency motors and wind turbine generators to precision-guided defence systems. This creates a strong and lasting structural tailwind for the sector, positioning it as a cornerstone of 21st-century industrial policy.

However, despite this surge in demand, expanding mining and processing capacity remains slow, complex, and capital-intensive. Bringing new projects online can take many years due to exploration, permitting, financing, and environmental hurdles. This lag between demand growth and supply development heightens the risk of price volatility, bottlenecks, and supply chain disruptions. In effect, global markets may face prolonged tightness, with potential for recurring "boom-and-bust" cycles tied to geopolitical shifts or production shortfalls.

Countries and companies that control more stages of the value chain, particularly processing, refining, and downstream manufacturing, gain significant strategic and economic advantages. These stages represent the highest value-add and the greatest leverage over global supply networks. As a result, nations such as the United States, Australia, Japan, and the European Union are pursuing industrial strategies to reduce dependence on single-source supply chains and to localise critical mineral processing capacity.

In sectors such as defence, renewable energy, and advanced manufacturing, rare earth and critical mineral security has evolved from an economic consideration into a national security priority. Reliable, diversified access to these materials is now viewed as vital to maintaining technological superiority and industrial resilience.

For the education and training sector, this shift underscores the importance of teaching learners about the full spectrum of risks within the critical minerals value chain—ranging from permitting delays and environmental regulations to geopolitical dependencies, export controls, and recycling limitations. Understanding these systemic pressures helps future professionals appreciate the necessity of robust regulation, targeted investment, and forward-thinking trade policy.

Finally, because the costs and pricing of rare earths directly affect downstream industries such as electric vehicles, wind turbines, and electronics, fluctuations in upstream markets have cascading effects across entire economies. This interdependence highlights why coordinated industrial policy, innovation in recycling and substitution, and international cooperation are essential to ensuring both economic stability and sustainable growth in the era of clean energy and advanced technology.

Government Strategies and Critical Minerals Policies

The Australian Government's Critical Minerals Strategy (2023–2030) establishes a national framework designed to expand Australia's leadership in the global critical minerals sector. This strategy goes beyond traditional mining activities to encompass processing, manufacturing, and integration into international supply chains. It reflects Australia's recognition that long-term economic growth, industrial capability, and national security depend on secure access to critical minerals and value-added production.

Key elements of the strategy include the identification of critical minerals and strategic materials essential for emerging technologies, energy systems, and defence applications. The framework outlines six focus areas through which the federal government will collaborate with industry, research institutions, communities, and international partners to build a sustainable and globally competitive critical minerals ecosystem. Targeted measures are being introduced to de-risk high-value projects, attract private investment, and develop advanced downstream processing and manufacturing capacity.

The strategy also places a strong emphasis on international cooperation, responsible supply chain development, and adherence to environmental, social, and governance (ESG) standards. To stimulate investment, the government is deploying financial incentives such as grants, tax offsets, and project guarantees to encourage private sector participation in refining, recycling, and advanced material production. Collectively, these measures aim to transition Australia from a raw materials exporter to a strategic player across the full critical minerals value chain.

In the United States, the federal government has identified a wide range of minerals as "critical" due to their economic importance, supply vulnerability, and relevance to national security. U.S. policy focuses on reducing dependence on foreign sources and strengthening domestic and allied supply chains through a combination of legislation, executive orders, and strategic investments.

Rare Earth and Critical Mineral Operations and Processing

Key policy tools include domestic initiatives targeting offshore and terrestrial mineral resources, particularly those found in the seabed and hard-rock environments, to enhance supply diversification. The United States has also entered into bilateral and multilateral agreements with like-minded partners—such as Australia, Canada, Japan, and the European Union—to foster investment, develop shared standards, and improve supply chain transparency.

A key example is the U.S.–Australia Climate, Critical Minerals and Clean Energy Transformation Compact, which commits both nations to coordinate on critical minerals supply, investment, and processing. Under this agreement, Australia has also announced the creation of a Strategic Minerals Reserve to safeguard national supply and reduce reliance on single-source suppliers.

Governments around the world are deploying a consistent set of policy tools to strengthen critical minerals resilience. Lists and classification systems formally identify critical and strategic minerals, focusing government action and private investment on materials most essential to technology and national defence. For instance, Australia's expanded list now includes minerals vital for renewable energy, semiconductors, and advanced manufacturing.

Incentives and financial support such as grants, concessional loans, export credit facilities, and tax offsets reduce investment risks and stimulate participation across both upstream (mining) and downstream (refining and manufacturing) activities. Meanwhile, international partnerships are being pursued to diversify supply chains and reduce dependence on any single dominant supplier. This includes collaboration in processing, recycling, and research and development to ensure secure, environmentally responsible, and competitive global supply.

Policies increasingly integrate standards and ESG governance, mandating transparency, ethical sourcing, and environmental protection across the supply chain. There is also a concerted push toward downstream processing and value-chain development, ensuring that resource-rich nations retain more economic value domestically. Furthermore, several governments—including Australia and the U.S.—are establishing strategic reserves and stockpiles to buffer against potential supply disruptions and market shocks.

Across nations, critical minerals policies share several overarching goals. They seek to stimulate economic development and job creation by transforming natural resource endowments into industrial and manufacturing opportunities. They also prioritise supply chain resilience and national security, particularly for defence, clean energy, and advanced technology sectors that rely on uninterrupted access to rare earths and strategic minerals.

Another key objective is reducing reliance on dominant global suppliers, most notably in refining and processing, by fostering diversified, trusted supply chains among allied nations. Governments are simultaneously focused on building domestic capabilities, moving beyond resource extraction to establish refining, alloying, and manufacturing capacity. At the same time, there is a growing emphasis on sustainability and governance, embedding ESG standards and community engagement—particularly with Indigenous and local groups—into policy design and project approval processes.

Finally, international engagement and cooperation have become central pillars of modern critical minerals policy. Through trade agreements, investment partnerships, and shared standards, governments are working collectively to strengthen global supply networks and advance shared industrial goals.

For educators and training professionals, understanding these government strategies provides critical context for developing workforce capabilities in the minerals and energy sectors. Training programs can integrate lessons on how regulatory, financial, and policy mechanisms shape real-world operations, helping learners connect technical mining knowledge with broader economic and strategic priorities.

Emphasising the value-chain transition—from mining and processing to refining, manufacturing, and recycling—aligns training content with current government priorities. Learners can explore how permitting, investment, export controls, and technology development influence each stage of the chain.

Recognising that national strategies are driven by industrial policy, national security, and sovereignty, not just market economics, helps frame why critical minerals are viewed as strategic assets. Moreover, understanding incentive structures, grants, and procurement policies enables learners to grasp how large-scale mining and processing projects are financed and regulated.

Finally, highlighting the importance of international cooperation and supply chain diversification helps learners appreciate the global nature of critical minerals governance—and the role that diplomacy, trade frameworks, and standard-setting play in securing sustainable, secure, and competitive resource futures.

Logistics, Exports, and Downstream Manufacturing Links

Material Logistics and Exports

The logistics of critical minerals involve moving raw ores, concentrates or processed intermediates from mines to refineries, separation plants, manufacturing hubs and then to final goods production. These logistics cover transport of large volumes (sometimes low value per tonne) but also specialised handling (hazardous chemicals, international shipping, regulatory export controls).

Exports are a key link: many resource-rich countries export raw or semi-processed materials to nations that do the bulk of refining or manufacturing. For example, one recent commentary with the International Energy Agency (IEA) notes that for 19 of 20 strategic minerals one country dominates the refining stage and export controls are increasing. IEA Export restrictions (for example on rare earths or battery-materials) suddenly raise shipping/handling complexity, cause supply disruptions and impact downstream manufacturing.

Rare Earth and Critical Mineral Operations and Processing

China remains the dominant player in refining rare earths and manufacturing permanent magnets – for instance, ~91 % of global separation and refining of rare earths is concentrated there [440]. That means logistics chains for many countries rely on shipping processed intermediates from China (or via China) to manufacturing hubs, which creates geostrategic vulnerability.

Another key point: Many governments are now implementing export controls, strategic stockpiling, and reviewing trade flows of critical minerals. These policy moves can affect export logistics (longer lead times, licensing, increased costs) and indirectly influence where manufacturing links are located.

For example:

- Australia is increasing its support for infrastructure and downstream processing to link mine output to global markets.

- Export-finance mechanisms exist to support Australian projects that can move further up the value chain (not just shipping raw ore).

Downstream Manufacturing and Value-Chain Links

The term "downstream" refers to the stages that occur after raw mineral extraction. This includes the separation and refinement of ores into usable minerals or oxides, followed by alloy production, component manufacturing (such as magnets, batteries, and electronic parts), and finally, the assembly of finished goods like electric vehicles, wind turbines, aerospace equipment, and defence systems. These downstream stages represent the most technologically advanced and value-intensive parts of the critical minerals supply chain.

Downstream links are crucial because the highest value and technological complexity exist at these later stages of production. For example, the manufacture of high-performance permanent magnets from rare earth alloys involves specialised knowledge, advanced equipment, and significant investment. Countries that focus primarily on mining raw ore but lack processing and manufacturing capabilities often lose much of the potential value-add to nations that control these downstream processes. This imbalance limits economic returns and can create dependency on foreign markets for refined and finished products.

According to the International Energy Agency (IEA) and other analysts, the processing and refinement stage remains a significant bottleneck in global supply chains. The number of players capable of refining rare earths and other critical minerals is limited, and the barriers to entry are high due to technical complexity, environmental risks, and substantial capital requirements. This concentration of capability means that disruptions in a few key locations can ripple across global industries.

The logistics of moving intermediate products—for instance, transferring rare earth oxides to alloy facilities and then to magnet manufacturers—also require efficient coordination and strong international supply chain relationships. When manufacturing hubs are geographically distant from resource sources, shipping costs, lead times, and trade risks increase, adding further complexity and vulnerability to global supply networks.

Recognising these challenges, governments are now prioritising domestic and allied downstream manufacturing as a cornerstone of industrial and economic policy. Nations such as Australia, through its Critical Minerals Strategy 2023–2030, are investing heavily in local processing and manufacturing capacity to improve control over supply chains, capture greater economic value, and create high-skilled jobs. Strengthening these downstream links not only supports industrial diversification and innovation but also enhances national resilience against supply disruptions in an increasingly competitive and geopolitically sensitive global market.

Key Risks, Strategic Considerations and Implications

Logistics risks represent one of the most immediate challenges in the critical minerals supply chain. Disruptions in shipping—whether caused by port congestion, export controls, or broader trade restrictions—can delay or even halt the movement of intermediate or finished products. These disruptions are particularly severe when imposed by dominant suppliers; for instance, export restrictions from China have historically caused major global ripple effects, as noted by the International Energy Agency (IEA) [440]. Maintaining predictable and secure logistics channels is therefore essential to ensuring continuous production and delivery throughout the supply chain.

Many countries also face export reliance risks, particularly those that primarily export raw ores or concentrates without domestic processing or manufacturing capabilities. This reliance leaves them vulnerable to global price fluctuations, trade restrictions, and downstream market dependencies. Developing domestic or regional downstream manufacturing capacity helps reduce this vulnerability by allowing nations to retain greater control over their resources and benefit from higher-value stages of production.

Another critical issue is manufacturing clustering, where much of the world's processing and manufacturing capacity is concentrated in a small number of countries. This concentration creates systemic risk: any geopolitical tension, policy shift, or environmental regulation change in those nations can disrupt entire global supply networks. The COVID-19 pandemic and subsequent trade disputes have demonstrated how such clustering can amplify vulnerability across interconnected industries.

Rare Earth and Critical Mineral Operations and Processing

To address these challenges, governments and industries alike are pursuing a diversification imperative—the need to distribute logistics and manufacturing capacity across multiple regions. Geographic diversification reduces chokepoints, enhances resilience, and allows for greater flexibility in responding to market or geopolitical shocks. Building new downstream capacity in allied and emerging economies is a key part of this strategy, fostering more stable and equitable global supply chains.

Equally important is value capture—the ability for countries to retain more economic benefit from their natural resources by developing domestic industries that extend beyond mining. Capturing more of the value chain creates jobs, stimulates industrial development, and enhances strategic independence by reducing reliance on foreign manufacturing.

Effective policy and infrastructure development underpin all these efforts. Downstream manufacturing requires reliable transport systems, affordable and sustainable energy, advanced processing facilities, a skilled workforce, and regulatory certainty. Logistics hubs, export terminals, and transport corridors must be planned alongside manufacturing investments. For example, Australia's Critical Minerals Strategy emphasises enabling infrastructure and coordinated planning to ensure that mining, processing, and export systems are fully integrated.

Finally, the recycling and circular economy dimension is becoming increasingly important. Downstream manufacturing links directly to recycling logistics, which involve collecting, processing, and reintegrating materials from end-of-life products back into the supply chain. Developing these closed-loop systems not only reduces waste and environmental impact but also adds resilience by supplementing primary mineral supplies with recycled inputs. Together, these measures form a comprehensive approach to strengthening global logistics and downstream manufacturing ecosystems for critical minerals.

Production Costs

Providing exact cost figures for mineral production is inherently difficult due to the wide variability in underlying factors. Costs can differ significantly depending on the deposit characteristics, ore grade, mining method, recovery rates, processing technology, energy and labour costs, environmental and regulatory compliance requirements, and geographic location. These variables make it challenging to establish a universal cost benchmark across projects or regions.

Moreover, many companies do not publicly disclose detailed cost breakdowns per tonne for individual elements, particularly for rare earths and high-value specialty materials. This lack of transparency limits the availability of granular data and complicates efforts to compare or model costs accurately.

Available cost models often aggregate multiple elements into bundled products, such as rare earth oxides (REO), rather than isolating costs for each element. For instance, one model estimates a blended cost of approximately US $70/kg (or US $70,000/tonne) for a combined REO product to achieve a 20% return on investment [441]. However, such models are illustrative and not universally applicable.

Historical data provides some insight but is often outdated or specific to individual projects. For example, operating cost estimates for REO production in 2013 ranged from US $8,480 to US $17,280 per tonne, depending on the mine [442]. These figures reflect past conditions and may not align with current market realities.

In contrast to production costs, price data for base and critical minerals—such as lithium, nickel, cobalt, and copper—is more readily available. For example, lithium hydroxide was forecast to reach around US $15,870/tonne in 2024 [442]. However, this represents market price, not production cost, and the two figures can diverge significantly.

Illustrative data points further highlight the complexity. One report cited the price of Terbium oxide at approximately 7.1 million yuan per tonne in 2025 [434]. Yet, for many rare earth mining projects, the cost of mining ore and converting it to REO varies widely based on throughput and ore grade, reinforcing the challenge of standardizing cost estimates.

Ultimately, while price forecasts offer some visibility, accurate and consistent production cost data remains elusive—especially at the level of individual elements.

Below is a stitched-together view of cost-of-production proxies for some of the better-reported critical minerals. This is not a single global cost curve — it mixes public company guidance, government/industry reviews, and analyst estimates (with flagged uncertainty where it's high). All figures are USD/tonne unless stated otherwise.

Table 6: Indicative Cost/Proxy Table.

Mineral / group	Typical product form in source	Indicative cost / proxy range	Notes on uncertainty / drivers	Main public sources
Rare earths (mixed REO)	Mixed rare earth oxide (REO)	8,000 – 20,000 $/t REO (opex, project-level, older studies); up to 70,000 $/t as an economic hurdle in newer modelling	Very wide range: ore grade, radioactivity handling, reagent cost, by-product credits, scale, Chinese vs ex-China operations. Separation into single REOs adds	Earlier project studies on REO mining/ separation; economic modelling for new non-Chinese supply

Rare Earth and Critical Mineral Operations and Processing

Mineral / group	Typical product form in source	Indicative cost / proxy range	Notes on uncertainty / drivers	Main public sources
			substantial extra cost.	
Nd/Pr (magnet REO focus)	Separated Nd/Pr oxide (China domestic ref.)	Price in 2025 at ≈ ¥445,000/t (~60–65k USD/t at 6.8–7.0 ¥/$); production cost typically well below price, but rarely disclosed.	Cost depends on having a mixed REO stream already. Non-Chinese plants face higher capex/ESG, so cost can be closer to 25–40k $/t.	SMM China rare earth pricing; analyst commentary on ex-China magnet REEs
Dy / Tb (heavy REEs)	Dy or Tb oxide	Production/cost proxies often >100,000 $/t and can run to several 100k $/t because ore bodies are small and processing is complex	Small volumes, few producers, often recovered as a minor stream → unit costs look very high; prices projected to rise strongly toward 2034.	Heavy REE market outlooks, 2025 supply-tightness commentary
Lithium	Lithium hydroxide monohydrate (LiOH·H$_2$O) / lithium carbonate (LCE)	Hard-rock cash costs commonly 4,000 – 8,000 $/t LCE; brine legacy assets can be 3,000 – 6,000 $/t; high-cost converters 8,000 – 10,000 $/t	Costs swing with reagent and energy prices; integrated mine-to-chemicals is cheaper than toll converting spodumene; ex-China conversion is higher. Market prices in 2024–25 (~15–16k $/t) sit well above low-cost brine opex.	WA Battery & Critical Minerals profile 2024; industry cost curves 2024–25
Nickel	Nickel in sulphate / Class-1 nickel (Ni metal equivalent)	Laterite HPAL: 14,000 – 20,000 $/t Ni opex on	Enormous spread: ore type (sulphide vs	Battery-supply assessments

Mineral / group	Typical product form in source	Indicative cost / proxy range	Notes on uncertainty / drivers	Main public sources
		difficult assets; good sulphide operations: 9,000 – 13,000 $/t Ni; Chinese NPI-to-sulphate route often higher	laterite), acid cost, by-products, power, carbon rules. Battery-grade Ni tends to the upper half.	2024–25; analyst nickel opex bands (multiple)
Cobalt	Cobalt hydroxide/intermediate, refined Co metal	DRC Cu-Co by-product (mine-site) often 3,000 – 6,000 $/t Co before logistics; refined ex-Africa into metal/sulphate more like 10,000 – 15,000 $/t	True standalone Co mines are rare; most Co is a by-product, so "cost" depends on how credits are allocated. Logistics/political risk pushes effective cost up.	Supply-chain commentary 2024–25 on Co for batteries (multiple)
Graphite (natural, battery anode)	Spherical purified graphite	Flake mining + spheronisation/ purification often 1,500 – 3,000 $/t; synthetic graphite anode much higher, 3,000 – 5,000 $/t+	Purification route (HF vs thermal), power price, and China vs ex-China location are the big cost levers.	Battery materials market outlooks 2024–25
Copper	Cathode	Large Latin American/US operations 5,000 – 7,000 $/t Cu C1; higher-cost or deeper orebodies 7,000 – 9,000 $/t	Treatment & refining charges, ore grade decline, and water/energy costs push costs up; by-products (Au, Mo) can lower net cash cost.	Global mining cost commentary 2025 (multiple)
Aluminium (from alumina)	Primary aluminium (smelter)	Power-driven: many smelters sit 1,500 – 2,000 $/t Al cash cost; high-	Electricity is 30–40%+ of cost. Hydro-powered	Industry aluminium cost

Rare Earth and Critical Mineral Operations and Processing

Mineral / group	Typical product form in source	Indicative cost / proxy range	Notes on uncertainty / drivers	Main public sources
		cost smelters >2,200 $/t	smelters are lower.	benchmarkin g 2024–25
Tungsten	APT equivalent	Many operations need 18,000 – 25,000 $/t APT to be economic; low-cost Chinese operations below that	Highly deposit-specific; by-products and grades matter a lot.	Critical minerals market reports 2024–25
Scandium	Sc_2O_3 (10–99.9% purity)	Very high: often >1,000,000 $/t equivalent for small streams recovered from laterite/processin g residues	Sc is mostly co-/by-product; very little dedicated capacity → costs are modelled, not observed.	Advanced materials and critical minerals analyses 2024–25
PGMs (Pt, Pd, Rh, Ru)	Refined metal	Costs usually reported per 4E/6E ounce basket, but back-converted, many South African PGM operations sit 25,000 – 40,000 $/t metal equivalent; Rh effectively higher	Energy, labour, deep shafts, and complex smelting drive costs. By-product credits (Ni, Cu, Co, Au) can reduce unit cost.	PGM cost commentary 2025 (SA & Russia)

These figures represent ranges rather than fixed costs. For example, a high-grade, efficiently run project like Mount Weld in Western Australia will typically operate near the lower end of the rare-earth cost band, whereas smaller, ex-China producers that face high environmental, social, and governance (ESG) compliance costs tend to fall toward the upper end.

Product form is another key variable. The cost of producing a raw oxide, a refined metal, an alloy, or a battery-grade chemical differs substantially even when all are derived from the same ore body. Each processing step adds energy use, chemical reagents, and technical complexity, which raise unit costs.

Geographical location also strongly affects operating expenses. China generally maintains lower costs because of its established industrial clusters, mature processing technologies, and economies of scale. In contrast, ex-China projects often need either higher selling prices or government support to remain commercially viable—precisely the challenge that recent cooperative frameworks among Australia, the United States, and Japan aim to address.

Finally, by-product credits can distort apparent production costs. Metals such as cobalt, scandium, certain rare earths, and tungsten may seem inexpensive only because their extraction is subsidised by revenue from a primary metal like copper or nickel. When the main metal carries most of the cost burden, the secondary material's "effective" cost appears artificially low.

Electricity Demand

It is not possible to assign a single, exact kilowatt-hour (kWh) value per tonne for every metal because electricity requirements vary greatly depending on the processing route, ore grade, geographic location, and the share of thermal versus electrical energy used. However, we can establish approximate electricity-intensity bands that reflect the typical energy consumption for each major production pathway and highlight where uncertainty is highest.

For rare earth elements (La–Lu and Y), most energy use occurs during crushing, beneficiation, acid or alkali cracking, and solvent extraction. A significant portion of this energy is thermal rather than electrical, and the electricity share differs from plant to plant. Light rare earth elements such as lanthanum, cerium, praseodymium, and neodymium produced from hard-rock deposits typically require 3,000–8,000 kWh of electricity per tonne of rare earth oxide (REO). Larger, integrated Chinese facilities with high-grade ores tend to operate near the lower end of this range, while smaller, ex-China operations with stricter environmental standards sit toward the upper end. The conversion of magnet REEs such as praseodymium and neodymium from oxide to metal (via electrolysis or metallothermic reduction) adds another 1,000–3,000 kWh per tonne of metal. Heavy rare earths like dysprosium and terbium generally demand 4,000–10,000 kWh/t, given their lower ore grades and longer separation processes. Promethium, however, is only produced in gram-scale quantities from nuclear by-products, making tonne-scale energy data meaningless.

In battery and energy transition minerals, lithium from spodumene ore requires around 200–400 kWh/t at the mining and concentration stage, and 1,000–2,500 kWh/t during roasting, leaching, and refining—resulting in a combined total of roughly 1–3 MWh per tonne of lithium chemical. Nickel's electricity consumption depends on ore type: sulphide operations consume 2,000–4,000 kWh/t Ni, while laterite projects using high-pressure acid leach (HPAL) can use 4,000–8,000 kWh/t Ni, with much of the energy supplied as heat or steam. Cobalt, typically a by-product of copper or nickel production, accounts for 1,000–3,000 kWh/t Co of allocated electricity use, with higher demand for high-purity sulphates. For graphite, natural flake processing uses 300–800 kWh/t, while producing battery-grade spherical graphite raises the requirement to 1,000–3,000 kWh/t due to purification steps.

Among base and industrial metals, copper requires 2–4 MWh/t of electricity for smelting and electro-refining, with electrolysis being the largest contributor. Aluminium is highly electricity-intensive, requiring 12–16 MWh/t in the Hall–Héroult process. Electrolytic manganese metal production typically uses 6–8 MWh/t, while titanium sponge production through the Kroll process demands around 10–15 MWh/t. Zirconium, refined from zircon, consumes between 5,000 and 10,000 kWh/t, reflecting its small-scale, highly specialised processing.

For technology and minor metals, energy use varies widely because of small production scales and complex chemistry. Gallium extraction from alumina or zinc residues can exceed 10,000 kWh/t, though total global production is only a few tonnes per year. Germanium follows a similar pattern, with several megawatt-hours per tonne of allocated electricity. Antimony, chromium, and vanadium smelting typically require 3,000–8,000 kWh/t, tungsten around 3,000–6,000 kWh/t, and difficult-to-process elements like beryllium, tantalum, and niobium consume between 5,000 and 12,000 kWh/t.

Finally, platinum group metals (PGMs) such as platinum, palladium, rhodium, and ruthenium are by-products of energy-intensive base-metal mining and refining chains. When electricity use is allocated back to the small volumes of precious metal produced, the figures can reach tens or even hundreds of megawatt-hours per tonne. For general reference, it is reasonable to describe PGMs as very high electricity metals, typically requiring well above 10 MWh per tonne, though the actual value depends entirely on the chosen energy allocation method.

Key Terms and Concepts	
Value Chain: The full sequence of activities involved in producing a product—from raw material extraction to processing, manufacturing, distribution, and end use.	

Key Terms and Concepts	

Supply Chain: The network of organisations, resources, and logistics involved in sourcing, processing, and delivering critical minerals to global markets.

Downstream Processing: The refining, separation, and manufacturing stages that add value to raw mineral concentrates and produce usable materials or components.

Upstream Operations: The early stages of the industry that include exploration, mining, and initial ore beneficiation.

Midstream Sector: The processing stage between mining and manufacturing, where minerals are refined, purified, and converted into industrial-grade materials.

Market Dynamics: The forces that influence the supply, demand, and pricing of commodities within global markets.

Geopolitics: The study of how political, geographic, and economic relationships between nations influence access to and control over critical mineral resources.

Strategic Reserve: A government-held stockpile of essential minerals or materials to mitigate supply disruptions and ensure national security.

Commodity Pricing: The system through which raw material prices are determined based on market demand, production costs, and global trade conditions.

Critical Minerals Strategy: A national or regional policy framework designed to secure supply chains, promote exploration, and encourage downstream processing.

Market Diversification: The practice of developing multiple supply sources, customers, or markets to reduce dependence on any single supplier or region.

Trade Restrictions: Government-imposed controls, such as export bans or tariffs, that affect the global flow of minerals and materials.

Processing Hub: A region or facility that specialises in refining and producing value-added critical mineral products for export.

Key Terms and Concepts

Sourcing Transparency: The practice of documenting and disclosing the origins and supply routes of materials to ensure ethical and sustainable procurement.

Responsible Sourcing: The procurement of materials in ways that meet environmental, social, and governance (ESG) standards and avoid human rights violations.

Logistics: The management of the movement and storage of materials, including transport networks, ports, and infrastructure supporting mineral exports.

Industrial Policy: A government's strategic plan to support specific sectors, such as critical minerals, to promote national economic and technological growth.

Chapter 11 Review Questions

Global Industry Structure and Value Chain

1. Short answer: Describe the main stages of the rare earth and critical minerals value chain.

2. Multiple choice: Which stage of the value chain adds the greatest technological and economic value?

(a) Exploration and Mining

(b) Separation and Refinement

(c) Alloying and Manufacturing

(d) Stockpiling and Transport

3. True/False: The majority of global refining and separation capacity for rare earths is located in the United States.

4. Matching: Match each stage of the value chain with its description:

Stage 1   Exploration & Mining

Stage 2   Separation & Refinement

Stage 3   Alloying & Intermediate Products

Stage 4   Manufacturing & Final Applications

Stage 5   Recycling / Secondary Supply

Stage 6   Logistics, Stockpiling, Strategic Reserves

Major Producing Countries and Companies

5. Short answer: Identify three leading countries in rare earth production and describe one major operation in each.

6. Multiple choice: Which of the following operations is the largest rare earth mine outside China?

(a) Mount Weld (Australia)

(b) Bayan Obo (China)

(c) Mountain Pass (USA)

(d) Ngualla (Tanzania)

7. Short answer: What role does Africa currently play in the global rare earth market?

Market Forces, Pricing, and Supply–Demand Trends

Rare Earth and Critical Mineral Operations and Processing

8. Multiple choice: Which factor most significantly affects rare earth oxide pricing?

(a) Ore grade

(b) Concentration of refining capacity

(c) Exchange rates

(d) Weather conditions

9. Short answer: What are two key reasons for price volatility in the rare earth market?

10. Data interpretation: In 2025, Praseodymium-Neodymium oxide was priced at ¥445,000 per tonne. What factors could cause this price to rise by 2034?

Geopolitical and Strategic Factors

11. Short answer: Explain how export controls can affect global supply chains for critical minerals.

12. True/False: The 2025 U.S.–Australia Critical Minerals Framework creates legally binding trade obligations.

13. Short answer: How do strategic alliances help mitigate geopolitical risks in mineral supply chains?

Regional Roles and Comparative Advantages

14. Essay question (extended response): Compare the roles of Australia, China, and the United States in the global critical minerals supply chain.

15. Short answer: Why is Australia's Critical Minerals Strategy 2023–2030 significant in the global context?

Government Policies and Strategies

16. Multiple choice: Which policy mechanism is *not* typically used by governments to promote critical minerals development?

(a) Tax offsets

(b) Strategic reserves

(c) Price ceilings on consumer goods

(d) Grants and concessional loans

17. Short answer: Name two ways the U.S. government supports domestic processing and manufacturing of critical minerals.

Logistics, Transport, and Exports

18. Short answer: Describe two logistics risks that can disrupt critical mineral supply chains.

19. True/False: Export reliance on raw ore strengthens a country's economic resilience.

20. Scenario question: A country exports only unprocessed ore to China for refining. What steps could it take to capture more value domestically?

Downstream Manufacturing and Recycling

21. Multiple choice: What is meant by "downstream" in the critical minerals value chain?

(a) Early-stage exploration and mining

(b) Midstream refining and separation

(c) Manufacturing and assembly of finished goods

(d) Stockpiling and transport

22. Short answer: Why is recycling considered an emerging stage in the value chain?

Market Trends and the Energy Transition

23. Short answer: Which industries are driving the largest increase in rare earth demand?

24. Multiple choice: The global rare earth market is expected to grow from USD 3.39 billion in 2023 to approximately USD 8.14 billion by:

(a) 2027

(b) 2030

(c) 2032

(d) 2040

25. Short answer: What is the main supply challenge associated with the transition to clean energy technologies?

Standards, Certification, and Responsible Sourcing

26. Short answer: What is the purpose of responsible sourcing initiatives?

27. Matching: Match the following terms with their definitions:

- Value Chain
- Supply Chain
- Responsible Sourcing
- Strategic Reserve
- Processing Hub

Definitions:

Rare Earth and Critical Mineral Operations and Processing

- o Steps from raw extraction to end use
- o Facility for refining critical minerals
- o Network of logistics and production links
- o Government stockpile of essential materials
- o Ethical and ESG-compliant procurement

28. Essay question (extended response): Discuss the importance of transparency, ESG standards, and certification schemes in ensuring sustainable global mineral supply chains.

Chapter 12

Future Skills, Technology, and Career Pathways

The global shift toward clean energy, advanced manufacturing, and resilient supply chains is transforming the rare-earth and critical-minerals industry. Once driven primarily by geological discovery and mechanical processing, today's sector operates at the intersection of digital technology, materials science, and sustainability. Automation, artificial intelligence (AI), green chemistry, and low-carbon refining are redefining how minerals are explored, extracted, and processed, while international policy frameworks are reshaping where value is created along the supply chain.

As the industry evolves, so too must its workforce. The emerging landscape demands professionals who can bridge traditional engineering with data analytics, automation, and environmental stewardship. Skills once confined to separate disciplines—metallurgy, chemistry, digital systems, and project management—are now converging. This chapter explores these transformations by examining the technologies driving digitalisation and low-carbon processing, the human capabilities required to implement them, and the new career pathways emerging across exploration, refining, manufacturing, and recycling. It concludes by outlining research and innovation priorities that will shape the sector's future, highlighting how education, collaboration, and lifelong learning will underpin a sustainable and technologically advanced minerals economy.

Learning Outcomes	
This chapter aims to give you the ability to: 1. Identify the emerging technological trends transforming the rare earth and critical minerals sector, including automation, artificial intelligence (AI), and digital process control.	

Learning Outcomes	

2. Explain how data analytics, machine learning, and remote sensing are improving exploration, extraction, and processing efficiency.
3. Describe the application of green chemistry, low-carbon technologies, and renewable energy integration in reducing the environmental footprint of mineral operations.
4. Recognise the increasing importance of cross-disciplinary skills that combine engineering, environmental science, data analysis, and project management in the modern mineral industry.
5. Discuss the workforce changes driven by digital transformation, decarbonisation, and automation, and their implications for training and professional development.
6. Identify key career pathways within the critical minerals industry, including roles in exploration, processing, quality control, research, and sustainability.
7. Explain the value of vocational education, higher education, and professional accreditation in supporting career progression and skill recognition across the sector.
8. Assess the need for continuous learning and upskilling in response to evolving technologies and global market demands.
9. Discuss how innovation, collaboration, and lifelong learning contribute to a resilient, future-ready critical minerals workforce.
10. Develop a personal or organisational plan for career growth and capability development aligned with emerging industry trends and sustainability goals.

Digitalisation, Automation, and AI in Mineral Processing

Digitalisation, automation, and artificial intelligence (AI) are transforming the way rare earth and critical minerals are discovered, processed, and refined. Digitalisation refers to the systematic collection, integration, and utilisation of data across operations. It involves the use of sensors, Internet of Things (IoT) devices, control systems, and software platforms that allow mining and processing facilities to monitor conditions in real time, analyse performance, and make data-driven decisions. This approach enables a seamless flow of information across the entire operation—from exploration and extraction to refining and product delivery—creating greater transparency and efficiency.

Automation involves the deployment of advanced machinery, robotics, control systems, and autonomous technologies to reduce manual intervention in mining, mineral processing, and refining. Through automated drilling rigs, robotic sampling, real-time control loops, and self-optimising circuits, automation not only enhances precision and consistency but also improves worker safety and operational uptime.

Richard Skiba

Artificial intelligence (AI) and machine learning build on this digital foundation. These technologies use algorithms that learn from data—whether supervised, unsupervised, or reinforcement-based—to detect patterns, predict outcomes, optimise process parameters, and support decision-making. In the mineral processing sector, AI can model complex chemical and physical processes, identify process inefficiencies, and recommend adjustments that maximise yield and minimise waste.

Together, these three technologies—digitalisation, automation, and AI—are reshaping every stage of rare earth and critical mineral value chains. Their impact extends from exploration and mine planning to ore processing, separation, refining, and downstream manufacturing, as well as to recycling and recovery of critical materials. This digital transformation is enabling smarter, cleaner, and more resilient operations, positioning technologically advanced producers at a strategic advantage in an increasingly competitive global minerals market.

The integration of digitalization, automation, and artificial intelligence (AI) throughout the rare earth and critical mineral value chain is transforming exploration, processing, and downstream manufacturing stages, enhancing overall efficiency, safety, and sustainability.

In the initial stages of exploration and mine planning, AI is increasingly applied to analyse vast geological datasets, leveraging machine learning algorithms to identify new deposits of rare earth elements (REEs) with higher accuracy and speed. For instance, the use of ambient noise tomography combined with AI has shown improvements in deposit delineation and enhances the understanding of subsurface structures which is crucial for effective mine planning [443]. Moreover, drones equipped with multispectral and hyperspectral sensors, along with remote sensing capabilities facilitated by Internet of Things (IoT) platforms have improved monitoring of environmental conditions and geological features, ensuring a more data-driven approach to drilling that minimizes ecological disturbances [444].

As operations transition to process optimization, digital twins emerge as transformative technologies. These virtual simulations of processing plants allow operators to model production stages—from crushing and beneficiation to extraction and separation—providing insights to optimize recovery rates and reduce operational downtime [443]. Additionally, AI-based ore sorting, utilizing advanced sensors like X-ray, laser, and optical systems, has been instrumental in differentiating high-grade ore from waste materials. Such systems substantively reduce energy consumption during downstream processing, cutting down on waste [169]. Predictive maintenance systems are now key components, relying on real-time sensor data to forecast mechanical failures proactively, further enhancing process reliability and worker safety [445].

In the context of downstream manufacturing and recycling, the incorporation of digital platforms is paramount for the production of essential components like magnets and batteries. Blockchain technology, when integrated with IoT and data analytics, ensures high transparency in tracking material provenance and compliance with environmental and social governance (ESG) standards [443]. This approach also aligns with circular economy practices, facilitating the recovery and reuse of critical materials throughout the value chain [444].

Rare Earth and Critical Mineral Operations and Processing

Sustainability has become a focal point, with digitalization playing a pivotal role in minimizing environmental impacts associated with rare earth extraction and processing. Automated monitoring systems enhance energy efficiency and control emissions during processing, which is vital given the energy-intensive nature of rare earth elements refining [169]. Furthermore, automation contributes to reducing worker exposure to hazardous processes, significantly advancing safety standards across the industry [446].

The adoption of digitalisation, automation, and artificial intelligence in rare earth and critical mineral processing brings a wide range of tangible benefits that enhance both operational performance and strategic positioning.

One of the most significant advantages is improved recovery and yield. Through the optimisation of crushing, separation, and sorting circuits, operators can achieve higher metal recovery rates, increase ore grade, and minimise waste generation. This results in greater extraction efficiency and better utilisation of valuable resources, particularly critical in deposits with low-grade or complex mineralogy.

Another key benefit is lower cost and energy use. Automated systems and AI-driven process optimisation streamline workflows, minimise manual labour requirements, and reduce production bottlenecks. These technologies enable fine-tuning of operations to achieve optimal energy consumption across both electrical and thermal systems, helping lower overall operating expenses while supporting sustainability goals.

Digital technologies also contribute to a faster time-to-market. Enhanced exploration targeting through AI-driven data analysis and the ability to rapidly model and optimise processing systems mean that new rare earth and critical-mineral projects can move from discovery to production more quickly. This accelerated project timeline is particularly important given the increasing global urgency to secure reliable critical-mineral supply chains.

Improved supply-chain transparency is another major value proposition. The integration of blockchain, IoT, and data analytics facilitates digital traceability throughout the production process. This ensures compliance with environmental, social, and governance (ESG) standards, supports responsible sourcing, and mitigates reputational and regulatory risks related to material provenance.

The use of predictive maintenance systems and digital twins significantly enhances risk reduction. By continuously monitoring plant performance, these tools can forecast equipment failures, prevent unplanned outages, and minimise safety and environmental incidents. This leads to more stable operations, improved safety outcomes, and reduced downtime.

Digital and data-driven processing capabilities enable downstream integration by strengthening refining and manufacturing competitiveness—particularly for producers outside traditional dominant regions such as China. By reducing processing costs, improving throughput, and enhancing consistency, these technologies help new and emerging supply chains become more resilient, efficient, and globally competitive.

Despite the clear advantages of digitalisation, automation, and artificial intelligence in mineral processing, several challenges and limitations affect their widespread adoption and effectiveness.

A major issue lies in data quality and integration. Many mineral-processing facilities—particularly older ones—lack sufficient sensor coverage, coherent data infrastructure, and consistent metadata. Since AI models rely heavily on accurate and comprehensive datasets, poor data quality can result in unreliable or misleading outputs. Without systematic data collection and integration, the potential of AI to optimise processing performance remains limited.

Scale and variability also pose difficulties. Rare earth and other critical-mineral operations often deal with small production volumes, variable ore grades, and complex metallurgical characteristics. This variability makes it difficult for AI models trained on one deposit or processing circuit to perform effectively on another. Research published in *Nature* notes that while AI can improve efficiency, it cannot fully overcome deeper structural investment shortfalls in critical minerals in the short to medium term.

The cost of technology and adoption presents another barrier. Implementing digital and automated systems requires significant upfront investment in sensors, control systems, software, workforce training, and workflow redesign. Smaller or emerging producers—especially those outside established mining regions—may find it difficult to finance these upgrades without government incentives or partnerships.

The workforce and skills gap is equally important. As mining and mineral processing become more digitally driven, demand is rising for professionals skilled in data science, automation engineering, and digitally enabled metallurgy. The *Minerals Council of Australia*'s "Digital Mine" review emphasises the urgent need for training and workforce transformation to support this technological shift.

There are also cybersecurity and system vulnerabilities associated with digital integration. Increased connectivity through IoT devices and networked control systems raises the risk of cyberattacks, system failures, or data breaches—issues of particular concern given the strategic importance of critical-mineral supply chains.

From a technical standpoint, metallurgical complexity remains a major limitation. Rare earth separation and refining processes are chemically intricate, often requiring customised approaches for specific deposits. While AI and automation can enhance control and efficiency, they cannot eliminate fundamental challenges such as low ore grades or high impurity levels.

Finally, geopolitical and regulatory risks continue to shape the operational environment. Even the most advanced digital systems can be undermined by trade restrictions, export controls, or sudden policy shifts affecting international mineral supply chains.

For producers, early adoption of digital and automated systems can provide a competitive advantage, reducing costs, improving recovery and grade, and accelerating project timelines—

particularly valuable in a market characterised by supply constraints and growing geopolitical urgency.

For countries seeking to build downstream processing and manufacturing capacity outside dominant regions such as China, digitalisation offers a means of overcoming scale and cost disadvantages. By integrating sensors, automation, and AI from the outset, new projects can develop "smart processing plants" that achieve higher efficiency, lower energy use, and stronger environmental, social, and governance (ESG) outcomes.

Digital tools also enhance supply-chain transparency through blockchain, IoT, and analytics, helping miners and refiners demonstrate material provenance and compliance with responsible-sourcing standards increasingly demanded by sectors like electronics, electric vehicles, and defence.

One widely used platform is NIAflow, a process-modelling and simulation software designed for mineral processing plants. Engineers use NIAflow to construct virtual flowsheets covering stages such as crushing, grinding, flotation, and sorting. The software allows users to simulate material flows and optimise plant layouts and equipment performance, reducing design time and improving overall process efficiency.

Another key tool is QEMSCAN (Quantitative Evaluation of Minerals by Scanning Electron Microscopy), an automated mineral-characterisation system that combines scanning electron microscopy (SEM), energy-dispersive X-ray spectroscopy (EDS), and advanced image processing. QEMSCAN provides detailed insights into mineral liberation, grain size, and gangue associations, helping engineers and geologists refine sorting and processing strategies for REE and other complex mineral deposits.

Sensor-based ore-sorting systems have also become an important innovation. These technologies integrate high-resolution optical, laser, and X-ray sensors with real-time data analytics and automated sorting mechanisms to distinguish high-grade material from waste. By removing low-value ore before fine grinding or chemical processing, sensor-based sorting reduces energy use, increases feed grade, and lowers downstream processing costs.

AI is also transforming the exploration phase through tools such as GAIA (Geoscience Artificial Intelligence and Assessment), developed by the U.S. National Energy Technology Laboratory (NETL). GAIA applies AI algorithms to large geological and geophysical datasets, enabling faster identification of potential critical-mineral deposits, including unconventional sources. This accelerates exploration efforts and improves targeting accuracy.

In processing and plant optimisation, digital twins—virtual replicas of physical operations—are increasingly being adopted. These models simulate entire process flows, from mine to mill and concentrator, incorporating particle-scale data to track material movement, evaluate scenarios, and optimise recovery and throughput. Digital twins allow operators to test changes virtually before implementation, improving efficiency and reducing operational risk.

Integration of IoT, automation, and drones is another frontier in REE and critical-mineral operations. Networks of sensors, remote-sensing devices, and drones collect real-time data

on parameters such as terrain conditions, overburden thickness, equipment health, and environmental indicators. These data streams feed into AI-driven analytics platforms that support predictive maintenance, optimise logistics, and improve safety and environmental performance.

A commercial example of AI in action is Weir Group's AI-based optimisation platform, which applies machine learning to mineral-processing circuits to reduce downtime, improve throughput, and predict equipment failures. Although not specific to REE processing, such systems are readily adaptable to critical-mineral operations that share similar equipment and process challenges.

Collectively, these tools demonstrate how digital transformation is delivering measurable value. They improve recovery and efficiency by optimising sorting and processing, reduce energy use by cutting unnecessary grinding or chemical treatment, and enable faster, data-informed decision-making through real-time monitoring and simulation. They also enhance traceability and responsible sourcing, integrating data from mine to market in support of environmental, social, and governance (ESG) compliance. Finally, by lowering barriers to entry, digital technologies help emerging producers build competitive downstream capacity, allowing smaller or newer projects to achieve operational performance comparable to that of established global refineries.

Education and training systems must evolve to support this transformation, developing digital-metallurgical skillsets across data analytics, AI model development, process automation, and system control. Meanwhile, governments and policymakers can play a key role by funding pilot projects, offering innovation incentives, and ensuring regulatory frameworks support safe and efficient digital integration.

Implementing digitalisation, automation, and artificial intelligence (AI) in mineral processing—particularly within rare earth element (REE) and critical-mineral industries—requires a highly skilled and multidisciplinary workforce. This workforce must combine expertise in traditional mining and metallurgy with new capabilities in data science, automation, and cybersecurity. The transition to smart and connected operations demands people who can bridge engineering, digital technologies, and real-time decision-making. The following sections outline the key human resources, skills, and knowledge areas required across all organisational levels.

At the leadership level, managers and planners must possess both technological awareness and business transformation capabilities. Successful digital integration depends on leaders who can design and implement a clear digital roadmap across exploration, processing, and supply chain operations.

Key leadership competencies include digital transformation management, the ability to oversee large-scale integration of digital tools; change management, to reskill teams and foster a digital-first culture; and systems integration, ensuring coordination among automation vendors, software developers, and engineers. Strategic leaders must also negotiate partnerships with technology providers, research institutions, and governments, aligning innovation funding and infrastructure with national and corporate priorities. Importantly, leaders require a strong understanding of ESG (environmental, social, and governance) obligations to ensure that digital and automated operations meet sustainability and compliance expectations.

Although automation and AI are reshaping mineral processing, they still rely heavily on core process, chemical, and mechanical engineering foundations. Modern engineers must complement these fundamentals with digital skills such as process modelling, systems optimisation, and control logic design.

Process engineers now combine metallurgical expertise with data analysis and statistical modelling to optimise recovery and throughput. Automation and control engineers manage programmable logic controllers (PLCs), SCADA systems, robotics, and distributed control systems (DCS) that enable remote or autonomous operations. Digital twin specialists create and maintain virtual plant models that replicate real-world conditions, allowing operators to simulate scenarios and improve efficiency. Sensor and instrumentation engineers install, calibrate, and maintain IoT devices, drones, and environmental sensors for real-time process monitoring. Meanwhile, maintenance and reliability engineers apply AI-based predictive analytics to anticipate failures and optimise equipment uptime.

Data professionals form the bridge between traditional engineering and data-driven decision-making. Data scientists and machine learning engineers design and train AI models for applications such as ore sorting, recovery optimisation, and predictive maintenance, using programming languages like Python or R and frameworks such as TensorFlow or PyTorch.

Data engineers and analysts ensure that sensor, laboratory, and production data are clean, structured, and integrated into usable formats. MLOps specialists deploy and maintain machine-learning models in live production environments, continuously refining them as new data emerge. Computational geoscientists apply AI and geostatistical models for geological mapping, resource estimation, and exploration targeting. In addition, automation software developers write code that connects industrial hardware to AI platforms and dashboards, often through industrial IoT or edge-computing environments.

At the operational level, the move to digital and automated plants has transformed the skills required for on-site personnel. Technicians and operators must now be comfortable working with semi-autonomous machinery, interpreting digital dashboards, and understanding data from sensors. Basic data literacy and familiarity with human–machine interfaces (HMI) are essential.

Remote operations staff control equipment, drones, and processing systems from centralised command centres, requiring hybrid knowledge of engineering systems and IT networks. Health, Safety, and Environment (HSE) professionals need to update their skills to include digital risk assessment, cybersecurity in safety systems, and protocols for remote or automated operations. Meanwhile, trainers and technical educators play a vital role in developing competency frameworks and retraining traditional workers for emerging digital roles.

With the expansion of connected systems, maintaining data integrity and system security has become a top priority. IT infrastructure engineers manage industrial networks, cloud computing, and secure data storage across geographically distributed sites. Cybersecurity specialists protect operational technology (OT) and IoT networks against cyberattacks, ensuring compliance with standards such as ISO/IEC 27001 and safeguarding intellectual property. Systems integration architects ensure seamless interoperability between enterprise and plant-level systems, linking ERP, MES, SCADA, and AI platforms into a unified digital ecosystem.

Ongoing research and innovation underpin competitiveness in digital mineral processing. R&D scientists develop advanced extraction and separation techniques using machine learning and computational chemistry. University and industry partnerships support the testing of pilot plants, sensors, and AI algorithms under real-world conditions. Innovation program coordinators bridge academia, government, and industry to accelerate the commercial adoption of emerging technologies and ensure that research translates into practical applications.

Across all job levels, professionals in digital mineral processing require hybrid knowledge that integrates science, technology, and systems thinking. Core knowledge areas include:

- Fundamentals of mineral processing and metallurgy

- Digital literacy and data interpretation

- Process automation and control theory

- Machine learning and AI principles

- Sensor technologies and industrial IoT

- Energy efficiency and emissions optimisation

- Systems thinking and supply-chain awareness

- Cybersecurity and data ethics

- Environmental sustainability and ESG compliance

Finally, the convergence of AI, electrification, and clean-energy technologies is creating a virtuous cycle within the critical-minerals sector. As demand for these materials rises, digital and automated processing plants become more valuable—driving lower costs, faster scaling, improved sustainability, and stronger supply security across the global critical-minerals value chain.

Green Chemistry and Low-Carbon Refining Technologies

Green chemistry is defined as the design, development, and implementation of chemical processes and materials aimed at reducing or eliminating hazardous substances, minimizing waste, and lowering energy and environmental footprints. This framework becomes particularly critical in the mining and minerals-processing sectors, where the goal is to develop extraction methods, separation processes, and refining techniques that yield lower pollution levels, minimize the use of toxic reagents, and reduce tailings [447, 448]. The implementation of the "12 principles of green chemistry," specifically in the context of metal recovery, emphasizes essential tenets such as waste prevention, utilization of less hazardous reagents, and the efficiency of processes [449, 450].

In the realm of low-carbon refining technologies, innovative methods are being explored to minimize greenhouse gas emissions and reliance on fossil fuels during the ore conversion to value-added materials. Techniques include utilizing renewable electricity, hydrogen substitution, electrified leaching and roasting, and recycling [451, 452]. These advances hold significant importance in the critical minerals sector, as many essential minerals—such as rare earth elements (REEs), lithium, nickel, and cobalt—are foundational to clean energy technologies yet present substantial carbon footprints in their production processes [448]. Furthermore, the integration of renewable energy sources fosters a sustainable critical-minerals supply chain aligning with broader climate and environmental objectives [448].

The combined approaches of green chemistry and low-carbon refining not only aim to enhance sustainability in the extraction of resources but also underpin the supply chain necessary for clean energy and advanced manufacturing [453]. Researchers suggest that embracing the principles of green chemistry—including methodologies such as molecular recognition technology (MRT)—can significantly improve the sustainability of the metals industry by reducing waste generation, energy consumption, and resource utilization in processes such as metal beneficiation and recycling from waste [454, 455]. Remarkably, these strategies indicate a move towards a more circular economy within the mining sector, which can capitalize on waste reduction and resource conservation [456, 457].

The overarching impact of green chemistry and low-carbon strategies reveals a promising trajectory towards a more sustainable mineral processing and extraction paradigm, crucial for meeting the growing demands of low-carbon technologies while adhering to environmental standards [447, 448, 452]. Therefore, an emphasis on innovative extraction processes, green metrics, and sustainable methodologies is vital as the global economy transitions to cleaner energy solutions.

Rare earth elements (REEs) and critical minerals are pivotal for the advancement of clean energy technologies, highlighting their strategic value within modern industries. The application of REEs in high-performance magnets, and lithium, cobalt, and nickel in battery technologies underscores their importance in facilitating the shift to renewable energy sources, such as electric vehicles (EVs) and wind turbines, thereby playing a crucial role in addressing climate change [458]. The International Energy Agency (IEA) has recognized that these minerals are integral to the transition towards sustainable energy solutions and acknowledges the increasing global demand for them [458].

However, the traditional mining and refining processes involved in extracting these minerals present significant environmental challenges. These processes often consume high levels of energy and generate substantial carbon emissions, alongside large volumes of hazardous chemicals and waste. Techniques such as acid leaching and solvent extraction not only carry an environmental cost but also introduce a paradox where the very technologies designed to promote sustainability are undermined by the harmful practices employed for their production [459]. A comprehensive review of the chemical processes involved in rare earth extraction reinforces the need for more sustainable practices that minimize environmental damage while improving efficiency [459].

In response to these multifaceted challenges, there is a growing demand from governments, investors, and consumers for supply chains that are low-carbon, ethical, and transparent. Stakeholders, including EV manufacturers and alternative energy organizations, are pushing industries to adopt more sustainable practices throughout the supply chain. Companies that invest in green technologies not only enhance their reputations but also improve their access to necessary financing and regulatory incentives [460]. As noted by researchers, the adoption of advanced technologies such as the Internet of Things (IoT) and blockchain can significantly enhance traceability and transparency in supply chains, which is essential for meeting these ethical and environmental demands [461, 462]. Blockchain technology, in particular, offers a promising avenue for verifying product provenance, thereby reassuring end-users and investors of the sustainable nature of these minerals [463, 464].

From a national strategic perspective, countries like the U.S., Australia, and those in Europe are increasingly focused on developing domestic supply chains for critical minerals. This approach is not only about securing supply but also about ensuring that the mining and refining processes employed are conducted with a lower ecological footprint than those typically found in existing markets, predominantly in Asia [458]. Efforts toward process innovation, recycling secondary materials, and exploring alternative feedstocks have emerged as essential pathways to enhance the sustainability of mineral processing [465]. These innovations are vital for aligning with global sustainability goals while maintaining competitive advantages in the global market.

A range of innovative technologies and process improvements are being applied to promote green chemistry and low-carbon refining in critical-mineral processing. These approaches collectively aim to lower emissions, reduce energy use, and minimise the environmental impact of extraction and refining operations across the supply chain.

Modern mineral processing increasingly favours hydrometallurgical routes over traditional high-temperature pyro-roasting or smelting. These methods use chemical leaching at lower temperatures—often through acid or alkaline solutions combined with solvent extraction—to recover metals more efficiently while reducing energy consumption. Another emerging approach is direct extraction technology, particularly in lithium production from brines or clays, where processes bypass large evaporation ponds or energy-intensive thermal operations, as demonstrated by firms like Future Bridge Mining. In addition, research into deep eutectic solvents (DES), ionic liquids, and other novel reagent systems offers safer, more sustainable alternatives to conventional chemicals. Studies using DES for tungsten and arsenic removal through electrodialytic systems show that such solvents can recover valuable materials at significantly lower energy inputs, offering promise for broader application in secondary resource recovery.

Another key pathway toward low-carbon refining involves replacing fossil-fuelled furnaces and roasters with electrified heating systems powered by renewable or low-carbon energy sources such as solar, wind, or hydroelectricity. Several emerging technologies are also exploring hydrogen-based reduction processes as substitutes for carbon-intensive fuels like coke. While these innovations have primarily been developed for steelmaking, they have direct analogues in mineral refining, where hydrogen can serve as a clean reducing agent. Complementary measures—such as capturing waste heat, improving thermal insulation, and integrating heat recovery systems—further reduce total energy demand across the refining process.

Recycling is a cornerstone of sustainable and low-carbon critical-mineral supply chains. By recovering rare earth elements (REEs) and other critical minerals from end-of-life products—including electronics, magnets, and batteries—companies can offset the need for virgin ore and significantly cut emissions associated with mining. Studies such as *Green Recovery of Rare Earth Elements from Secondary Resources* (ScienceDirect) show that hydrometallurgical and electrochemical recycling can achieve high recovery rates using less energy and fewer harmful reagents. Designing closed-loop supply chains—where recycled feedstocks are reintroduced into production—further enhances sustainability by minimising waste and capitalising on the fact that secondary materials are often already partially refined.

Advances in separation and purification are another major driver of green processing. Techniques such as selective solvent extraction, ion exchange, membrane separation, and electrodialysis/electrowinning can replace traditional, chemical-heavy methods with cleaner and more efficient systems. These approaches reduce reagent use, heat load, and overall environmental impact. The principle of process intensification—designing systems that achieve more in fewer steps—also reduces the physical and energy footprint of refining plants. As highlighted in *Mining Technology*, there is a growing push to replace sulphur-based collectors and frothers with less hazardous chemical reagents, thereby aligning metallurgical processes with green-chemistry principles.

Finally, digitalisation and AI-driven optimisation play a critical enabling role in achieving low-carbon outcomes. Advanced process control systems can monitor real-time plant conditions, adjust reagent dosages, optimise leaching times, and improve separation efficiency. By integrating data analytics, automation, and predictive modelling, these systems reduce chemical waste, energy use, and tailings generation—all of which contribute to a smaller carbon footprint. Digital twins, machine-learning models, and IoT-based monitoring allow continuous improvement of process parameters, ensuring that critical-mineral refining operations become progressively more sustainable and efficient over time.

Training, Education, and Career Development in the Sector

The global rare earth element (REE) and critical-mineral supply chain, spanning mining, refining, magnet manufacturing, and recycling, is undergoing an unprecedented surge in hiring as nations race to secure supplies essential to clean energy, defence, and advanced technology sectors. For decades, China's dominance in this field rested not only on its vast resources and refining capacity but also on a deep, technically skilled workforce, giving it what many analysts have called an "absolute advantage" in rare earth production [466]. However, as geopolitical tensions and supply-chain risks grow, countries such as the United States and Australia are rebuilding domestic REE industries and expanding their workforces to reduce dependence on China. U.S. employment in rare earths—once negligible—has now reached several thousand workers and is expected to grow dramatically as new facilities come online .

While precise figures are hard to determine, U.S. employment in the rare earth supply chain is estimated at between 3,000 and 7,000 direct and indirect jobs, with thousands more projected as new operations scale up [466]. Companies such as MP Materials and USA Rare Earth are leading this expansion. MP Materials alone has created roughly 150 direct jobs and 1,300 indirect jobs at its Texas magnet facility, while its California operations employ more than 800 workers [466]. The company expects further growth as it ramps up refining and magnet manufacturing. Similarly, USA Rare Earth plans to add at least 100 employees at its Stillwater, Oklahoma magnet plant. Analysts suggest that building a fully resilient U.S. supply chain may require 5,000 to 7,000 highly skilled professionals, from mining engineers to materials scientists, to support exploration, refining, and advanced manufacturing [466].

The upstream stage of the REE supply chain, including exploration, mining, and primary processing, is witnessing strong hiring as new projects come online across the U.S., Australia, Africa, and Southeast Asia. Companies need geologists, mining engineers, metallurgists, and field technicians to locate, extract, and process ore, alongside environmental and safety specialists to ensure regulatory compliance. For example, Lynas Rare Earths in Western Australia has been hiring mine geologists, metallurgical technicians, and processing supervisors for its Mount Weld operations and new Kalgoorlie facilities. These roles demand expertise in mineral processing, ore-grade control, flotation, and XRF analysis [466].

In the U.S., MP Materials—operator of the Mountain Pass mine in California, the country's only active rare earth mine—has expanded its workforce significantly. Its job postings in 2025 include roles such as junior mine engineer, mill technician, and process engineer. Other firms, such as Ramaco Resources, are diversifying into rare earths, hiring senior executives and project leaders to launch new extraction initiatives. Globally, new mines in Texas, Wyoming, Canada, and Africa are expected to add thousands of jobs as countries invest billions in domestic mining capacity. The upstream workforce is therefore expanding beyond geologists and engineers to include specialists in health and safety, environmental management, and community engagement to support sustainable development [466].

The refining and separation stage—transforming mined ore into purified oxides—is one of the most critical and labour-intensive parts of the rare earth value chain. Historically concentrated in China, this stage is now being rebuilt in the West. Demand is high for chemical engineers, process engineers, and metallurgists skilled in hydrometallurgy, solvent extraction, and separation technologies.

In Australia, Lynas is recruiting processing supervisors, chemical technicians, and health and safety officers for its new Kalgoorlie refinery, while in Malaysia, the company employs hundreds of engineers at its established plant. In the U.S., Lynas USA—supported by the Department of Defense (DoD)—is constructing a separation facility in Texas that will require dozens of process engineers, quality managers, and plant operators. MP Materials is also expanding midstream operations at Mountain Pass to include oxide processing and reagent production, employing process engineers and chemists to optimise chemical extraction systems [466].

The shortage of skilled professionals in this segment is acute. Western countries are essentially building refining capacity from scratch, and the available expertise is limited. Consequently, governments and companies are funding training programs and university partnerships to develop the necessary workforce. Startups and R&D labs are also hiring laboratory chemists, pilot plant operators, and process technologists to advance next-generation separation methods, including membrane and bioleaching techniques.

Downstream manufacturing, particularly the production of neodymium-iron-boron (NdFeB) magnets, is the final step toward rare earth independence. These magnets are essential for electric vehicles, wind turbines, and defense systems. The U.S., once entirely dependent on Asia, is now establishing domestic magnet factories.

MP Materials' new Fort Worth, Texas facility, backed by the U.S. government, aims to produce 10,000 tons of NdFeB magnets annually by 2028 [466]. The company is hiring process engineers, commissioning engineers, coating specialists, and senior magnet technologists. Similarly, USA Rare Earth's Stillwater, Oklahoma plant is building a team of electrical, materials, and quality engineers, while Noveon Magnetics in San Marcos, Texas has transitioned to pilot-scale production, hiring machinists, press operators, and quality inspectors.

However, the skills gap in this area is significant. The U.S. lacks a deep talent base in magnet metallurgy and powder processing. Required roles include materials scientists, magnetic engineers, CNC machinists, and furnace technicians experienced in sintering and alloy fabrication. The challenge extends beyond the U.S.; even global magnet manufacturers like Vacuumschmelze (Germany) are investing in new facilities, such as their $506 million South Carolina plant, expected to employ 300 workers, to meet surging demand. Training programs in powder handling, furnace operations, and quality control will be critical to fill these roles [466].

Recycling rare earths from end-of-life products, such as electric vehicle motors and wind turbine magnets, is becoming one of the fastest-growing employment areas. These projects blend chemical engineering, materials science, and logistics expertise. Companies like ReElement Technologies in Indiana expect to hire over 300 workers for leaching, separation, and purification operations [466]. Meanwhile, India's Attero Recycling is investing ₹100 crore to expand its REE recycling capacity thirtyfold, from 300 to 30,000 tonnes per year, creating hundreds of new jobs in process engineering, waste collection, and plant operations [466].

In the U.S., startups like Phoenix Tailings and Cyclic Materials are recruiting for multi-disciplinary teams spanning R&D, process optimisation, operations, and quality assurance. These ventures require senior engineers, shift supervisors, and plant technicians to bridge laboratory research with full-scale industrial operations. However, competition for talent is fierce, as recycling companies often compete with mining and refining firms for the same pool of metallurgists and process engineers.

Across all segments—mining, refining, manufacturing, and recycling—the rare earth industry is seeing robust hiring and escalating competition for skilled labour. The most sought-after professionals include mining engineers, metallurgists, chemical and process engineers, materials scientists, and manufacturing engineers. These experts must understand both traditional metallurgy and modern digital and automation tools [466].

Employers also value expertise in sustainability, ESG compliance, recycling technologies, and supply-chain transparency—skills increasingly linked to corporate and regulatory requirements. Knowledge of international trade, export controls, and risk management is now desirable, especially as companies build supply chains resilient to geopolitical disruptions [466].

Globally, rare earth demand is expected to nearly triple by 2035, according to the Boston Consulting Group, translating into tens of thousands of new jobs. The U.S. industry alone is growing rapidly—employment rose by roughly 27% within a year, and projects such as e-VAC's 300-job magnet plant in South Carolina are adding momentum [466]. Countries like Australia, Canada, and the European Union are also funding training programs and apprenticeships in minerals engineering, processing, and manufacturing to address future shortages [466].

The rare earth element (REE) and critical mineral processing sectors are currently confronting significant workforce shortages across various skilled professions, in 2025. There is a high demand for professionals such as mining engineers, processing engineers, metallurgists, geologists, and geotechnical experts, yet there is a marked deficit in the supply of these essential roles. For instance, Australia's Critical Minerals Strategy explicitly identifies "a national shortage of key professions" in this sector as a pressing issue [467], which reflects a broader trend observed in various mining and mineral sectors.

In the United States, concerns about declining academic interest in mining and mineral engineering have been documented. Reports indicate that the number of graduates in this field has decreased, with only 327 students graduating with degrees in mining and mineral engineering in 2020, dropping from a peak of 25 academic programs in 1982 to just 14 by 2023 [468].

Furthermore, the sector grapples with demographic challenges, including an aging workforce and decreasing enrolment in relevant academic programs. This shift poses a threat to the future workforce capabilities in both the United States and Australia [469]. The importance of integrating training in digital technologies and advanced processes, such as automation and data analytics, has been highlighted by industry leaders, stressing that these skills are essential for maintaining competitiveness in a rapidly evolving technological landscape [470].

Additionally, there are specific skill shortages in midstream and downstream segments of the REE supply chain, such as hydrometallurgy and rare-earth separation chemistry. Industry reports suggest that there is a significant unmet demand for specialists in these areas, highlighting the importance of addressing these shortages to enhance the operational efficacy of the REE processing industry [471]. Moreover, workforce diversity remains a pressing concern, with female representation in large-scale mining workforces in Australia reported as being less than 15%, indicating challenges related to inclusivity in these sectors [472].

These workforce shortages significantly impact project execution and overall operational efficiency. Lack of critical technical personnel can lead to project delays and cost overruns, as well as hinder the construction and operation of new mines and processing facilities. The competitive landscape is also shaped by workforce dynamics; for example, China's leading position in the rare earth sector is bolstered by a well-established pool of skilled labour [473]. Without addressing these workforce challenges, countries aiming to enhance their critical mineral sectors may face severe technological and operational hurdles that could impede innovation and sustainability initiatives in the long run.

Workforce bottlenecks are particularly evident in midstream refining and downstream manufacturing segments that demand specialized skills. As the industry transforms due to automation and digitalization, the challenge of recruiting adequately skilled professionals intensifies [474]. New entrants into mining and processing often experience labour supply challenges, partly due to underdeveloped training pipelines [475].

To mitigate these pressing workforce challenges, targeted interventions are essential. Governments and industries are starting to invest in training programs and scholarships aimed at transitioning existing workers into critical mineral roles [476]. Educational institutions are encouraged to expand mining and mineral engineering curricula to attract more students and incorporate training on modern processing techniques [477]. Collaborative efforts between universities and mining companies, providing hands-on learning opportunities such as internships and apprenticeships, can pave the way for equipping future professionals with necessary skills.

The critical mineral processing sectors face complex workforce shortages that threaten the viability and competitiveness of the industry. Urgent actions, including workforce development initiatives, educational program expansions, and diverse recruitment strategies, are critical to addressing these shortages and fostering a sustainable future for the REE processing sector.

Emerging Research and Innovation Priorities

The transition toward cleaner energy systems has intensified interest in the research and innovation priorities surrounding rare earth elements (REEs) and critical minerals. This shift is primarily motivated by the need for sustainable energy production and enhanced supply chain resilience in a geopolitically sensitive landscape. Emerging priorities include technological innovations, circular economy solutions, and policy initiatives that increasingly focus on sustainability and industrial competitiveness.

One of the most pressing areas of research focuses on sustainable and low-carbon processing methods for REEs. Researchers are exploring a variety of techniques aimed at minimizing environmental impacts and energy consumption. Notable advancements include hydrometallurgical processes and direct extraction methods utilizing deep eutectic solvents, as well as electrified operations supported by renewable energy. For example, recent experimental work has highlighted bio-adsorptive separation processes that recover REEs from phosphogypsum waste, demonstrating the practical viability of secondary resource recovery with lower environmental footprints [478-481]. This innovative approach not only enhances the economic viability of resource recovery but aligns with broader sustainability goals by repurposing waste streams usually destined for landfills.

The principles of a circular economy are increasingly being integrated into the discourse surrounding critical minerals. Escalating demand for REEs, lithium, nickel, and cobalt has prompted a paradigm shift toward recycling metals from end-of-life products, including spent batteries and electronic devices. For instance, innovations in closed-loop systems are being pursued to minimize reliance on virgin materials through effective reclaiming processes [482, 483]. However, despite substantial progress, significant hurdles remain in scaling pilot technologies to commercially viable operations that can address global supply challenges. Factors impeding commercialization include the fragmented recycling capabilities and the complex processing routes required for efficient resource recovery [484, 485].

Value-chain diversification has emerged as a critical priority in response to the geographic concentration of REE processing, particularly in China [486]. There are concerted efforts to establish midstream and downstream capabilities in allied nations through the development of modular processing units and advanced manufacturing techniques for magnets and alloys [487]. These initiatives not only serve to diversify supply but also contribute to local industrial development and enhance energy security. Furthermore, research into advanced materials is producing next-generation alternatives, such as high-entropy alloys and perovskites, which could reduce dependence on specific critical minerals while optimizing performance in various applications [488, 489].

In exploring resource availability, advances in artificial intelligence (AI) and other digital technologies are being harnessed for resource mapping and identifying alternative feedstocks, such as deep-sea nodules and phosphogypsum [490-492]. The integration of AI-driven modeling tools is aiding in evaluating supply chain vulnerabilities and simulating trade dependencies, which is essential for developing resilient systems that can mitigate risks associated with critical minerals supply [490]. Such tools are crucial in understanding global trends and responding proactively to shifts in supply demands.

Rare Earth and Critical Mineral Operations and Processing

In terms of policy engagement, international cooperation and government-led initiatives are pivotal in underpinning these research efforts. Programs like Australia's Critical Minerals Research & Development Hub are fostering partnerships across academia, industry, and government to tackle key challenges [493]. Similarly, the European Union's Critical Raw Materials Act is mandating investments into domestic processing and recycling infrastructures, fortifying supply chain security [494].

Nonetheless, persistent research gaps and challenges, such as the scaling of green processing technologies and workforce shortages in specialized engineering fields, are constraining the sector's growth [483, 485, 495]. Rigorous life-cycle assessments (LCAs) and techno-economic evaluations are urgently needed to validate the true environmental and economic benefits of emerging technologies in this domain.

Key Terms and Concepts	

Automation: The use of technology and control systems to operate machinery and processes with minimal human intervention, improving efficiency and safety.

Digitalisation: The integration of digital technologies—such as data analytics, IoT (Internet of Things), and cloud computing—into mineral industry operations.

Artificial Intelligence (AI): The application of computer algorithms that enable machines to perform tasks such as prediction, optimisation, and decision-making in exploration and processing.

Machine Learning (ML): A subset of AI where systems learn from data patterns to improve performance in tasks like mineral identification, equipment diagnostics, and process optimisation.

Remote Sensing: The use of satellite or aerial sensor data to detect and map geological features, aiding in exploration and environmental monitoring.

Green Chemistry: The design of chemical processes and products that minimise the use and generation of hazardous substances, supporting sustainability goals.

Low-Carbon Technology: Equipment or processes that reduce greenhouse gas emissions through improved energy efficiency, electrification, or renewable energy use.

Key Terms and Concepts	

Predictive Maintenance: A data-driven maintenance strategy that uses sensors and analytics to anticipate equipment failures before they occur.

Interdisciplinary Skills: The combination of knowledge from multiple disciplines—such as engineering, environmental science, and data analytics—essential for modern mineral professionals.

Vocational Education and Training (VET): Skills-based learning that prepares individuals for specific technical roles in the mining and processing industries.

Professional Accreditation: Formal recognition of qualifications and competencies by professional bodies within the minerals, engineering, or environmental sectors.

Lifelong Learning: The continuous development of skills and knowledge throughout a professional career to remain current with evolving technologies and industry needs.

Decarbonisation: The process of reducing carbon dioxide emissions from industrial operations through efficiency improvements, renewable energy use, or carbon capture technologies.

Innovation Ecosystem: The interconnected network of research institutions, industry partners, and government bodies that drive technological advancement and workforce development.

Career Pathway: The progression of roles and specialisations available within the critical minerals industry, from technical operations to leadership and research positions.

Future-Ready Workforce: A workforce equipped with adaptable skills, digital literacy, and sustainability awareness to meet emerging challenges in the minerals sector.

Chapter 12 Review Questions

Emerging Technological Trends

1. Multiple Choice: Which of the following best describes *digitalisation* in the minerals sector?

(a) The use of manual control systems to monitor mining sites

(b) The systematic capture, integration, and analysis of data across operations

(c) The physical automation of drilling machinery only

(d) The conversion of mining operations into purely virtual environments

2. Short Answer: Define *automation* and explain one benefit it brings to mineral processing.

3. True or False: Artificial intelligence (AI) in mineral processing only performs routine calculations and cannot support decision-making.

4. Extended Response: Describe how automation, artificial intelligence, and digital process control are transforming exploration, extraction, and refining stages in the rare-earth and critical-minerals industries.

Data Analytics and Process Efficiency

5. Matching: Match each technology to its main function:

Technology	Function
A. Digital Twin	1. Uses sensor data to detect faults before breakdowns
B. AI Ore Sorting	2. Simulates a full processing plant to test performance
C. Predictive Maintenance	3. Uses imaging sensors to separate ore from waste

6. Short Answer: How do data analytics and remote sensing contribute to more efficient exploration programs?

7. Multiple Choice: Which technology provides virtual simulation of mine-to-mill systems for optimisation?

(a) QEMSCAN

(b) NIAflow

(c) Digital Twin

(d) Hydrometallurgical model

Green Chemistry and Low-Carbon Refining

8. Short Answer: What is the main goal of *green chemistry* in mineral refining?

9. Multiple Choice: Which of the following practices aligns with low-carbon refining?

(a) Using fossil-fuel roasters only

(b) Applying hydrogen reduction powered by renewable electricity

(c) Expanding evaporation ponds

(d) Increasing thermal smelting temperatures

10. Extended Response: Discuss how green chemistry and renewable-energy integration reduce the environmental footprint of critical-mineral operations.

Cross-Disciplinary and Digital Skills

11. True or False: Future minerals professionals require both engineering and data-analysis knowledge to remain competitive.

12. Short Answer: List three interdisciplinary skills that are increasingly essential in the modern mineral industry.

13. Short Answer: Why is cybersecurity becoming a key competency in digitally connected mineral operations?

Workforce Change and Training

14. Multiple Choice: Which trend is driving the most significant workforce transformation in the critical-minerals sector?

(a) Declining demand for digital skills

(b) Digitalisation and decarbonisation of operations

(c) Return to manual processing methods

(d) Reduction in training investment

15. Short Answer: Identify one challenge companies face when recruiting for automation and AI-related roles.

16. Extended Response: Explain how digital transformation and decarbonisation are reshaping training needs and professional development in the sector.

Career Pathways and Education

17. Matching: Match each education type to its primary outcome:

Education Type	Outcome
A. Vocational Education and Training (VET)	1. Formal recognition of professional competencies
B. Higher Education	2. Applied technical skills for specific occupations
C. Professional Accreditation	3. Theoretical and research-based knowledge

18. Short Answer: Identify two key career pathways available in the critical-minerals industry.

19. Multiple Choice: Which area offers emerging opportunities due to recycling and circular-economy growth?

(a) Traditional smelting

(b) REE recycling

(c) Coal processing

(d) Petroleum refining

Continuous Learning and Innovation

20. Short Answer: What is meant by *lifelong learning* in the context of the critical-minerals workforce?

21. True or False: Continuous upskilling is only necessary for new graduates entering the industry.

22. Extended Response: Discuss how innovation, collaboration, and lifelong learning contribute to building a resilient and future-ready minerals workforce.

Career Development Planning

23. Short Answer: Why is aligning a career-growth plan with sustainability goals becoming essential in this sector?

24. Extended Response: Develop a brief personal or organisational plan outlining how you (or your company) could build capabilities in digitalisation, automation, or sustainability over the next five years.

Review Question Sample Answers

Chapter 1 Introduction to Rare Earth and Critical Minerals

Definitions

1. REEs: The 17 metallic elements (15 lanthanides + Sc, Y) with similar chemistry and distinctive magnetic/optical/catalytic properties.

2. Critical minerals: Materials essential to economic/strategic sectors that face elevated supply risk (concentrated production, processing bottlenecks, limited substitutes).

3. Difference: REEs are a chemical group; "critical minerals" is an economic/strategic designation that can include REEs plus others (Li, Co, Ni, graphite, PGMs).

Classification & Identification

4. LREE: Nd, La, Sm. HREE: Dy, Tb, Yb.

5. Any four: Lithium, cobalt, nickel, graphite, manganese (also PGMs).

6. Lithium–(i); Cobalt–(ii); Graphite–(iii); PGMs–(iv).

Applications

7. (a) REEs: Nd-Pr-Dy magnets in wind/EV motors.
(b) Lithium: Li-ion batteries (EVs/grid).
(c) PGMs: Electrolysers/fuel cells; catalytic converters.

8. Examples: LED/display phosphors (Eu, Tb, Y); fiber-optic amplifiers (Er); optical lenses/polishing (La, Ce).

9. Primary: Stainless steel (~70%). Growing: Battery cathodes (NMC/NCA).

Deposits & Geology

10. C

11. B (Ion-adsorption clay regolith)

12. B (Pegmatites & brine salars)

Supply Chain & Criticality

13. Exploration → Resource definition → Mining → Beneficiation/concentration → Chemical refining/separation → Intermediate materials (e.g., precursors/magnets) → Component manufacturing → End-use products → Collection/reuse/recycling.

14. Refining/separation—technically complex, capital-intensive, geographically concentrated (notably in China).

15. Criticality: Intersection of economic importance and supply risk (drivers: concentration of mining/refining, geopolitics, substitutability, recyclability).

16. Examples: Co-investment/loans/grants; strategic stockpiles; allied trade partnerships; permitting reform; recycling mandates; R&D support.

Geography & Producers

17. DRC (over 60% of supply; ESG/child labour concerns).

18. Examples: Indonesia (laterite), Philippines (laterite), Russia (sulphide), Canada (sulphide), Australia (both; major sulphide historically).

19. Strengths: World-leading hard-rock lithium; major REE feedstock (Mount Weld); strong governance. Expanding: Onshore refining/precursor manufacturing (e.g., Kwinana lithium hydroxide; Nickel West; Alpha HPA).

20. Australia: Major hard-rock REE producer/exporter, growing refining. China: Dominant in refining/separation and magnet manufacturing; significant HREE from ion-adsorption clays.

Strategy, Risk & Sustainability

21. Reasons: (i) Reduces reliance on concentrated primary supply; (ii) Lowers environmental footprint and stabilises availability/costs.

22. Metrics: (i) HHI concentration for mining/refining by country; (ii) Stock-to-use/days-of-cover; optional: recycling rate, price volatility, ESG/permits risk.

23. Technical: Shift to high-Ni/low-Co chemistries; deploy LFP where suitable; improve recycling.
Supply-chain: Diversify sources beyond DRC; secure long-term offtakes; invest in traceable/ethical supply; build precursor/refining capacity in allied regions.

24. Cooperation levers: Common ESG/traceability standards, joint funding of refining/recycling hubs, strategic stockpiles, transparent data sharing, technology transfer for cleaner processing, coordinated permitting—maintaining open but resilient trade.

Chapter 2 Geology, Occurrence, and Exploration

Geological Processes

1. Rare earth and critical mineral deposits form through magmatic, hydrothermal, sedimentary, and lateritic processes that concentrate dispersed elements into mineable accumulations.

2. Magmatic differentiation in carbonatite or alkaline magma concentrates REEs in residual melts, forming minerals like bastnäsite and monazite.

3. Weathering in tropical climates leaches soluble elements while enriching REEs in clays such as kaolinite and halloysite, creating ion-adsorption deposits.

4. Brine deposits develop in closed arid basins through evaporation of lithium-rich groundwater, while pegmatites crystallise lithium minerals from late-stage granitic melts.

Host Rocks and Ore Minerals

5. Bastnäsite occurs mainly in carbonatite and alkaline igneous rocks rich in carbonate and fluorine.

6. Monazite is a phosphate mineral found in granitic pegmatites and metamorphic rocks; it is resistant to weathering and accumulates in placers.

7. Xenotime (YPO_4) is rich in heavy rare earth elements (HREEs) like yttrium and dysprosium, making it valuable for high-tech applications.

8. Spodumene ($LiAl(Si_2O_6)$) forms in granitic pegmatites and is a primary source of lithium for battery compounds.

Key Deposit Types

9. Carbonatites host minerals such as bastnäsite, monazite, and pyrochlore, rich in REEs and niobium.

10. Pegmatites display coarse-grained textures, zoning, and contain lithium, niobium, and beryllium minerals such as spodumene and beryl.

11. Lateritic deposits form by intense tropical weathering, which leaches mobile ions and concentrates nickel and cobalt in limonite and saprolite zones.

12. Placer deposits are secondary accumulations of dense, resistant minerals (e.g., monazite, xenotime, ilmenite) concentrated by water or wind action.

Global Examples

13. Mount Weld (Australia) and Bayan Obo (China) are world-class carbonatite REE deposits.

14. Major pegmatites occur at Greenbushes and Pilgangoora (Australia) and Tanco (Canada).

15. Southern China produces most of the world's HREEs from ion-adsorption clays.

16. The Bushveld Complex (South Africa) hosts vast PGM, vanadium, and titanium resources, critical for alloys and clean energy tech.

Exploration Methods

17. Mapping defines lithology, structures, and alteration zones linked to mineralisation.

18. Geochemical sampling measures trace element anomalies that reveal concealed ore bodies.

19. Geophysical methods (magnetic, gravity, radiometric) detect subsurface features such as dense carbonatites or magnetic mafic intrusions.

20. Core logging records rock type, structure, and mineral content, helping to build 3D models and plan further drilling.

Assay and Geochemical Data

21. Assays measure the concentration of elements (e.g., % Li_2O or ppm REEs) in rock or soil samples.

22. Anomalously high values indicate zones of enrichment and guide further drilling.

23. Multiple analyses ensure grade consistency and help define ore boundaries within the deposit.

Ore Grade, Tonnage, and Cut-Off Grade

24. Ore grade is the metal concentration in the rock; tonnage is the total ore volume.

25. Cut-off grade is the lowest grade that can be mined profitably under current conditions.

26. A deposit with a high grade but low metallurgical recovery may yield less payable metal than a lower-grade ore with better recovery.

27. New extraction technologies (e.g., hydrometallurgical leaching, DLE) reduce costs, making lower-grade resources viable.

Environmental, Social, and Regulatory Considerations

28. Early environmental planning prevents erosion, water contamination, and habitat loss.

29. Engaging local and Indigenous communities builds social licence and ensures equitable benefits.

30. Regulations require impact assessments, rehabilitation plans, and waste-handling controls.

31. Impacts may include deforestation, soil degradation, and radiation concerns (especially from monazite or bastnäsite mining).

Chapter 3 Mineral Characterisation and Ore Testing

Physical and Chemical Properties of Ores

1. Key physical properties include hardness, density, magnetism, electrical conductivity, cleavage, grain size, and texture. Chemical properties include oxidation state, solubility, acid–base reactivity, and elemental composition.

2. Harder ores (e.g., spodumene, cassiterite) require more energy to grind and wear-resistant liners; softer ores (e.g., bastnäsite) grind more easily but risk over-liberation.

3. Density differences between ore and gangue enable gravity separation using jigs, spirals, or shaking tables.

4. Magnetic susceptibility and conductivity guide both exploration (geophysical surveys) and separation (magnetic or electrostatic).

5. Substitution of REE^{3+} ions (e.g., La, Ce, Nd) affects mineral chemistry and separation selectivity.

6. Ce^{4+} is harder to dissolve than Ce^{3+}, requiring reducing conditions during leaching for efficient recovery.

7. Acid–base reactivity dictates whether ores are processed by acid leaching (bastnäsite), alkaline cracking (monazite/xenotime), or mild salt leaching (ion-adsorption clays).

Mineral Characterisation and Processing Routes

8. Mineral characterisation defines mineral species, chemistry, and texture, enabling selection of efficient beneficiation and extraction routes.

9. Fine-grained or interlocked textures require finer grinding for liberation, increasing energy costs.

10. Mineralogical data ensure that recovery models and process flowsheets reflect actual mineral behaviour, not just bulk assays.

Mineral Identification and Liberation Studies

11. Key techniques: optical microscopy, SEM-EDS, XRD, EMPA, MLA/QEMSCAN, and LA-ICP-MS.

12. Reflected light microscopy identifies opaque minerals (monazite, xenotime); transmitted light identifies transparent phases (spodumene, quartz).

13. SEM-EDS provides micro-scale images and elemental composition, revealing intergrowths and impurities.

14. MLA quantifies mineral abundance, associations, and liberation using automated imaging.

15. Different size fractions simulate grinding stages, showing how liberation changes with particle size.

Analytical Tools for Mineral and Elemental Analysis

16. XRD identifies crystalline structure and mineral phases (e.g., α–β spodumene).

17. XRF measures bulk elemental composition, providing oxide percentages for geochemical classification.

18. ICP-MS quantifies trace and rare earth elements with very high sensitivity (ppm–ppb).

19. Qualitative methods identify what is present (e.g., EDS, microscopy); quantitative methods measure how much (e.g., XRF, ICP-MS).

20. Combining XRD, SEM-EDS, MLA, and ICP-MS yields a complete picture of mineralogy, texture, and chemistry.

Bench-Scale Testing

21. Bench-scale testing validates laboratory findings under semi-controlled, small-scale conditions before pilot operations.

22. Flotation tests assess surface chemistry; leaching tests measure dissolution behaviour and reagent efficiency.

23. Data include recovery rates, reagent use, pH control, and energy needs.

24. Different mineralogical zones may respond differently to the same process, requiring tailored parameters.

Pilot-Scale Testing and Process Design

25. Pilot-scale testing replicates continuous operation, confirming process stability and scalability.

26. Results define equipment size, residence times, and flow sequencing for PFDs.

27. Reagent, energy, and throughput data enable accurate CAPEX/OPEX projections.

28. Waste composition and emissions data feed into EIAs for compliance and sustainability design.

Application to Process Flowsheets

29. Bastnäsite: acid leaching; Monazite/Xenotime: roasting or caustic cracking; Ion-adsorption clays: ammonium-salt leaching.

30. Hard-rock deposits require crushing, flotation, and leaching; clay deposits rely on in-situ ion exchange.

31. Integrated data reduce uncertainty in design, investment, and environmental outcomes.

Chapter 4 Comminution, Classification, and Particle Preparation

Fundamental Principles of Comminution

1. Comminution is the mechanical reduction of ore particle size through crushing and grinding to liberate valuable minerals from gangue.

2. b) To reduce particle size for mineral liberation.

3. True. Comminution consumes 50–70% of total plant energy.

4. Impact breaks by striking force; compression squeezes particles between surfaces; attrition abrades particles through friction.

Major Types of Comminution Equipment

5. b – a – a – d – c (Jaw = b; Cone = e; Ball = a; Rod = d; HPGR = c).

6. A jaw crusher compresses ore between a fixed and moving jaw driven by an eccentric shaft, fracturing the material until it passes through the discharge opening.

7. c) Vertical or stirred mill.

8. Cone crushers produce uniform feed for ball mills; ball mills then achieve fine liberation for leaching.

Classification Equipment

9. Classification separates coarse and fine particles, sending fines forward and returning coarse material for regrinding.

10. c) Hydrocyclone.

11. True. Air classifiers are used where water use is undesirable.

12. A hydrocyclone uses centrifugal force to drive coarse particles outward and down through the underflow while fine particles exit via the overflow.

Relationship between Energy Input, Particle Size, and Grind Efficiency

13. Bond's Work Index quantifies ore hardness and estimates the energy required for grinding; it supports mill sizing and energy efficiency analysis.

14. b) Using pre-classification and recycling of oversize material.

15. False. Overgrinding wastes energy and can reduce recovery.

16. A finer grind increases liberation but raises energy use; the goal is the smallest size that achieves required recovery without excessive power consumption.

Importance of Liberation before Separation

17. Liberation exposes valuable minerals so that separation processes can act selectively.

18. a) Loss of valuable minerals to tailings due to incomplete liberation.

19. True. Overgrinding damages graphite flakes and lowers value.

20. Bastnäsite flotation requires fine ($\approx$ P80 100 μm) grinding for surface exposure; spodumene leaching requires post-roast grinding to medium size for reactivity.

Rare Earth and Critical Mineral Operations and Processing

Particle Size distribution (PSD) Data

21. P80 is the particle size at which 80 % of the material passes through the specified screen aperture.

22. c) 75–150 µm.

23. True. Narrow PSD improves separation efficiency.

24. If overflow is too coarse, the mill is under-grinding; if underflow contains excessive fines, the circuit is over-grinding and wasting energy.

Ore Hardness and Texture

25. Hard ores favour SAG/AG mills for coarse, energy-efficient reduction; softer ores may need ball mills for finer control.

26. a) Mineral grain size and texture.

27. True. Coarse-grained ores liberate at larger particle sizes.

28. Bastnäsite: jaw/cone → ball → hydrocyclone (fine). Xenotime: jaw/cone → stirred mill (<50 µm) → fine classification.

Safety, Maintenance, and Operational Best Practices

29. Isolate power (lock-out/tag-out) and ensure guards are in place before maintenance.

30. b) Act as a safety release under overload.

31. True. Lock-out/tag-out is essential for rotating mills.

32. Regular liner and lubrication checks prevent mechanical failure, reduce downtime, and protect personnel from hazards associated with equipment wear or overheating.

Chapter 5 Physical Separation Processes

Principles of Physical Separation

1. (b) Exploiting differences in physical properties such as density, magnetism, and conductivity.

2. Physical separation separates valuable minerals from gangue by exploiting measurable physical property differences. It increases recovery, reduces reagent use, and enhances overall efficiency.

3. Gravity separation relies on density contrasts; magnetic separation exploits variations in magnetic susceptibility; and electrostatic separation uses conductivity differences—together forming the foundation of REE and critical-mineral beneficiation.

Gravity Separation

4. (c) Specific-gravity (density) difference.

5. - Jigs: Use pulsating water currents to stratify particles by density.

- Spirals: Employ helical troughs where gravity and centrifugal forces separate heavy minerals.

- Shaking tables: Combine vibration and thin-film flow for fine-particle stratification.

6. Jigs 2 mm – 0.1 mm; Spirals 2 mm – 37 µm; Shaking tables 3 mm – 37 µm.

7. Jigs treat coarse, dense minerals such as cassiterite; spirals handle mid-size heavy minerals in beach sands; shaking tables refine fine concentrates (e.g., gold, tungsten, monazite) requiring precise separation.

Magnetic Separation

8. (c) Low-Intensity Magnetic Separator (LIMS).

9. - LIMS: < 0.2 T; strongly magnetic minerals (e.g., magnetite).

- HIMS: 0.8 – 2 T; weakly magnetic minerals (e.g., hematite, ilmenite).

- HGMS: > 1000 T/m gradient; ultrafine, weakly magnetic minerals (e.g., monazite, xenotime).

10. Xenotime is fine-grained and only weakly magnetic; high-gradient fields generate sufficient magnetic force for its recovery.

11. Magnetic separation removes ferromagnetic and paramagnetic minerals before electrostatic or flotation stages. Desliming eliminates fine slimes that hinder selectivity and reduce concentrate quality.

Electrostatic Separation

12. (b) Electrical conductivity.

13. In an HTR separator, corona discharge charges dry particles; conductive minerals discharge quickly and adhere to a grounded roll, while non-conductors retain charge and are repelled into separate bins.

14. Moisture prevents efficient charging and causes particle clumping; dry, free-flowing feed is essential.

15. Magnetic separation removes magnetic minerals (e.g., ilmenite); electrostatic separation then divides conductors (monazite, rutile) from non-conductors (zircon, xenotime) to produce high-purity concentrates.

Froth Flotation

16. (b) Collector.

17. - Collectors: Render mineral surfaces hydrophobic.

- Frothers: Control bubble size and froth stability.

- Modifiers/Regulators: Adjust pulp pH and selectivity.

- Depressants: Prevent gangue minerals from floating.

18. Over-stabilised froth traps unwanted gangue and lowers concentrate grade.

19. Collectors selectively attach to target minerals; frothers stabilise the froth; modifiers regulate pulp chemistry. REE circuits use hydroxamate collectors under mildly acidic pH, whereas lithium flotation uses fatty-acid collectors under alkaline conditions.

Recovery, Selectivity, and Process Efficiency

20. (b) Density contrast between minerals.

21. Particle-size distribution, surface chemistry (hydrophobicity or charge), and pulp pH.

22. Efficient separation depends on suitable particle liberation and method matching. Fine particles require flotation or high-gradient magnetic fields; coarse particles respond better to gravity methods. Surface chemistry and fluid dynamics also govern recovery and selectivity.

Process Flowsheets and Integration

23. Typical REE flowsheet: (1) Crushing and grinding for liberation; (2) Desliming/classification; (3) Magnetic separation (LIMS/HIMS); (4) Electrostatic separation (by conductivity); (5) Flotation or leaching for final upgrading.

24. Comminution liberates spodumene crystals; classification removes slimes and produces a uniform particle size suitable for collector adsorption and flotation efficiency.

25. Gravity (e.g., DMS or spirals) removes coarse gangue; flotation then upgrades spodumene by surface-chemistry selectivity, improving grade and reducing reagent use.

Safety and Environmental Considerations

26. (b) High-voltage discharge and grounding.

27. Recycle process water and treat tailings to remove residual reagents before release.

28. Safety: Guard high-voltage equipment, isolate magnetic systems during maintenance, and use PPE for reagent handling. Environmental: Employ closed-loop water systems, dust suppression for dry circuits, and secure tailings management to prevent contamination.

Chapter 6 Hydrometallurgical Techniques

Principles and Role of Hydrometallurgy

1. Extraction and purification of metals through aqueous chemical reactions at low temperature.

2. Leaching → Solution Purification (SX/IX) → Metal Recovery (precipitation or electrowinning).

3. Provides selective, low-temperature recovery with higher purity and lower energy use.

Leaching Methods and Applicability

4. Acid, Alkaline, Pressure (autoclave), and Bioleaching.

5. C) High-Pressure Acid Leaching (HPAL).

6. Acid leaching dissolves oxides, phosphates, and carbonates using H_2SO_4 or HCl; alkaline leaching uses NaOH or Na_2CO_3 for amphoteric oxides such as alumina and tungsten.

7. Roasting converts α-spodumene to β-spodumene, increasing its reactivity.

8. Advantages – High recovery and broad applicability; Limitations – Acid waste and corrosion issues.

Parameters Controlling Leaching Efficiency

9. Temperature, pH, reagent concentration, redox potential, and particle size.

10. High temperature increases reaction rate; controlled Eh oxidises U^{4+} to U^{6+} for solubility.

11. Smaller particles increase surface area and contact efficiency.

12. Adjusting pH and acid strength optimises metal solubility and selectivity.

Solvent Extraction (SX)

13. To selectively bind metal ions forming organo-metallic complexes.

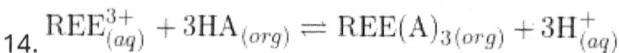

14. $$REE^{3+}_{(aq)} + 3HA_{(org)} \rightleftharpoons REE(A)_{3(org)} + 3H^+_{(aq)}$$

15. Extraction, Scrubbing, Stripping, and Regeneration.

16. pH and extractant concentration.

17. REEs form stable complexes with acidic extractants such as D2EHPA and PC-88A, allowing fine separation.

Ion Exchange (IX)

18. Resins exchange charged ions with metal ions in solution, capturing them for later elution.

19. IX – higher selectivity, lower throughput; SX – cost-effective for bulk operations.

20. Strong-acid cation resins (e.g., Dowex 50) used to capture REE^{3+} or Co^{2+} ions.

Precipitation and Purification

21. Conversion of dissolved metal ions into insoluble solids by chemical or pH change.

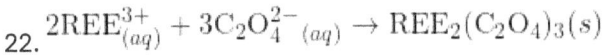

22. $$2REE^{3+}_{(aq)} + 3C_2O_4^{2-}{}_{(aq)} \rightarrow REE_2(C_2O_4)_3(s)$$

23. To remove impurities (e.g., Fe, Al) before target metal precipitation.

24. Hydroxide, Carbonate, Oxalate.

25. Higher temperatures increase crystal growth but may reduce purity through co-precipitation.

Process Flow Interpretation

26. Leaching, Solvent Extraction / Ion Exchange, and Precipitation + Calcination.

27. Leaching with acid → SX for purification → precipitation of lithium carbonate or hydroxide.

28. Leach reactors, mixer–settler SX units, and filter presses or crystallisers.

Environmental and Safety

29. Acidic effluents and heavy-metal residues.

30. Neutralisation with lime and closed-loop water recycling.

31. Over-pressure hazard controlled by pressure relief systems and remote operation.

Bench and Pilot Scale

32. To verify chemical behaviour and optimise conditions before scale-up.

33. To test continuous operation and equipment performance under industrial conditions.

34. Pilot-autoclave for HPAL or mixer-settler SX pilot units.

Sustainability and Comparison

35. Lower energy consumption, reagent recycling, and minimal air emissions.

36. Aqueous chemistry allows dissolution of metals locked in complex matrices.

37. Enables recovery from wastes and end-of-life products, supporting a circular economy.

Chapter 7 Pyrometallurgy and Thermal Processing

Principles of Pyrometallurgy

1. Pyrometallurgy uses heat to drive chemical transformations that extract or purify metals. It is essential for converting ores into metallic or oxide forms for further refining.

2. Endothermic reactions absorb heat (e.g., calcination), while exothermic reactions release heat (e.g., sulphide oxidation).

3. High temperatures overcome activation energy barriers, promote phase changes, and accelerate diffusion-controlled reactions.

Major Pyrometallurgical Processes

4. Roasting oxidises sulfides or carbonates (e.g., bastnäsite → CeO_2 + CO_2 + F_2).

5. Calcination removes volatiles in limited oxygen, whereas roasting requires an oxidising atmosphere.

6. Reduction uses agents such as carbon or hydrogen to remove oxygen from oxides.

7. Smelting melts the ore and flux together, separating metal from slag—e.g., nickel laterites reduced in electric furnaces.

Chemical and Thermodynamic Reactions

8. Ellingham diagrams show Gibbs free energy ($\Delta G°$) of oxide formation versus temperature, indicating which reactions are spontaneous.

9. Carbon reduces Fe_2O_3 to Fe because its reaction line lies below Fe_2O_3, but MgO's line lies lower, making it thermodynamically stable.

10. $2ZnS + 3O_2 → 2ZnO + 2SO_2$.

Equipment and Operating Parameters

11. Rotary kilns provide continuous flow for coarse materials; multiple-hearth furnaces offer staged heating for fine feeds.

12. Fluidised-bed roasters maintain intense mixing and even heat transfer, ideal for fine or uniform feeds.

13. Electric and induction furnaces use resistive or electromagnetic heating, ensuring tight temperature control.

14. Feed rate, rotation speed, slope, and burner control all affect residence time and uniformity.

Process Control

15. Ramp-and-soak programming gradually raises temperature and holds it to avoid thermal shock and complete reaction.

16. Thermocouple arrays measure temperatures along the chamber to detect hot/cold zones and adjust heat distribution.

17. Gas composition affects oxidation or reduction outcomes, controlling product phase and purity.

18. PLCs and PID loops automatically adjust air–fuel ratios, gas flows, and power inputs for stable operation.

Pyro- vs. Hydro-Processing

19. Pyro uses heat and solid–gas reactions; hydro uses aqueous chemistry at low temperature.

20. Bastnäsite is first roasted to remove volatiles and then leached in acid.

21. Hydro advantages: high selectivity and low temperature; disadvantages: slow kinetics and liquid waste management.

Integrated Processing

22. Example: Bastnäsite → Roasting (650 °C) → Leaching → Solvent extraction → REE oxides.

23. Pre-roasting removes volatiles and converts refractory minerals into reactive oxides, enhancing leach efficiency.

Environmental and Safety Aspects

24. CO_2, SO_2, and HF are major pollutants.

25. Heat stress, fume inhalation, and fire hazards are key occupational risks.

26. Use of insulated shielding, local exhaust ventilation, and gas scrubbers reduces exposure.

Emerging Technologies

27. Hydrogen-based reduction forms water instead of CO_2, lowering emissions.

25. Microwave heating provides volumetric energy absorption, reducing energy loss.

29. Plasma smelting uses ionised gas at very high temperatures for cleaner, faster reactions.

Process Optimisation

30. Waste-heat recovery and improved insulation reduce fuel use.

31. Uniform particle size improves heat transfer and reaction kinetics.

32. Data logging enables trend analysis and automated correction, improving yield and consistency.

Chapter 8 Plant Design, Process Flows, and Equipment

Fundamentals of Plant Design

1. To achieve efficient, safe, and environmentally responsible transformation of raw ore into marketable products through optimised layout, workflow, and integration of unit operations.

2. Answer: C) Randomised equipment placement.

3. Answer: True.

4. Integrating stages allows continuous material flow, reduces handling losses, balances capacity between circuits, and improves recovery and energy efficiency through coordinated control.

PFDs and P&IDs

5. A PFD shows the major unit operations and material flows, while a P&ID adds detailed instrumentation, valves, control loops, and safety systems.

6. Answer: C) Pressure transmitter.

7. PFDs provide a visual overview of the process, helping engineers identify bottlenecks, plan modifications, and ensure mass balance.

8. Instrumentation is represented with symbols and tags connected to control loops that interface with DCS or PLC systems.

Equipment and Systems

9. a-2, b-3, c-1, d-4.

10. To separate fine particles based on size and density, ensuring correctly sized material advances while oversize returns for regrinding.

11. Flotation uses air bubbles and reagents to selectively attach target minerals, while magnetic separation exploits magnetic properties—both increase concentrate grade.

12. Answer: D) Electric arc furnace.

Design Considerations and Capacity

13. Ore characteristics (hardness, grade), equipment size and configuration, and operational constraints (power, water, maintenance).

14. Options include increasing filter area, optimizing feed concentration, using higher vacuum pressure, or adding parallel units.

15. Answer: B) Tonnes of material processed per hour.

Process Integration and Control

16. Answer: True.

17. A DCS is a computer-based system that monitors and controls plant operations by integrating sensors, actuators, and process data in real time.

18. XRF analysers continuously measure elemental composition in slurries, allowing automatic adjustment of reagent dosing and air flow.

19. Feedback automation uses sensors and PLCs to detect deviations and correct them instantly, ensuring stable temperature, flow, and composition for consistent product quality.

Quality Control, Sampling, and Product Specification

20. To collect samples that accurately represent the bulk material for reliable analytical and decision-making outcomes.

21. Answer: A) ISO 3082.

22. Moisture analysis determines drying energy requirements and prevents incomplete calcination or material degradation.

23. CoAs verify batch compliance, while LIMS link analytical data, sampling points, and production batches, providing transparency and traceability across the supply chain.

Safety and Environmental Factors

24. Gas scrubbers and baghouse dust collectors.

25. Answer: B) Increase water recovery and minimize waste.

26. Answer: False – they are usually segregated for safety and environmental compliance.

27. Effective tailings and water systems reduce environmental footprint, conserve resources, and support compliance with sustainability and community standards.

Applied Design Evaluation

28. Because brine operations rely on evaporation ponds and solution chemistry, while hard-rock plants require comminution, roasting, and acid leaching.

29. REE flowsheets involve multiple minerals and closely related chemical behaviours, requiring more complex separation and purification stages.

30. Local ore composition, water availability, energy costs, and environmental laws dictate whether hydrometallurgical or pyrometallurgical routes—and corresponding layouts—are most suitable.

Chapter 9 Environmental Management and Safety

Environmental and Safety Challenges

1. REE and critical mineral extraction pose challenges such as radiation exposure from thorium/uranium, dust inhalation, chemical contamination, tailings stability, and water/air pollution.

2. Answer: b) Monazite

3. Answer: True

4. Poor tailings management can cause environmental contamination, groundwater pollution, dust emissions, and potential dam failures—endangering workers, communities, and ecosystems.

Radiation and Chemical Hazards

5. NORM stands for *Naturally Occurring Radioactive Material*, which includes thorium and uranium found in REE ores; it requires strict handling and monitoring.

6. Answer: a) Alpha

7. ALARA ("As Low As Reasonably Achievable") means all radiation exposure must be minimized considering social and economic factors.

8. Control measures include radiation surveys, personal dosimeters, HEPA filters, radon monitoring, shielding, PPE, and water containment systems.

Environmental Management Systems (EMS)

9. Answer: c) Environmental Management Systems

10. Core components: (1) Environmental policy and objectives, (2) Monitoring and measurement, (3) Continuous improvement and audits.

11. Apply water recycling, process optimization, and waste treatment systems within ISO 14001's "Plan–Do–Check–Act" framework.

Tailings and Waste Management

12. Tailings are fine-grained waste residues from ore processing containing residual chemicals, metals, and radionuclides.

13. Answer: b) Prevent groundwater contamination

14. Steps: Covering and sealing tailings, re-vegetation, monitoring radiation and water quality, and long-term stability assessments.

Water Use and Emissions Control

15. Recycling reduces freshwater demand, minimizes discharge, and supports sustainable operation in arid regions.

16. Answer: b) Wet scrubber

17. Answer: False

18. Systems like thickeners, neutralisation plants, scrubbers, and baghouses reduce pollution and ensure compliance with environmental permits.

Occupational Health and Safety (OHS)

19. Hazard identification, risk assessment, and incident reporting or control measures.

20. Answer: c) ISO 45001

21. Isolate the area, provide first aid, remove the worker from exposure, notify supervisors, and investigate for gas leaks or ventilation failure.

Regulatory and Legislative Frameworks

22. Answer: Australian Radiation Protection and Nuclear Safety Agency (ARPANSA).

23. Answer: b) Tailings dam failures worldwide

24. Australia regulates under ARPANSA and the Radiation Protection Act; the U.S. under the Nuclear Regulatory Commission (NRC) and Department of Energy (DOE). Both follow IAEA and ICRP exposure limits (20 mSv/year).

ESG and Community Engagement

25. Environmental – waste and emissions control; Social – Indigenous engagement; Governance – transparent reporting and compliance.

26. Answer: c) Ongoing acceptance and approval by stakeholders

27. FPIC ensures Indigenous communities consent freely and are fully informed before project commencement, protecting their rights.

28. ESG frameworks improve transparency, reduce environmental harm, promote equity, and enhance reputation, leading to sustainable and ethical operations.

Chapter 10 Circular Economy, Recycling, and Secondary Sources

Concepts of the Circular Economy

1. A circular economy aims to eliminate waste by designing products and systems that enable reuse, recycling, and regeneration; unlike the linear model, it maintains material value throughout multiple life cycles.

2. Answer: b) Minimising resource use and closing material loops.

3. It ensures long-term access to critical materials by reducing dependence on primary mining and mitigating environmental impacts.

Recycling and Resource Recovery

4. **Environmental:** Reduced landfill waste, lower carbon emissions, decreased mining impacts. **Economic:** Lower production costs, improved supply stability.

5. Answer: b) Electronic waste (e-waste).

6. Recycling these materials reduces greenhouse gas emissions from mining and refines domestic supply chains for renewable-energy technologies.

Secondary Sources and Recovery Challenges

7. Common secondary sources: e-waste, spent batteries, industrial residues, tailings, catalysts.

8. Answer: False.

9. Challenges include material heterogeneity, contamination, safety risks, and economic feasibility.

Recycling Processes

10. a → Mechanical; b → Hydrometallurgical; c → Pyrometallurgical; d → Direct.

11. Advantage: Efficient metal recovery from complex feeds. Limitation: High energy consumption and CO_2 emissions.

12. Steps: Pre-treatment → Acid leaching → Solvent extraction/Ion exchange → Precipitation or electrowinning → Effluent treatment.

Primary vs. Secondary Production

13. Answer: b) Consumes less energy and generates lower emissions.

14. It diversifies material supply, reducing exposure to geopolitical risks and resource shortages.

Life-Cycle Assessment (LCA)

15. LCA assesses environmental impacts from extraction to disposal; stages: extraction → processing → use → end-of-life/recycling.

16. Answer: b) ISO 14040 and ISO 14044.

17. LCAs identify high-impact processes, guide cleaner technology investment, and support evidence-based sustainability reporting.

Policy and International Frameworks

18. EPR holds manufacturers responsible for product recovery and recycling, promoting eco-design and resource efficiency.

19. Answer: b) Critical Raw Materials Act (2023).

20. They fund R&D, establish recycling targets, and foster industrial collaboration for circular supply chains.

Case Studies and Industry Examples

21. Example: Redwood Materials recovers lithium, cobalt, nickel, and copper from used batteries and reintroduces them into new battery production.

22. Europe enforces recycling efficiency via legislation; China focuses on large-scale "urban mining" initiatives—both enhance resource sustainability.

Future Trends and Sustainability

23. Innovations: Bioleaching; ionic-liquid and deep-eutectic-solvent extraction; AI-based resource tracking.

24. Answer: b) Continuous reuse and recycling of materials within production.

25. Recycling reduces emissions, conserves resources, and supports renewable-energy transitions, aligning with global sustainability and net-zero objectives.

Chapter 11 - Industry Operations and Global Market Dynamics

Global Industry Structure and Value Chain

1. The main stages are: exploration & mining, separation & refinement, alloying & intermediate products, manufacturing & final application, recycling, and logistics/stockpiling.

2. Answer: c) Alloying and Manufacturing

3. Answer: False – Most refining and separation capacity is located in China.

4. Answer:

- Stage 1 → Identifying and extracting ores
- Stage 2 → Producing oxides and concentrates
- Stage 3 → Producing alloys/metals
- Stage 4 → Manufacturing final products
- Stage 5 → Recycling end-of-life materials
- Stage 6 → Managing transport and reserves

Major Producing Countries and Companies

5. China (Bayan Obo Mine), Australia (Mount Weld Mine), USA (Mountain Pass Mine).

6. Answer: c) Mountain Pass (USA)

7. Africa holds large, high-grade deposits (e.g., Ngualla, Kangankunde) but remains project-heavy with limited operational tonnage.

Market Forces, Pricing, and Supply–Demand Trends

8. Answer: b) Concentration of refining capacity

9. Price volatility arises from export restrictions, supply bottlenecks, environmental regulations, and limited refining capacity.

10. Rising demand for EVs and renewables, geopolitical restrictions, and insufficient new refining capacity could all drive prices upward.

Geopolitical and Strategic Factors

11. Export controls can disrupt supply chains, cause price spikes, and shift manufacturing dependencies.

12. Answer: False – The U.S.–Australia framework outlines policy intentions but is non-binding.

13. Alliances diversify supply sources, share technology, and build resilient "trusted partner" networks.

Regional Roles and Comparative Advantages

14. China dominates refining; Australia leads in mining and is building processing; the U.S. focuses on defence-linked downstream manufacturing.

15. It expands onshore processing and strengthens Australia's role as a secure, sustainable critical minerals supplier.

Government Policies and Strategies

16. Answer: c) Price ceilings on consumer goods

17. DoD-backed funding for processing plants (e.g., Texas, California) and strategic mineral stockpiling initiatives.

Logistics, Transport, and Exports

18. Export restrictions, shipping delays, and port congestion are key risks.

19. Answer: False – Export reliance increases vulnerability to foreign markets.

20. Develop refining capacity, attract manufacturing investment, and build enabling infrastructure.

Downstream Manufacturing and Recycling

21. Answer: c) Manufacturing and assembly of finished goods

22. Recycling closes the loop, reduces waste, and supplements supply from end-of-life products.

Market Trends and the Energy Transition

23. EVs, wind turbines, defence, and advanced electronics.

24. Answer: c) 2032

25. The supply of refined materials is not keeping pace with demand from new technologies.

Standards, Certification, and Responsible Sourcing

26. To ensure ethical, transparent, and environmentally sound sourcing practices.

27. Answers:

- Value Chain → Steps from raw extraction to end use

- Supply Chain → Network of logistics and production links

- Responsible Sourcing → Ethical and ESG-compliant procurement

- Strategic Reserve → Government stockpile of essential materials

- Processing Hub → Facility for refining critical minerals

28. Transparency and certification ensure traceability, prevent environmental harm, and promote sustainable, socially responsible global mineral trade.

Chapter 12 Future Skills, Technology, and Career Pathways

Emerging Technological Trends

1. b – It is the capture, integration, and use of data across mining and processing operations.

2. Automation uses machines, robotics, and control systems to perform tasks with minimal human input; benefits include higher precision, safety, and efficiency.

3. False – AI supports decision-making through pattern recognition and predictive modelling.

4. Automation and AI improve exploration by analysing geological data, enhance processing via self-optimising circuits and digital twins, and increase refining accuracy while improving safety and energy efficiency.

Data Analytics and Process Efficiency

5. A→2, B→3, C→1

6. They allow large-scale data collection and analysis for identifying new deposits and monitoring operations efficiently, reducing costs and environmental impact.

7. c – Digital Twin.

Green Chemistry and Low-Carbon Refining

8. To design chemical processes that minimise hazardous substances, waste, and energy use.

9. b – Hydrogen reduction powered by renewables.

10. Green chemistry replaces toxic reagents with safer alternatives; electrification and renewable-energy use lower emissions, making refining cleaner and supporting sustainability goals.

Cross-Disciplinary and Digital Skills

11. True.

12. Examples: data analytics, environmental management, project coordination, systems thinking, digital literacy.

13. Because interconnected IoT and control systems increase vulnerability to cyberattacks, requiring protection of operational data and equipment.

Workforce Change and Training

14. b – Digitalisation and decarbonisation of operations.

15. A shortage of skilled professionals with both engineering and digital expertise.

16. They require new training in automation, AI, and ESG practices; professional development must integrate digital skills and sustainability awareness.

Career Pathways and Education

17. A→2, B→3, C→1

18. Exploration geology, processing engineering, quality control, research, and sustainability roles.

19. b – REE recycling.

Continuous Learning and Innovation

20. It means continuously updating skills and knowledge to stay relevant as technologies evolve.

21. False – All professionals must upskill continuously.

22. They encourage adaptive thinking, innovation in processing and sustainability, and collaboration across disciplines, ensuring workforce resilience.

Career Development Planning

23. Because aligning personal or corporate growth with sustainability ensures long-term competitiveness and compliance with global decarbonisation goals.

Rare Earth and Critical Mineral Operations and Processing

24. Plans should include training in automation and data analytics, partnerships with education providers, investment in green technologies, and clear performance milestones to build future-ready capability.

References

1.   Deady, E., et al., *Volcanic-Derived Placers as a Potential Resource of Rare Earth Elements: The Aksu Diamas Case Study, Turkey.* Minerals, 2019. **9**(4): p. 208.
2.   Beard, C.D., et al., *Alkaline-Silicate REE-HFSE Systems.* Economic Geology, 2023. **118**(1): p. 177-208.
3.   Goodenough, K., et al., *Europe's Rare Earth Element Resource Potential: An Overview of REE Metallogenetic Provinces and Their Geodynamic Setting.* Ore Geology Reviews, 2016. **72**: p. 838-856.
4.   Silva, A.C., et al., *Classification of Rare Earths Mineral Resources Using the Jequié/Ba Complex Rock Weathering Chemical Index.* Concilium, 2024. **24**(11): p. 15-29.
5.   Borst, A., et al., *Adsorption of Rare Earth Elements in Regolith-Hosted Clay Deposits.* Nature Communications, 2020. **11**(1).
6.   Vahidi, E., J. Navarro, and F. Zhao, *An Initial Life Cycle Assessment of Rare Earth Oxides Production From Ion-Adsorption Clays.* Resources Conservation and Recycling, 2016. **113**: p. 1-11.
7.   Goodenough, K., F. Wall, and D. Merriman, *The Rare Earth Elements: Demand, Global Resources, and Challenges for Resourcing Future Generations.* Natural Resources Research, 2017. **27**(2): p. 201-216.
8.   Peacock, J., K.M. Denton, and D.A. Ponce, *Magnetotelluric Imaging of a Carbonatite Terrane in the Southeast Mojave Desert, California and Nevada.* Aseg Extended Abstracts, 2016. **2016**(1): p. 1-5.
9.   Taylor, R.D., et al., *Geochemistry and Geophysics of Iron Oxide-Apatite Deposits and Associated Waste Piles With Implications for Potential Rare Earth Element Resources From Ore and Historical Mine Waste in the Eastern Adirondack Highlands, New York, USA.* Economic Geology, 2019. **114**(8): p. 1569-1598.
10.   Wall, F. and R. Pell, *Responsible Sourcing of Critical Metals.* Applied Earth Science Transactions of the Institutions of Mining and Metallurgy Section B, 2017. **126**(2): p. 103-104.
11.   McLemore, V.T., *Rare Earth Elements (REE) Deposits Associated With Great Plain Margin Deposits (Alkaline-Related), Southwestern United States and Eastern Mexico.* Resources, 2018. **7**(1): p. 8.
12.   Hussain, A., et al., *Geochronology, Mineral Chemistry and Genesis of REE Mineralization in Alkaline Rocks From the Kohistan Island Arc, Pakistan.* Ore Geology Reviews, 2020. **126**: p. 103749.
13.   Aide, M., *The Soil Chemistry of Cerium With an Emphasis on the Formation Of Ion-Adsorption Rare Earth Element Deposits.* 2024.
14.   Dostál, J. and O. Gerel, *Rare Earth Element Deposits in Mongolia.* Minerals, 2023. **13**(1): p. 129.
15.   El-Sayed, O., et al., *Significant Enrichment of Rare Earth Element Concentrations in Stream Sediments of Sharm El-Sheikh Area, Southern Sinai-Egypt:*

Geochemical Prospecting and Heavy Mineral Survey. Iraqi Geological Journal, 2023: p. 1-15.

16. Taggart, R.K., et al., *Trends in the Rare Earth Element Content of U.S.-Based Coal Combustion Fly Ashes.* Environmental Science & Technology, 2016. **50**(11): p. 5919-5926.

17. Deng, H. and A. Kendall, *Life Cycle Assessment With Primary Data on Heavy Rare Earth Oxides From Ion-Adsorption Clays.* The International Journal of Life Cycle Assessment, 2019. **24**(9): p. 1643-1652.

18. Liu, H., et al., *Geochemical Signatures of Rare Earth Elements and Yttrium Exploited by Acid Solution Mining Around an Ion-Adsorption Type Deposit: Role of Source Control and Potential for Recovery.* The Science of the Total Environment, 2022. **804**: p. 150241.

19. Melegari, M., et al., *Tailoring the Use of 8-Hydroxyquinolines for the Facile Separation of Iron, Dysprosium and Neodymium.* Chemsuschem, 2024. **17**(21).

20. Zou, K. and H. Zhao, *Ligand-Induced Tetrad Effect in Coordination Leaching of Ion-Adsorption Rare Earth Ores: Enhanced Recovery of High-Value Low-Lanthanum Rare Earth Concentrates.* Green Chemistry, 2025. **27**(27): p. 8174-8187.

21. Watts, K.E., D.M. Miller, and D.A. Ponce, *Mafic Alkaline Magmatism and Rare Earth Element Mineralization in the Mojave Desert, California: The Bobcat Hills Connection to Mountain Pass.* Geochemistry Geophysics Geosystems, 2024. **25**(1).

22. Shahbaz, A., *Process Development to Recover Rare Earth Metals From Selected Primary Ores: A Review.* Pakistan Journal of Analytical & Environmental Chemistry, 2022. **23**(1): p. 1-20.

23. Chen, W., et al., *Geochemistry of Monazite Within Carbonatite Related REE Deposits.* Resources, 2017. **6**(4): p. 51.

24. Xu, C., et al., *Origin of Heavy Rare Earth Mineralization in South China.* Nature Communications, 2017. **8**(1).

25. Sergeev, N. and T. Collins, *Regolith-Hosted Rare Earth Element Mineralization in the Esperance Region, Western Australia: Major Characteristics and Potential Controls.* 2024.

26. Marin, R., G. Brunet, and M. Murugesu, *Shining New Light on Multifunctional Lanthanide Single-Molecule Magnets.* Angewandte Chemie International Edition, 2020. **60**(4): p. 1728-1746.

27. Larochelle, T., et al., *Recovery of Rare Earth Element From Acid Mine Drainage Using Organo-Phosphorus Extractants and Ionic Liquids.* Minerals, 2022. **12**(11): p. 1337.

28. Aravena, D., M. Atanasov, and F. Neese, *Periodic Trends in Lanthanide Compounds Through the Eyes of Multireference Ab Initio Theory.* Inorganic Chemistry, 2016. **55**(9): p. 4457-4469.

29. Kavanagh, L., et al., *Global Lithium Sources—Industrial Use and Future in the Electric Vehicle Industry: A Review.* Resources, 2018. **7**(3): p. 57.

30. Peiró, L.T., G. Villalba, and R.U. Ayres, *Lithium: Sources, Production, Uses, and Recovery Outlook.* Jom, 2013. **65**(8): p. 986-996.

31. Vera, M.L., et al., *Environmental Impact of Direct Lithium Extraction From Brines.* Nature Reviews Earth & Environment, 2023. **4**(3): p. 149-165.

32. Bing, P., et al., *Characteristics and Material Sources of Global Brine-Type Lithium Deposits.* Geological Journal, 2024. **60**(8): p. 1956-1973.

33. Song, J., et al., *Lithium Extraction From Chinese Salt-Lake Brines: Opportunities, Challenges, and Future Outlook.* Environmental Science Water Research & Technology, 2017. **3**(4): p. 593-597.

34. Li, L., et al., *Lithium Recovery From Aqueous Resources and Batteries: A Brief Review.* Johnson Matthey Technology Review, 2018. **62**(2): p. 161-176.

35. Stringfellow, W.T. and P. Dobson, *Technology for the Recovery of Lithium From Geothermal Brines.* Energies, 2021. **14**(20): p. 6805.

36. Handayani, S.S., et al., *Mass Balance of Nickel Manganese Cobalt Cathode Battery Recycle Process.* Journal of Bioresources and Environmental Sciences, 2024. **3**(3): p. 161-165.

37. Avarmaa, K., et al., *Battery Scrap and Biochar Utilization for Improved Metal Recoveries in Nickel Slag Cleaning Conditions.* Batteries, 2020. **6**(4): p. 58.

38. Cruz, K.A.M.L., et al., *Multi-Energy Calibration for Determining Critical Metals in Nickel-Metal Hydride Battery Residues by Microwave-Induced Plasma Atomic Emission Spectrometry.* Analytical Methods, 2023. **15**(30): p. 3675-3682.

39. Liu, F., et al., *Synergistic Recovery of Valuable Metals From Spent Nickel–Metal Hydride Batteries and Lithium-Ion Batteries.* Acs Sustainable Chemistry & Engineering, 2019. **7**(19): p. 16103-16111.

40. Ma, Z., et al., *The Forming Age and the Evolution Process of the Brine Lithium Deposits in the Qaidam Basin Based on Geochronology and Mineral Composition.* Frontiers in Earth Science, 2021. **9**.

41. Kim, K., et al., *Selective Cobalt and Nickel Electrodeposition for Lithium-Ion Battery Recycling Through Integrated Electrolyte and Interface Control.* Nature Communications, 2021. **12**(1).

42. Agusdinata, D.B., et al., *Socio-Environmental Impacts of Lithium Mineral Extraction: Towards a Research Agenda.* Environmental Research Letters, 2018. **13**(12): p. 123001.

43. Sinha, R. and B.C. Raymahashay, *Evaporite Mineralogy and Geochemical Evolution of the Sambhar Salt Lake, Rajasthan, India.* Sedimentary Geology, 2004. **166**(1-2): p. 59-71.

44. Filho, W.L., et al., *Understanding Rare Earth Elements as Critical Raw Materials.* Sustainability, 2023. **15**(3): p. 1919.

45. Alonso, E., et al., *Evaluating Rare Earth Element Availability: A Case With Revolutionary Demand From Clean Technologies.* Environmental Science & Technology, 2012. **46**(6): p. 3406-3414.

46. Diallo, M.S., M.R. Kotte, and M. Cho, *Mining Critical Metals and Elements From Seawater: Opportunities and Challenges.* Environmental Science & Technology, 2015. **49**(16): p. 9390-9399.

47. Roelich, K., et al., *Assessing the Dynamic Material Criticality of Infrastructure Transitions: A Case of Low Carbon Electricity.* Applied Energy, 2014. **123**: p. 378-386.

48. Hernández-Ávila, J., et al., *Use of Porous No Metallic Minerals to Remove Heavy Metals, Precious Metals and Rare Earths, by Cationic Exchange.* 2021.

49. McLellan, B., G. Corder, and S.H. Ali, *Sustainability of Rare Earths—An Overview of the State of Knowledge.* Minerals, 2013. **3**(3): p. 304-317.

50. Ambrose, H. and A. Kendall, *Understanding the Future of Lithium: Part 2, Temporally and Spatially Resolved Life-cycle Assessment Modeling.* Journal of Industrial Ecology, 2019. **24**(1): p. 90-100.

51. Yang, Y., *Impacts of the U.S.-China Trade War on Lithium-Ion Battery Supply Chains: A Network Analysis Approach.* International Journal of Global Economics and Management, 2024. **5**(2): p. 362-381.

52. Sanchez-Lopez, M.D., *Geopolitics of the Li-ion Battery Value Chain and the Lithium Triangle in South America.* Latin American Policy, 2023. **14**(1): p. 22-45.

53. Gibson, C.E., et al., *The Recovery and Concentration of Spodumene Using Dense Media Separation.* Minerals, 2021. **11**(6): p. 649.

54. Dhurandhar, A.P., *Lithium: A Discourse on Global Reserves, India's Ascendancy, and Substitutes for a Verdant Future.* International Journal for Multidisciplinary Research, 2025. **7**(5).

55. Bowell, R., et al., *Classification and Characteristics of Natural Lithium Resources.* Elements, 2020. **16**(4): p. 259-264.

56. Paranthaman, M., et al., *Recovery of Lithium From Geothermal Brine With Lithium–Aluminum Layered Double Hydroxide Chloride Sorbents.* Environmental Science & Technology, 2017. **51**(22): p. 13481-13486.

57. Wei, F., et al., *A Critical Bottleneck in Energy Transition: Quantitative Predictions and Potential Strategies for Lithium Resource Depletion.* 2025.

58. Montes, T., *Lithium Governance in the Lithium Triangle and the Goal of Achieving Sustainable Development: Where Is It Heading?* The Law and Development Review, 2025.

59. Skiba, R., *Battery Powered: The Social, Economical, and Environmental Impacts of the Lithium Ion Battery.* 2024: After Midnight Publishing.

60. Rodríguez, N.G.P., et al., *Cobalt Metal: Overview of Deposits, Reserves, Processing, and Recycling.* 2023.

61. Subagja, R., et al., *A Preliminary Study of Cobalt Solvent Extraction From Nickel Sulphate Solution Using Organic Extractant-Pc-88a.* Metalurgi, 2023. **38**(1): p. 33.

62. Domènech, C., et al., *Co–Mn Mineralisations in the Ni Laterite Deposits of Loma Caribe (Dominican Republic) and Loma De Hierro (Venezuela).* Minerals, 2022. **12**(8): p. 927.

63. Gleeson, S.A., C.R.M. Butt, and M. Eliás, *Nickel Laterites: A Review.* Seg Discovery, 2003(54): p. 1-18.

64. Peiseler, L., et al., *Carbon Footprint Distributions of Lithium-Ion Batteries and Their Materials.* 2024.

65. Pell, R., et al., *Towards Sustainable Extraction of Technology Materials Through Integrated Approaches.* Nature Reviews Earth & Environment, 2021. **2**(10): p. 665-679.

66. Dunn, J., et al., *Circularity of Lithium-Ion Battery Materials in Electric Vehicles.* Environmental Science & Technology, 2021. **55**(8): p. 5189-5198.

67.   Dai, Q., et al., *Life Cycle Analysis of Lithium-Ion Batteries for Automotive Applications.* Batteries, 2019. **5**(2): p. 48.

68.   Li, X.H., et al., *Experimental Study on Recovery of Nickel From Nickel-Bearing Laterite.* Advanced Materials Research, 2014. **881-883**: p. 1611-1615.

69.   Ang, C.A., Z. Feixiong, and G. Azimi, *Waste Valorization Process: Sulfur Removal and Hematite Recovery From High Pressure Acid Leach Residue for Steelmaking.* Acs Sustainable Chemistry & Engineering, 2017. **5**(9): p. 8416-8423.

70.   Walvekar, H., et al., *Implications of the Electric Vehicle Manufacturers' Decision to Mass Adopt Lithium-Iron Phosphate Batteries.* Ieee Access, 2022. **10**: p. 63834-63843.

71.   Manthiram, A., *A Reflection on Lithium-Ion Battery Cathode Chemistry.* Nature Communications, 2020. **11**(1).

72.   Pickles, C.A. and R. Elliott, *Thermodynamic Analysis of Selective Reduction of Nickeliferous Limonitic Laterite Ore by Carbon Monoxide.* Mineral Processing and Extractive Metallurgy Transactions of the Institutions of Mining and Metallurgy Section C, 2015. **124**(4): p. 208-216.

73.   Lee, S., D.-S. Kang, and J.-S. Roh, *Bulk Graphite: Materials and Manufacturing Process.* Carbon Letters, 2015. **16**(3): p. 135-146.

74.   Lower, L., et al., *Catalytic Graphitization of Biocarbon for Lithium-Ion Anodes: A Minireview.* Chemsuschem, 2023. **16**(24).

75.   Zhang, J., C. Liang, and J.B. Dunn, *Graphite Flows in the U.S.: Insights Into a Key Ingredient of Energy Transition.* Environmental Science & Technology, 2023. **57**(8): p. 3402-3414.

76.   Jurkiewicz, K., et al., *Sucrose-Based Dense, Pure, and Highly-Crystalline Graphitic Materials for Lithium-Ion Batteries.* Advanced Functional Materials, 2024. **34**(51).

77.   Banek, N.A., et al., *Sustainable Conversion of Biomass to Rationally Designed Lithium-Ion Battery Graphite.* Scientific Reports, 2022. **12**(1).

78.   Sen, A., et al., *Reviving Graphite Anode From Spent Li-Ion Batteries via Acid Leaching and Carbonization Methodology.* Acs Sustainable Resource Management, 2025. **2**(4): p. 642-650.

79.   Nyathi, M.S., C.B. Clifford, and H.H. Schobert, *Effect of Petroleum Feedstock and Reaction Conditions on the Structure of Coal-Petroleum Co-Cokes and Heat-Treated Products.* Energy & Fuels, 2012. **26**(7): p. 4413-4419.

80.   Frankenstein, L., et al., *Revealing the Impact of Different Iron-Based Precursors on the 'Catalytic' Graphitization for Synthesis of Anode Materials for Lithium Ion Batteries.* Chemelectrochem, 2023. **10**(5).

81.   Gani, S.A., et al., *Designing a Silicon-Dominant Anode With Graphitic Carbon Coating From Biomass for High-Capacity Li-Ion Batteries.* Chemelectrochem, 2025. **12**(17).

82.   Kulkarni, S., et al., *Prospective Life Cycle Assessment of Synthetic Graphite Manufactured via Electrochemical Graphitization.* Acs Sustainable Chemistry & Engineering, 2022. **10**(41): p. 13607-13618.

83.   Xia, X., *Modification of Graphite Anode for Lithium Ion Battery.* Applied and Computational Engineering, 2024. **60**(1): p. 159-164.

84.   Barkov, A.Y. and F. Zaccarini, *Editorial for the Special Issue "Platinum-Group Minerals: New Results and Advances in PGE Mineralogy in Various Ni-Cu-Cr-Pge Ore Systems".* Minerals, 2019. **9**(6): p. 365.

85.   Wilson, A.H. and M.D. Prendergast, *Platinum-Group Element Mineralisation in the Great Dyke, Zimbabwe, and Its Relationship to Magma Evolution and Magma Chamber Structure.* South African Journal of Geology, 2001. **104**(4): p. 319-342.

86.   O'Connor, C.T. and T.N. Alexandrova, *The Geological Occurrence, Mineralogy, and Processing by Flotation of Platinum Group Minerals (PGMs) in South Africa and Russia.* Minerals, 2021. **11**(1): p. 54.

87.   Yakoumis, I., *PROMETHEUS: A Copper-Based Polymetallic Catalyst for Automotive Applications. Part I: Synthesis and Characterization.* Materials, 2021. **14**(3): p. 622.

88.   Mvokwe, S.A., et al., *A Critical Review of the Hydrometallurgy and Pyrometallurgical Recovery Processes of Platinum Group Metals From End-of-Life Fuel Cells.* Membranes, 2025. **15**(1): p. 13.

89.   Firmansyah, M.L., et al., *Application of Ionic Liquids in Solvent Extraction of Platinum Group Metals.* Solvent Extraction Research and Development Japan, 2020. **27**(1): p. 1-24.

90.   Oberthür, T., et al., *Geochemistry and Mineralogy of Platinum-Group Elements at Hartley Platinum Mine, Zimbabwe.* Mineralium Deposita, 2003. **38**(3): p. 344-355.

91.   Sun, X., *Supply Chain Risks of Critical Metals: Sources, Propagation, and Responses.* Frontiers in Energy Research, 2022. **10**.

92.   Popov, V.V., et al., *Powder Bed Fusion Additive Manufacturing Using Critical Raw Materials: A Review.* Materials, 2021. **14**(4): p. 909.

93.   Maulida, S., et al., *Multivariate Analysis of Geochemical Data to Determine Critical Minerals in Prigi Beach Sand, Trenggalek.* Journal of Physics Conference Series, 2024. **2866**(1): p. 012071.

94.   Liu, W., et al., *Research Trend and Dynamical Development of Focusing on the Global Critical Metals: A Bibliometric Analysis During 1991–2020.* Environmental Science and Pollution Research, 2021. **29**(18): p. 26688-26705.

95.   Parbhakar-Fox, A., *The Critical Importance of Mine Waste- An Australian Perspective.* 2023.

96.   Du, X. and G.C. Thakur, *The Growing Importance of Critical Minerals for the Energy Transition and U.S. Supply Chain Security.* 2025.

97.   Pawar, G. and R.C. Ewing, *Recent Advances in the Global Rare-Earth Supply Chain.* Mrs Bulletin, 2022. **47**(3): p. 244-249.

98.   Nygaard, A., *The Geopolitical Risk and Strategic Uncertainty of Green Growth After the Ukraine Invasion: How the Circular Economy Can Decrease the Market Power of and Resource Dependency on Critical Minerals.* Circular Economy and Sustainability, 2022. **3**(2): p. 1099-1126.

99.   Ali, A., F.A. Shah, and H. Raza, *The Geopolitical Implications of Rare Earth Mineral Dependencies and Technological Rivalries.* Wah Academia Journal of Social Sciences, 2024. **3**(2): p. 732-746.

100. Chowdhury, M.O.S. and D. Talan, *From Waste to Wealth: A Circular Economy Approach to the Sustainable Recovery of Rare Earth Elements and Battery Metals From Mine Tailings.* Separations, 2025. **12**(2): p. 52.

101. Patil, A.B., R.P.W.J. Struis, and C. Ludwig, *Opportunities in Critical Rare Earth Metal Recycling Value Chains for Economic Growth With Sustainable Technological Innovations.* Circular Economy and Sustainability, 2022. **3**(2): p. 1127-1140.

102. Mohamed, D., et al., *Transforming Waste Into Wealth: The Role of E-Waste in Sustainable Mineral Resource Management.* 2025.

103. Laudal, D., et al., *Rare Earth Elements in North Dakota Lignite Coal and Lignite-Related Materials.* Journal of Energy Resources Technology, 2018. **140**(6).

104. Agusdinata, D.B., H. Eakin, and W. Liu, *Critical Minerals for Electric Vehicles: A Telecoupling Review.* Environmental Research Letters, 2022. **17**(1): p. 013005.

105. Government of Canada, *Lithium facts.* 2025, Government of Canada,.

106. Australian Government, *Critical Minerals Strategy 2023–2030.* 2023.

107. Chen, G., *The Impact and Recovery of Covid-19 on the Chinese Economy: A Supply Chain Perspective as an Example.* Transactions on Economics Business and Management Research, 2024. **14**: p. 377-383.

108. Wang, Q., H. Zhou, and X. Zhao, *The Role of Supply Chain Diversification in Mitigating the Negative Effects of Supply Chain Disruptions in COVID-19.* International Journal of Operations & Production Management, 2023. **44**(1): p. 99-132.

109. Abbott, P., *Where Do We Stop for First Nations Cultural Heritage on the Critical Mineral-Paved Road to Net-Zero in Australia?* Asia Pacific Journal of Environmental Law, 2025. **28**(1): p. 11-36.

110. Goreczky, P., *Decoupling or Diversification? Dilemmas of India, Japan, and Australia in Shaping Economic Relations With China.* 2021.

111. Elizabeth, A.O., *Effect of Supply Chain Agility on Business Resilience in Manufacturing Firms in Australia.* Journal of Business and Strategic Management, 2025. **10**(1): p. 59-67.

112. Han, Y., *Rare Earth as Strategic Leverage: China's Export Restrictions, Global Responses, and Conceptual Framework for Analysis.* Advances in Economics Management and Political Sciences, 2025. **222**(1): p. 193-199.

113. Smith, J.A. and C. Williams, *Bridging Boundaries: Health Promotion Leadership in the Context of Health-In-All-Policies.* Health Promotion Journal of Australia, 2021. **32**(3): p. 369-371.

114. Anenburg, M., S. Broom-Fendley, and W. Chen, *Formation of Rare Earth Deposits in Carbonatites.* Elements, 2021. **17**(5): p. 327-332.

115. Simandl, G.J. and S.J. Paradis, *Carbonatites: Related Ore Deposits, Resources, Footprint, and Exploration Methods.* Applied Earth Science Transactions of the Institutions of Mining and Metallurgy, 2018. **127**(4): p. 123-152.

116. Zheng, X., et al., *Carbonatitic Magma Fractionation and Contamination Generate Rare Earth Element Enrichment and Mineralization in the Maoniuping Giant REE Deposit, SW China.* Journal of Petrology, 2023. **64**(6).

117. Dowman, E., et al., *Rare-Earth Mobility as a Result of Multiple Phases of Fluid Activity in Fenite Around the Chilwa Island Carbonatite, Malawi.* Mineralogical Magazine, 2017. **81**(6): p. 1367-1395.

118. Nikolenko, A., et al., *Crystallization of Bastnäsite and Burbankite From Carbonatite Melt in the System La(CO3)F-CaCO3-Na2CO3 at 100 MPa.* American Mineralogist, 2022. **107**(12): p. 2242-2250.

119. Watts, K.E. and A.K. Andersen, *Complex Carbonate Ore Mineralogy in the Mountain Pass Carbonatite Rare Earth Element Deposit, USA.* American Mineralogist, 2025.

120. Tamayo-Soriano, D.A., et al., *Acid Leaching of La and Ce From Ferrocarbonatite-Related REE Ores.* Minerals, 2024. **14**(5): p. 504.

121. Zapp, P., et al., *Environmental Impacts of Rare Earth Production.* Mrs Bulletin, 2022. **47**(3): p. 267-275.

122. Bui, T.T., et al., *Rare-Earth Mineralization in Dong Pao Deposit, Laichau Province, Vietnam.* Resource Geology, 2025. **75**(1).

123. Silva, Y.J.A.B.d., et al., *Rare Earth Element Geochemistry During Weathering of S-type Granites From Dry to Humid Climates of Brazil.* Journal of Plant Nutrition and Soil Science, 2018. **181**(6): p. 938-953.

124. Cressey, G., F. Wall, and B.A. Cressey, *Differential <i>REE</i> Uptake by Sector Growth of Monazite.* Mineralogical Magazine, 1999. **63**(6): p. 813-828.

125. Dostál, J., *Rare Earth Element Deposits of Alkaline Igneous Rocks.* Resources, 2017. **6**(3): p. 34.

126. Watt, G.R., *High-Thorium Monazite-(Ce) Formed During Disequilibrium Melting of Metapelites Under Granulite-Facies Conditions.* Mineralogical Magazine, 1995. **59**(397): p. 735-743.

127. Weng, Z., et al., *Assessing Rare Earth Element Mineral Deposit Types and Links to Environmental Impacts.* Applied Earth Science Transactions of the Institutions of Mining and Metallurgy Section B, 2013. **122**(2): p. 83-96.

128. Agarwal, V., M.S. Safarzadeh, and J. Galvin, *A Comparative Study of the Solvent Extraction of Lanthanum(III) From Different Acid Solutions.* Mineral Processing and Extractive Metallurgy Transactions of the Institutions of Mining and Metallurgy, 2019. **130**(2): p. 90-97.

129. Manikyamba, C., et al., *Gold, Uranium, Thorium, and Rare Earth Mineralization in the Kadiri Volcanic Province of Eastern Dharwar Craton, India: An Evaluation of Mineralogical, Textural, and Geochemical Attributes.* Geological Journal, 2020. **56**(1): p. 359-381.

130. Balaram, V., *Potential Future Alternative Resources for Rare Earth Elements: Opportunities and Challenges.* 2023.

131. Abdellah, W.M., et al., *Cerium Sulfate Preparation From Egyptian Monazite's Rare Earth Cake for Its Application as Corrosion Inhibitor of Aluminum Alloy AA6061.* Egyptian Journal of Chemistry, 2021. **0**(0): p. 0-0.

132. Ölmez, İ., et al., *Rare Earth Elements in Sediments Off Southern California: A New Anthropogenic Indicator.* Environmental Science & Technology, 1991. **25**(2): p. 310-316.

133. Hassan, E.-S.R.E., *Rare Earths in Egypt: Occurrence, Characterization and Physical Separation, a Review.* International Journal of Materials Technology and Innovation, 2023. **0**(0): p. 0-0.

134. Hentschel, F., et al., *Corona Formation Around Monazite and Xenotime During Greenschist-Facies Metamorphism and Deformation.* European Journal of Mineralogy, 2020. **32**(5): p. 521-544.

135. Lu, L., et al., *Geochemical and Geochronological Constraints on the Genesis of Ion-Adsorption-Type REE Mineralization in the Lincang Pluton, SW China.* Minerals, 2020. **10**(12): p. 1116.

136. Nazzareni, S., et al., *Characterization of Gold of the Murcielago Fluvial Placer (Central Honduras) and Its Possible Primary Sources.* Geosciences, 2023. **13**(6): p. 175.

137. Zglinicki, K., K. Szamałek, and S. Wołkowicz, *Critical Minerals From Post-Processing Tailing. A Case Study From Bangka Island, Indonesia.* Minerals, 2021. **11**(4): p. 352.

138. Natasya, Z.A. and N.S. Abdullah, *Malaysia's Rare Earth Element Story: Characterizing the "Amang".* Key Engineering Materials, 2022. **908**: p. 494-502.

139. Beurlen, H., et al., *Evaluation of the Potential for Rare Earth Element (Ree) Deposits Related to the Borborema Pegmatite Province in Northeastern Brazil.* Estudos Geológicos, 2020. **29**(2).

140. Cook, N.J., et al., *Mineralogy and Distribution of REE in Oxidised Ores of the Mount Weld Laterite Deposit, Western Australia.* Minerals, 2023. **13**(5): p. 656.

141. Lyalina, L.M., et al., *Beryllium Mineralogy of the Kola Peninsula, Russia—A Review.* Minerals, 2018. **9**(1): p. 12.

142. Yang, Z.-Y., et al., *Paragenesis of Li Minerals in the Nanyangshan Rare-Metal Pegmatite, Northern China: Toward a Generalized Sequence of Li Crystallization in Li-Cs-Ta-Type Granitic Pegmatites.* American Mineralogist, 2022. **107**(12): p. 2155-2166.

143. Zhang, X., et al., *Solubility and Modeling of $Li_2SO_4 \cdot H_2O$ in Aqueous H_2SO_4–$MgSO_4$ Solutions for Lithium Extraction From Spodumene.* Journal of Chemical & Engineering Data, 2022. **67**(4): p. 919-931.

144. Asif, A.H., et al., *Australia's Spodumene: Advances in Lithium Extraction Technologies, Decarbonization, and Circular Economy.* Industrial & Engineering Chemistry Research, 2024. **63**(5): p. 2073-2086.

145. Feng, Y., et al., *Environmental Impacts of Lithium Supply Chains From Australia to China.* Environmental Research Letters, 2024. **19**(9): p. 094035.

146. Ahmed, S., A.K.M.R. Reddy, and K. Zaghib, *Transformations of Critical Lithium Ores to Battery-Grade Materials: From Mine to Precursors.* Batteries, 2024. **10**(11): p. 379.

147. Phelps-Barber, Z., A. Trench, and D.I. Groves, *Recent Pegmatite-Hosted Spodumene Discoveries in Western Australia: Insights for Lithium Exploration in Australia and Globally.* Applied Earth Science Transactions of the Institutions of Mining and Metallurgy, 2022. **131**(2): p. 100-113.

148. Fosu, A.Y., et al., *Literature Review and Thermodynamic Modelling of Roasting Processes for Lithium Extraction From Spodumene*. Metals, 2020. **10**(10): p. 1312.

149. Aylmore, M., et al., *Assessment of a Spodumene Ore by Advanced Analytical and Mass Spectrometry Techniques to Determine Its Amenability to Processing for the Extraction of Lithium*. Minerals Engineering, 2018. **119**: p. 137-148.

150. Han, S., et al., *Direct Extraction of Lithium From A-Spodumene by Salt Roasting–Leaching Process*. Acs Sustainable Chemistry & Engineering, 2022. **10**(40): p. 13495-13504.

151. Припачкин, П.В., et al., *Lithium in Pegmatites of the Fennoscandian Shield and Operation Prospects for the Kolmozero Deposit on the Kola Peninsula (Russia)*. Applied Earth Science Transactions of the Institutions of Mining and Metallurgy, 2022. **131**(4): p. 179-192.

152. Santos, L.L.d., R.M. Nascimento, and S.B.C. Pergher, *Beta-Spodumene:Na2CO3:NaCl System Calcination: A Kinetic Study of the Conversion to Lithium Salt*. Chemical Engineering Research and Design, 2019. **147**: p. 338-345.

153. Zhu, Y., et al., *Lithium Recovery From Pretreated <i>α</I>-spodumene Residue Through Acid Leaching at Ambient Temperature*. The Canadian Journal of Chemical Engineering, 2023. **101**(8): p. 4360-4373.

154. Agarwal, S., et al., *Ceria Nanoshapes—Structural and Catalytic Properties*. 2015: p. 31-70.

155. Silva, Y.J.A.B.d., et al., *Rare Earth Element Concentrations in Brazilian Benchmark Soils*. Revista Brasileira De Ciência Do Solo, 2016. **40**(0).

156. Verni, E.R., et al., *REE Profiling in Basic Volcanic Rocks After Ultrasonic Sample Treatment and ICPMS Analysis With Oxide Ion Formation in ICP Enriched With O 2*. Microchemical Journal, 2017. **130**: p. 14-20.

157. Sadeghi, M., et al., *Rare Earth Element Distribution and Mineralization in Sweden: An Application of Principal Component Analysis to FOREGS Soil Geochemistry*. Journal of Geochemical Exploration, 2013. **133**: p. 160-175.

158. Silva, C.M.C.A.C., et al., *Geochemistry and Spatial Variability of Rare Earth Elements in Soils Under Different Geological and Climate Patterns of the Brazilian Northeast*. Revista Brasileira De Ciência Do Solo, 2018. **42**(0).

159. Tyler, G., *Rare Earth Elements in Soil and Plant Systems - A Review*. Plant and Soil, 2004. **267**(1-2): p. 191-206.

160. Akçıl, A., et al., *Hydrometallurgical Recycling Strategies for Recovery of Rare Earth Elements From Consumer Electronic Scraps: A Review*. Journal of Chemical Technology & Biotechnology, 2021. **96**(7): p. 1785-1797.

161. Rodríguez-Rastrero, M., et al., *Geochemical Anomalies in Soils and Surface Waters in an Area Adjacent to a Long-Used Controlled Municipal Landfill*. Sustainability, 2023. **15**(23): p. 16280.

162. Varvoutis, G., et al., *Recent Advances on Fine-Tuning Engineering Strategies of CeO2-Based Nanostructured Catalysts Exemplified by CO2 Hydrogenation Processes*. Catalysts, 2023. **13**(2): p. 275.

163. Turgeon, K., J.F. Boulanger, and C. Bazin, *Simulation of Solvent Extraction Circuits for the Separation of Rare Earth Elements*. Minerals, 2023. **13**(6): p. 714.

164. Romano, P., et al., *Simulation and Economical Analysis of Hydro-Nd Process for the Recovery of Rare Earth From End-of-Life Permanent Magnets: NEW-RE and INSPIREE Projects.* Detritus, 2024(28): p. 141-149.

165. Luo, X., et al., *Review on the Development and Utilization of Ionic Rare Earth Ore.* Minerals, 2022. **12**(5): p. 554.

166. Chai, X., et al., *Leaching Kinetics of Weathered Crust Elution-Deposited Rare Earth Ore With Compound Ammonium Carboxylate.* Minerals, 2020. **10**(6): p. 516.

167. Abbasalizadeh, A., et al., *Electrochemical Extraction of Rare Earth Metals in Molten Fluorides: Conversion of Rare Earth Oxides Into Rare Earth Fluorides Using Fluoride Additives.* Journal of Sustainable Metallurgy, 2017. **3**(3): p. 627-637.

168. Yang, H., et al., *Recovery of Trace Rare Earths From High-Level Fe^{3+}and Al^{3+}Waste of Oil Shale Ash (Fe–Al–OSA).* Industrial & Engineering Chemistry Research, 2010. **49**(22): p. 11645-11651.

169. Huang, S., et al., *Technology Development for Rare Earth Cleaner Hydrometallurgy in China.* Rare Metals, 2015. **34**(4): p. 215-222.

170. Zhu, S., et al., *Resource Recovery of Waste Nd–Fe–B Scrap: Effective Separation of Fe as High-Purity Hematite Nanoparticles.* Sustainability, 2020. **12**(7): p. 2624.

171. Zhuang, J., et al., *Monodispersed B-NaYF$_4$ Mesocrystals: In Situ Ion Exchange and Multicolor Up- And Down-Conversions.* Crystal Growth & Design, 2013. **13**(6): p. 2292-2297.

172. Xu, Z., et al., *Rare Earth Fluorides Nanowires/Nanorods Derived From Hydroxides: Hydrothermal Synthesis and Luminescence Properties.* Crystal Growth & Design, 2009. **9**(11): p. 4752-4758.

173. Nghipulile, T., S. Nkwanyana, and N. Lameck, *The Effect of HPGR and Conventional Crushing on the Extent of Micro-Cracks, Milling Energy Requirements and the Degree of Liberation: A Case Study of UG2 Platinum Ore.* Minerals, 2023. **13**(10): p. 1309.

174. Pedrosa, F.J.B., et al., *HPGR as Alternative to Fused Alumina Comminution Route: An Assessment of Circuit Simplification Potential.* Rem - International Engineering Journal, 2019. **72**(3): p. 543-551.

175. Rashidi, S., R.K. Rajamani, and D.W. Fuerstenau, *A Review of the Modeling of High Pressure Grinding Rolls.* Kona Powder and Particle Journal, 2017. **34**(0): p. 125-140.

176. Zan, Y., et al., *The Study on Properties Change of Bauxite Crushed by High Pressure Grinding Roller.* Advanced Materials Research, 2013. **826**: p. 148-151.

177. Oliveira, R., H. Delboni, and M.G. Bergerman, *Performance Analysis of the HRCTM HPGR in Pilot Plant.* Rem Revista Escola De Minas, 2016. **69**(2): p. 227-232.

178. Hassanzadeh, A., M. Safari, and D.H. Hoang, *Fine, Coarse and Fine-Coarse Particle Flotation in Mineral Processing With a Particular Focus on the Technological Assessments.* 2021.

179. Thatipamula, S. and S. Devasahayam, *Energy-Efficient Gold Flotation via Coarse Particle Generation Using VSI and HPGR Comminution.* Materials, 2025. **18**(15): p. 3553.

180. Li, S., et al., *Improving Separation Efficiency of Galena Flotation Using the Aerated Jet Flotation Cell.* Physicochemical Problems of Mineral Processing, 2020. **56**(3): p. 513-527.

181. Beneventi, D., et al., *Optimization and Management of Flotation Deinking Banks by Process Simulation.* Industrial & Engineering Chemistry Research, 2009. **48**(8): p. 3964-3972.

182. Galas, J. and D. Litwin, *Machine Learning Technique for Recognition of Flotation Froth Images in a Nonstable Flotation Process.* Minerals, 2022. **12**(8): p. 1052.

183. Wang, X., et al., *Effect of Two-Stage Energy Input on Enhancing the Flotation Process of High-Ash Coal Slime.* Physicochemical Problems of Mineral Processing, 2024.

184. Kumar, A., R. Sahu, and S.K. Tripathy, *Energy-Efficient Advanced Ultrafine Grinding of Particles Using Stirred Mills—A Review.* Energies, 2023. **16**(14): p. 5277.

185. Li, H., et al., *Adaptive Decoupling Control of Pulp Levels in Flotation Cells.* Asian Journal of Control, 2013. **15**(5): p. 1434-1447.

186. Abidi, A., et al., *Entrainment and True Flotation of a Natural Complex Ore Sulfide.* Journal of Mining Science, 2014. **50**(6): p. 1061-1068.

187. Ya, K.Z., et al., *Thermodynamics and Electrochemistry of the Interaction of Sphalerite With Iron (II)-Bearing Compounds in Relation to Flotation.* Resources, 2022. **11**(12): p. 108.

188. Abkhoshk, E., M. Kor, and B. Rezai, *A Study on the Effect of Particle Size on Coal Flotation Kinetics Using Fuzzy Logic.* Expert Systems With Applications, 2010. **37**(7): p. 5201-5207.

189. Bahrami, A., et al., *Untitled.* Rudarsko-Geološko-Naftni Zbornik, 2023. **38**(1).

190. Hosseini, S.H. and E. Forssberg, *Studies on Selective Flotation of Smithsonite From Silicate Minerals Using Mercaptans and One Stage Desliming.* Mineral Processing and Extractive Metallurgy Transactions of the Institutions of Mining and Metallurgy Section C, 2011. **120**(2): p. 79-84.

191. Santana, R.C.d., et al., *Evaluation of the Influence of Process Variables on Flotation of Phosphate.* Materials Science Forum, 2010. **660-661**: p. 555-560.

192. Polat, M., H. Polat, and S. Chander, *Physical and Chemical Interactions in Coal Flotation.* International Journal of Mineral Processing, 2003. **72**(1-4): p. 199-213.

193. Xu, M., et al., *Effect of Ultrasonic Pretreatment on Oxidized Coal Flotation.* Energy & Fuels, 2017. **31**(12): p. 14367-14373.

194. Li, X., et al., *Flotation Recovery of Barite From High-Density Waste Drilling Fluid Using B-Cyclodextrin as a Novel Depressant and Its Mechanism.* Plos One, 2024. **19**(3): p. e0298626.

195. Wei, Y. and R.F. Sandenbergh, *Effects of Grinding Environment on the Flotation of Rosh Pinah Complex Pb/Zn Ore.* Minerals Engineering, 2007. **20**(3): p. 264-272.

196. Oleksik, K., D. Saramak, and A. Młynarczykowska, *Evaluation of Flotation Process Course on the Example of Sulphide Ores.* E3s Web of Conferences, 2017. **18**: p. 01006.

197. Wang, X., et al., *Effects of Sec-Octanol and Terpineol on Froth Properties and Flotation Selectivity Index for Microcrystalline Graphite.* Minerals, 2023. **13**(9): p. 1231.

198. Batjargal, K., et al., *Correlation of Flotation Recoveries and Bubble–Particle Attachment Time for Dodecyl Ammonium Hydrochloride/Frother/Quartz Flotation System*. Minerals, 2023. **13**(10): p. 1305.

199. Ebrahimi, H. and M. Karamoozian, *Effect of Ultrasonic Irradiation on Particle Size, Reagents Consumption, and Feed Ash Content in Coal Flotation*. International Journal of Coal Science & Technology, 2020. **7**(4): p. 787-795.

200. Calisaya, D.A., et al., *A Strategy for the Identification of Optimal Flotation Circuits*. Minerals Engineering, 2016. **96-97**: p. 157-167.

201. Lee, S., C.E. Gibson, and A. Ghahreman, *The Separation of Carbonaceous Matter From Refractory Gold Ore Using Multi-Stage Flotation: A Case Study*. Minerals, 2021. **11**(12): p. 1430.

202. Radmehr, V., et al., *Optimizing Flotation Circuit Recovery by Effective Stage Arrangements: A Case Study*. Minerals, 2018. **8**(10): p. 417.

203. Bu, X., et al., *Discrimination of Six Flotation Kinetic Models Used in the Conventional Flotation and Carrier Flotation of −74 Mm Coal Fines*. Acs Omega, 2020. **5**(23): p. 13813-13821.

204. Norgren, A. and C. Anderson, *Ultra-Fine Centrifugal Concentration of Bastnaesite Ore*. Metals, 2021. **11**(10): p. 1501.

205. Eckert, K., et al., *Carrier Flotation: State of the Art and Its Potential for the Separation of Fine and Ultrafine Mineral Particles*. Materials Science Forum, 2019. **959**: p. 125-133.

206. Chapleski, R.C., et al., *A Molecular-Scale Approach to Rare-Earth Beneficiation: Thinking Small to Avoid Large Losses*. Iscience, 2020. **23**(9): p. 101435.

207. Jonglertjunya, W. and T. Rubcumintara, *Titanium and Iron Dissolutions From Ilmenite by Acid Leaching and Microbiological Oxidation Techniques*. Asia-Pacific Journal of Chemical Engineering, 2012. **8**(3): p. 323-330.

208. Sahu, K.K., et al., *An Overview on the Production of Pigment Grade Titania From Titania-Rich Slag*. Waste Management & Research the Journal for a Sustainable Circular Economy, 2006. **24**(1): p. 74-79.

209. Shen, L., et al., *Kinetics of Scheelite Conversion in Sulfuric Acid*. Jom, 2018. **70**(11): p. 2499-2504.

210. Jonglertjunya, W., S. Rattanaphan, and P. Tipsak, *Kinetics of the Dissolution of Ilmenite in Oxalic and Sulfuric Acid Solutions*. Asia-Pacific Journal of Chemical Engineering, 2013. **9**(1): p. 24-30.

211. Shen, L., et al., *Wolframite Conversion in Treating a Mixed Wolframite–Scheelite Concentrate by Sulfuric Acid*. Jom, 2017. **70**(2): p. 161-167.

212. Nie, W., et al., *Leaching Behaviors of Impurities in Titanium-Bearing Electric Furnace Slag in Sulfuric Acid*. Processes, 2020. **8**(1): p. 56.

213. Wang, W.J., et al., *Kinetic Study on Sulfuric Acid Dissolution of Na$_x$H$_{2-X}$TiO$_3$ From Sodium Hydroxide Molten Method*. Advanced Materials Research, 2014. **881-883**: p. 1545-1548.

214. Thambiliyagodage, C., R. Wijesekera, and M.G. Bakker, *Leaching of Ilmenite to Produce Titanium Based Materials: A Review*. Discover Materials, 2021. **1**(1).

215. Jiang, X., T. Herricks, and Y. Xia, *Monodispersed Spherical Colloids of Titania: Synthesis, Characterization, and Crystallization.* Advanced Materials, 2003. **15**(14): p. 1205-1209.

216. Dubenko, A.V., et al., *Mechanism, Thermodynamics and Kinetics of Rutile Leaching Process by Sulfuric Acid Reactions.* Processes, 2020. **8**(6): p. 640.

217. Shen, L., et al., *Sustainable and Efficient Leaching of Tungsten in Ammoniacal Ammonium Carbonate Solution From the Sulfuric Acid Converted Product of Scheelite.* Journal of Cleaner Production, 2018. **197**: p. 690-698.

218. Baba, A.A., J.O. Kayode, and M.A. Raji, *Low-Energy Feasibility for Leaching an Indigenous Scheelite Ore for Industrial Applications.* Journal of Sustainable Metallurgy, 2020. **6**(4): p. 659-666.

219. Silva, C.N.d., et al., *Sulphuric Acid Digestion of Anatase Concentrate.* Mining, 2024. **4**(1): p. 79-90.

220. Premaratne, W.A.P.J. and N.A. Rowson, *Microwave Assisted Dissolution of Sri Lankan Ilmenite: Extraction and Leaching Kinetics of Titanium and Iron Metals.* Journal of Science of the University of Kelaniya, 2015. **9**: p. 01-14.

221. Li, D., et al., *Green and Efficient Recovery of Tungsten From Spent SCR Denitration Catalyst by Na$_2$S Alkali Leaching and Calcium Precipitation.* Advanced Sustainable Systems, 2025. **9**(4).

222. Zhao, L., et al., *Production of Rutile TiO$_2$ Pigment From Titanium Slag Obtained by Hydrochloric Acid Leaching of Vanadium-Bearing Titanomagnetite.* Industrial & Engineering Chemistry Research, 2013. **53**(1): p. 70-77.

223. Liu, L., et al., *Complex Leaching Process of Scheelite in Hydrochloric and Phosphoric Solutions.* Jom, 2016. **68**(9): p. 2455-2462.

224. Zhang, Y., et al., *Sulfur-Doped TiO$_2$ Anchored on a Large-Area Carbon Sheet as a High-Performance Anode for Sodium-Ion Battery.* Acs Applied Materials & Interfaces, 2019. **11**(47): p. 44170-44178.

225. Nasab, M.H., M. Noaparast, and H. Abdollahi, *Dissolution Optimization and Kinetics of Nickel and Cobalt From Iron-rich Laterite Ore, Using Sulfuric Acid at Atmospheric Pressure.* International Journal of Chemical Kinetics, 2020. **52**(4): p. 283-298.

226. Smailov, K. and Y. Nuruly, *Complex Ni and Co Extraction From Leached Nontronitized Serpentinite via Hydrometallurgical Process at Atmospheric Pressure.* 2023: p. 62-71.

227. Li, J., et al., *Applications of Rietveld-Based QXRD Analysis in Mineral Processing.* Powder Diffraction, 2014. **29**(S1): p. S89-S95.

228. Altansukh, B., K. Haga, and A. Shibayama, *Recovery of Nickel and Cobalt From a Low Grade Laterite Ore.* Resources Processing, 2014. **61**(2): p. 100-109.

229. Desaulty, A.-M., et al., *Tracing the Origin of Lithium in Li-Ion Batteries Using Lithium Isotopes.* Nature Communications, 2022. **13**(1).

230. Barbosa, L., et al., *Lithium Extraction From B-Spodumene Through Chlorination With Chlorine Gas.* Minerals Engineering, 2014. **56**: p. 29-34.

231. Mystrioti, C., et al., *Counter-Current Leaching of Low-Grade Laterites With Hydrochloric Acid and Proposed Purification Options of Pregnant Solution.* Minerals, 2018. **8**(12): p. 599.

232. Başlayıcı, S., et al., *Hydrometallurgical Nickel and Cobalt Production From Lateritic Ores: Optimization and Comparison of Atmospheric Pressure Leaching and Pug-Roast-Leaching Processes.* Acta Metallurgica Slovaca, 2021. **27**(1): p. 17-22.

233. Kaya, Ş., et al., *Concentration and Separation of Scandium From Ni Laterite Ore Processing Streams.* Metals, 2017. **7**(12): p. 557.

234. Lindgren, M., *Influence of the Amount of Ferric/Ferrous Ions on the Corrosion Resistance of a NiCrMo Alloy in High-temperature Sulfuric Acid Solutions.* Materials and Corrosion, 2016. **67**(9): p. 952-957.

235. Zunaidi, M.A., et al., *Iron Removal Process From Nickel Pregnant Leach Solution Using Sodium Hydroxide.* Metalurgi, 2022. **37**(3).

236. Haruna, B., et al., *Selective Separation of Lithium From Leachate of Spent Lithium-Ion Batteries by Zirconium Phosphate/Polyacrylonitrile Composite: Leaching and Sorption Behavior.* Batteries, 2024. **10**(7): p. 254.

237. Nguyen, T.T.H., T.T. Tran, and M.S. Lee, *A Modified Process for the Separation of Fe(III) and Cu(II) From the Sulfuric Acid Leaching Solution of Metallic Alloys of Reduction Smelted Spent Lithium-Ion Batteries.* Resources Recycling, 2022. **31**(1): p. 12-20.

238. Matsumoto, M., T. Yamaguchi, and Y. Tahara, *Extraction of Rare Earth Metal Ions With an Undiluted Hydrophobic Pseudoprotic Ionic Liquid.* Metals, 2020. **10**(4): p. 502.

239. Bao, S., et al., *Recovery and Separation of Metal Ions From Aqueous Solutions by Solvent-Impregnated Resins.* Chemical Engineering & Technology, 2016. **39**(8): p. 1377-1392.

240. Xu, Z., et al., *A Green and Efficient Recycling Strategy for Spent Lithium-Ion Batteries in Neutral Solution Environment.* Angewandte Chemie International Edition, 2025. **64**(17).

241. Kim, J.S. and M.A. Keane, *Ion Exchange of Divalent Cobalt and Iron With Na–Y Zeolite: Binary and Ternary Exchange Equilibria.* Journal of Colloid and Interface Science, 2000. **232**(1): p. 126-132.

242. Gmar, S., et al., *Lithium-Ion Battery Recycling: Metal Recovery From Electrolyte and Cathode Materials by Electrodialysis.* Metals, 2022. **12**(11): p. 1859.

243. Sahu, S., et al., *Synergistic Approach for Selective Leaching and Separation of Strategic Metals From Spent Lithium-Ion Batteries.* Acs Omega, 2024. **9**(9): p. 10556-10565.

244. Alkhadra, M.A. and M.Z. Bazant, *Continuous and Selective Separation of Heavy Metals Using Shock Electrodialysis.* Industrial & Engineering Chemistry Research, 2022.

245. Roy, J.J., B. Cao, and M. Srinivasan, *A Review on the Recycling of Spent Lithium-Ion Batteries (LIBs) by the Bioleaching Approach.* Chemosphere, 2021. **282**: p. 130944.

246. Zhu, X.H., et al., *Recycling Valuable Metals From Spent Lithium-Ion Batteries Using Carbothermal Shock Method.* Angewandte Chemie International Edition, 2023. **135**(15).

247. Fan, P., et al., *Separation and Purification of Light Rare Earth Elements From Chloride Media Using P204 and Cyanex272 in Sulfonated Kerosene Under Non-Saponification Conditions.* Physicochemical Problems of Mineral Processing, 2023.

248. Kostanyan, A.E., et al., *Separation of Rare Earth Elements in Multistage Extraction Columns in Chromatography Mode: Experimental Study and Mathematical Simulation.* Processes, 2023. **11**(6): p. 1757.

249. Sposato, C., et al., *Towards the Circular Economy of Rare Earth Elements: Lanthanum Leaching From Spent FCC Catalyst by Acids.* Processes, 2021. **9**(8): p. 1369.

250. Chen, Y., et al., *Synthesis of a Novel Water-Soluble Polymer Complexant Phosphorylated Chitosan for Rare Earth Complexation.* Polymers, 2022. **14**(3): p. 419.

251. Jeon, S., et al., *Glycopolymer-Mediated Selective Separation of Middle Rare Earth Elements.* Angewandte Chemie International Edition, 2024. **137**(6).

252. Bian, Y., et al., *Extraction of Rare Earth Elements From Permanent Magnet Scraps by VIM-HMS Method.* Materials Science Forum, 2016. **863**: p. 139-143.

253. Zhou, M., et al., *Synthesis of Acidic Phosphonic Chitosan and the Complexation of La(III) in Acidic Aqueous Solution.* Polymers, 2025. **17**(10): p. 1341.

254. Tan, H., et al., *Nitrogen-Doped Nanoporous Graphene Induced by a Multiple Confinement Strategy for Membrane Separation of Rare Earth.* Iscience, 2021. **24**(1): p. 101920.

255. Li, G., et al., *Preparation and Properties of C=X (X: O, N, S) Based Distillable Ionic Liquids and Their Application for Rare Earth Separation.* Acs Sustainable Chemistry & Engineering, 2016. **4**(12): p. 6258-6262.

256. Wang, L., et al., *Eliminating Ammonia Emissions During Rare Earth Separation Through Control of Equilibrium Acidity in a HEH(EHP)-Cl System.* Green Chemistry, 2013. **15**(7): p. 1889.

257. Sinclair, L., et al., *Rare Earth Element Extraction From Pretreated Bastnäsite in Supercritical Carbon Dioxide.* The Journal of Supercritical Fluids, 2017. **124**: p. 20-29.

258. Tjhia, M.A., M.N.F. Yahya, and R.M. Ulum, *Characteristics of Treated Monazite in Different Particle Sizes to Upgrade the Rare Earth Elements Content by Using Mechanochemical and Roasting Processes.* International Journal of Technology, 2024. **15**(2): p. 463.

259. Vibbert, H.B., A.W.S. Ooi, and A.H.A. Park, *Selective Recovery of Cerium as High-Purity Oxides via Reactive Separation Using CO_2-Responsive Structured Ligands.* Acs Sustainable Chemistry & Engineering, 2024. **12**(13): p. 5186-5196.

260. Anticoi, H., et al., *Ore Processing Technologies Applied to Industrial Waste Decontamination: A Case Study.* Minerals, 2022. **12**(6): p. 695.

261. Li, J., et al., *Sustainable and Efficient Recovery of Tungsten From Wolframite in a Sulfuric Acid and Phosphoric Acid Mixed System.* Acs Sustainable Chemistry & Engineering, 2020. **8**(36): p. 13583-13592.

262. Orefice, M., et al., *Solvometallurgical Process for the Recovery of Tungsten From Scheelite*. Industrial & Engineering Chemistry Research, 2021. **61**(1): p. 754-764.

263. Alguacil, F.J., et al., *Strategies for the Recovery of Tungsten From Wolframite, Scheelite, or Wolframite–Scheelite Mixed Concentrates of Spanish Origin*. Metals, 2025. **15**(8): p. 819.

264. Gaur, R.P.S., T.A. Wolfe, and S.A. Braymiller, *Sodium Carbonate-Roasting-Aqueous-Leaching Method to Process Flot-Grade Scheelite-Sulfide Tungsten Ore Concentrates*. Industrial & Engineering Chemistry Research, 2025. **64**(16): p. 8339-8358.

265. Baimbetov, B., et al., *Prospects of Processing Tungsten Ores From the Akchatau Deposit*. Processes, 2023. **12**(1): p. 77.

266. Wan, L., et al., *Study on Process of Wolframite and Calcium Carbonate Synthesizing Scheelite*. Advanced Materials Research, 2014. **962-965**: p. 847-851.

267. Zeng, D., V. Shcherbina, and J. Li, *Thermal Efficiency Analysis of the Rotary Kiln Based on the Wear of the Lining*. International Journal of Applied Mechanics and Engineering, 2023. **28**(2): p. 125-138.

268. Li, S., et al., *A Mathematical Model of Heat Transfer in a Rotary Kiln Thermo-Reactor*. Chemical Engineering & Technology, 2005. **28**(12): p. 1480-1489.

269. Song, X. and Q. Fan, *The Analysis of Rotary Kiln Thermal Characteristics Based on ANSYS and FLUENT*. Advanced Materials Research, 2013. **834-836**: p. 1523-1528.

270. Du, W., B. Wang, and L. Cheng, *Experimental Research and Numerical Analysis on Thermal Dynamic Characteristics of Rotary Kiln*. The Canadian Journal of Chemical Engineering, 2019. **97**(4): p. 1022-1032.

271. Zhang, B., et al., *Experimental and Numerical Simulation Study on Co-Incineration of Solid and Liquid Wastes for Green Production of Pesticides*. Processes, 2019. **7**(10): p. 649.

272. Zhong, Q., et al., *Thermal Behavior of Coal Used in Rotary Kiln and Its Combustion Intensification*. Energies, 2018. **11**(5): p. 1055.

273. Akram, N., et al., *Improved Waste Heat Recovery Through Surface of Kiln Using Phase Change Material*. Thermal Science, 2018. **22**(2): p. 1089-1098.

274. Kilbourn, B.T., *Cerium and Cerium Compounds*. 2011: p. 1-23.

275. Nawab, A. and R. Honaker, *Pilot Scale Testing of Lignite Adsorption Capability and the Benefits for the Recovery of Rare Earth Elements From Dilute Leach Solutions*. Minerals, 2023. **13**(7): p. 921.

276. Patkowski, W., et al., *Lanthanide Oxides in Ammonia Synthesis Catalysts: A Comprehensive Review*. Catalysts, 2023. **13**(12): p. 1464.

277. Emil-Kaya, E., et al., *NdFeB Magnets Recycling Process: An Alternative Method to Produce Mixed Rare Earth Oxide From Scrap NdFeB Magnets*. Metals, 2021. **11**(5): p. 716.

278. Whitty-Léveillé, L., N. Reynier, and D. Larivière, *Selective Removal of Uranium From Rare Earth Leachates via Magnetic Solid-Phase Extraction Using Schiff Base Ligands*. Industrial & Engineering Chemistry Research, 2018. **58**(1): p. 306-315.

279. Schreiber, A., et al., *Environmental Impacts of Rare Earth Mining and Separation Based on Eudialyte: A New European Way*. Resources, 2016. **5**(4): p. 32.

280. Norgate, T. and N. Haque, *Energy and Greenhouse Gas Impacts of Mining and Mineral Processing Operations.* Journal of Cleaner Production, 2010. **18**(3): p. 266-274.

281. Han, B., et al., *Development of Copper Recovery Process From Flotation Tailings by a Combined Method of High–pressure Leaching–solvent Extraction.* Journal of Hazardous Materials, 2018. **352**: p. 192-203.

282. Hawker, W., et al., *The Synergistic Copper Process Concept.* Mineral Processing and Extractive Metallurgy Transactions of the Institutions of Mining and Metallurgy, 2017. **127**(4): p. 210-220.

283. Tungpalan, K., E. Wightman, and E. Manlapig, *Relating Mineralogical and Textural Characteristics to Flotation Behaviour.* Minerals Engineering, 2015. **82**: p. 136-140.

284. Jassim, H.M., H.Z.A. Toma, and L.S. Oudah, *Solvent Extraction and Electro-Wining From Copper Leaching Product of Mawat Sulfide Ore Using Taguchi Method.* Ukh Journal of Science and Engineering, 2017. **1**(1): p. 53-59.

285. Dreisinger, D., T. Glück, and J. Lü, *The Recovery of Cobalt From the Boleo Deposit Using Leach, SX and EW.* 2012: p. 169-180.

286. Wang, Z., et al., *Recovery of High-Purity Silver From Spent Silver Oxide Batteries by Sulfuric Acid Leaching and Electrowinning.* Acs Sustainable Chemistry & Engineering, 2020. **8**(41): p. 15573-15583.

287. Wang, J., et al., *Gallium Electrowinning in Complex Solutions Produced by a Zinc Hydrometallurgy System.* Acs Omega, 2024. **9**(52): p. 51421-51430.

288. Godirilwe, L.L., et al., *Copper Recovery and Reduction of Environmental Loading From Mine Tailings by High-Pressure Leaching and SX-EW Process.* Metals, 2021. **11**(9): p. 1335.

289. Calvo, G., et al., *Decreasing Ore Grades in Global Metallic Mining: A Theoretical Issue or a Global Reality?* Resources, 2016. **5**(4): p. 36.

290. Yuan, Z., et al., *Mineralogical Characterization and Comprehensive Utilization of Micro-Fine Tantalum–niobium Ores From Songzi.* Rare Metals, 2015. **34**(4): p. 282-290.

291. Wang, Y.T., *The Application and Development of Microbubble Column Flotation Technology in China.* Advanced Materials Research, 2010. **136**: p. 194-201.

292. Cvetkovski, V., et al., *Construction of Isotherms in Solvent Extraction of Copper.* Hemijska Industrija, 2009. **63**(4): p. 309-312.

293. Annan, C., *Radon Risks in the Rare Earth Industry: A Critical Review of Exposure Pathways, Health Impacts and Policy Gaps.* Advances in Research, 2025. **26**(4): p. 458-467.

294. Dekusar, V.M., et al., *Mineral Reserves of Naturally Radioactive Thorium-Bearing Raw Materials.* Atomic Energy, 2012. **111**(3): p. 185-194.

295. Kotb, N.A., M.S.A.E. Ghany, and A.A. El-Sayed, *Radiological Assessment of Different Monazite Grades After Mechanical Separation From Black Sand.* Scientific Reports, 2023. **13**(1).

296. Hak, C.R.C., et al., *Extraction of Thorium Oxide (ThO$_2$) From Malaysian Monazite Through Alkali Digestion: Physical and Chemical Characterization Using X-Ray Analysis.* Key Engineering Materials, 2022. **908**: p. 509-514.

297. Silva, A.L.M.A.d., et al., *Radon in Brazilian Underground Mines.* Journal of Radiological Protection, 2018. **38**(2): p. 607-620.
298. Seo, S., et al., *Health Effects of Exposure to Radon: Implications of the Radon Bed Mattress Incident in Korea.* Epidemiology and Health, 2019. **41**: p. e2019004.
299. Kim, B.-G., K.-H. Jeong, and H. Shin, *Evaluation of Dose in Sleep by Mattress Containing Monazite.* Radiation Protection Dosimetry, 2019. **187**(3): p. 286-299.
300. Borai, E.H., M.M. Hamed, and A.M.S. El-Din, *A New Method for Processing of Low-Grade Monazite Concentrates.* Journal of the Geological Society of India, 2017. **89**(5): p. 600-604.
301. Abdel-Azeem, M.M., *Genesis of the Rare Metals Mineralisation in Um Safi Acidic Volcanics, Central Eastern Desert, Egypt.* Applied Earth Science Transactions of the Institutions of Mining and Metallurgy, 2024. **133**(1): p. 67-82.
302. Medeiros, F.F.d., et al., *Rare Earth Elements: A Review of Primary Sources, Applications, Business Investment, and Characterization Techniques.* Applied Sciences, 2025. **15**(20): p. 10949.
303. Claudia, D., et al., *Separation of Thorium (Th) From Monazite Sand of Bangka Island Using Primene JMT Solvent Extraction Method.* Eksplorium, 2024. **44**(2): p. 67-74.
304. Kim, M.S., et al., *Geochemical Origins and Occurrences of Natural Radioactive Materials in Borehole Groundwater in the Goesan Area.* The Journal of Engineering Geology, 2014. **24**(4): p. 535-550.
305. Alnour, I.A., et al., *Determination of the Elemental Concentration of Uranium and Thorium in the Products and by-Products of Amang Tin Tailings Process.* 2017. **1802**: p. 030003.
306. Chen, W.T. and M.F. Zhou, *Mineralogical and Geochemical Constraints on Mobilization and Mineralization of Rare Earth Elements in the Lala Fe-Cu-(Mo, Ree) Deposit, SW China.* American Journal of Science, 2015. **315**(7): p. 671-711.
307. Al-Ani, T., et al., *Geology and Mineralogy of Rare Earth Elements Deposits and Occurrences in Finland.* Minerals, 2018. **8**(8): p. 356.
308. Navaranjan, G., et al., *Uncertainties Associated With Assessing Ontario Uranium Miners' Exposure to Radon Daughters.* Journal of Radiological Protection, 2019. **39**(1): p. 136-149.
309. McDiarmid, M.A. and K.S. Squibb, *Uranium and Thorium.* 2001.
310. Rosenthal, I., P.R. Kleindorfer, and M.R. Elliott, *Predicting and Confirming the Effectiveness of Systems for Managing Low-probability Chemical Process Risks.* Process Safety Progress, 2006. **25**(2): p. 135-155.
311. Peron, M., et al., *Risk Assessment for Handling Hazardous Substances Within the European Industry: Available Methodologies and Research Streams.* Risk Analysis, 2022. **43**(7): p. 1434-1462.
312. Markowski, A.S., et al., *Process Safety Management Quality in Industrial Corporation for Sustainable Development.* Sustainability, 2021. **13**(16): p. 9001.
313. Tong, R., et al., *Process Safety Management in China: Progress and Performance Over the Last 10 years and Future Development.* Process Safety Progress, 2020. **39**(4).

314. Zhou, J. and X. Wang, *Analysis of the Progress of Chemical Process Safety Management in China.* Process Safety Progress, 2024. **43**(3): p. 425-435.

315. Yang, D., et al., *Characteristics and Statistical Analysis of Large and Above Hazardous Chemical Accidents in China From 2000 to 2020.* International Journal of Environmental Research and Public Health, 2022. **19**(23): p. 15603.

316. Ștefănescu, L., C. Botezan, and I. Crăciun, *Vulnerability Analysis for Two Accident Scenarios at an Upper-Tier Seveso Establishment in Romania.* Geographia Technica, 2018. **13**(1).

317. Möldri, M., et al., *Integration of the SMS to IMS in Estonian Seveso II Establishments: Selected Case Studies.* 2012.

318. Maria, G., D. Dinculescu, and H.H.S. Khwayyir, *Proximity Risk Assessment for Two Sensitive Chemical Plants Based on the Accident Scenario Consequence Analysis.* Asia-Pacific Journal of Chemical Engineering, 2013. **9**(1): p. 146-158.

319. Bernatík, A., *Industry Process Safety.* 2022: p. 960-995.

320. Chen, W., et al., *Reshaping Heavy Rare Earth Supply Chains Amidst China's Stringent Environmental Regulations.* Fundamental Research, 2025. **5**(2): p. 505-513.

321. Ge, Z., et al., *Revealing the Material, Cost, and Monetary Value Flows of Rare Earth Elements in China.* Environmental Research Communications, 2025. **7**(9): p. 095009.

322. Ширази, А., et al., *Investigation of Environmental and Biological Effects of Rare Earth Elements (REEs) With a Special Focus on Industrial and Mining Pollutions in Iran: A Review.* Advances in Geological and Geotechnical Engineering Research, 2021. **4**(1): p. 1-10.

323. Wolkersdorfer, C., et al., *Guidance for the Integrated Use of Hydrological, Geochemical, and Isotopic Tools in Mining Operations.* Mine Water and the Environment, 2020. **39**(2): p. 204-228.

324. Demirkan, C.P., N. Smith, and Ş. Düzgün, *A Quantitative Sustainability Assessment for Mine Closure and Repurposing Alternatives in Colorado, USA.* Resources, 2022. **11**(7): p. 66.

325. Lansbury, N. and T. Jeanneret, *Social Licence to Operate.* Corporate Communications an International Journal, 2015. **20**(2): p. 213-227.

326. Vanclay, F. and P. Hanna, *Conceptualizing Company Response to Community Protest: Principles to Achieve a Social License to Operate.* Land, 2019. **8**(6): p. 101.

327. Mei, L.C., *Logging and Indigenous Peoples' Well-Being: An Overview of the Relevant International Human Rights Jurisprudence.* International Forestry Review, 2023. **25**(1): p. 17-27.

328. Merino, R., *Unraveling a Political Technology: Free, Prior, and Informed Consent in Peruvian Oil and Mining Sectors.* Journal of Politics in Latin America, 2024. **17**(1): p. 131-154.

329. Phiri, O., E. Mantzari, and P. Gleadle, *Stakeholder Interactions and Corporate Social Responsibility (CSR) Practices.* Accounting Auditing & Accountability, 2018. **32**(1): p. 26-54.

330. Boiral, O., I.H. Saizarbitoria, and M.C. Brotherton, *Corporate Sustainability and Indigenous Community Engagement in the Extractive Industry.* Journal of Cleaner Production, 2019. **235**: p. 701-711.

331. Diantini, A., et al., *Is This a Real Choice? Critical Exploration of the Social License to Operate in the Oil Extraction Context of the Ecuadorian Amazon.* Sustainability, 2020. **12**(20): p. 8416.

332. Demajorovic, J., V. Pisano, and A.A.F. Pimenta, *Reframing the Social Acceptance of Mining Projects: The Contribution of Social Impact Assessment in the Brazilian Amazon.* Current Sociology, 2023. **72**(4): p. 649-671.

333. Ruwhiu, D. and L. Carter, *Negotiating "Meaningful Participation" for Indigenous Peoples in the Context of Mining.* Corporate Governance, 2016. **16**(4): p. 641-654.

334. Rodhouse, T.S.G.H. and F. Vanclay, *Is Free, Prior and Informed Consent a Form of Corporate Social Responsibility?* Journal of Cleaner Production, 2016. **131**: p. 785-794.

335. Ronalter, L.M., M. Bernardo, and J. Romaní, *Quality and Environmental Management Systems as Business Tools to Enhance ESG Performance: A Cross-Regional Empirical Study.* Environment Development and Sustainability, 2022. **25**(9): p. 9067-9109.

336. OECD, *Handbook on Environmental Due Diligence in Mineral Supply Chains.* 2023.

337. Fransen, L. and G. LeBaron, *Big Audit Firms as Regulatory Intermediaries in Transnational Labor Governance.* Regulation & Governance, 2018. **13**(2): p. 260-279.

338. Bohinc, R., *Corporate Directors' Sustainability Due Diligence.* 2024: p. 179-205.

339. Voland, T. and S. Daly, *The EU Regulation on Conflict Minerals: The Way Out of a Vicious Cycle?* Journal of World Trade, 2018. **52**(Issue 1): p. 37-63.

340. Sarfaty, G.A. and R. Deberdt, *Supply Chain Governance at a Distance.* Law & Social Inquiry, 2023. **49**(2): p. 1036-1059.

341. Pelaudeix, C., E.M. Basse, and N. Loukacheva, *Openness, Transparency and Public Participation in the Governance of Uranium Mining in Greenland: A Legal and Political Track Record.* Polar Record, 2017. **53**(6): p. 603-616.

342. Wood, J., *Seeking Community Consent for Resource Development in Greenland.* Nordicum-Mediterraneum, 2022. **17**(2).

343. Singh, N., et al., *Characterizing the Materials Composition and Recovery Potential From Waste Mobile Phones: A Comparative Evaluation of Cellular and Smart Phones.* Acs Sustainable Chemistry & Engineering, 2018. **6**(10): p. 13016-13024.

344. Das, S., G. Natarajan, and Y.P. Ting, *Bio-Extraction of Precious Metals From Urban Solid Waste.* 2017. **1805**: p. 020004.

345. Parlar, E.D., M.A. Sarı, and M. Can, *Selective Recovery of Gold From Electronic Circuit Board Waste With Pyrogallol-Formaldehyde Polymer.* 2024.

346. Godigamuwa, K. and N. Okibe, *Gold Leaching From Printed Circuit Boards Using a Novel Synergistic Effect of Glycine and Thiosulfate.* Minerals, 2023. **13**(10): p. 1270.

347. Gurung, M., et al., *Selective Recovery of Precious Metals From Acidic Leach Liquor of Circuit Boards of Spent Mobile Phones Using Chemically Modified*

Persimmon Tannin Gel. Industrial & Engineering Chemistry Research, 2012. **51**(37): p. 11901-11913.

348.   W, l.K., et al., *Hydrometallurgical Treatment for Mixed Waste Battery Material.* Iop Conference Series Materials Science and Engineering, 2017. **170**: p. 012024.

349.   Nguyen, V.N.H. and M.S. Lee, *Separation of <scp>Co(II), Ni(II), Mn(II) and Li(I)</Scp> From Synthetic Sulfuric Acid Leaching Solution of Spent Lithium Ion Batteries by Solvent Extraction.* Journal of Chemical Technology & Biotechnology, 2020. **96**(5): p. 1205-1217.

350.   Shafique, M., et al., *Global Material Flow Analysis of End-of-Life of Lithium Nickel Manganese Cobalt Oxide Batteries From Battery Electric Vehicles.* Waste Management & Research the Journal for a Sustainable Circular Economy, 2022. **41**(2): p. 376-388.

351.   Cucchiella, F., et al., *Recycling of WEEEs: An Economic Assessment of Present and Future E-Waste Streams.* Renewable and Sustainable Energy Reviews, 2015. **51**: p. 263-272.

352.   Khaliq, A., et al., *Metal Extraction Processes for Electronic Waste and Existing Industrial Routes: A Review and Australian Perspective.* Resources, 2014. **3**(1): p. 152-179.

353.   Gurumurthy, K. and M. Annamalai, *Microbiological Leaching of Metals and Its Recovery From Waste Electrical and Electronic Equipment: A Review.* World Review of Science Technology and Sustainable Development, 2019. **15**(1): p. 1.

354.   Arora, M., *Resources Recovery From Electronic Waste.* 2020.

355.   Ghodrat, M., M. Rashidi, and B. Samali, *Investigation Into the Recovery of Valuable Metals From Waste Mobile Phone Printed Circuit Boards (PCBs): An Australian Case Study.* International Journal of Waste Resources, 2018. **08**(04).

356.   Wan, X., et al., *Behavior of Waste Printed Circuit Board (WPCB) Materials in the Copper Matte Smelting Process.* Metals, 2018. **8**(11): p. 887.

357.   Harvey, J.P., et al., *Greener Reactants, Renewable Energies and Environmental Impact Mitigation Strategies in Pyrometallurgical Processes: A Review.* Mrs Energy & Sustainability, 2022. **9**(2): p. 212-247.

358.   Ghimire, H. and P.A. Ariya, *E-Wastes: Bridging the Knowledge Gaps in Global Production Budgets, Composition, Recycling and Sustainability Implications.* Sustainable Chemistry, 2020. **1**(2): p. 154-182.

359.   Nurjaman, F., et al., *Utilisation of Biomass Waste as a Reductant in the Smelting of Saprolitic Nickel Ore Using a DC-arc Furnace.* Mineral Processing and Extractive Metallurgy Transactions of the Institutions of Mining and Metallurgy, 2024. **133**(4): p. 160-170.

360.   Yuan, X., et al., *Dynamic Environmental–Economic Impacts Analysis of Upgrading E-Waste Recycling From Crude Disassembly to Precise Disassembly.* Acs Sustainable Chemistry & Engineering, 2024. **12**(19): p. 7591-7602.

361.   Yao, Y., et al., *Hydrometallurgical Processes for Recycling Spent Lithium-Ion Batteries: A Critical Review.* Acs Sustainable Chemistry & Engineering, 2018. **6**(11): p. 13611-13627.

362. Gerold, E., C. Schinnerl, and H. Antrekowitsch, *Critical Evaluation of the Potential of Organic Acids for the Environmentally Friendly Recycling of Spent Lithium-Ion Batteries.* Recycling, 2022. **7**(1): p. 4.

363. Or, T., et al., *Recycling of Mixed Cathode Lithium-ion Batteries for Electric Vehicles: Current Status and Future Outlook.* Carbon Energy, 2020. **2**(1): p. 6-43.

364. Lerchbammer, R., E. Gerold, and H. Antrekowitsch, *Gluconic Acid Leaching of Spent Lithium-Ion Batteries as an Environmentally Friendly Approach to Achieve High Leaching Efficiencies in the Recycling of NMC Active Material.* Metals, 2023. **13**(8): p. 1330.

365. Chen, X. and T. Zhou, *Hydrometallurgical Process for the Recovery of Metal Values From Spent Lithium-Ion Batteries in Citric Acid Media.* Waste Management & Research the Journal for a Sustainable Circular Economy, 2014. **32**(11): p. 1083-1093.

366. Yuliusman, Y., et al., *Recovery of Cobalt and Nickel From Spent Lithium Ion Batteries With Citric Acid Using Leaching Process: Kinetics Study.* E3s Web of Conferences, 2018. **67**: p. 03008.

367. Tanong, K., et al., *Recovery of Zn From Unsorted Spent Batteries Using Solvent Extraction and Electrodeposition.* Journal of Environmental Engineering, 2018. **144**(6).

368. Dong, Y., et al., *Trends of Sustainable Recycling Technology for Lithium-ion Batteries: Metal Recovery From Conventional Metallurgical Processes to Innovative Direct Recycling.* Metalmat, 2023. **1**(1).

369. Swain, B., et al., *Hydrometallurgical Process for Recovery of Cobalt From Waste Cathodic Active Material Generated During Manufacturing of Lithium Ion Batteries.* Journal of Power Sources, 2007. **167**(2): p. 536-544.

370. Lei, S., et al., *Strengthening Valuable Metal Recovery From Spent Lithium-Ion Batteries by Environmentally Friendly Reductive Thermal Treatment and Electrochemical Leaching.* Acs Sustainable Chemistry & Engineering, 2021. **9**(20): p. 7053-7062.

371. Clyde, D.A., et al., *Upscaled Recycling of Lithium Nickel Manganese Cobalt Oxide (NMC) Cathode Using an Automated Electrochemical System Towards Low-Carbon Utilization of Waste Lithium-Ion Battery (LIB).* Acs Es&t Engineering, 2025. **5**(4): p. 979-990.

372. Pang, D., et al., *Sustainable Recycling of Lithium-Ion Battery Cathodes: Life Cycle Assessment, Technologies, and Economic Insights.* Nanomaterials, 2025. **15**(16): p. 1283.

373. Inaba, Y., A.C. West, and S. Banta, *Enhanced Microbial Corrosion of Stainless Steel by <i>Acidithiobacillus Ferrooxidans</i> Through the Manipulation of Substrate Oxidation and Overexpression of <i>rus</i>.* Biotechnology and Bioengineering, 2020. **117**(11): p. 3475-3485.

374. Pathak, A., L. Morrison, and M.G. Healy, *Catalytic Potential of Selected Metal Ions for Bioleaching, and Potential Techno-Economic and Environmental Issues: A Critical Review.* Bioresource Technology, 2017. **229**: p. 211-221.

375. Yang, C., et al., *Bioleaching of Copper From Metal Concentrates of Waste Printed Circuit Boards by a Newly Isolated Acidithiobacillus Ferrooxidans Strain Z1.* Journal of Material Cycles and Waste Management, 2015. **19**(1): p. 247-255.

376. Gao, X.Y., et al., *Novel Strategy for Improvement of the Bioleaching Efficiency of Acidithiobacillus Ferrooxidans Based on the AfeI/R Quorum Sensing System.* Minerals, 2020. **10**(3): p. 222.

377. Madrigal-Arias, J.E., et al., *Bioleaching of Gold, Copper and Nickel From Waste Cellular Phone PCBs and Computer Goldfinger Motherboards by Two ≪italic>Aspergillus Niger</Italic>strains.* Brazilian Journal of Microbiology, 2015. **46**(3): p. 707-713.

378. Islam, M.M. and Y.P. Ting, *Fungal Bioleaching of Spent Hydroprocessing Catalyst: Effect of Decoking and Particle Size.* Advanced Materials Research, 2009. **71-73**: p. 665-668.

379. Willner, J. and A. Fornalczyk, *Extraction of Metals From Electronic Waste by Bacterial Leaching.* Environment Protection Engineering, 2013. **39**(1).

380. Watling, H.R., *Review of Biohydrometallurgical Metals Extraction From Polymetallic Mineral Resources.* Minerals, 2014. **5**(1): p. 1-60.

381. Giese, E.C. and P.M. Vaz, *Bioleaching of Primary Nickel Ore Using Acidithiobacillus Ferrooxidans LR Cells Immobilized in Glass Beads.* Orbital - The Electronic Journal of Chemistry, 2015. **7**(2).

382. Samani, L.D., et al., *Studies of the Anaerobic Leaching of Iron And Copper Containing Dump Material and Comparison to Chemical Leaching.* Chemie Ingenieur Technik, 2023. **95**(12): p. 2022-2029.

383. Quach, N.T., et al., *Bioleaching Potential of Indigenous Bacterial Consortia From Gold-Bearing Sulfide Ore Of Ta Nang Mine in Vietnam.* Polish Journal of Environmental Studies, 2022. **31**(1): p. 803-813.

384. Pinchuk, A.A., Н. Ткаленко, and V. Marhasova, *Implementation of Circular Economy Elements in the Mining Regions.* E3s Web of Conferences, 2019. **105**: p. 04048.

385. Withanage, S.V. and K. Habib, *Life Cycle Assessment and Material Flow Analysis: Two Under-Utilized Tools for Informing E-Waste Management.* Sustainability, 2021. **13**(14): p. 7939.

386. Tarek, A. and S.M. El-Haggar, *Sustainable Guideline for Developing the E-Waste Sector in Egypt.* Journal of Environmental Protection, 2019. **10**(08): p. 1043-1071.

387. Hossain, R. and V. Sahajwalla, *Current Recycling Innovations to Utilize E-Waste in Sustainable Green Metal Manufacturing.* Philosophical Transactions of the Royal Society a Mathematical Physical and Engineering Sciences, 2024. **382**(2284).

388. Gönen, Ç. and E. Kaplanoğlu, *Environmental and Economic Evaluation of Solar Panel Wastes Recycling.* Waste Management & Research the Journal for a Sustainable Circular Economy, 2019. **37**(4): p. 412-418.

389. Jk, P., et al., *Effects of Electronic Waste on Developing Countries.* Advances in Recycling & Waste Management, 2017. **02**(02).

390. Patil, A. and W.A. Vonk, *Stakeholder Perspectives on EU Regulatory Frameworks: Navigating Critical Raw Materials, Battery Innovation, and Recycling Challenges.* Open Research Europe, 2025. **5**: p. 104.

391. Zhang, X., et al., *Assessing the GHG Emissions and Savings During the Recycling of NMC Lithium-Ion Batteries Used in Electric Vehicles in China.* Processes, 2022. **10**(2): p. 342.

392. Yu, X., et al., *Current Challenges in Efficient Lithium-Ion Batteries' Recycling: A Perspective.* Global Challenges, 2022. **6**(12).

393. Donnelly, L., et al., *The Recycling of End-of-Life Lithium-Ion Batteries and the Phase Characterisation of Black Mass.* Recycling, 2023. **8**(4): p. 59.

394. Wijayanti, W., et al., *Developing a Strategic Plan for Efficient Management of Industrial Waste in the Machinery Sector at United Tractors Company.* Eastern-European Journal of Enterprise Technologies, 2024. **3**(10 (129)): p. 70-83.

395. Sheth, R.P., et al., *The Lithium-Ion Battery Recycling Process From a Circular Economy Perspective—A Review and Future Directions.* Energies, 2023. **16**(7): p. 3228.

396. Oh, J., et al., *Forecasting the Recycled Material Content of EV Batteries In Korea.* Korean Journal of Life Cycle Assessment, 2025. **26**(1): p. 1-6.

397. Huang, X., et al., *Protecting the Environment and Public Health From Rare Earth Mining.* Earth S Future, 2016. **4**(11): p. 532-535.

398. Chingwaru, S.J., B.P.v.d. Heyden, and M. Tadie, *An Underexploited Invisible Gold Resource in the Archean Sulphides of the Witwatersrand Tailings Dumps.* Scientific Reports, 2023. **13**(1).

399. Chingwaru, S.J., B.P.v.d. Heyden, and M. Tadie, *Invisible Gold in the Archean Detrital Sulphides of the Witwatersrand Tailings Dumps: A Large and Under-Exploited Gold Resource.* 2022.

400. Kootbodien, T., et al., *Environmental Silica Dust Exposure and Pulmonary Tuberculosis in Johannesburg, South Africa.* International Journal of Environmental Research and Public Health, 2019. **16**(10): p. 1867.

401. Laker, M.C., *Environmental Impacts of Gold Mining—With Special Reference to South Africa.* Mining, 2023. **3**(2): p. 205-220.

402. Utembe, W., et al., *Hazards Identified and the Need for Health Risk Assessment in the South African Mining Industry.* Human & Experimental Toxicology, 2015. **34**(12): p. 1212-1221.

403. Maest, A.S., *Remining for Renewable Energy Metals: A Review of Characterization Needs, Resource Estimates, and Potential Environmental Effects.* Minerals, 2023. **13**(11): p. 1454.

404. Pollmann, O., et al., *Mine Tailings: Waste or Valuable Resource?* Waste and Biomass Valorization, 2010. **1**(4): p. 451-459.

405. Zaimes, G.G., et al., *Environmental Life Cycle Perspective on Rare Earth Oxide Production.* Acs Sustainable Chemistry & Engineering, 2015. **3**(2): p. 237-244.

406. Marx, J., et al., *Comparative Life Cycle Assessment of NdFeB Permanent Magnet Production From Different Rare Earth Deposits.* Acs Sustainable Chemistry & Engineering, 2018. **6**(5): p. 5858-5867.

407. Jouini, M., et al., *Sustainable Production of Rare Earth Elements From Mine Waste and Geoethics.* Minerals, 2022. **12**(7): p. 809.

408. Lai, C., et al., *Unleashing the Power of Closed-Loop Supply Chains: A Stackelberg Game Analysis of Rare Earth Resources Recycling.* Sustainability, 2024. **16**(12): p. 4899.

409. Chen, W., D. Zhou, and B. Xue, *LCA-Based Carbon Footprint Accounting of Mixed Rare Earth Oxides Production From Ionic Rare Earths.* Processes, 2022. **10**(7): p. 1354.

410. Ouaneche, T., et al., *Challenges in the Direct Lithiation of Spent LFP Cathodes: The Crucial Role of Reducing Agents.* Ees Batteries, 2025. **1**(5): p. 1068-1082.

411. Rabbani, M., et al., *Advancing Circular Economy Through Phytomining: Critical Mineral Recovery From Mine Tailings and Environmental Impact Assessment.* Acs Sustainable Chemistry & Engineering, 2025. **13**(29): p. 11335-11347.

412. Baars, J., et al., *Circular Economy Strategies for Electric Vehicle Batteries Reduce Reliance on Raw Materials.* Nature Sustainability, 2020. **4**(1): p. 71-79.

413. Shen, H., et al., *Impact of Urban Mining on Energy Efficiency: Evidence From China.* Sustainability, 2022. **14**(22): p. 15039.

414. Amalia, D., et al., *The Effect of a Molasses Reductant on Acetic Acid Leaching of Black Mass From Mechanically Treated Spent Lithium-Ion Cylindrical Batteries.* Sustainability, 2023. **15**(17): p. 13171.

415. Losa, G. and S. Bindschedler, *Enhanced Tolerance to Cadmium in Bacterial-Fungal Co-Cultures as a Strategy for Metal Biorecovery From E-Waste.* Minerals, 2018. **8**(4): p. 121.

416. Julapong, P., et al., *Rare Earth Elements Recovery From Primary and Secondary Resources Using Flotation: A Systematic Review.* Applied Sciences, 2023. **13**(14): p. 8364.

417. Nickless, E. and N. Yakovleva, *Resourcing Future Generations Requires a New Approach to Material Stewardship.* Resources, 2022. **11**(8): p. 78.

418. Lapko, Y., et al., *In Pursuit of Closed-Loop Supply Chains for Critical Materials: An Exploratory Study in the Green Energy Sector.* Journal of Industrial Ecology, 2018. **23**(1): p. 182-196.

419. Richa, K., C.W. Babbitt, and G. Gaustad, *Eco-Efficiency Analysis of a Lithium-Ion Battery Waste Hierarchy Inspired by Circular Economy.* Journal of Industrial Ecology, 2017. **21**(3): p. 715-730.

420. Pagliaro, M. and F. Meneguzzo, *Lithium Battery Reusing and Recycling: A Circular Economy Insight.* Heliyon, 2019. **5**(6): p. e01866.

421. Khan, S., et al., *South Africa's Mineral Resource Availability as a Potential Driver for Transitioning to a Circular Economy.* Journal of the Southern African Institute of Mining and Metallurgy, 2025. **125**(2): p. 61-68.

422. Deravian, B. and C.N. Mulligan, *Sustainable Recovery of Critical Minerals From Wastes by Green Biosurfactants: A Review.* Molecules, 2025. **30**(11): p. 2461.

423. Raugei, M. and P.H. Winfield, *Prospective LCA of the Production and EoL Recycling of a Novel Type of Li-Ion Battery for Electric Vehicles.* Journal of Cleaner Production, 2019. **213**: p. 926-932.

424. Je, J. and S. Lee, *A Review on Process Mineralogical Characteristics and Current State of Beneficiation Process of Mount Weld Rare Earths Mine in Western*

Australia. Journal of the Korean Society of Mineral and Energy Resources Engineers, 2023. **60**(3): p. 172-180.

425. Rozaila, Z.S., et al., *Environmental Monitoring Through Use of Silica-Based TLD.* Journal of Radiological Protection, 2017. **37**(3): p. 761-779.

426. Mudd, G.M. and S.M. Jowitt, *Rare Earth Elements From Heavy Mineral Sands: Assessing the Potential of a Forgotten Resource.* Applied Earth Science Transactions of the Institutions of Mining and Metallurgy Section B, 2016. **125**(3): p. 107-113.

427. Kang, X., L. Csetényi, and G.M. Gadd, *Monazite Transformation Into Ce‑and La‑containing Oxalates by <i>Aspergillus Niger</I>.* Environmental Microbiology, 2020. **22**(4): p. 1635-1648.

428. Goodenough, K. and F. Wall, *Critical Metal Mineralogy: Preface to the Special Issue of Mineralogical Magazine.* Mineralogical Magazine, 2016. **80**(1): p. 1-4.

429. Chowdhury, N.A., et al., *Sustainable Recycling of Rare-Earth Elements From NdFeB Magnet Swarf: Techno-Economic and Environmental Perspectives.* Acs Sustainable Chemistry & Engineering, 2021. **9**(47): p. 15915-15924.

430. Vaughan, J., et al., *Toward Closing a Loophole: Recovering Rare Earth Elements From Uranium Metallurgical Process Tailings.* Jom, 2020. **73**(1): p. 39-53.

431. Jordan, B.W., et al., *Thorium: Crustal Abundance, Joint Production, and Economic Availability.* Resources Policy, 2015. **44**: p. 81-93.

432. Zhang, T., et al., *Material Flow and Supply–demand Feature of Thulium in China.* Journal of Industrial Ecology, 2024. **28**(6): p. 1952-1964.

433. Song, W., *Mine-on-a-Chip: Megascale Opportunities for Microfluidics in Critical Materials and Minerals Recovery.* Lab on a Chip, 2025. **25**(18): p. 4461-4472.

434. Zadeh, J., *Rare Earth Market and Price Trends: Stability Amid Supply-Demand Tensions.* 2025, Discovery Alert,.

435. Global Mining Review, *Reshaping the rare earths supply chain amid soaring demand and strategic risks.* 2025, Global Mining Review.

436. Grand View Research, *Rare Earth Elements Market (2025 - 2030).* 2025, Grand View Research,.

437. Fortune Business Insights, *Rare Earth Elements Market Size, Share & Industry Analysis, By Type (Lanthanum, Cerium, Neodymium, Praseodymium, Samarium, Europium, Others), By Application (Magnets, Metallurgy, Batteries, Polishing Agent, Glass and Ceramics, Catalyst, Phosphors, and Others) and Regional Forecast, 2024-2032.* 2025.

438. Adamas Intelligence, *The industry's go-to reference for magnet market intelligence* 2024.

439. Ramon Barua Costa, *Outlook 2025: Reshaping the rare earth elements supply chain amid soaring demand and strategic risks.* 2025, Canadian Mining Journal.

440. Tae-Yoon Kim, S.D., Amrita Dasgupta, Alessio Scanziani, , *With new export controls on critical minerals, supply concentration risks become reality.* 2025, IEA.

441. Thunder Said Energy, *Rare Earth mining and refining: the economics?* 2025.

442. Government of Western Australia, *Western Australia Battery and Critical Minerals Profile – August 2024.* 2024.

443. Sachdev, D., *Enabling Data Democracy in Supply Chain Using Blockchain and Iot.* Journal of Management, 2019. **6**(1).

444. Ardagna, C.A., et al., *From Trustworthy Data to Trustworthy IoT.* Acm Transactions on Cyber-Physical Systems, 2020. **5**(1): p. 1-26.

445. Li, S., et al., *The Effect of a New Process on the Environment of Soil in Ion Adsorption Rare Earth Ores.* E3s Web of Conferences, 2024. **490**: p. 01009.

446. Navarro, J. and F. Zhao, *Life-Cycle Assessment of the Production of Rare-Earth Elements for Energy Applications: A Review.* Frontiers in Energy Research, 2014. **2**.

447. Izatt, R.M., et al., *Industrial Applications of Molecular Recognition Technology to Separations of Platinum Group Metals and Selective Removal of Metal Impurities From Process Streams.* Green Chemistry, 2015. **17**(4): p. 2236-2245.

448. Nansai, K., et al., *Global Mining Risk Footprint of Critical Metals Necessary for Low-Carbon Technologies: The Case of Neodymium, Cobalt, and Platinum in Japan.* Environmental Science & Technology, 2015. **49**(4): p. 2022-2031.

449. Hayler, J., D.K. Leahy, and E.M. Simmons, *A Pharmaceutical Industry Perspective on Sustainable Metal Catalysis.* Organometallics, 2018. **38**(1): p. 36-46.

450. Sheldon, R.A., *Metrics of Green Chemistry and Sustainability: Past, Present, and Future.* Acs Sustainable Chemistry & Engineering, 2017. **6**(1): p. 32-48.

451. Zanoletti, A., A. Cornelio, and E. Bontempi, *A Post-Pandemic Sustainable Scenario: What Actions Can Be Pursued to Increase the Raw Materials Availability?* Environmental Research, 2021. **202**: p. 111681.

452. Huntington, V.E., F. Coulon, and S. Wagland, *Innovative Resource Recovery From Industrial Sites: A Critical Review.* Sustainability, 2022. **15**(1): p. 489.

453. Hessel, V., et al., *Continuous-Flow Extraction of Adjacent Metals—A Disruptive Economic Window for In Situ Resource Utilization of Asteroids?* Angewandte Chemie International Edition, 2020. **60**(7): p. 3368-3388.

454. Izatt, R.M., et al., *Green Chemistry Molecular Recognition Processes Applied to Metal Separations in Ore Beneficiation, Element Recycling, Metal Remediation, and Elemental Analysis.* 2018: p. 189-240.

455. Nguyen, V.T., S. Riaño, and K. Binnemans, *Separation of Precious Metals by Split-Anion Extraction Using Water-Saturated Ionic Liquids.* Green Chemistry, 2020. **22**(23): p. 8375-8388.

456. Zeng, X. and J. Li, *Emerging Anthropogenic Circularity Science: Principles, Practices, and Challenges.* Iscience, 2021. **24**(3): p. 102237.

457. Ylä-Mella, J. and É. Pongrácz, *Drivers and Constraints of Critical Materials Recycling: The Case of Indium.* Resources, 2016. **5**(4): p. 34.

458. Dou, S. and D. Xu, *The Security of Critical Mineral Supply Chains.* Mineral Economics, 2022. **36**(3): p. 401-412.

459. Demol, J., et al., *The Sulfuric Acid Bake and Leach Route for Processing of Rare Earth Ores and Concentrates: A Review.* Hydrometallurgy, 2019. **188**: p. 123-139.

460. Reynolds, S., *Unveiling Supply Chain Transparency and Traceability in the Renewable Energy Sector: Challenges and Opportunities.* 2024.

461. Rejeb, A., J.G. Keogh, and H. Treiblmaier, *Leveraging the Internet of Things and Blockchain Technology in Supply Chain Management.* Future Internet, 2019. **11**(7): p. 161.

462. Hastig, G.M. and M.S. Sodhi, *Blockchain for Supply Chain Traceability: Business Requirements and Critical Success Factors.* Production and Operations Management, 2020. **29**(4): p. 935-954.

463. Boissieu, E.d., et al., *The Use of Blockchain in the Luxury Industry: Supply Chains and the Traceability of Goods.* Journal of Enterprise Information Management, 2021. **34**(5): p. 1318-1338.

464. Francisco, K. and D. Swanson, *The Supply Chain Has No Clothes: Technology Adoption of Blockchain for Supply Chain Transparency.* Logistics, 2018. **2**(1): p. 2.

465. Malik, M., H. Ghaderi, and A. Andargoli, *A Resource Orchestration View of Supply Chain Traceability and Transparency Bundles for Competitive Advantage.* Business Strategy and the Environment, 2021. **30**(8): p. 3866-3881.

466. Rare Earth Exchanges, *Rare Earth Supply Chain Hiring Trends.* 2025, Rare Earth Exchanges.

467. Souza, A.S.C.d. and L. Debs, *Identifying Emerging Technologies and Skills Required for Construction 4.0.* Buildings, 2023. **13**(10): p. 2535.

468. Ercik, C. and K. Kardaş, *Reflections of Digital Technologies on Human Resources Management in the Tourism Sector.* Worldwide Hospitality and Tourism Themes, 2024. **16**(5): p. 646-663.

469. Mishra, P.C. and P.K. Mishra, *Challenges and Opportunities of Big Data Analytics for Human Resource Management in Mining and Metal Industries.* Journal of Mines Metals and Fuels, 2023: p. 1747-1753.

470. Borst, R.T., R. Blom, and W. Vandenabeele, *Comparing the Employability of Public and Private Employees: The Role of PSM and Red Tape in Linking Employability and Work Engagement.* Public Personnel Management, 2025. **54**(3): p. 331-360.

471. Perini, S., et al., *Increasing Middle School Students' Awareness and Interest in Manufacturing Through Digital Game -based Learning (DGBL).* Computer Applications in Engineering Education, 2017. **25**(5): p. 785-799.

472. Schlegel, D. and P. Kraus, *Skills and Competencies for Digital Transformation – A Critical Analysis in the Context of Robotic Process Automation.* International Journal of Organizational Analysis, 2021. **31**(3): p. 804-822.

473. Satzger, M. and R. Vogel, *Do Inclusive Workplace Policies Foster Employer Attractiveness? Comparative Evidence From an Online Employer Review Platform.* Public Personnel Management, 2023. **52**(4): p. 566-589.

474. Gurieva, L. and A. Dzhioev, *Digital Skills Supply and Demand on the Russia Regional Labor Markets.* E3s Web of Conferences, 2023. **413**: p. 05017.

475. Popoola, O.A., et al., *The Impact of Automation on Maritime Workforce Management: A Conceptual Framework.* International Journal of Management & Entrepreneurship Research, 2024. **6**(5): p. 1467-1488.

476. Akyazi, T., et al., *Skills Requirements for the European Machine Tool Sector Emerging From Its Digitalization.* Metals, 2020. **10**(12): p. 1665.

477. Jakobsen, M.M. and F. Homberg, *First Impressions: An Analysis of Professional Stereotypes and Their Impact on Sector Attraction.* Public Administration Review, 2024. **85**(4): p. 1134-1149.

478. Moshynskyi, V., et al., *Investigation of Technogenic Deposits of Phosphogypsum Dumps.* E3s Web of Conferences, 2021. **280**: p. 08008.

479. Góralczyk, S. and E. Uzunow, *The Recovery of Yttrium and Europium Compounds From Waste Materials*. Archives of Environmental Protection, 2013. **39**(3): p. 107-114.

480. Chernysh, Y., et al., *Phosphogypsum Recycling: A Review of Environmental Issues, Current Trends, and Prospects*. Applied Sciences, 2021. **11**(4): p. 1575.

481. Cánovas, C.R., et al., *Mobility of Rare Earth Elements, Yttrium and Scandium From a Phosphogypsum Stack: Environmental and Economic Implications*. The Science of the Total Environment, 2018. **618**: p. 847-857.

482. Pak, S.J., et al., *Rare Earth Elements and Other Critical Metals in Deep Seabed Mineral Deposits: Composition and Implications for Resource Potential*. Minerals, 2018. **9**(1): p. 3.

483. Maina, L., et al., *Chemical and Radiochemical Characterization of Phosphogypsum From Poland*. Nukleonika, 2024. **69**(2): p. 113-117.

484. Yasukawa, K., et al., *Tracking the Spatiotemporal Variations of Statistically Independent Components Involving Enrichment of Rare-Earth Elements in Deep-Sea Sediments*. Scientific Reports, 2016. **6**(1).

485. Li, J., et al., *Integrated Mineral Carbonation of Ultramafic Mine Deposits—A Review*. Minerals, 2018. **8**(4): p. 147.

486. Kering, M.K., A. Rahemi, and V.W. Temu, *Effect of Harvest Management on Biomass Yield, Forage Quality, and Nutrient Removal by Bioenergy Grasses in Mid-Central Virginia*. Agronomy, 2024. **14**(4): p. 825.

487. Deng, Y., et al., *Rare Earth Element Geochemistry Characteristics of Seawater and Porewater From Deep Sea in Western Pacific*. Scientific Reports, 2017. **7**(1).

488. Joseph, L., et al., *Bio-Accelerated Weathering of Ultramafic Minerals With Gluconobacter Oxydans*. 2024.

489. Rozelle, P.L., et al., *The Mercer Clay in Pennsylvania as a Polymetallic Mineral Resource: Review and Update*. Mining Metallurgy & Exploration, 2021. **38**(5): p. 2037-2054.

490. Pourret, O. and J. Tuduri, *Continental Shelves as Potential Resource of Rare Earth Elements*. Scientific Reports, 2017. **7**(1).

491. Malanchuk, Z., et al., *Results of Research Into the Content of Rare Earth Materials in Man-Made Phosphogypsum Deposits*. Key Engineering Materials, 2020. **844**: p. 77-87.

492. Kuhn, T. and C. Rühlemann, *Exploration of Polymetallic Nodules and Resource Assessment: A Case Study From the German Contract Area in the Clarion-Clipperton Zone of the Tropical Northeast Pacific*. Minerals, 2021. **11**(6): p. 618.

493. Thompson, K.F., et al., *Urgent Assessment Needed to Evaluate Potential Impacts on Cetaceans From Deep Seabed Mining*. Frontiers in Marine Science, 2023. **10**.

494. Shao, M., W. Song, and X. Zhao, *Polymetallic Nodule Resource Assessment of Seabed Photography Based on Denoising Diffusion Probabilistic Models*. Journal of Marine Science and Engineering, 2023. **11**(8): p. 1494.

495. Baciocchi, R. and G. Costa, *CO2 Utilization and Long-Term Storage in Useful Mineral Products by Carbonation of Alkaline Feedstocks*. Frontiers in Energy Research, 2021. **9**.

Index

A

Acid Leaching, 20, 41, 57, 87, 90, 107, 111, 145, 157, 244, 246, 247, 248, 249, 250, 251, 252, 253, 255, 260, 261, 267, 276, 278, 279, 282, 284, 285, 286, 287, 292, 295, 296, 299, 305, 311, 312, 315, 316, 319, 325, 326, 328, 329, 346, 347, 360, 381, 390, 400, 450, 466,467, 472, 477, 490, 493, 495, 498, 499, 500, 503, 507, 508, 511

Activated Carbon, 260

Adsorption, 7, 8, 33, 36, 37, 57, 58, 90, 94, 99, 110, 111, 113, 145, 152, 183, 184, 222, 223, 224, 227, 228, 229, 230, 231, 233, 235, 246, 248, 261, 414, 462, 464, 467, 471, 486, 487, 494, 502, 513

Advanced Manufacturing, 3, 14, 26, 28, 31, 335, 349, 402, 409, 410, 418, 421, 423, 440, 449, 452, 456

Advanced Materials, 28, 261, 275, 456, 490, 496, 498, 499, 502, 503, 509

Air Classification, 181, 387

Alumina, 247, 248, 287, 288, 289, 316, 400, 430, 433, 472, 496

Aluminium Smelting, 248

Aluminum, 489, 493

Anode Materials, 29, 490

Antimony, 282, 433

Arsenic Removal, 451

B

Ball Mill, 153, 155, 157, 167, 168, 169, 172, 173, 174, 184, 191, 192, 233, 340, 468, 469

Barite, 49, 65, 125, 126, 133, 137, 157, 173, 234, 253, 254, 497

Base Metals, 39, 63, 65, 73, 76, 169, 175, 221, 289, 329, 341, 358, 393, 396

Beneficiation, 5, 81, 96, 103, 104, 105, 106, 107, 108, 117, 119, 121, 143, 144, 151, 153, 157, 158, 160, 163, 164, 166, 167, 170, 173, 174, 177, 181, 182, 185, 187, 194, 195, 197, 202, 209, 211, 220, 221, 231, 233, 234, 238, 239, 242, 244, 252, 286, 341, 344, 346, 349, 351,

360, 361, 362, 396, 397, 398, 399, 400, 411, 412, 432, 434, 442, 449, 462, 466, 469, 498, 511, 513

Beryllium, 38, 51, 52, 54, 92, 433, 464, 494

Bioleaching, 88, 242, 276, 390, 391, 392, 396, 403, 453, 472, 481, 500, 508, 509

Blast Furnace, 289

Boron, 13, 50, 54, 236, 453

Brine Extraction, 11

By-Product, 6, 17, 24, 25, 39, 41, 43, 44, 76, 98, 145, 146, 285, 290, 309, 327, 332, 362, 367, 396, 399, 400, 402, 405, 406, 428, 429, 430, 431, 432, 433

By-Product Recovery, 405

C

Calcination, 255, 261, 272, 280, 281, 286, 287, 288, 295, 296, 297, 300, 301, 303, 305, 307, 312, 313, 315, 316, 317, 319, 320, 321, 322, 323, 324, 325, 327, 328, 329, 331, 332, 333, 345, 349, 358, 370, 389, 473, 474, 477, 495

Calcium Carbonate, 234, 286, 502

Carbon Footprint, 185, 330, 331, 384, 398, 400, 449, 451, 489, 511

Catalysts, 2, 11, 13, 14, 30, 33, 45, 62, 100, 102, 254, 259, 319, 345, 347, 350, 384, 420, 480, 495, 502

Cerium, 3, 10, 13, 14, 40, 46, 60, 92, 98, 99, 100, 102, 110, 113, 114, 124, 125, 126, 147, 253, 254, 260, 324, 325, 329, 367, 399, 400, 432, 486, 493, 501, 502, 512

Cerium (Ce), 3, 10, 13, 98, 102, 110, 126, 147, 260, 399, 400

Chemical Precipitation, 61, 346, 366, 369

Chromium, 44, 47, 166, 433

Circular Economy, 27, 83, 91, 274, 279, 367, 383, 384, 388, 390, 393, 394, 395, 397, 401, 402, 403, 404, 405, 406, 408, 420, 427, 442, 449, 456, 473, 479, 491, 492, 494, 498, 501, 507, 508, 509, 510, 511

Clean Energy Transition, 19, 26, 45, 92

Coarse Particle Flotation, 496

Cobalt, 2, 3, 4, 6, 8, 11, 13, 14, 17, 18, 19, 25, 26, 27, 28, 29, 30, 33, 34, 37, 38, 39, 45, 46, 55, 56,

57, 58, 61, 62, 63, 64, 65, 70, 73, 75, 76, 81, 86, 88, 91, 92, 94, 170, 173, 175, 178, 182, 183, 184, 243, 244, 246, 249, 250, 251, 252, 256, 257, 259, 260, 261, 264, 266, 268, 269, 270, 271, 273, 278, 280, 281, 290, 292, 296, 297, 299, 300, 301, 305, 308, 311, 313, 315, 319, 322, 326, 329, 333, 335, 337, 342, 344, 346, 347, 348, 350, 352, 358, 377, 384, 385, 386, 388, 389, 390, 393, 395, 396, 401, 402, 403, 404, 406, 428, 430, 432, 433, 449, 456, 462, 464, 480, 488, 489, 499, 500, 503, 507, 508, 513

Cobalt (Co), 17, 56, 322, 388

Comminution, 98, 104, 106, 119, 120, 121, 143, 153, 154, 157, 158, 162, 167, 173, 182, 183, 184, 185, 186, 187, 188, 190, 191, 193, 195, 233, 234, 235, 236, 241, 245, 253, 336, 339, 340, 343, 344, 346, 356, 467, 468, 471, 477, 496

Continuous Improvement, 353, 360, 368, 369, 451

Copper, 17, 24, 25, 38, 40, 46, 47, 50, 62, 63, 64, 65, 73, 98, 106, 124, 170, 213, 221, 223, 224, 227, 228, 229, 230, 244, 250, 251, 257, 271, 281, 283, 284, 300, 346, 347, 350, 385, 386, 388, 390, 391, 393, 428, 430, 432, 433, 480, 491, 503, 507, 509

Corporate Social Responsibility (CSR), 374, 505

Critical Raw Materials, 32, 352, 375, 399, 403, 408, 457, 480, 488, 491, 509

Crushing, 5, 87, 96, 97, 104, 106, 107, 116, 119, 125, 133, 146, 148, 149, 151, 153, 154, 155, 156, 157, 158, 159, 160, 161, 162, 163, 164, 166, 167, 169, 182, 183, 184, 185, 186, 188, 189, 190, 193, 233, 234, 253, 336, 339, 340, 346, 351, 354, 356, 360, 386, 387,389, 404, 407, 432, 442, 443, 445, 467, 471, 496

D

Data Analytics, 440, 441, 442, 443, 445, 446, 451, 455, 457, 458, 459, 483, 484, 485, 514

Data-Driven Maintenance, 458

Deep Eutectic Solvents (DES), 451

Defence Applications, 5, 422

Dense Media Separation, 221, 489

Dewatering, 175, 337, 341, 359

Digital Literacy, 458, 484

Digital Twin, 442, 443, 445, 459, 483

Digitalisation, 5, 421, 440, 441, 442, 443, 444, 445, 446, 451, 457, 459, 460, 461, 484

Dissolution Kinetics, 249

Downstream Processing, 8, 28, 32, 156, 158, 160, 170, 171, 179, 187, 220, 226, 259, 264, 309, 410, 422, 423, 425, 434, 442, 445

Dysprosium, 4, 10, 12, 13, 14, 43, 57, 60, 76, 90, 99, 101, 116, 144, 147, 265, 266, 315, 329, 385, 399, 400, 417, 419, 432, 464, 487

Dysprosium (Dy), 4, 10, 12, 13, 57, 76, 90, 147, 265, 399, 400, 417

E

Economic Geology, 83, 486

Electrolysis, 20, 24, 236, 290, 347, 400, 432, 433

Electrowinning, 244, 245, 327, 337, 342, 347, 356, 388, 389, 390, 451, 472, 480, 503

Energy Consumption, 122, 153, 154, 157, 158, 167, 173, 185, 188, 191, 192, 194, 198, 215, 244, 265, 273, 290, 304, 307, 320, 328, 330, 337, 353, 372, 388, 389, 393, 397, 401, 432, 442, 443, 449, 450, 456, 473, 480

Environmental Compliance, 145, 148, 304, 310, 314, 331, 342, 344, 348, 351, 365, 389, 477

Environmental Management Systems (EMS), 359, 380, 399, 478

Environmental Monitoring, 125, 138, 141, 361, 362, 372, 379, 457, 512

Erbium, 4, 10, 13, 43, 60, 101, 399

Europium, 13, 101, 102, 110, 274, 324, 385, 398, 399, 400, 401, 512, 515

Europium (Eu), 13, 102, 110, 398, 399, 400, 401

Exploration, 2, 3, 5, 6, 8, 9, 23, 26, 27, 29, 30, 31, 32, 35, 36, 40, 48, 50, 52, 54, 55, 62, 63, 65, 69, 71, 72, 74, 75, 77, 80, 82, 85, 86, 93, 94, 95, 96, 97, 105, 107, 109, 112, 116, 122, 125, 126, 130, 133, 134, 138, 140, 141, 142, 151, 409, 410, 415, 419, 421,434, 436, 438, 440, 441, 442, 443, 445, 446, 447, 452, 457, 459, 462, 463, 464, 466, 481, 483, 484, 492, 494, 495, 506, 515

Exploration Geophysics, 93

Extraction Efficiency, 111, 112, 251, 260, 311, 315, 337, 443

F

Filtration Systems, 146, 273

Financial Modelling, 147

Flotation Reagents, 224, 227, 230, 231, 236, 240

Froth Flotation, 105, 195, 221, 233, 240, 470

Functional Materials, 490

Furnace Design, 300

G

Gadolinium, 4, 10, 101, 399
Gadolinium (Gd), 4, 10, 399
Gallium, 25, 26, 39, 396, 433, 503
Gallium (Ga), 25
Gas Scrubbing, 302, 308, 342
Geochronology, 60, 67, 486, 488
Geological Survey, 27, 70
Geophysical Exploration, 105
Geopolitical Risk, 416, 419, 437, 480, 491
Germanium, 433
Graphite, 2, 3, 4, 6, 11, 12, 14, 20, 21, 22, 23, 26, 28, 29, 30, 33, 34, 37, 38, 39, 45, 62, 64, 65, 66, 67, 68, 72, 104, 109, 110, 114, 167, 170, 171, 173, 175, 177, 178, 181, 182, 183, 184, 186, 192, 216, 218, 219, 221, 222, 223, 224, 227, 229, 231, 246, 312, 430, 433, 462, 468, 490, 497
Graphite (C), 20, 45, 109, 110
Green Chemistry, 440, 441, 449, 450, 457, 460, 483, 487, 501, 513
Greenhouse Gas Emissions, 290, 330, 373, 393, 449, 457, 479
Grinding Efficiency, 151, 156, 170, 188

H

Hafnium, 47, 49
Hard-Rock Mining, 6, 11, 16, 57, 77, 78
Heap Leaching, 246
Heavy Rare Earth Elements (Hrees), 4, 6, 8, 14, 37, 43, 57, 60, 90, 99, 108, 464
High-Pressure Acid Leach (HPAL), 249, 251, 433
High-Purity Alumina, 29
Holmium, 10, 101, 399
Holmium (Ho), 10, 399
Hydrogen Reduction, 289, 322, 460
Hydrometallurgical Recycling, 404, 407, 495
Hydrometallurgy, 81, 187, 242, 243, 244, 247, 252, 255, 257, 262, 274, 276, 277, 278, 279, 281, 298, 325, 326, 327, 328, 329, 330, 331, 389, 390, 411, 453, 455, 471, 491, 496, 503, 513
Hydroxide Precipitation, 269

I

Ilmenite, 39, 43, 44, 47, 48, 50, 59, 60, 61, 62, 103, 104, 105, 106, 109, 110, 112, 113, 114, 184, 196, 197, 200, 202, 207, 209, 211, 212, 216, 218, 219, 220, 234, 246, 341, 464, 470, 498, 499
Indium (In), 25
Industrial Ecology, 489, 511, 512
Instrumentation, 305, 335, 336, 337, 343, 353, 354, 356, 447, 476
International Cooperation, 3, 34, 422, 424, 457
Ion-Adsorption Clays, 6, 8, 10, 14, 57, 58, 89, 97, 98, 107, 112, 113, 167, 182, 252, 261, 262, 266, 285, 344, 412, 414, 463, 464, 466, 486, 487
Ionic Liquids, 275, 451, 487, 491, 501, 513
Iron Ore Beneficiation, 207, 209, 215, 397
Isotopic Analysis, 119, 142

J

Job Creation, 423

K

Kaolinite, 57, 58, 113, 145, 464
Kinetics Of Leaching, 250

L

Lanthanum, 3, 10, 13, 14, 31, 40, 46, 60, 92, 99, 100, 102, 113, 126, 131, 253, 254, 260, 265, 367, 399, 400, 432, 487, 493, 501, 512
Lanthanum (La), 3, 10, 13, 14, 31, 102, 126, 260, 399, 400
Leaching, 6, 7, 8, 33, 36, 37, 56, 57, 87, 89, 90, 91, 96, 97, 98, 106, 108, 109, 110, 111, 113, 114, 115, 116, 125, 133, 141, 143, 144, 145, 146, 147, 148, 152, 153, 154, 156, 157, 164, 167, 168, 169, 170, 171, 173, 174, 175, 182, 183, 185, 186, 187, 191, 192, 233, 234, 241, 242, 243, 244, 245, 246, 247, 248, 249, 250, 251, 252, 253, 254, 255, 257, 258, 265, 266, 269, 276, 277, 278, 279, 281, 282, 283, 284, 285, 286, 287, 290, 295, 296, 299, 300, 305, 308, 309, 315, 316, 325, 327, 328, 329, 332, 334, 336, 337, 342, 344, 345, 347, 349, 351, 354, 360, 362, 363, 365, 367, 368, 369, 370, 378, 388, 389, 391, 396, 400, 404, 407, 411, 432, 449, 450, 451, 453, 465, 466, 467, 468, 471,

472, 473, 475, 480, 487, 495, 496, 498, 499, 500, 501, 502, 503, 506, 507, 508, 509
Leaching Efficiency, 133, 242, 278, 316, 472
Legislative Frameworks, 381, 479
Lifecycle, 362, 374, 376, 395, 402
Life-Cycle Assessment (LCA), 383, 384, 397, 399, 404, 407, 480
Light Rare Earth Elements (Lrees), 3, 14, 40, 46, 60, 99, 253, 260
Lithium, 2, 3, 4, 6, 8, 11, 12, 13, 14, 15, 16, 17, 19, 21, 23, 25, 26, 27, 28, 30, 33, 34, 37, 38, 39, 43, 44, 45, 50, 51, 52, 53, 54, 55, 58, 70, 71, 73, 74, 75, 76, 77, 78, 80, 81, 82, 84, 85, 86, 87, 88, 90, 91, 92, 94, 103, 106, 108, 110, 111, 112, 113, 114,116, 124, 125, 126, 128, 129, 133, 144, 145, 157, 166, 167, 169, 170, 173, 174, 175, 177, 178, 180, 181, 182, 183, 186, 194, 195, 196, 221, 224, 227, 229, 231, 233, 234, 235, 236, 240, 241, 242, 243, 244, 246, 247, 251, 252, 256, 257, 260, 261, 262, 266, 268, 270, 273, 278, 279, 280, 282, 284, 287, 292, 295, 296, 299, 300, 304, 305, 308, 311, 313, 315, 319, 328, 329, 335, 336, 341, 342, 344, 345, 346, 347, 348, 350, 352, 357, 358, 369, 370, 371, 376, 377, 380, 384, 385, 388, 389, 390, 393, 395, 396, 401, 402, 403, 404, 406, 407, 410, 428, 429, 432, 449, 450, 456, 462, 463, 464, 471, 473, 480, 487, 488, 489, 490, 492, 494, 495, 499, 500, 507, 508, 510, 511
Lithium (Li), 15, 52, 126, 388
Lithium-Ion Batteries, 11, 12, 13, 15, 17, 19, 21, 23, 25, 43, 45, 51, 58, 256, 319, 388, 389, 390, 393, 395, 403, 407, 488, 489, 490, 500, 507, 508, 510
Low-Carbon Refining, 440, 449, 450, 451, 460, 483
Lutetium, 4, 10, 14, 31, 102, 399
Lutetium (Lu), 4, 10, 14, 31, 102, 399

M

Magnet Manufacturing, 30, 266, 400, 413, 414, 452, 463
Magnet Recycling, 383, 401
Magnetic Separation, 61, 87, 133, 143, 147, 153, 157, 166, 167, 170, 171, 173, 175, 183, 184, 186, 195, 207, 209, 210, 211, 216, 220, 237, 239, 240, 252, 275, 386, 396, 469, 470, 476
Manganese, 4, 13, 17, 19, 25, 26, 28, 29, 38, 46, 59, 61, 62, 67, 68, 159, 166, 197, 220, 258, 385, 388, 393, 433, 462, 488, 507, 508

Manganese (Mn), 25, 258, 388
Market Pricing, 409, 417
Material Flow Analysis, 507, 509
Mechanical Separation, 106, 173, 503
Membrane Filtration, 90
Metallurgical Accounting, 354
Metallurgical Testwork, 137
Milling, 107, 156, 158, 183, 189, 236, 349, 397, 496
Mine Closure, 380, 505
Mine Planning, 75, 80, 81, 83, 85, 442
Mine Safety, 372
Mineral Characterisation, 96, 97, 143, 147, 148, 151, 465, 466
Mineral Economics, 513
Mineral Sands, 42, 43, 60, 61, 92, 103, 112, 196, 197, 200, 202, 207, 209, 216, 218, 234, 341, 512
Modular Processing Units, 456
Molybdenum (Mo), 128

N

Nanomaterials, 275, 320, 508
Ndfeb, 320, 324, 401, 402, 413, 418, 453, 502, 510, 512
Ndfeb Magnets, 324, 401, 402, 418, 453
Neodymium, 3, 10, 12, 13, 14, 15, 40, 46, 60, 73, 76, 81, 82, 90, 92, 99, 100, 116, 126, 144, 147, 253, 254, 260, 263, 265, 266, 315, 324, 329, 350, 367, 385, 386, 395, 399, 400, 401, 417, 432, 437, 453, 487, 512, 513
Neodymium (Nd), 3, 10, 12, 13, 76, 90, 126, 147, 260, 265, 399, 400, 417
Neodymium-Iron-Boron, 401
Nickel, 2, 3, 4, 6, 8, 11, 13, 14, 17, 19, 20, 24, 25, 26, 28, 29, 30, 33, 34, 37, 38, 39, 45, 46, 47, 50, 55, 56, 57, 58, 61, 62, 63, 64, 65, 70, 73, 75, 76, 81, 86, 91, 92, 94, 98, 170, 173, 178, 182, 183, 184, 221, 243, 244, 246, 249, 250, 251, 252, 256, 257,259, 261, 264, 266, 268, 269, 270, 271, 272, 273, 278, 280, 281, 283, 284, 290, 292, 299, 300, 305, 308, 310, 311, 313, 315, 316, 319, 326, 329, 333, 337, 342, 347, 350, 352, 377, 384, 385, 388, 389, 390, 391, 393, 395, 402, 403, 404, 428, 429, 430, 432, 449, 456, 462, 463, 464, 474, 480, 488, 489, 490, 499, 500, 507, 508, 509
Nickel (Ni), 19, 56, 388
Niobium, 4, 5, 14, 25, 26, 30, 39, 45, 46, 47, 49, 51, 52, 53, 54, 55, 70, 92, 107, 109, 111, 115,

116, 126, 157, 158, 182, 183, 184, 196, 220, 413, 433, 464, 503
Niobium (Nb), 25, 52, 126

O

Offtake Agreements, 419
Open-Pit Mining, 400
Operational Efficiency, 185, 256, 259, 266, 268, 304, 315, 337, 351, 455
Ore Deposits, 91, 92, 492
Ore Sorting, 442, 447, 459
Oxalate Precipitation, 270, 279, 400
Oxidative Roasting, 311

P

Palladium, 23, 24, 47, 385, 433
Palladium (Pd), 23
Particle-Size Distribution, 182
Permanent Magnets, 2, 4, 12, 13, 14, 58, 62, 76, 90, 102, 208, 210, 320, 350, 385, 386, 403, 412, 420, 425, 496
Perovskites, 456
Physical Separation, 106, 143, 147, 194, 195, 234, 237, 238, 239, 242, 363, 469, 494
Pilot Plant, 88, 97, 144, 148, 152, 200, 316, 319, 448, 453, 496
Platinum, 2, 3, 5, 13, 23, 24, 26, 30, 44, 47, 50, 73, 75, 128, 132, 202, 221, 385, 433, 491, 496, 513
Platinum (Pt), 23
Policy Frameworks, 409, 421, 440
Polymetallic Nodules, 515
Potassium, 37, 47, 49, 156, 222, 223, 266, 271
Praseodymium, 3, 10, 13, 14, 76, 90, 100, 144, 254, 260, 265, 266, 324, 329, 350, 399, 417, 432, 437, 512
Praseodymium (Pr), 3, 10, 13, 76, 90, 260, 265, 399, 417
Precious Metals, 73, 75, 198, 205, 221, 385, 397, 489, 506, 513
Predictive Analytics, 447
Predictive Maintenance, 338, 443, 446, 447, 458, 459
Price Volatility, 417, 421, 437, 463
Primary Production, 384, 396, 398, 403
Process Automation, 305, 351, 446, 514
Process Intensification, 451
Process Simulation, 497
Processing Plant Design, 335, 336, 356
Production Capacity, 314, 365, 419

Promethium, 98, 100, 399, 432
Promethium (Pm), 98, 399
Purification, 16, 183, 184, 211, 215, 216, 218, 242, 243, 244, 251, 254, 255, 256, 260, 262, 263, 265, 266, 267, 268, 269, 270, 271, 273, 274, 275, 276, 278, 279, 281, 292, 296, 325, 326, 327, 328, 329, 331, 336, 337, 345, 346, 347, 356, 370, 389, 393, 396, 430, 433, 451, 453, 472, 473, 477, 499, 501
Pyrometallurgy, 243, 278, 279, 280, 281, 288, 325, 326, 327, 328, 329, 330, 331, 332, 388, 473

R

Radiation Safety, 366, 380, 381
Radioactive, 33, 41, 42, 48, 60, 87, 100, 145, 360, 361, 362, 363, 365, 366, 367, 368, 369, 373, 378, 379, 406, 478, 503, 504
Radioactive Waste, 368, 369
Rare Earth Carbonate, 287, 300
Rare Earth Oxides (Reos), 144, 254, 266, 350
Rare Earth Recycling, 395, 398
Rare Earth Separation, 272, 274, 275, 277, 501
Rare Earth Supply Chains, 505
Recycling Logistics, 427
Reducing Agents, 269, 327, 331, 511
Regulatory Frameworks, 35, 352, 359, 368, 369, 446, 509
Remote Sensing, 52, 55, 62, 70, 366, 368, 441, 442, 457, 459
Renewable Energy Integration, 373, 441
Responsible Sourcing, 18, 352, 410, 435, 438, 443, 446, 483, 486
Rhodium, 23, 24, 47, 128, 140, 433
Rhodium (Rh), 23, 128
Risk Management, 98, 371, 374, 454
Rotary Kilns, 280, 281, 285, 287, 301, 302, 303, 304, 305, 307, 311, 317, 319, 321, 328, 333, 337
Ruthenium, 23, 24, 433
Ruthenium (Ru), 23

S

Safety Management Systems, 371, 381
Samarium, 3, 10, 12, 14, 100, 274, 324, 399, 401, 512
Samarium (Sm), 3, 10, 12, 399
Scandium, 3, 9, 11, 14, 31, 99, 324, 367, 396, 399, 400, 401, 402, 413, 431, 432, 500, 515

Scandium (Sc), 3, 9, 14, 324, 399, 400

Separation, 10, 87, 96, 102, 104, 105, 107, 108, 111, 112, 119, 121, 122, 133, 137, 143, 149, 151, 154, 155, 157, 160, 164, 169, 172, 174, 175, 176, 177, 178, 179, 181, 182, 183, 184, 185, 186, 187, 188, 189, 192, 194, 195, 196, 198, 199, 200, 201, 202, 203, 206, 207, 208, 209, 210, 213, 214, 215, 216, 217, 218, 219, 220, 221, 222, 227, 229, 230, 233, 234, 235, 237, 238, 239, 240, 241, 243, 245, 249, 251, 252, 254, 255, 256, 257, 258, 260, 261, 262, 263, 265, 266, 267, 268, 271, 272, 273, 274, 275, 276, 281, 284, 309, 315, 326, 327, 328, 329, 336, 337, 339, 340, 341, 343, 344, 345, 346, 356, 365, 366, 367, 386, 387, 389, 390, 396, 402, 403, 411, 412, 413, 415, 416, 417, 418, 419, 420, 424, 425, 428, 432, 434, 436, 438, 442, 443, 444, 448, 449, 451, 452, 453, 455, 456, 462, 463, 466, 468, 469, 470, 471, 472, 477, 481, 487, 495, 496, 497, 498, 500, 501, 502, 504, 507, 513

Separation Chemistry, 455

Sintering, 285, 287, 288, 293, 295, 297, 298, 299, 308, 316, 319, 320, 321, 322, 324, 453

Smelting, 40, 57, 109, 110, 243, 280, 281, 282, 284, 285, 286, 287, 288, 300, 305, 308, 312, 313, 314, 315, 325, 326, 327, 329, 331, 332, 333, 334, 337, 342, 370, 388, 390, 393, 396, 404, 407, 431, 433, 450, 460, 461, 474, 475, 507

Solvent Extraction, 88, 109, 111, 112, 114, 141, 143, 146, 148, 235, 242, 243, 244, 245, 246, 249, 251, 252, 254, 255, 256, 257, 258, 259, 260, 264, 265, 266, 267, 268, 269, 273, 274, 275, 276, 278, 287, 295, 296, 299, 305, 311, 315, 319, 326, 327, 329, 334, 342, 347, 369, 388, 389, 390, 396, 398, 399, 400, 404, 407, 414, 432, 450, 451, 453, 472, 473, 481, 489, 491, 493, 495, 503, 504, 507, 508

Solvent Extraction (SX), 254, 255, 256, 258, 259, 265, 266, 267, 268, 269, 273, 276, 278, 299, 388, 389, 414, 472

Sustainability Assessment, 505

Sustainability Reporting, 379, 399, 407, 480

Sustainable Mining, 83, 397

Sustainable Supply Chains, 12, 393

Systems Thinking, 448, 484

T

Tailings, 43, 61, 79, 83, 87, 88, 125, 133, 141, 143, 148, 187, 192, 196, 197, 198, 199, 200, 201, 202, 203, 204, 205, 208, 209, 210, 211, 212, 213, 214, 215, 221, 222, 238, 245, 249, 273, 342, 343, 348, 352, 358, 359, 361, 362, 363, 364, 365, 366, 367, 368, 369, 370, 373, 378, 379, 380, 381, 383, 396, 397, 398, 399, 401, 406, 408, 414, 415, 449, 451, 454, 468, 471, 477, 478, 479, 480, 492, 503, 504, 510, 511, 512

Tailings Dam, 363, 369, 370, 378

Tailings Management, 342, 352, 358, 365, 366, 368, 369, 373, 379, 381, 471, 478

Tantalum, 14, 38, 39, 45, 46, 47, 51, 52, 53, 54, 55, 92, 107, 109, 111, 115, 126, 157, 158, 183, 196, 197, 200, 202, 205, 211, 215, 218, 220, 433, 503

Tantalum (Ta), 52, 126

Technology Transfer, 417, 463

Terbium, 10, 13, 14, 57, 76, 90, 101, 102, 144, 265, 324, 385, 399, 417, 428, 432

Terbium (Tb), 10, 13, 57, 76, 90, 102, 265, 399, 417

Thermal Efficiency, 334, 502

Thulium, 10, 101, 399, 512

Thulium (Tm), 10, 399

Titanium, 6, 14, 25, 37, 39, 44, 47, 50, 60, 61, 62, 103, 110, 182, 183, 209, 211, 246, 290, 313, 316, 341, 367, 396, 400, 433, 464, 498, 499

Trade Policies, 7, 9, 409

Transition Metals, 13, 264, 266

Tungsten, 4, 62, 63, 64, 65, 67, 128, 131, 196, 197, 200, 202, 205, 207, 209, 211, 215, 218, 220, 223, 224, 229, 231, 246, 247, 248, 288, 289, 290, 296, 299, 300, 305, 313, 316, 319, 322, 431, 432, 433, 451, 470, 472, 499, 501, 502

Tungsten (W), 64, 128, 220, 322

U

Underground Mining, 79

Uranium, 9, 39, 61, 67, 106, 243, 244, 246, 247, 257, 258, 259, 261, 262, 266, 268, 273, 276, 278, 312, 359, 360, 361, 362, 363, 365, 366, 376, 378, 379, 400, 477, 478, 493, 502, 504, 506, 512

V

Value-Chain Development, 423
Vanadium, 5, 25, 26, 29, 39, 44, 47, 50, 59, 61,
  246, 268, 312, 433, 464, 499
Vanadium (V), 25
Ventilation, 336, 342, 360, 362, 475, 479
Vibration Analysis, 337

W

Waste Heat Recovery, 399, 502
Wastewater Treatment, 114
Water Management, 359, 363, 370, 401
Wind Energy, 373
Workforce Development, 455, 458

X

Xenotime, 6, 7, 10, 11, 14, 35, 37, 42, 43, 59, 60,
  61, 62, 67, 87, 94, 97, 98, 99, 103, 104, 105,
  106, 107, 108, 109, 110, 111, 112, 113, 115,
  116, 117, 119, 121, 122, 124, 125, 134, 135,
141, 143, 145, 147, 152, 154, 157, 160, 167,
168, 171, 173, 182, 183, 193, 200, 202, 207,
209, 211, 212, 215, 216, 218, 219, 220, 223,
227, 229, 231, 233, 234, 235, 240, 284, 285,
344, 349, 360, 363, 411, 464, 466, 467, 469,
470, 494
X-Ray Diffraction (XRD), 97, 117, 122, 151, 349
X-Ray Fluorescence (XRF), 49, 73, 97, 126, 151,
  349, 351, 357

Y

Ytterbium, 4, 10, 14, 43, 102, 399
Ytterbium (Yb), 4, 10, 102, 399
Yttrium, 3, 9, 11, 13, 14, 31, 43, 60, 99, 100, 103,
  113, 265, 285, 324, 396, 399, 400, 401, 402,
  464, 487, 515
Yttrium (Y), 3, 9, 13, 14, 324, 399, 400, 401

Z

Zirconium, 37, 47, 49, 61, 246, 290, 316, 367, 396,
  433, 500

www.ingramcontent.com/pod-product-compliance
Lightning Source LLC
Chambersburg PA
CBHW082119210326
41599CB00031B/5813